CURRICULUM STANDARDS FOR GRADES K–4

The following curriculum standards were taken from *Curriculum and Evaluation Standards for School Mathematics*, published by the National Council of Teachers of Mathematics, 1989.

STANDARD 1: Mathematics as Problem Solving

In grades K–4, the study of mathematics should emphasize problem solving so that students can—

- use problem-solving approaches to investigate and understand mathematical content;
- formulate problems from everyday and mathematical situations;
- develop and apply strategies to solve a wide variety of problems;
- verify and interpret results with respect to the original problem;
- acquire confidence in using mathematics meaningfully.

STANDARD 2: Mathematics as Communication

In grades K–4, the study of mathematics should include numerous opportunities for communication so that students can—

- relate physical materials, pictures, and diagrams to mathematical ideas;
- reflect on and clarify their thinking about mathematical ideas and situations;
- relate their everyday language to mathematical language and symbols;
- realize that representing, discussing, reading, writing, and listening to mathematics are a vital part of learning and using mathematics.

STANDARD 3: Mathematics as Reasoning

In grades K–4, the study of mathematics should emphasize reasoning so that students can—

- draw logical conclusions about mathematics;
- use models, known facts, properties, and relationships to explain their thinking;
- justify their answers and solution processes;
- use patterns and relationships to analyze mathematical situations;
- believe that mathematics makes sense.

STANDARD 4: Mathematical Connections

In grades K–4, the study of mathematics should include opportunities to make connections so that students can—

- link conceptual and procedural knowledge;
- relate various representations of concepts or procedures to one another;
- recognize relationships among different topics in mathematics;
- use mathematics in other curriculum areas
- use mathematics in their daily lives.

STANDARD 5: Estimation

In grades K–4, the curriculum should include estimation so students can—

- explore estimation strategies;
- recognize when an estimate is appropriate;
- determine the reasonableness of results;
- apply estimation in working with quantities, measurement, computation, and problem solving.

STANDARD 6: Number Sense and Numeration

In grades K–4, the mathematics curriculum should include whole number concepts and skills so that students can—

- construct number meanings through real-world experiences and the use of physical materials;
- understand our numeration system by relating counting, grouping, and place-value concepts;
- develop number sense;
- interpret the multiple uses of numbers encountered in the real world.

STANDARD 7: Concepts of Whole Number Operations

In grades K–4, the mathematics curriculum should include concepts of addition, subtraction, multiplication, and division of whole numbers so that students can—

- develop meaning for the operations by modeling and discussing a rich variety of problem situations;
- relate the mathematical language and symbolism of operations to problem situations and informal language;
- recognize that a wide variety of problem structures can be represented by a single operation;
- develop operation sense.

STANDARD 8: Whole Number Computation

In grades K–4, the mathematics curriculum should develop whole number computation so that students can—

- model, explain, and develop reasonable proficiency with basic facts and algorithms;
- use a variety of mental computation and estimation techniques;
- use calculators in appropriate computational situations;
- select and use computation techniques appropriate to specific problems and determine whether the results are reasonable.

STANDARD 9: Geometry and Spatial Sense

In grades K–4, the mathematics curriculum should include two- and three-dimensional geometry so that students can—

- describe, model, draw, and classify shapes;
- investigate and predict the results of combining, subdividing, and changing shapes;
- develop spatial sense;
- relate geometric ideas to number and measurement ideas;
- recognize and appreciate geometry in their world.

STANDARD 10: Measurement

In grades K–4, the mathematics curriculum should include measurement so that students can—

- understand the attributes of length, capacity, weight, area, volume, time, temperature, and angle;
- develop the process of measuring and concepts related to units of measurement;
- make and use estimates of measurement;
- make and use measurements in problem and everyday situations.

STANDARD 11: Statistics and Probability

In grades K–4, the mathematics curriculum should include experiences with data analysis and probability so that students can—

- collect, organize, and describe data;
- construct, read, and interpret displays of data;
- formulate and solve problems that involve collecting and analyzing data;
- explore concepts of chance.

STANDARD 12: Fractions and Decimals

In grades K–4, the mathematics curriculum should include fractions and decimals so that students can—

- develop concepts of fractions, mixed numbers, and decimals;
- develop number sense for fractions and decimals;
- use models to relate fractions to decimals and to find equivalent fractions;
- use models to explore operations on fractions and decimals;
- apply fractions and decimals to problem situations.

STANDARD 13: Patterns and Relationships

In grades K–4, the mathematics curriculum should include the study of patterns and relationships so that students can—

- recognize, describe, extend, and create a wide variety of patterns;
- represent and describe mathematical relationships;
- explore the use of variables and open sentences to express relationships.

Curriculum Standards for Grades 5–8 are listed inside the back cover.

GEOMETRY

An Investigative Approach

Second Edition

Phares G. O'Daffer
Illinois State University
Normal, Illinois

Stanley R. Clemens
Bluffton College
Bluffton, Ohio

GEOMETRY

An Investigative Approach

Second Edition

Phares G. O'Daffer
Illinois State University
Normal, Illinois

Stanley R. Clemens
Bluffton College
Bluffton, Ohio

Addison-Wesley Publishing Company
Reading, Massachusetts   Menlo Park, California   New York
Don Mills, Ontario   Wokingham, England   Amsterdam   Bonn
Sydney   Singapore   Tokyo   Madrid   San Juan   Milan   Paris

Library of Congress Cataloging-in-Publication Data

O'Daffer, Phares G.

Geometry : an investigative approach / by Phares G. O'Daffer and Stanley R. Clemens.

p. cm.

Includes bibliographical references and index.

ISBN 0-201-21795-3

1. Geometry. I. Clemens, Stanley R. II. Title.

QA445.042 1991

516—dc20 91-29153

CIP

Photo Credits appear on page 451.

4 5 6 7 8 9 10-DOC-00999

PREFACE

The first edition of *Geometry: An Investigative Approach* has been widely used by preservice and inservice teachers in elementary, middle, and secondary schools and colleges. From this use, we have received a great deal of feedback reflecting a variety of philosophies about teaching geometry. These comments and some of the current ideas about teaching geometry have played an important role in our preparation of this revised edition.

The basic investigative philosophy of this text, described in the preface of the 1976 edition, has been positively affirmed by the feedback we have received. Consequently, it has been retained and expanded. It is our hope that the philosophy and approach of the first edition will continue to influence directions and emphasis in teaching geometry in grades K–12 in decades to come. Here are some excerpts from the 1976 preface:

> Students are encouraged to explore geometrical ideas using constructions, laboratory materials, and a variety of other investigative techniques. This emphasis on investigations—designed to set the stage for discovery of key geometrical relationships—is central in each of the chapters of the text and helps teachers gain confidence in their search for patterns in geometry.
>
> The central role of the "investigation" facilitates a student-centered course, rather than the more traditional instructor-centered course. We have found that students react positively toward working on the investigations in groups of from two to four persons; they can test ideas and get immediate feedback from each other. As they become more familiar with the small group format, we find that it is a natural setting for the experience of sharing mathematical thoughts—an experience too often missed.
>
> As the students work on the investigations, it is the job of the instructor (a) to ask penetrating questions and to give leading suggestions when necessary, (b) to

communicate that learning sometimes involves making mistakes, (c) to emphasize the idea that a "good" question is often as valuable as a "good" answer, and (d) to encourage the student to look at a situation with an eye toward making a conjecture about it.

We believe that teachers often tend to use teaching methods similar to the ones employed by their teachers. Consequently, we have attempted to develop this book so that its use will provide a good example of sound pedagogical technique. One assumption which has guided us in our writing is that teachers are best prepared to direct discovery experiences for pupils if teachers themselves have had many similar experiences while in the student role.

We feel that the material in this text provides a model for helping teachers implement the geometry standards described by the Commission on Standards of the National Council of Teachers of Mathematics in the book *Curriculum and Evaluation Standards for School Mathematics* (1989). Here are some excerpts from the focus statements for the geometry standards that we find especially pertinent:

> In learning geometry, children need to investigate, experiment, and explore with everyday objects and other physical materials . . . . Although a facility with the language of geometry is important, it should not be the focus of the geometry program but rather should grow naturally out from exploration and experimentation. Explorations can range from simple activities to challenging problem solving situations that develop useful mathematical thinking skills. [K–4 Standard, p. 48]
>
> Students discover relationships and develop spatial sense by constructing, drawing, measuring, visualizing, comparing, transforming, and classifying geometric figures. Discussing ideas, conjecturing, and testing hypotheses precede the development of more formal summary statements [5–8 Standard, p. 112]
>
> Physical models and other real-world objects should be used to provide a strong base for the development of students' geometric intuition so that they can draw on these experiences in their work with abstract ideas. [9–12 Standard, p. 157]

We believe this book helps teachers implement the NCTM Standards by addressing the role of asking questions in order to encourage the development of critical thinking skills. It also encourages the use of technology to explore geometric relationship and the use of geometry as a medium for problem solving. Finally, it stresses the importance of applications in geometry.

Throughout the text, Critical Thinking activities and problems appear within the Exercise Sets, while numerous supplemental activities designed to develop critical thinking skills are included in Appendix C. Each chapter highlights use of the computer to explore geometric patterns and relationships. While the LOGO computer language is developed and primarily used in the text, opportunities are also provided for using other software to discover geometric generalizations. Within each chapter, the Exercise Sets contain a final section on Problem Solving to develop skills and strategies, and Applications Problems are included in appropriate chapters.

Each chapter also contains numerous Investigation sections. These activities are developed and sequenced according the Van Hiele levels for learning geometry:

Level 0: Visualization
Level 1: Analysis
Level 2: Informal Deduction
Level 3: Deduction
Level 4: Rigor

The primary emphasis in this text is on Levels 0 through 2, which include the following ability goals:

Visualize geometric figures: manipulate (color, cut out, fold, and so on) geometric figures; identify a figure or relation in a set of objects, in a drawing, in the real world, and within other figures; create geometric figures by copying or tracing, by joining dots, by drawing, with manipulatives, and by construction using ruler, compass, Mira, and computer; describe figures verbally using everyday words and by using geometry words.

Classify geometric figures: Sort figures according to selected attributes; compare and contrast figures; identify a minimum set of properties for a figure; identify a figure from an oral description of its properties, from a written description of its properties, and from visual clues; develop and use a definition for a figure; describe inclusion relationships for figures.

Analyze geometric figures/relationships: Explain how figures and parts of figures are related; interpret the meaning of a geometric situation.

Discover/formulate generalizations: Answer and pose "what if" questions; make and test hypotheses; continue patterns; discover generalizations from empirical consideration of examples.

"Proving" and disproving generalizations: Give simple reasons to convince someone that something is true; present informal, convincing arguments that something is true (using cut-outs, measurement, construction, diagrams, computers, and so on); follow informal arguments that show that something is true; show a counterexample for a false generalization.

We have purposely included more material than can ordinarily be used in a one-semester course. Chapter 1 is of an introductory nature and may be assigned for reading. It is intended to broaden the reader's conception of geometry. Chapter 2 contains a more thorough introduction of the basic ideas of geometry than was developed in the first edition. It could be used selectively with students who have a strong background in geometry. Chapter 3, a new chapter in this edition, extends the basic ideas of geometry and focuses on discovery of generalizations. It is an important chapter for all students and might be chosen as the first or second chapter for classes having a better background in geometry. Chapters 4 and 5 focus on tessellations and analysis of figures in the plane and in three dimensions. Chapter 6 is a new chapter that develops measurement concepts in greater depth. Chapters 7 and 8 deal with transformations and the important ideas of congruence and similarity. The remaining chapters extend the basis concepts and relationships in geometry.

The chapters are related to each other, yet are independent enough so that it would be feasible to use chapters as individual units. Several different sequence options are possible, a few of which are described here:

Option 1: Follow the sequence of chapters as presented and proceed through as much of the text as possible.

Option 2: Begin with Chapter 2 and continue with Chapters 3 and 4, parts of Chapters 5, 6, and 7, part of Chapter 8, and Chapter 10.

Option 3: Use selected parts of Chapters 1, 2, and 3, the first part of Chapters 5, 6, and 7, and selected topics from the remaining chapters to complete the course.

Option 4 (for classes with a better background in geometry): Chapters 1, 10, 3, 4, and 5, followed by the more challenging parts of Chapters 6, 7, 8, and 9.

A flexible approach is encouraged, with sequence choices based on student background and instructor interest.

An option also exists for various degrees of pedagogical emphasis in a course using this text. On one end of the scale, the content and the more difficult, starred exercises might be emphasized, omitting the chapter Teaching Notes and the end-of-chapter Pedagogy for Teachers sections. At the other end, methods of teaching geometry might be emphasized by focussing primarily on these pedagogical elements and the Investigations. Our experience suggests that even in a more content-oriented course, it is important and motivational to allow the pedagogical material to play some role. A few experiences with some of the suggested pedagogical activities appear to better equip teachers to use new approaches in their own teaching.

Patience is necessary when using an investigation approach. Many students want to be told what to do and whether what they have done is correct. They expect direct factual items on tests. They are not accustomed to investigating, hypothesizing, experimenting, discussing alternatives, and so on.

When these different goals are explained and these processes are nurtured carefully, it is quite satisfying to see the progress students make in developing their thinking abilities. And that is what this book is all about.

SUPPLEMENTS

For the Instructor

Instructor's Solutions Manual: This valuable supplement contains completely worked out solutions to selected exercises in the text as well as teaching hints, solutions to applications problems, and information on commercially available geometry software packages and programming languages.

For the Student and Instructor

Cooperative Learning Activities for Geometry: This lab manual extends the philosophy of the text—to thoroughly teach geometry principles while teaching how to teach these concepts. It contains classroom activities, including the use of cut-outs and manipulatives, laboratory exercises, and group investigations, all of which are intended to foster group interaction and learning. It also contains investigations to be used with geometry exploration software.

Euclid's Toolbox Software: This user-friendly software program allows students to explore geometry, collect data, make conjectures, and test hypotheses. Students can construct a shape and measure its attributes; then with one keystroke, they can repeat the same measurements with a shape of a different size. Available for Macintosh and IBM/compatibles.

ACKNOWLEDGMENTS

We wish to express our appreciation to the elementary and secondary preservice and inservice teachers at Illinois State University and other colleges and universities around the world who have used this manuscript in a variety of situations. We gratefully acknowledge valuable feedback and revision ideas provided us by the many professors who have used this text.

We appreciate the work of Beverly Rich, who helped prepare some of the LOGO activities and instruction.

We are grateful to the authors of the many excellent articles and books on geometry that sparked our ideas and to the reviewers of our manuscript for their helpful reactions and suggestions.

The dedicated editorial effort of Tina Ashton in the initial stages of this revision is also greatly appreciated, as is the quality work of Jerome Grant and the AW College Division editorial staff in the final stages.

A special thanks to Stuart Brewster, long-time Director of the Addison-Wesley Innovative Division for his untiring encouragement and support during our entire careers as authors.

Finally we wish to thank our wives, Harriet and Jo, for their continuing affirmation and helpfulness during the revising of this text.

Phares G. O'Daffer
Stanley R. Clemens

Note to Persons Using This Book

I hear and it helps a little.
I see and ideas begin to form.
I do and the ideas become real to me.
I do, see, and hear and I understand.
I talk about it and I understand more.
I apply it and I see it's value.

This modification of an oft-quoted Chinese proverb indicates what we hope is the spirit of this text. The book has been designed to encourage you to become involved in exploring the ideas of geometry. These ideas relate to the world of nature, the world of art, and to the world of mathematical thought. Geometry can be exciting to study, and it is our hope that your experiences with this text will help you sense this excitement and develop confidence in extending your knowledge of the subject.

Involvement in investigation throughout the book may be a somewhat different approach to learning than that to which you are accustomed. At first it may seem strange to have a mathematics course that involves more than someone explaining how to do some skill and then testing you on that skill. Initially, it may also seem unsettling to be asked to explore a situation to see "what you can learn about it" rather than be asked to arrive at a single answer to a problem.

Our experience has been that as people become involved in the investigations and work with laboratory materials such as a ruler, compass, mirrors, dot paper, geoboards, styrofoam solids, and so on, the ideas of geometry that were previously unclear and uninteresting begin to have real meaning and to arouse a curiosity that leads to further exploration. We also have evidence that the investigations foster a new attitude toward what geometry is all about and a new confidence in a person's approach to a mathematical situation or problem. We urge you to "let yourself go" and follow the dictates of your curiosity and interest in the investigation situations. We also recommend working in groups whenever possible. Involvement with others in a setting where you can talk about your ideas, write about them, and give and get feedback can be a very valuable learning experience.

It has been said that a "good" question in mathematics is sometimes as valuable as a "good" answer. Real involvement means reaching the stage where you begin to ask questions and even suggest conjectures. This type of involvement often takes you beyond the basic assignment and into some articles, books, computer software, or other experiences related to your interest. It is our hope that you can find the satisfaction of this type of involvement from this text.

We encourage you to enter the experience of the text with an open mind and with the curiosity of a small child. If your experience translates into an approach to teaching that places even a small premium on helping your students develop a spirit of inquiry, a confidence in approaching new situations, and a clear knowledge and deep appreciation of geometry, then our goals for writing the text will be in the process of realization.

Phares G. O'Daffer
Stanley R. Clemens

Contents

Chapter One **A Panoramic View of Geometry** **1**
Introduction 1
Geometry in the Physical World 2
Geometry as a Mathematical System 8
Geometry as a Formal Axiomatic Structure 9
Aesthetic and Recreational Aspects of Geometry 12
Summary 21
Pedagogy for Teachers 22

Chapter Two **Basic Ideas of Geometry** **24**
Introduction 24
Points, Lines, and Their Relationships 25
Basic Geometric Figures and Their Measures 31
Line and Angle Relationships 38
Theorems about Intersecting Lines 45
More about Triangles and Quadrilaterals 47
Pedagogy for Teachers 54

Chapter Three **Discovering Polygon Relationships** **56**
Introduction 56
More About Polygons 56
Symmetry of Polygons 59
Construction of Regular Polygons 68
Star Polygons 74
Discovering Theorems About Polygons 78
Pedagogy for Teachers 84

Chapter Four **Patterns of Polygons: Tessellations** **86**
Introduction 86
Tessellations of Polygons 90
Tessellations with Regular Polygons 101
Analysis of Semiregular Tessellations 106
Three-Mirror Kaleidoscope 112
Tessellations of Curved Patterns 114
Pedagogy for Teachers 116

Chapter Five **Geometry in Three Dimensions 118**
Introduction 118
Regular Polyhedra 122
Symmetry in Space 132
Semiregular Polyhedra 140
Tessellations of Space 148
Pedagogy for Teachers 155

Chapter Six **Measurement—Length, Area, and Volume 158**
Introduction 158
Need for Standard Units in Measuring 162
Metric System of Measurement Units 163
Measuring Length 165
Measuring Area 172
Measuring Volume and Capacity 180
Measuring Mass and Weight 188
Pedagogy for Teachers 194

Chapter Seven **Motions in Geometry 196**
Introduction 196
Slides or Translations 197
Turns or Rotations 204
Flips or Reflections 213
Combining Slides, Turns, and Flips 222
Motions and Congruence 229
Pedagogy for Teachers 233

Chapter Eight **Magnification and Similarity 236**
Introduction 236
Magnification 238
Properties of Magnifications 242
Similarity 248
Pedagogy for Teachers 254

Chapter Nine **Topology 256**
Introduction 256
Topological Equivalence 259
Networks 267
Jordan Curve Theorem 272
Contiguous Regions 276
Four-Color Problem 277
Pedagogy for Teachers 280

Chapter Ten **Number Patterns in Geometry 282**
Introduction 282

Patterns of Points 282
Patterns of Lines 294
Pattern and Proof 302
Pascal's Triangle and Related Patterns 304
Pedagogy for Teachers 308

Appendix A **Using LOGO in Geometry 309**

Appendix B **Constructions 324**

Appendix C **Metric Units 343**

Appendix D **Critical Thinking 364**

Appendix E **Computer Explorations in Geometry 391**

Appendix F **Glossary, Postulates, and Theorems 421**

Bibliography 437

Selected Answers 446

Index 453

CHAPTER ONE A Panoramic View of Geometry

INTRODUCTION

In the fourth century B.C., the Greek philosopher and teacher Plato (429–348 B.C.) carved the inscription "Let no one ignorant of geometry enter my doors" above the entrance to his academy. From that time to the present, the question, "What is geometry?" has been appropriate for both students and teachers. The material in subsequent sections of this chapter provides information that sheds light on an answer to this question. Persons interested in teaching geometry might also find it valuable to analyze their own ideas and views about the nature of geometry and to explore the ideas of others. Investigation 1.1 suggests group activities that might be used.

TEACHING NOTE

Many teachers are surprised to find that geometry can be interesting and fun for students of all ages. Beyond this, there are also many other practical reasons for teaching geometry at every level. For some of these reasons for teaching geometry, see NCTM (1973, pp. 28–50) and Lindquist and Schulte (1987).

INVESTIGATION 1.1

Join a small group, and together engage in one or more of the following activities. Don't feel you must know all the answers to participate, but instead think of the group situation as a safe place to share untested ideas and to ask questions.

A. *Activity 1:* Write not more than five key words that best suggest your ideas of a general answer to the question, "What is geometry?" Then write as many words as you can that suggest basic concepts that are studied in geometry.
B. *Activity 2:* Draw a picture that expresses symbolically your group's answer to the question, "What is geometry?" The quality of the art isn't important, but the components of the picture should convey the intended message.
C. *Activity 3:* Describe at least one specific example for each of the following aspects of geometry. If your group thinks no examples exist, mark N. If you think examples exist, but you can't think of any, mark C.

    1. Geometry in nature
    2. Geometry in art
    3. A modern-day practical use of geometry
    4. Geometry as a theory
    5. An historic practical use of geometry
    6. A recreational use of geometry
    7. A scientific application of geometry

In the next few sections, we shall look at geometry in the physical world, geometry as a mathematical system, and the aesthetic and recreational aspects of geometry. It is hoped that after reading these sections, the reader will have made progress in formulating an answer to the question, "What is geometry?"

GEOMETRY IN THE PHYSICAL WORLD

As we consider the nature of geometry, we cannot help but imagine when, where, why, and how humans first became involved with geometric ideas. When we become sensitive to such objects as those shown in Figures 1.1 through 1.6, we begin to see that geometric form and structure permeate the universe and that we were immersed in a geometrical setting from the beginning. It remained for us to notice it, to appreciate it, to abstract ideas from it, and to use it.

FIGURE 1.1

FIGURE 1.2

FIGURE 1.3

FIGURE 1.4

TEACHING NOTE

At all age levels, a person's environment is a rich source of geometrical ideas. Preschool children should be encouraged to notice balls, blocks, cans, cones, dials, and a whole universe of other shapes. Older students might be challenged to investigate the geometry of sunflowers, snail shells, starfish, bridges, architecture, Wankel engines, and other intriguing situations. For a valuable classroom resource, see NCTM (1987).

FIGURE 1.5

FIGURE 1.6

TEACHING NOTE

A study of the Golden Ratio provides an interesting setting for enrichment activities for older students. Ideas involved are ratios, similarity, sequences, constructions, and other concepts of algebra and geometry. A book by Runion (1990) provides some interesting extensions of this idea and contains a useful bibliography.

It is difficult to pinpoint the persons, the times, or the places where the intricate geometry of nature was first noticed by humans. For example, it has been observed that the ratio of a to b for each of the objects in Figures 1.7 through 1.9 is close to 1.618. This ratio, called the Golden Ratio, is the ratio of the length to the width of what is said to be one of the most aesthetically pleasing rectangular shapes. This rectangle, called the Golden Rectangle, appears in nature and is used by humans in both art and architecture.

FIGURE 1.7

FIGURE 1.8

FIGURE 1.9

As shown in Figures 1.10 and 1.11, the Golden Ratio can be perceived in the proportions of both human and animal bodies. It can be seen in the way trees grow and even in the frequency of rabbit births. Whoever first discovered these intriguing manifestations of geometry in nature must have been very excited about the discovery.

A B E D F G H C J L N O M K a b

$$\phi = \frac{BC}{AB} = \frac{BD}{ED} = \frac{DC}{FC} = \frac{ED}{DF} = \frac{DG}{DF}$$

$$\phi = \frac{FC}{GC} = \frac{HC}{JK} = \frac{JK}{LM} = \frac{LM}{NO}$$

FIGURE 1.10

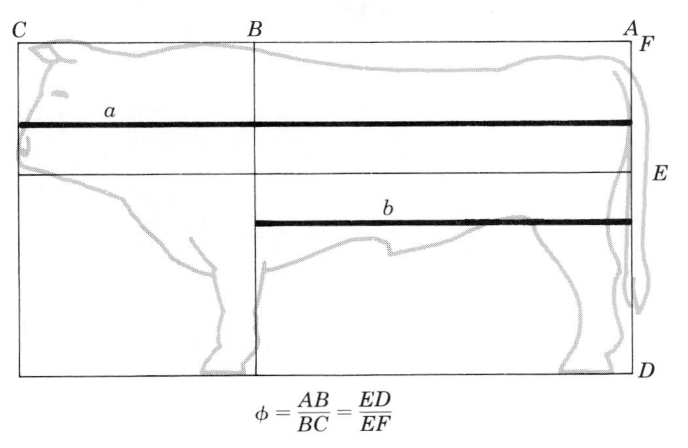

$$\phi = \frac{AB}{BC} = \frac{ED}{EF}$$

FIGURE 1.11

FIGURE 1.12

The geometry of the physical world can be touched and seen. In a very real sense, one *experiences* physical geometry. It shapes our aesthetic views and influences the work we do. Architects and designers, for example, experience it as they contemplate and create new works (Figure 1.12); engineers and carpenters experience it as they execute design and construction plans (Figure 1.13). Children experience physical geometry as they observe and handle objects in their environments, and this experience is an extremely important component of the learning process.

Our early ancestors' experience with geometric forms in nature was translated into useful solutions to everyday problems. They appear to have reduced nature's shapes to simpler forms, which they used in various ways. For example, early humans used geometric forms derived from nature to invent a method of pattern making that was then used to decorate utensils and to make weapons and useful tools (Figs. 1.14–1.15).

Later, as people's understanding of physical geometry grew, they developed even more sophisticated uses of this knowledge. For example, since the Nile River annually overflowed its banks, the Egyptians used empirical "rule of thumb" geometric techniques to re-mark the land boundaries that had washed away. The word **geometry**, which means "earth measure," originated from this setting.

FIGURE 1.13

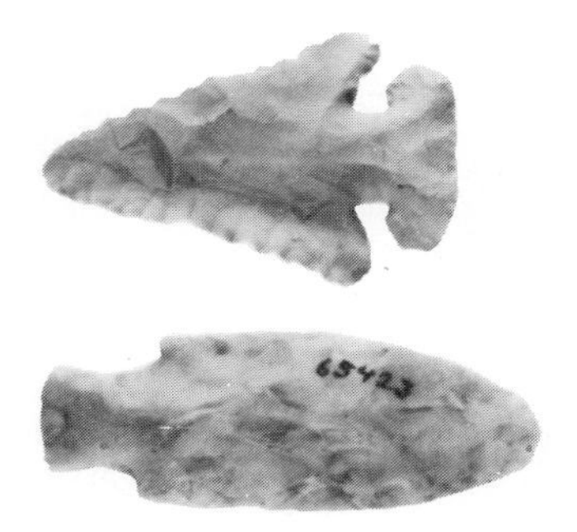

FIGURE 1.14

FIGURE 1.15

Perhaps one of the important early discoveries was that a triangle with side lengths 3, 4, 5 is a right triangle (Fig. 1.16). Thus, knots in a rope spaced 3, 4, and 5 units apart provided a useful method of determining a square corner for a field or for the base of a pyramid (Fig. 1.17).

Historically, the experience with physical geometry was of great value. It served to provide practical methods, usually involving measurement, for solving the problems of the day. Perhaps even more importantly, it provided the foundation for the more logical approach to geometry developed by the Greeks that was ultimately refined to become useful in scientific applications. In a similar way, experience with physical objects plays an important role in the child's learning of geometry, for this experience provides the necessary foundation for later abstraction and generalization of geometrical ideas.

Clearly, then, the geometry of actual experience—physical geometry—cannot be overlooked when answering the question, "What is geometry?" for it has played a major role in both the history of geometry and in the curriculum for the modern elementary and junior high school student.

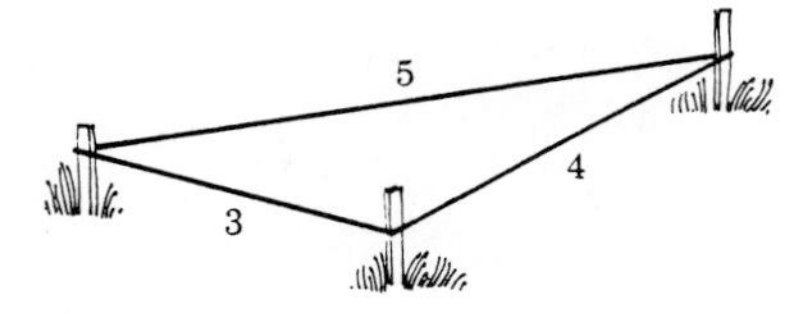

FIGURE 1.16

FIGURE 1.17

EXERCISE SET 1.1

1. Describe at least one example of each of the following that hasn't been discussed in this chapter.
   a. Geometry in nature
   b. Geometry in architecture
   c. Geometry in sports or games
   d. Geometry in design
2. Is a 3×5 card a Golden Rectangle? Explain. Describe examples of approximations of the Golden Ratio that you can find in your classroom.
3. Figure 1.9 asserts that the ratio of a person's height to the height of the person's navel is approximately equal to the Golden Ratio. Collect some data and decide whether you think this is an accurate statement.
4. A booklet entitled *I Can Count the Petals on a Flower* (Wahl, 1976) shows flowers with different numbers of petals. If you draw lines connecting the tips of the petals on a picture of starfire flower, for example, you form a 7-sided figure called a septagon. Look for flowers or pictures of flowers and name the figures that would be produced by "connecting" the tips.
5. Look at Appendix B to review the definitions of some common geometric figures and relationships. Take a walk around your home or school and list as many objects as you can that suggest these geometric ideas.

Problem Solving: Developing Skills and Strategies

Problem solving has always been an important part of the study of geometry. In this and all subsequent sections entitled "Problem Solving: Developing Skills and Strategies," ideas that will help improve your problem-solving abilities will be presented.

A problem-solving strategy called **guess and revise** is often useful when solving problems in geometry. This strategy is illustrated in the following problem.

Problem: The line segment in Figure 1.18 is divided so that $x - y = 1$ and $xy = 1$. This division of a segment was of special interest to early Greek geometers. Use a calculator to find two decimals, to the nearest thousandth, for x and y. What did you discover about this pair of numbers?

To solve this problem with the guess-and-revise strategy, simply guess a value for x, see how close it is, revise your guess, and try again. Continue this process until the correct pair of numbers has been found. Try it.

Can you think of another way to solve the problem? Explain.

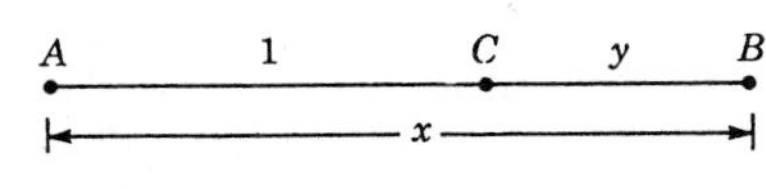

FIGURE 1.18

GEOMETRY AS A MATHEMATICAL SYSTEM

Geometry Organized Logically

In physical geometry, points, lines, and geometric figures are considered in much the same way as we consider physical objects. We often rely on the appearance of figures or on approximate measurements in making decisions about geometric properties and relationships. The situations in Figure 1.19a through 1.19f suggest that appearance is sometimes deceiving. Try Investigation 1.2.

Perhaps it was situations like these, as well as a new emphasis on the value of reason, that caused Greek philosophers to develop a geometry that relied less upon visual information and more upon logical reasoning. Regardless of the motivation, a new view of geometry emerged in the first half of the sixth century B.C. Thales of Miletus (640–550 B.C.) was one of the first Greeks to insist that geometric facts be established not by trial-and-error observation and experimentation, but by logical reasoning. His efforts set the stage for the monumental work of Euclid (300 B.C.) called *The Elements of Geometry*.

Thales, Euclid, and other Greeks helped geometry evolve from the purely physical level to the more abstract logical level. Euclid's work has so influenced the teaching of geometry that until quite recently, most secondary school students have thought of "geometry" and "proof" as essentially synonymous.

The spirit of this Greek view of geometry is illustrated by the story of Euclid's student who, when he learned the first theorem, asked Euclid, "But what advantage shall I get by learning these things?" Euclid called his assistant and said, "Give him threepence, since he must need to make a profit out of what he learns."

TEACHING NOTE

It is appropriate to ask young children why a particular geometric relationship is true. While their response will likely involve a reference to measurement or other experimentation with physical materials, these "proofs," based on the physical world, provide a valuable background for a later understanding of proofs that rely on postulates and logic.

INVESTIGATION 1.2

With members of your group, look at Figures 1.19a through 1.19f.

A. Answer the questions posed in the figures, relying on visual information only to make your group decision. No measuring or other methods allowed! Then find ways to check your answers.
B. Discuss the following question: Can accurate conclusions be drawn about geometric figures using only visual information?

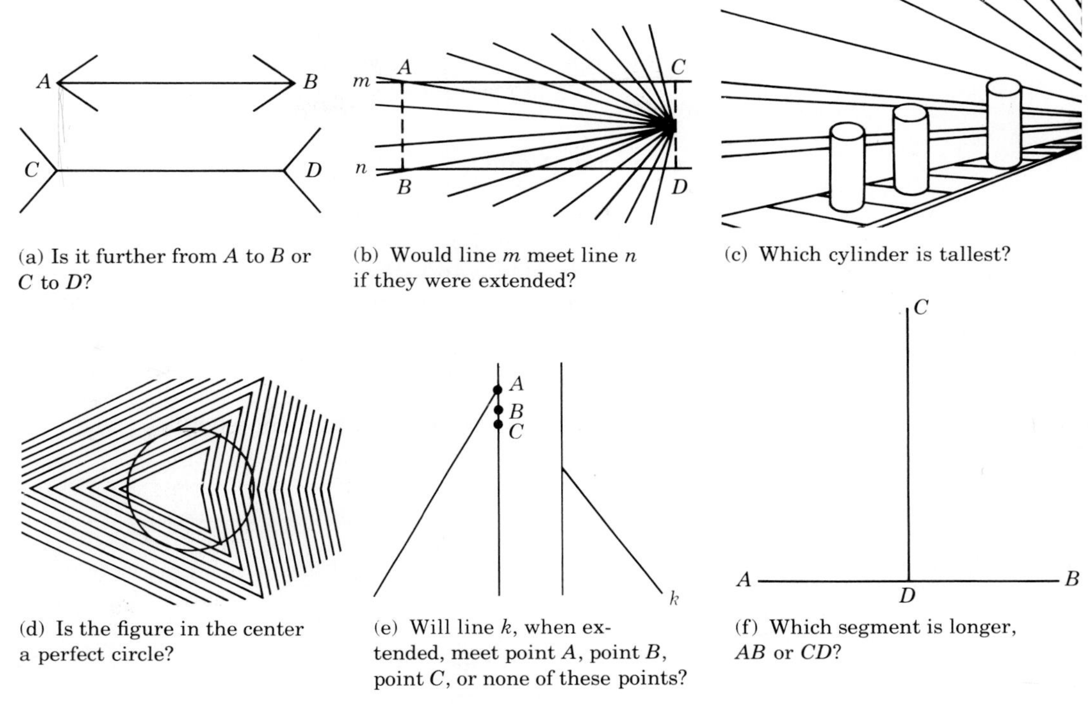

(a) Is it further from A to B or C to D?

(b) Would line m meet line n if they were extended?

(c) Which cylinder is tallest?

(d) Is the figure in the center a perfect circle?

(e) Will line k, when extended, meet point A, point B, point C, or none of these points?

(f) Which segment is longer, AB or CD?

FIGURE 1.19

GEOMETRY AS A FORMAL AXIOMATIC STRUCTURE

The development of geometry by Euclid, which seemed to disregard the practical, was essentially a logically derived chain of theorems based on initial postulates and common notions. **Postulates** are statements that are assumed to be true.

Postulates

1. A straight line can be drawn from any point to any point.
2. A finite straight line can be produced continuously in a straight line.
3. A circle may be described with any center and distance.
4. All right angles are equal to one another.
5. If a straight line falling on two straight lines makes the interior angles on the same side together less than two right angles, the two straight lines, if produced indefinitely, meet on that side on which the angles are together less than two right angles.

Postulates, or assumptions, are important in everyday life, as well as in a formal approach to geometry. Think about the role of assumptions as you try Investigation 1.3.

Common Notions

1. Things which are equal to the same thing are also equal to one another.
2. If equals be added to equals, the wholes are equal.
3. If equals be subtracted from equals, the remainders are equal.
4. Things that coincide with one another are equal to one another.
5. The whole is greater than the part.

INVESTIGATION 1.3

Critical Thinking. Critical thinking is a process that helps one decide what to believe or to do in any situation. Critical thinking activities like the one in this investigation will be given throughout the text and in Appendix D to help you develop and apply useful critical thinking skills.

A. Accept the following story as an accurate account of something that actually happened.

   Little Jack Horner sat in a corner
   eating his Christmas pie.
   He put in his thumb, pulled out a plum
   and said, "What a good boy am I."

B. Have each member of your group individually decide which of these conclusions they have accepted (perhaps without being aware of it) about the story.

   1. Jack was eating a plum pie.
   2. It was Christmas day.
   3. Jack felt he was a good boy because he pulled out the plum.
   4. Jack was sitting in the corner because he was being punished.
   5. Jack was a child.
   6. There were other people in the room with Jack.

C. Now, discuss with your group which of the members' conclusions are based on assumptions that are not necessarily confirmed by the story.

D. Use Critical Thinking Activities 1 through 3 in Appendix D as a follow-up to this investigation.

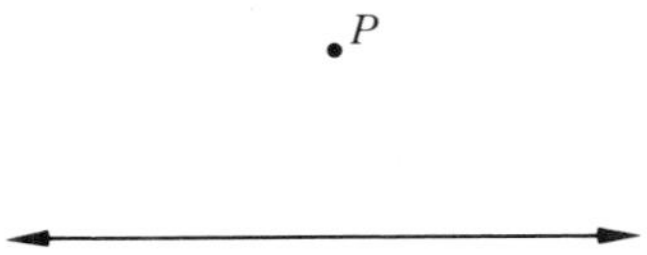

$\bullet P$

FIGURE 1.20

For two thousand years after Euclid listed these five postulates and common notions, mathematicians were concerned about the need for the fifth postulate. The version of this postulate usually encountered is as follows (see Fig. 1.20):

> Through a given point not on a given line, there can be drawn only one line parallel to the given line.

Since Euclid's original statement of the fifth postulate, is so much more complex than the first four and since its converse can be proved, many mathematicians felt it could be proved using the other postulates.

The story of the futile attempts to find a proof, along with the discoveries resulting from these attempts, presents one of the more exciting accounts in the history of mathematics. In the nineteenth century, Gauss, Bolyai, and Lobachevsky discovered independently that new, logically valid, non-Euclidean geometries resulted when Euclid's fifth postulate was *replaced* by a contradictory postulate. From two thousand years of belief that there could be only one system of geometry to describe space evolved the conclusion that different sets of postulates, chosen to be consistent, could be used, and new legitimate geometries could be developed.

This discovery of non-Euclidean geometries, often classified as one of the most significant events to happen in mathematics, marked the beginning of an interpretation of geometry as a formal axiomatic (or postulational) structure,* rather than a formal study of the physical world, and contributed greatly to an expanded view of the nature and use of geometry. In fact, Einstein in his 1921 lecture "Geometry and Experience" asserted, "To this interpretation of geometry I attach great im-

TEACHING NOTE

Providing a good intuitive background in geometry at the elementary and junior high level makes it possible for secondary-level students to explore new topics. A book by Posamentier (1983) includes, among many other ideas, some material on non-Euclidian geometry for secondary-level students.

* As a formal axiomatic structure, a system of geometry can roughly be described as consisting of the following:

— A set of elements and a collection of undefined terms
— A set of statements called axioms (or postulates) that establish relations between the undefined terms and that are accepted without proof
— A set of definitions that utilizes undefined terms or previously defined terms to specify additional concepts
— A system of logic that provides a means for determining the truth or falsity of new statements within the system
— A chain of statements called theorems, each of which is logically deduced from axioms, definitions, or previously proven theorems

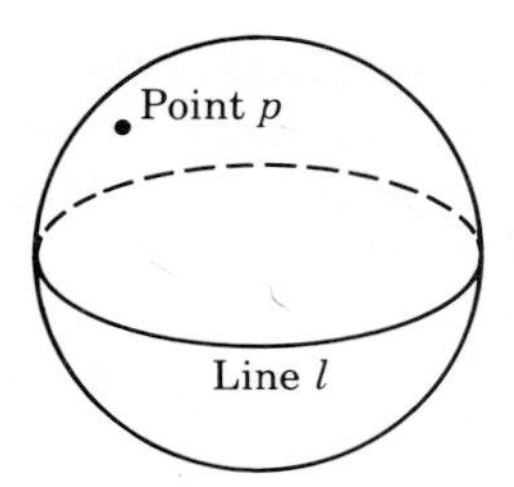

FIGURE 1.21

portance, for should I have not been acquainted with it, I would never have been able to develop the theory of relativity." Not only has the axiomatic view of geometry provided many useful models for scientific applications, but it has also had a developmental influence on many other areas of mathematics as well.

EXERCISE SET 1.2

1. Without looking at a reference list, state a postulate, a definition, and a theorem that you remember from your previous study of geometry. (See Appendix F for a list of postulates for geometry.)
2. The idea of a formal axiomatic structure was inherent in the development of non-Euclidean geometry. Refer to Figure 1.21. If the "plane" in a certain non-Euclidean geometry is the surface of a sphere and a "line" is always a great circle of a sphere, how many lines will there be through point P not intersecting line l?
3. **Critical Thinking.** Logic is an important part of everyday reasoning, as well as the basis for an axiomatic development of geometry. Test your understanding of logical deduction by deciding which of the following items are valid patterns of reasoning.
   a. If I can get the keys, then I will open the door. I can get the keys.
   Therefore, I will open the door.
   b. If pigeons can fly, then pigeons have wings. Pigeons have wings.
   Therefore, pigeons can fly.
   c. If you solve this logic problem, then you will get supper. You don't solve the problem.
   Therefore, you don't get supper.
   d. If you get an A on the test, then you will pass the course. You do not pass the course.
   Therefore, you did not get an A in the course. test

   Use Critical Thinking Activities 4 through 8 in Appendix D to follow up on these ideas.

AESTHETIC AND RECREATIONAL ASPECTS OF GEOMETRY

Geometry As an Art

Throughout the centuries philosophers, mathematicians, and some artists have recognized and emphasized the artistic aspects of mathematics. G. H. Hardy (1877–1947), a well-known mathematician who taught many years at Cambridge University, said:

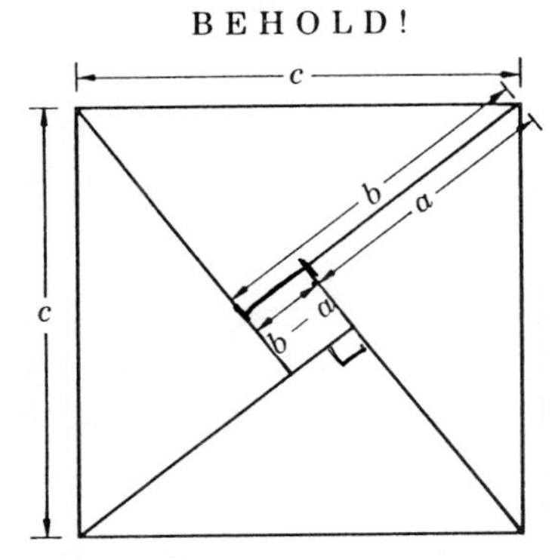

FIGURE 1.22

> A mathematician, like a painter or a poet, is a maker of patterns. If his patterns are more permanent than theirs, it is because they are made with ideas.
>
> The mathematician's patterns, like the painter's or the poet's, must be beautiful; the ideas, like the colors or the words, must fit together in a harmonious way. Beauty is the first test; there is no permanent place in the world for ugly mathematics.

When the Hindu mathematician Bhaskara (1114–1185) drew the picture in Figure 1.22 and offered no further explanation than the word "Behold," he was not suggesting that the lines in his diagram were particularly appealing to the eye, but rather that this particular arrangement suggested a "beautiful" proof of the Pythagorean Theorem*—one that probably gave him a feeling of delight and satisfaction.

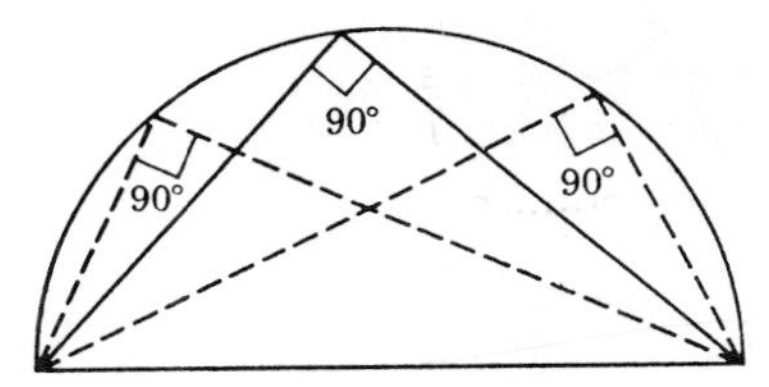

FIGURE 1.23

From the time that Thales was believed to have sacrificed a bull in joy over the discovery and proof that any angle inscribed in a semicircle is a right angle (Fig. 1.23), mathematicians have found an inspiring beauty in the structure of mathematics.

As mentioned earlier, the discovery of non-Euclidean geometry and the resulting axiomatic approach to geometry brought this structure into clear focus. It is interesting to note that Johann Bolyai, one of the "codiscoverers" of non-Euclidean geometry, wrote a letter at that time to his father in which he said, "Out of nothing I have created a strange new universe." Perhaps it was the reference to the purity of geometry as an axiomatic structure that prompted the philosopher Bertrand Russell (1872–1969) to say:

> Mathematics possesses not only truth, but supreme beauty—a beauty cold and austere, like that of sculpture, without any appeal to our weaker nature . . . sublimely pure, and capable of stern perfection such as only the greatest art can show.

If humans differ from other animals in being able to create and admire beautiful objects, then mathematicians join the ranks of artists in that they are able to admire and create beautiful ideas. This beauty of geometrical ideas and the opportunity to participate, at an appropriate level, in the creation or discovery of "new" relationships is what has interested many children and adults in the study of geometry.

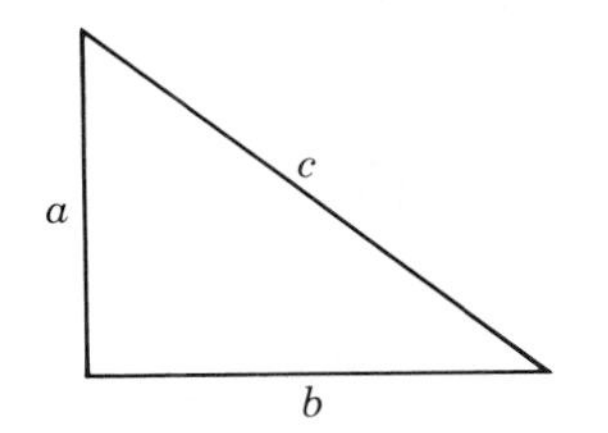

FIGURE 1.24

* From consideration of the area of the large square (c^2) in Figure 1.22 as formed by the area of the 4 triangles ($4 \cdot \frac{1}{2}ab$) and the area of the center square $(b - a)^2$, we get the equation $2ab + (b - a)^2 = c^2$, which easily simplifies to $a^2 + b^2 = c^2$ (Fig. 1.24).

Geometry in Art and Design

Another aspect of geometry is the explicit use of geometric principles or figures in the creation of paintings, sculptures, and other art objects, as well as in the design of useful objects and interesting buildings. Perhaps pictures, as presented in Figures 1.25 through 1.31, better than words, can be used to illustrate this aspect. While our geometrical artistic creations cannot equal those found in nature, it seems unthinkable that one would omit the aesthetic aspect when asking, "What is geometry?"

TEACHING NOTE

Students often find it interesting and valuable to create drawings and artistic expressions of geometric ideas. Curve stitching or ruler and compass designs provide useful activities in this regard. See the books *Curve Stitching* (Millington, 1989) and *How to Enrich Geometry Using String Designs* (Pohl, 1986).

FIGURE 1.25

St. Jerome, DaVinci (Note the Golden Rectangle shape into which the main figure fits.)

FIGURE 1.26

FIGURE 1.28

FIGURE 1.29

FIGURE 1.27

FIGURE 1.30

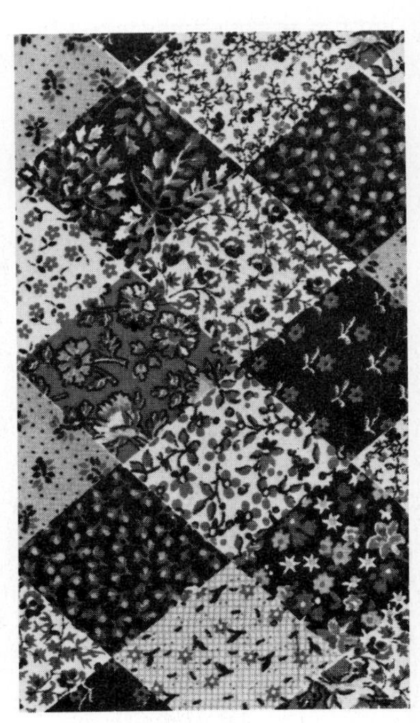

FIGURE 1.31

EXPLORING WITH THE COMPUTER 1.1

Read the introduction and instructions in Appendix A, Using LOGO in Geometry. Do Investigations 1 and 2 in Appendix A to become familiar with LOGO.

1. Just for fun, have someone knowledgeable about LOGO define the following procedure (Thornburg,1983) and help you run it to see what it draws.

```
TO WHIRL :S
IF :S < 1 [STOP]
REPEAT 4[FD :S RT 90]
FD :S RT 90 FD :S
WHIRL :S*0.61803
END
```

2. Then, have them help you execute the following LOGO commands. What is the result?

```
CS PU SETPOS [-120-60] PD HT WHIRL 150
```

When you complete this exercise, execute the ST command so that the turtle reappears ready for additional exploration.

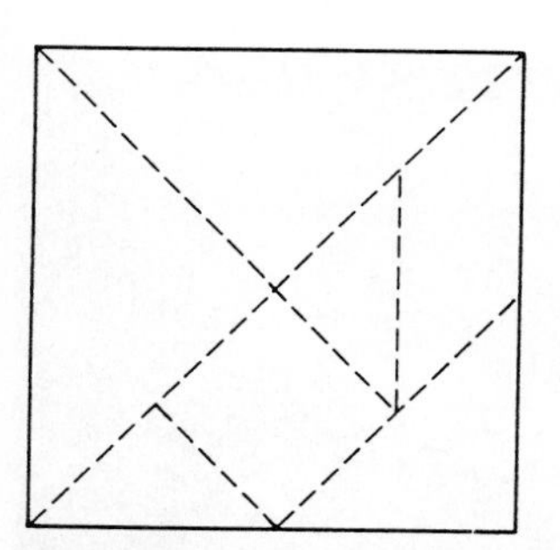

FIGURE 1.32

FIGURE 1.33

Geometry for Recreation

The recreational potential in geometry has been known and pleasurably exploited for centuries. Simple dissection puzzles in the plane and in space, for example, have long been a source of enjoyment. One example is the famous Chinese tangram puzzle, which is said to be at least 4000 years old. The book *Tangrams—330 Puzzles* (Read, 1965) is an interesting source. The puzzle consists of seven pieces, as shown in Figure 1.32, and the challenge is to put these pieces together to form interesting shapes, like the "portrait" shown in Figure 1.33.

Other recreations have involved searching for convincing proofs of important theorems. Henry Perigal, a London stockbroker and amateur astronomer, discovered the paper-and-scissors "proof" of the Pythagorean Theorem (Euclid's 47th theorem) illustrated in Figure 1.34. This proof shows that the area of the squares on the legs of a right triangle is equal to the area of the square on the hypotenuse, as in the following equation: $a^2 + b^2 = c^2$.

You can solve the "puzzle" in Figure 1.34 by copying and cutting out the triangle and the two squares on the legs. Next draw two lines through the center of the larger square, intersecting at right angles with one of the lines perpendicular to

TEACHING NOTE

To develop concepts fully, important ideas are introduced early in simple form and are reintroduced and broadened at successive levels in the child's development. The puzzle shown here can give young children an opportunity to experience the beginning aspects of the Pythagorean Theorem.

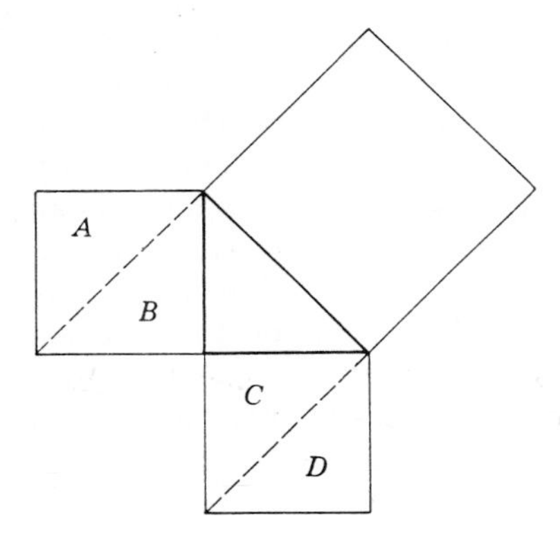

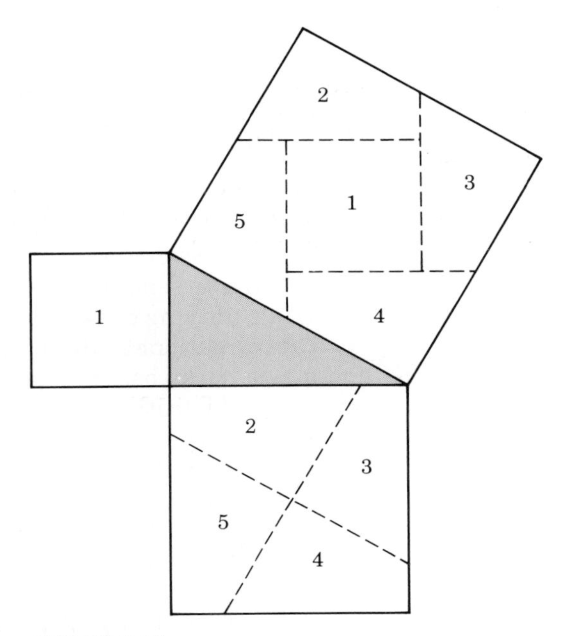

FIGURE 1.34

the hypotenuse of the right triangle. Cut out the four pieces of the larger square, and put them with the smaller square to form a large square on the hypotenuse.

In a more formal way, many nonmathematicians have found a hobby in searching for proofs of the Pythagorean Theorem. James Garfield, the twentieth President of the United States, devised an ingenious proof for the theorem, which appears along with others suggested by both mathematicians and nonmathematicians in the book *The Pythagorean Proposition* (Loomis, 1940). Other puzzles, such as the topological puzzles illustrated in Figures 1.35 and 1.36 and those appearing in

TEACHING NOTE

Recreational geometry can be motivational. But teachers should also be aware that important concepts can sometimes be developed through recreational experiences. The book *Mathematical Puzzles* (Kohl, 1987) contains some interesting geometric puzzles.

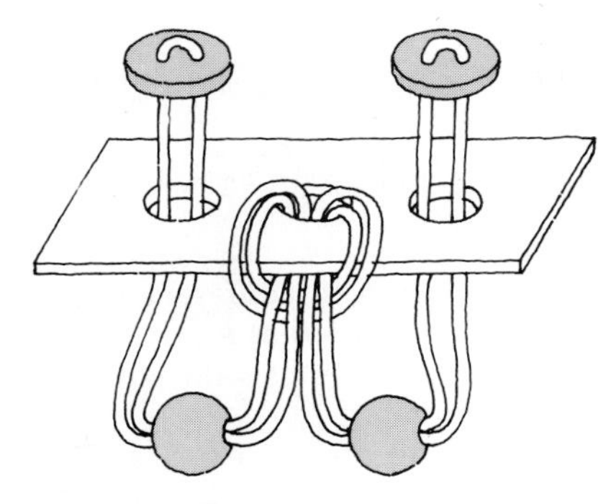

FIGURE 1.35

FIGURE 1.36

subsequent chapters of this book, suggest the variety of recreations available.

The puzzle in Figure 1.35 is made from string, cardboard, two beads, and two buttons. The object is to undo the loop and get the two beads together without cutting or untying the string, tearing the cardboard, or taking off the buttons. With the puzzle in Figure 1.36, each person ties a piece of string to each wrist in such a way that the strings are looped as shown. The object is to separate from the other person without cutting the string, untying the knots, or taking the string off the wrists.

The recreational value of the problems and the many games and magic tricks based on geometric ideas is well known. The preceding illustrations are only a few examples of the areas in which geometry has provided recreation for many. The literally hundreds of mathematical recreation books that have been published in the past 50 years, with such titles as *Play Mathematics, 536 Puzzles and Curious Problems,* and *Mathemagic*, attest to the popularity of mathematics in general and to geometry in particular as a recreational activity.

The recreation aspect of geometry is not only "just plain fun," but can also be used as motivational material for a serious study of geometry.

EXPLORING WITH THE COMPUTER 1.2

The following LOGO procedure (Lindquist and Shulte, 1987, p. 91) can be used to generate a variety of designs. When the procedure is run, there are 5 different input values needed. Have someone knowledgeable about LOGO help you enter and run this procedure, first for the values 40, 10, 40, 50, 20 and then for the values 10, 20, 30, 40, 50. Then answer the questions.

```
TO TEST.FIVE :A :B :C :D :E
RT 90
REPEAT 4 [FD:A RT 90 FD:B RT 90 FD:C
   RT 90 FD:D RT 90 FD:E RT 90]
PU HOME PD HT
END
```

1. How many different patterns can you find using this procedure with different input values?
2. Which input values result in a pattern that is a dissection of a square?
3. Alter the procedure and create new patterns.

APPLICATIONS

Suitcase Problem: Suppose you could take your fishing rod on a trip only if you could carry it inside your suitcase. The rod collapses to a length of 80 cm. You want to order a new suitcase from a catalog. It has interior dimensions 66 cm long, 18 cm wide, and 46 cm high.

A. Discuss some ways to decide whether you can take the fishing rod.

B. Draw a picture showing your solution to prove whether or not it can be done.

EXERCISE SET 1.3

1. Make a design using a compass and a straight edge. Color it in an interesting way.
2. Fold paper to make a geometric design. Color it in an interesting way.
3. Copy the square in Figure 1.32 that forms the famous Chinese tangram puzzle. Cut your square into 7 pieces, as shown. Can you place all 7 pieces together to form the following shapes?

   a. triangle c. parallelogram e. pentagon
   b. rectangle d. trapezoid f. hexagon

4. *Curve stitching designs* are designs made entirely with straight lines of string or thread (Fig. 1.37). To get an idea of how a simple string design is made, copy the angle shown in Figure 1.38 and mark it as shown. Use a fine-point colored pen or pencil to connect points with the same number. Why do you think this is called curve stitching?

FIGURE 1.37

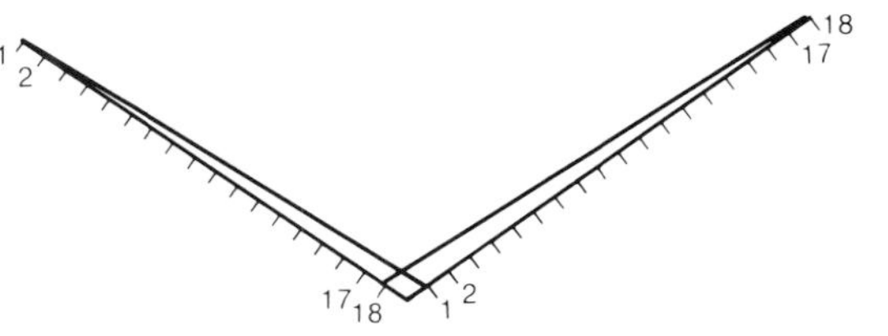

FIGURE 1.38

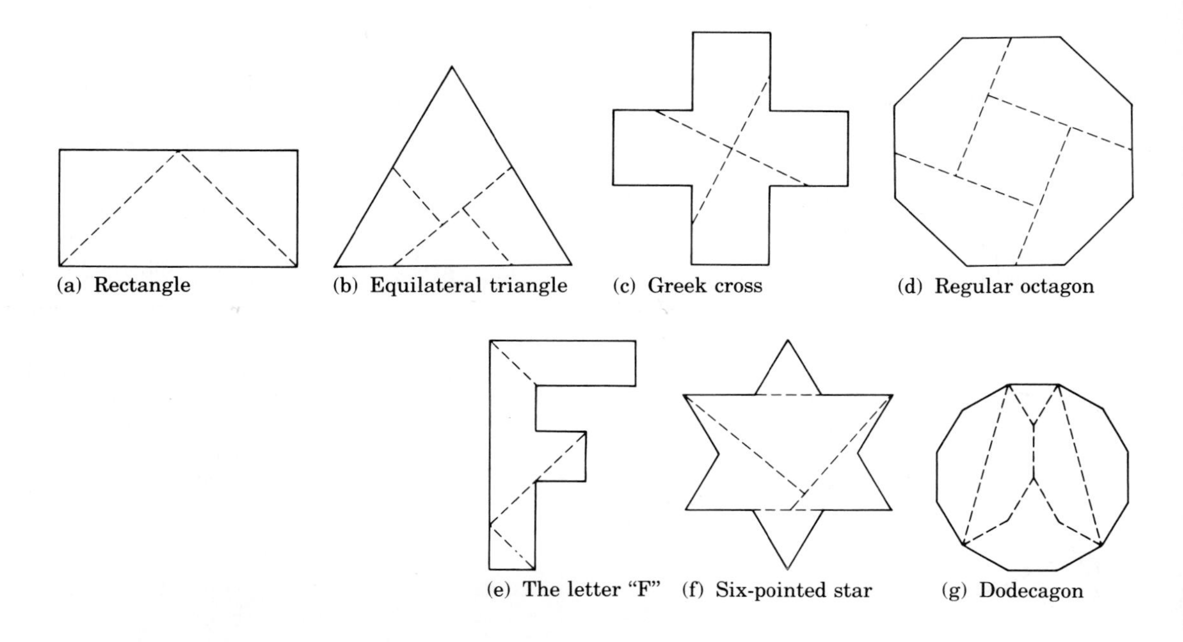

FIGURE 1.39

5. Can you place all 7 pieces of your tangram puzzle together to form the following shapes?
   a. The "portrait" (Fig. 1.33)
   b. A double arrow
   c. A digit
   d. A letter of the alphabet
   e. An interesting figure
6. Copy each of the shapes in Figure 1.39, and cut them into pieces as shown. Form single squares using all the pieces in each figure.
7. Try the puzzle illustrated by the two people in Figure 1.36. Be prepared to demonstrate your solutions.
8. Suppose an Egyptian used a knotted rope, as shown in Figure 1.40, to establish a square corner for building a pyramid. Would it work? Explain.

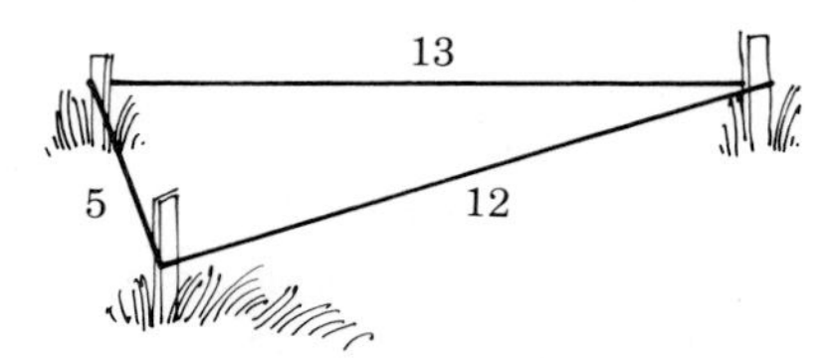

FIGURE 1.40

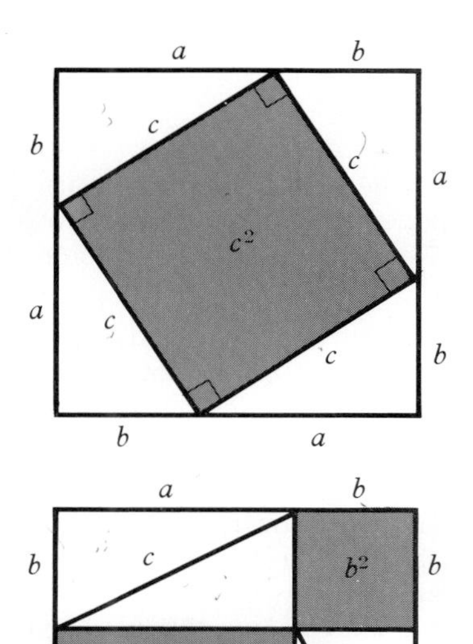

FIGURE 1.41

9. **Critical Thinking.** How could you use the areas of the pair of congruent squares marked as shown in Figure 1.41 to convince someone that the Pythagorean Theorem is true? Do you think your convincing argument was a proof of the theorem? Give reasons to support your answer.

SUMMARY

We began this chapter by asking, "What is geometry?" Although it is an oversimplification, we like the following answer: Geometry is the study of space, spatial relationships, and spatial patterns.

Some choose to emphasize in this study those relationships in the physical world that can be discovered through the handling and observation of physical objects and through the application of inductive reasoning. Others prefer to create an abstract mathematical model (system) and to study these relationships through the logic of the abstract system. This approach employs deductive reasoning and is the more traditional approach to geometry in the schools. Still others are content to study and experience these relationships in a much less structured way—through the aesthetic appeal of art objects.

Regardless of the approach chosen, prospective teachers of geometry who study the subject begin to realize that it has many dimensions. They realize that a great deal of geometry emanates from certain physical situations and experiences. They realize that geometry can be logically challenging, that it involves patterns and can be beautiful, that it is a fertile vehicle for creativity, and that it can be fun. It becomes clear that to really understand geometry, one must become deeply involved in these various aspects of the subject. Any one of the above aspects could easily be the central goal of an entire course. Ptolemy (?–168) asked Euclid if there was an easier way to an understanding of geometry—a "king's shortcut"—than studying Euclid's "Elements." Euclid replied, "There is no royal road to geometry."

The chapters that follow do not constitute a "royal road" to geometry, but a particular path designed to provide teachers of geometry with important experiences and valuable insights into the subject that they may not have encountered previously. Along this path the reader will explore such ideas as the Golden Ratio, star polygons, tessellations of the plane and of space, motion geometry, topology, and many other aspects of geometry. It is hoped that through this exploration each person will be motivated to provide his or her own answer to the basic question of this chapter.

TEACHING NOTE

It is important for teachers at all levels to realize the crucial effect their attitude toward a subject has on their students. An adequate perception of the breadth and scope of geometry can positively affect your attitude toward geometry and, hence, affect the attitudes of your students.

PEDAGOGY FOR TEACHERS

ACTIVITIES

1. Create a project assignment for students at a level of your choice. One goal of the project should be to help the students become more aware of geometry in their environment. Write a project description that you could mimeograph for the students or that you would use to describe the project to the students.

2. Devise a simple diagnostic interview that can be used to find out what a student at a given level of your choice knows about the ideas of geometry. You may wish to use geometric models, pictures, or devices and verbal instructions as part of the interview. Interview a student or group of students and summarize your findings.

3. Make a set of photographic slides depicting geometry in the environment. Possible settings for pictures are architecture, sidewalk patterns, nature, window designs, quilts, and so on. You may wish to tape music and synchronize this music with your slides. Present your slide-music program to a group of students. What was their reaction?

4. Geometric recreations can be valuable for motivating students. Refer to books that include geometric recreations and start a card file containing at least 5 such recreations that you feel confident would be useful to motivate students at a level of your choice.

COOPERATIVE LEARNING GROUP DISCUSSION

Teachers are often asked, "Why should we teach geometry?" Here are five reasons that can be given students:

1. To acquire visualization skills
2. To acquire reasoning skills
3. To acquire skills at translating visual description to verbal description
4. To see how geometrical relationships are applied to physical world problems
5. To experience geometrical problem solving

Discuss each reason. What does it mean? Can you give examples? Do you believe each is a good reason for teaching geometry? What other good reasons can you find for teaching geometry? Be ready to present and defend your reasons to the class.

CHAPTER REFERENCES

Kohl, Herbert. 1987. *Mathematical Puzzlements*. New York: Schocken Books.

Lindquist, Mary M., Ed. 1980. *Selected Issues in Mathematics Education*. Berkeley, CA: McCutchan Publishing Company.

Lindquist, Mary M., and Schulte, Albert P., Eds. 1987. *Learning and Teaching Geometry, K–12* (1987 Yearbook). Reston, VA: National Council of Teachers of Mathematics.

Loomis, Elisha Scoot. 1940. *The Pythagorean Proposition*. © 1968 NCTM, Reston, VA.

Millington, Jon. 1989. *Curve Stitching*. Palo Alto, CA: Dale Seymour Publications.

NCTM. 1987. *Geometry in Our World*. (Includes slides.) Reston, VA: National Council of Teachers of Mathematics.

NCTM. 1973. *Geometry in the Mathematics Curriculum* (Thirty-Sixth Yearbook). Reston, VA: National Council of Teachers of Geometry.

Pohl, Victoria. 1986. *How to Enrich Geometry Using String Designs*. Palo Alto, CA: Dale Symour Publications.

Posamentier, Alfred. 1983. *Investigations in Geometry*. Reading, MA: Addison-Wesley Publishing Company.

Read, Ronald C. 1965. *Tangrams—330 Puzzles*. New York: Dover Books.

Runion, Garth. 1990. *The Golden Section and Related Curiosa*. Palo Alto, CA: Dale Seymour Publications.

Thornburg, David D. 1983. *Discovering Aple LOGO*. Reading, MA: Addison-Wesley Publishing Company.

Wahl, John and Stacey. 1976. *I Can Count the Petals on a Flower*. Reston, VA: National Council of Teachers of Mathematics.

SUGGESTED READINGS

Consult the following sources listed in the Bibliography for additional related readings: Beamer (1989), Dunkels (1990), and Millman and Speranza (1991).

CHAPTER TWO Basic Ideas of Geometry

INTRODUCTION

Throughout history, people have used their creativity either to discover in the physical world or to create for themselves geometrical patterns and designs. The reader likely has seen many interesting point, line, and polygon patterns similar to the ones pictured in Figure 2.1. It is assumed that in addition to these types of patterns, the reader has had many experiences with geometry through handling, measuring, and drawing. We now look more carefully at the geometrical ideas that evolve from these earlier physical experiences.

Pattern of prime numbers from 1 to 10,000

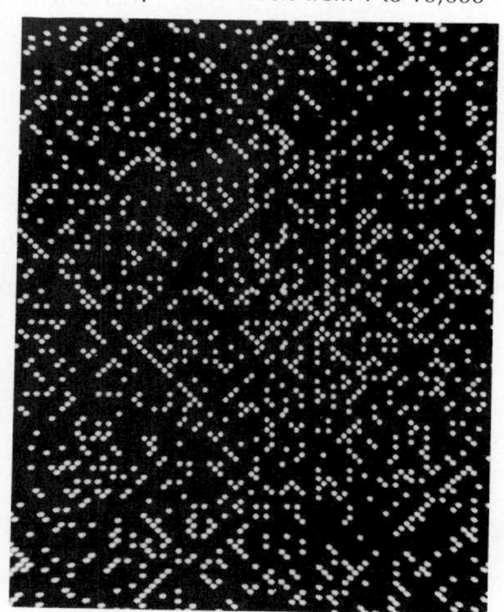

Alltair design by Ensor Holiday based on investigations of architecture and pattern development in the Middle East

Patterns of all lines connecting 24 evenly spaced points on a circle

FIGURE 2.1

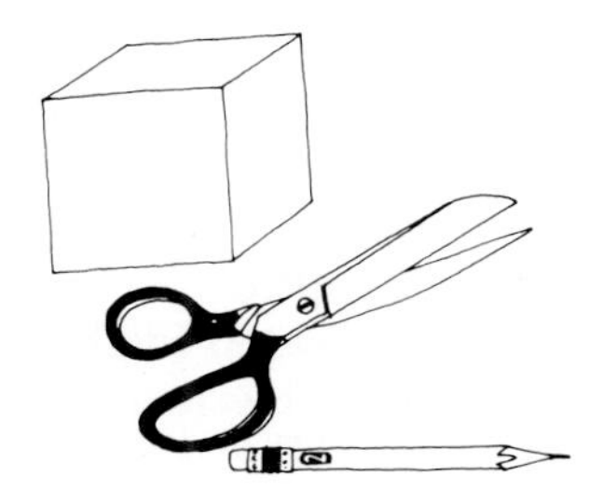
FIGURE 2.2

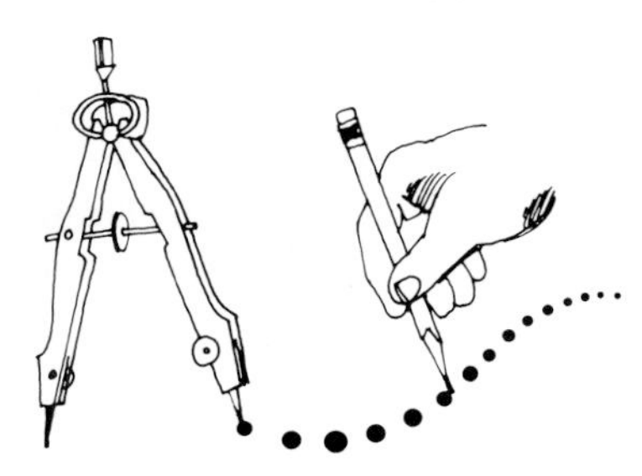
FIGURE 2.3

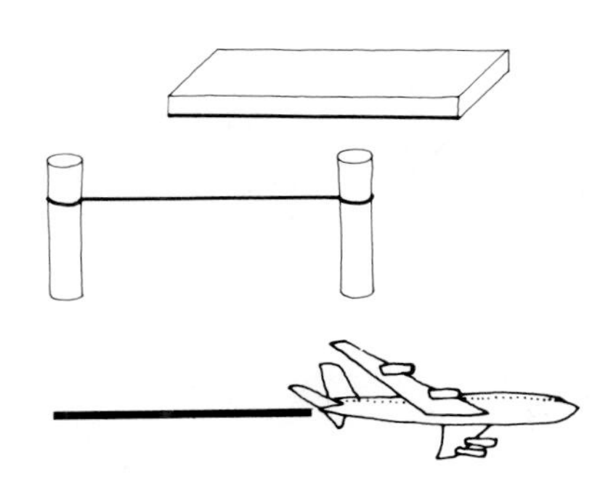
FIGURE 2.4

POINTS, LINES, AND THEIR RELATIONSHIPS

The concept of **point** as it is used in geometry is an abstraction, and it evolves out of experiences with physical objects. In the beginning, objects such as the tip of pencil, the corner of a block, or the forbidden point of a pair of scissors suggest the notion of point to the child (Fig. 2.2). These important early experiences provide the basis for later abstraction of the *idea* of point—a dot of any size—that marks *a location and has no length, width, or height.* They also provide the basis for a mature view of the concept of point as it is used in a geometric setting (Fig. 2.3).

Just as a number is an abstraction that cannot be seen or touched, so is a point. In a formal study of geometry, no attempt is made to define this abstraction (a point) using previously defined simpler terms. Since the idea of point is so basic that it is used without definition, it is called an **undefined term.**

In most formal developments of geometry, **line,** like point, is an undefined term. Sometimes a line is described as a set of points that satisfies a certain list of postulates. Examples from such a list of postulates are

1. Given any two different points, there is exactly one line containing them.
2. Every line contains at least two points.

Investigation 2.1 explores some relationships between points and lines.

As was true for point, the notion of "line" or "line segment" can be cultivated in the mind of a child through experiences with physical objects. The edge of a piece of lumber, a string or wire held taut, or the paths of certain moving objects all suggest the idea of line segments or lines (Fig. 2.4). These

---

INVESTIGATION 2.1

These figures show that 3 points can be arranged to determine 1 line or to determine 3 lines.

A. Can you arrange 6 points to determine 1 line? 6 lines? 7 lines? 8 lines? 9 lines?
B. How many other collections of lines can be determined by 6 points?
C. What is the largest number of lines that can be determined by 6 points?

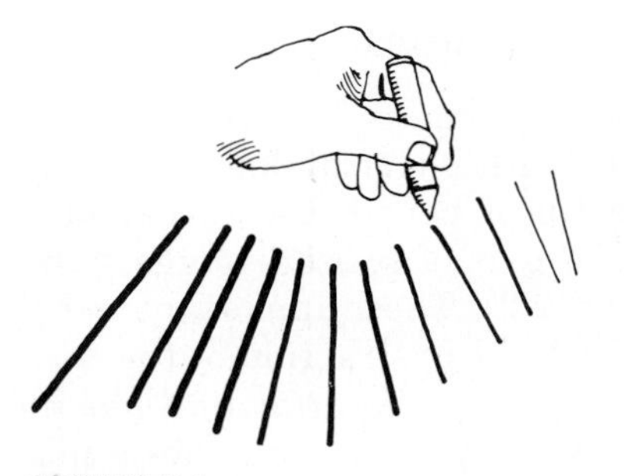

FIGURE 2.5

physical models of lines provide the basis for later abstraction of the *idea* of line as *an unlimited length that is straight, with no thickness and no endpoints*. They also provide the basis for the concept of line as it is used in a geometric setting (Fig. 2.5).

Just as point and line are undefined terms, **plane** is an undefined term. The idea of "plane" can be cultivated in the mind of a child through experiences with physical objects such as a table top or as the thinnest slice you can cut (Fig. 2.6). A plane *has no boundary, continues in all directions, is flat, and has no thickness*.

The undefined term **space** can be thought of as the collection of all points in a system. In a physical sense, the notion of space can be shown to a child as all the points inside, on, and outside of an inflated balloon and as what remains after the balloon is burst (Fig. 2.7). Space is an *idea or abstraction that has no boundary, length, width, or height and is the set of all points*.

We often use the word "point" to describe a dot or inkspot on a piece of paper. This dot may represent any number of

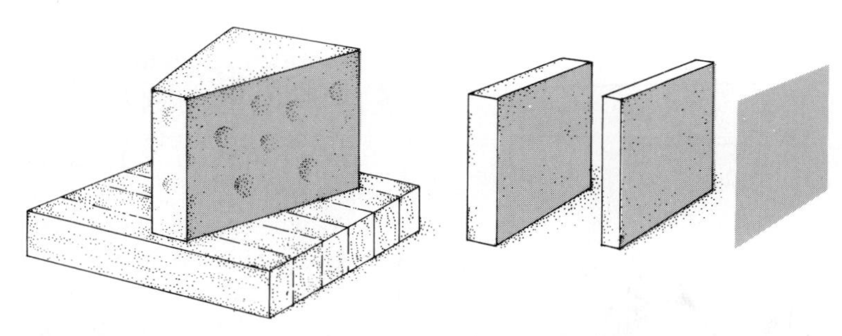

FIGURE 2.6

TEACHING NOTE

According to the studies by Piaget, children are not ready to think about lines, triangles, squares, etc., as sets of points before eleven to twelve years of age. The early notions of points, lines, and planes should probably be developed by looking at natural three-dimensional objects in the child's environment and making special notice of the corners, edges, and surfaces.

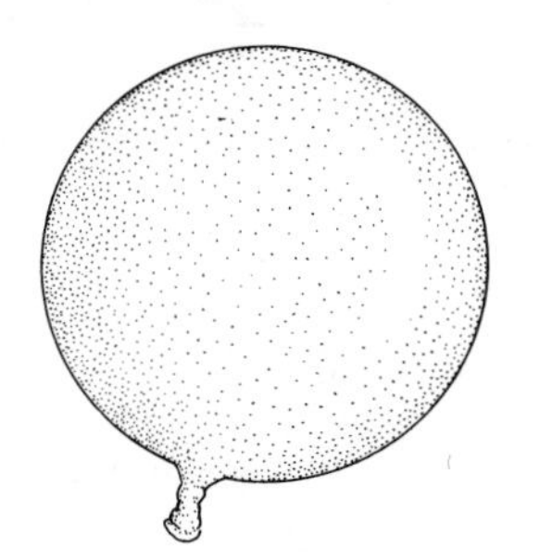

FIGURE 2.7

FIGURE 2.8

FIGURE 2.9

physical situations that suggest points; consequently, it conveys the notion of point for each of the objects represented. We say "draw" or "mark a point" when we mean draw a dot to represent a point. When we draw dots on paper to represent points, we use capital letters beside them to name the points. We call the points A, B, and C (Fig. 2.8).

We can think of a line as a set of points. By labeling a pair of points, we can name the line in terms of two points. For example, points A and B are on the line, so we call it *line AB*. We assume that only one line goes through both A and B. Another way of saying this is, *Two points determine a line.* Sometimes a line is labeled by using one small letter (Fig. 2.9). Here, line AB can also be called *line* ℓ. We write: $\overleftrightarrow{AB}$.

A plane can also be thought of as a set of points. A plane is named either by placing a single letter by the plane or by naming a set of three points on the plane that are not all on one line. Points A, B, and C are on plane N (Fig. 2.10). We say *plane N* or *plane ABC*.

We assume that only one plane contains the three points. We say that *three points not all on one line determine a plane.*

When thinking of line ℓ as a set of points, we can say point A is *on* line ℓ and point A is an *element* of line ℓ to describe the same situation. We can also say line ℓ *contains* point A.

If A, B, and C are points on line ℓ, as shown in Figure 2.11, we say point B is *between* points A and C. If A, B, and C are not all on the same line, we do not use "between" to describe their relationship.

FIGURE 2.10

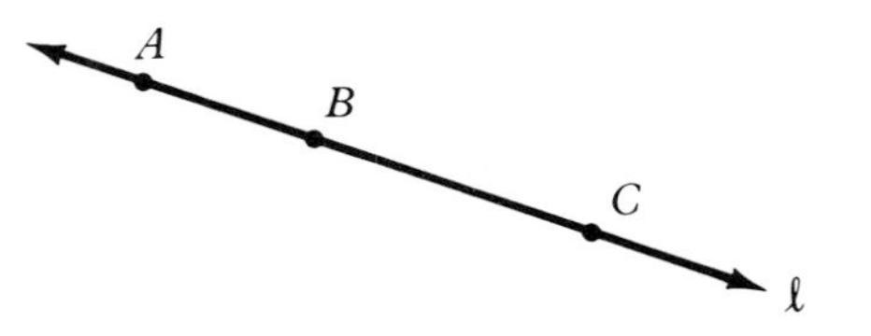

FIGURE 2.11

Some of the basic relationships for points and lines in a plane are described in Table 2.1.

TABLE 2.1

Physical Model, Figure	Description, Symbol	Definition
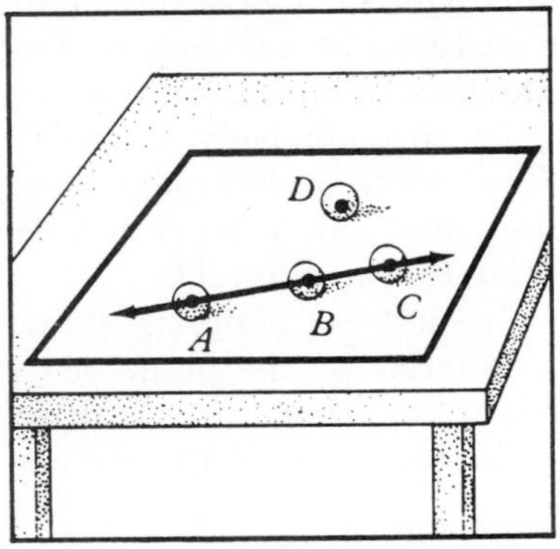	A, B, and C are *collinear*. A, D, and C are *noncollinear*. A, B, C, and D all lie in the same plane. They are *coplanar* points. Points not all in the same plane are *noncoplanar* points.	**Collinear points** are points that lie on the same line. **Coplanar points** are points that all lie in one plane.
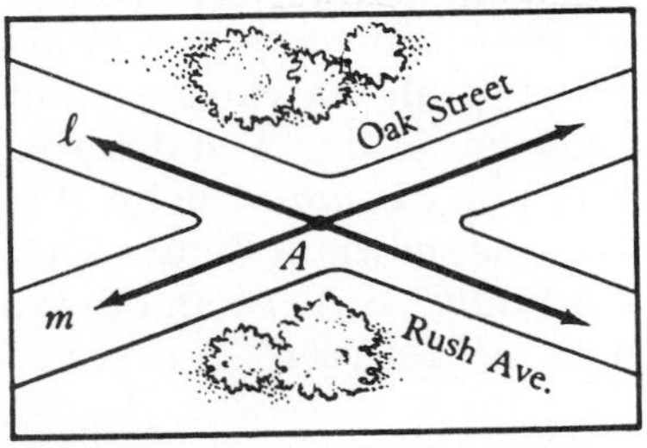	Lines ℓ and m *intersect* at point A.	**Intersecting lines** are two lines with a point in common.
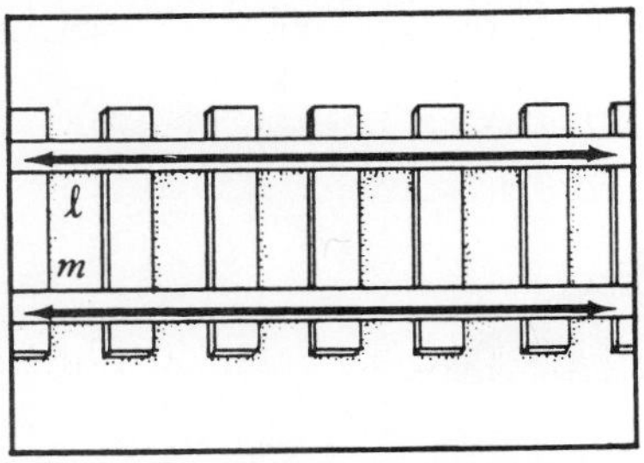	Lines ℓ and m do not have a common point. ℓ is *parallel* to m. We write: $\ell \parallel m$.	**Parallel lines** are lines in the same plane that do not intersect.
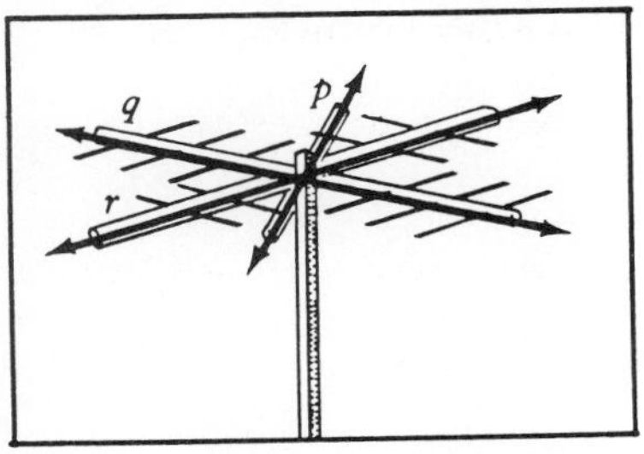	Lines p, q, and r have exactly one point in common. They are *concurrent lines*.	**Concurrent lines** are three or more coplanar lines that have a point in common.

TABLE 2.1 Continued

Physical Model, Figure	Description, Symbol	Definition
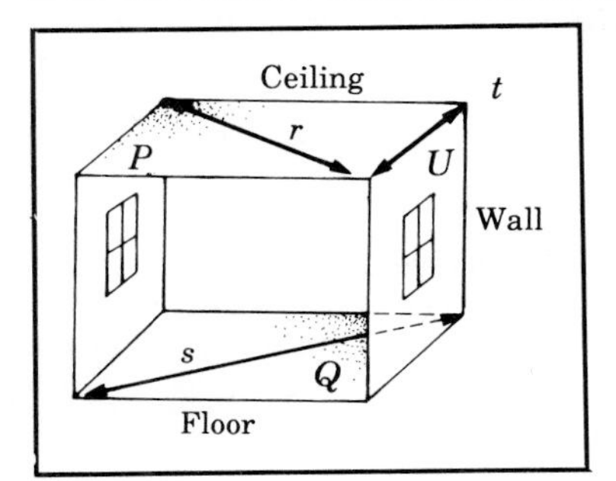	Planes P and Q do not have any points in common. P is parallel to Q. We write: $P \parallel Q$. Lines r and s are not in the same plane. They are skew lines. Planes P and U intersect in a line t.	**Parallel planes** are planes that have no points in common. **Skew lines** are two lines that are not in the same plane. **Intersecting planes** are planes that have a line in common.

EXERCISE SET 2.1

1. Name objects that are shaped so that all or part of them suggests the following:

   a. A point
   b. A part of a line
   c. A part of a plane
   d. Intersecting lines
   e. Parallel lines
   f. Concurrent lines
   g. Skew lines
   h. Intersecting planes
   i. Parallel planes

2. Refer to Figure 2.12 and answer the following:

   a. Give 4 different names for this line.
   b. Are there any other points on the line besides points A, B, and C? Explain.
   c. Do you think line AB is the same as line BA? Explain.

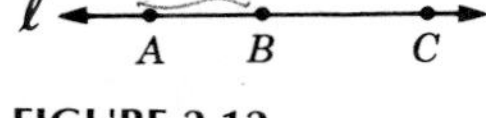

FIGURE 2.12

EXPLORING WITH THE COMPUTER 2.1

Use LOGO computer software. Give instructions to teach the turtle to draw the following:

1. Two lines that intersect to make a square corner
2. Two lines that are parallel
3. Three lines that determine as many points of intersection as possible
4. Four lines that intersect in as few points as possible
5. A figure with 2 parallel lines, 2 intersecting lines, and 3 concurrent lines
6. An interesting design that uses points and lines.

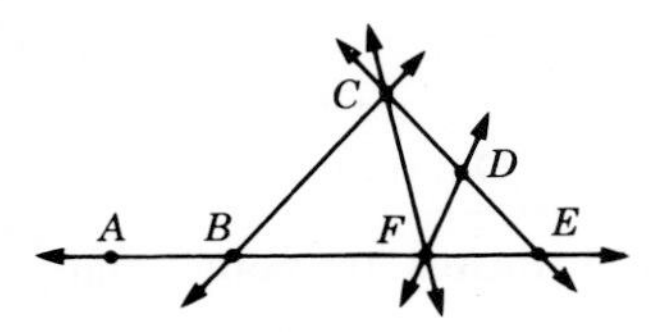

FIGURE 2.13

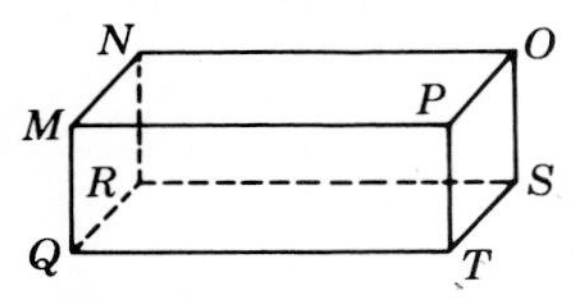

FIGURE 2.14

3. Refer to Figure 2.13 and answer the following:
   a. Name sets of 3 collinear points.
   b. Name sets of 3 noncollinear points.
   c. Name pairs of intersecting lines. Give the points of intersection.
   d. Name 3 lines that are concurrent.
   e. Do you think line *BC* intersects line *FD*? Why or why not?
4. Refer to Figure 2.14 and answer the following:
   a. Name sets of points that are coplanar.
   b. Name sets of points that are noncoplanar.
   c. Use 3 points in each plane to name pairs of intersecting planes. Name the lines of intersection.
   d. Name 3 pairs of parallel planes.
   e. Name a pair of skew lines.
   f. Name some pairs of intersecting planes.
5. Draw and label a picture for each of the following:
   a. Two points, *R* and *S*
   b. A line containing points *P* and *Q*
   c. A plane *S* containing points *C* and *D*
   d. Three collinear points, *G*, *H*, and *I*
   e. Three lines that intersect in points *R*, *S*, and *T*
   f. A point *P* on $\overleftrightarrow{AC}$, but not on $\overleftrightarrow{BC}$
6. Which of the following statements do you think are *always true*? Explain.
   a. Three collinear points are always coplanar.
   b. Three coplanar points are always collinear.
   c. Any 3 points are coplanar.
   d. It is possible to have 4 coplanar points, no 3 of which are collinear.
7. If 2 parallel planes *P* and *Q* intersect a third plane *R* in 2 lines ℓ and *m*, are ℓ and *m* necessarily parallel? Explain your answer.
8. Copy each of the figures in Figure 2.15. In your copy, use dashed lines for segments that would not be seen and use solid lines for segments that would be seen.

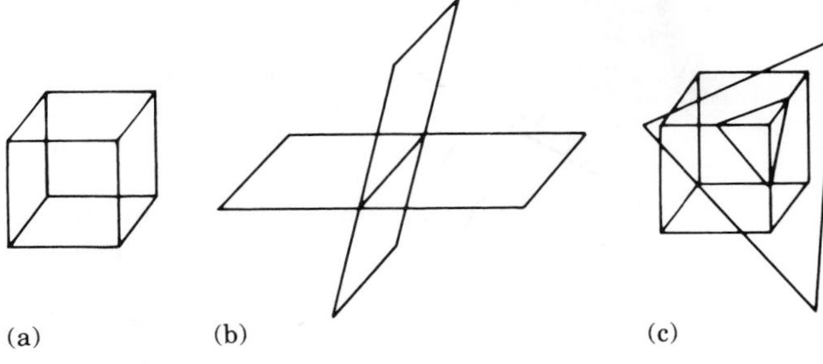
(a) (b) (c)

FIGURE 2.15

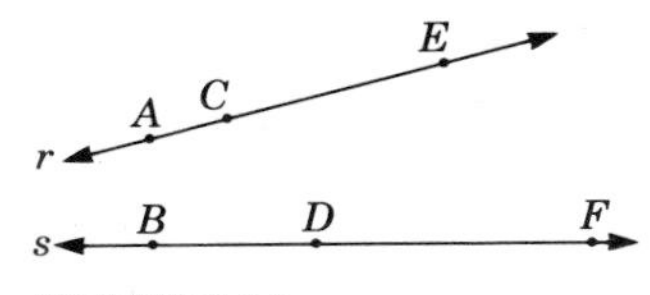

FIGURE 2.16

9. Four lines can be drawn to intersect in only 1 point. How can 4 lines be drawn to intersect in only
   a. 2 points? b. 3 points?
   c. 4 points? d. 5 points?
   e. 6 points?
10. **Critical Thinking.** Refer to Figure 2.16. A student said, "If you draw any 2 lines r and s and choose points A, B, C, D, E, and F alternately on r and s in any position, the 3 points of intersection of lines AB and DE, and lines BC and EF, and lines CD and AF will always be collinear."
    a. What does this statement mean?
    b. Do you think it is probably true?
    c. How did you decide?

Problem Solving: Developing Strategies and Skills

The problem-solving strategy **draw a picture** is very useful for solving geometric problems. Problem 1 illustrates the use of this strategy.

1. How many lines are determined by 9 points arranged in 3 rows of 3 points, each determining a square grid? Two pictures may be better than one. You may wish to draw one 3×3 array of dots and investigate the number of 3-point lines. Then draw another array and investigate the number of 2-point lines.
2. Three lines can be arranged to enclose at most one region in the plane. What is the greatest number of regions that can be enclosed by 6 lines?

TEACHING NOTE

Geoboards are an excellent aid for exploring geometric ideas at all ages. Activities that involve cooperative learning groups in a sustained exploration are quite useful. The activities in *Junior High Cooperative Problem Solving with Geoboards* (Goodnow, 1991) and other similar activities for primary and intermediate grades are examples.

BASIC GEOMETRIC FIGURES AND THEIR MEASURES

In Investigation 2.2, a geoboard is used to explore two important geometric figures: segments and triangles. A rubber band around two nails shows a segment. A rubber band around three noncollinear nails shows a triangle.

INVESTIGATION 2.2

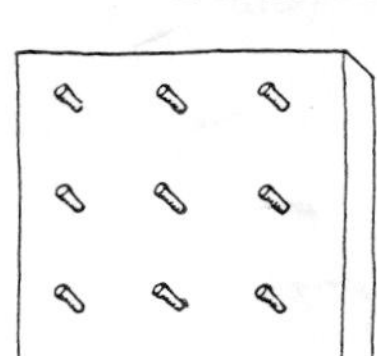

Explore the following questions on a geoboard such as the one shown. Record the results on dot paper.

A. How many different-length segments can be shown on a geoboard with 3 nails on a side? 4 nails on a side?
B. How many different-shaped triangles can be shown on a geoboard with 3 nails on a side? 4 nails on a side?

Some basic geometry figures are shown in Table 2.2. Symbols used to represent them are described, and the figures are defined.

TABLE 2.2

Physical Model, Figure	Description, Symbol	Definition
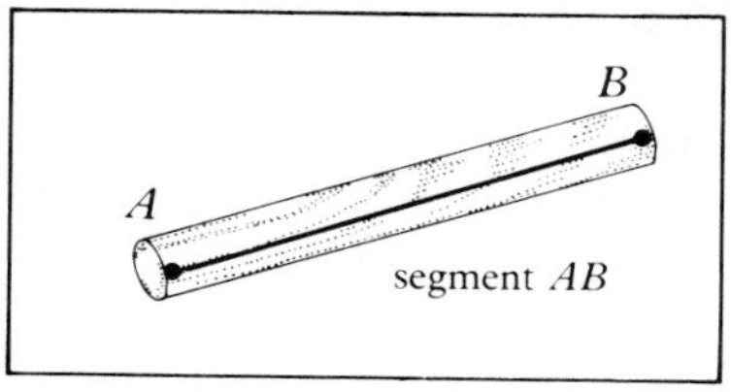	A and B are *endpoints*. We write $\overline{AB}$.	A **segment,** $\overline{AB}$, is the set of points A and B and all the points between A and B.
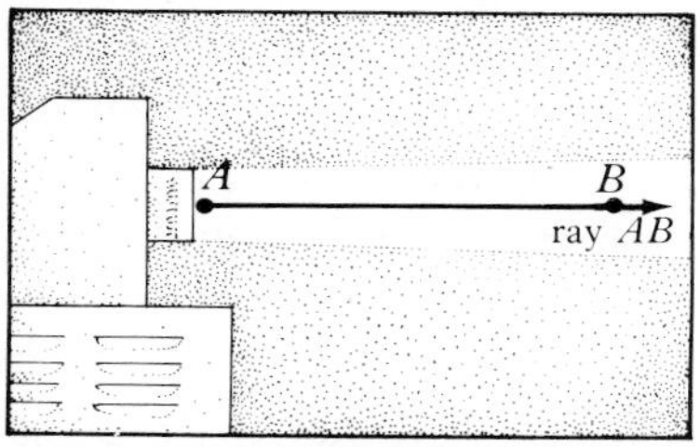	A is the endpoint. We write $\overrightarrow{AB}$.	A **ray** $\overrightarrow{AB}$, is a subset of a line that contains a given point A and all points on the same side of A as B.
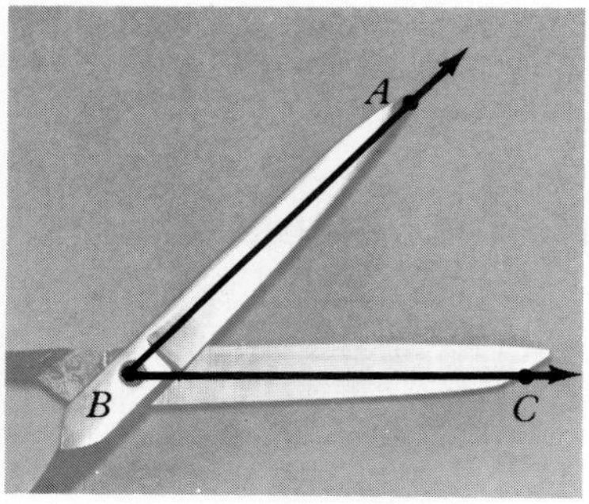	B is the *vertex*. $\overrightarrow{BA}$ and $\overrightarrow{BC}$ are the *sides*. The interior of $\angle ABC$ is the intersection of the points on A's side of $\overleftrightarrow{BC}$ with the points on C's side of $\overleftrightarrow{AB}$.	An **angle**, $\angle ABC$, is the union of two noncollinear rays that have the same endpoint. An **interior angle** of a figure is an angle formed by the vertex and two adjacent sides.
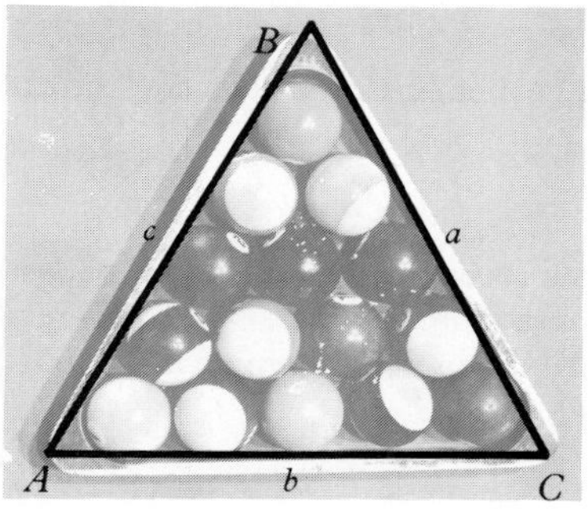	A, B, and C are vertices. $\overline{AB}$, $\overline{BC}$, and $\overline{AC}$ are sides. We write $\triangle ABC$.	A **triangle**, $\triangle ABC$, is the union of three segments determined by three noncollinear points.

TABLE 2.2 Continued

Physical Model, Figure	Description, Symbol	Definition
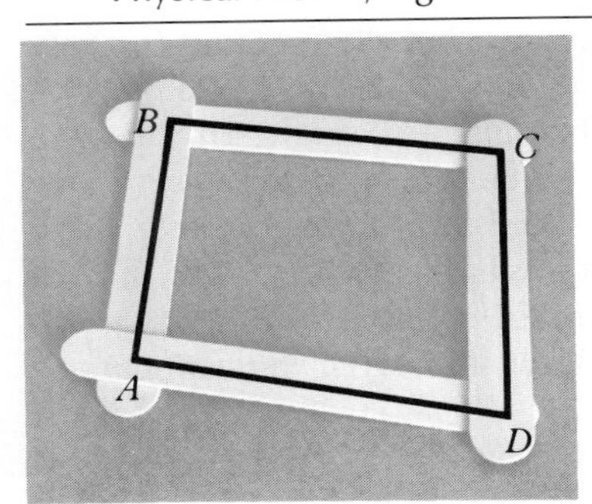	A, B, C, and D are vertices. $\overline{AB}$, $\overline{BC}$, $\overline{CD}$, and $\overline{AD}$ are sides. We write quadrilateral $ABCD$.	A **quadrilateral** is the union of four segments determined by four points, no three of which are collinear. The segments intersect only at their endpoints.
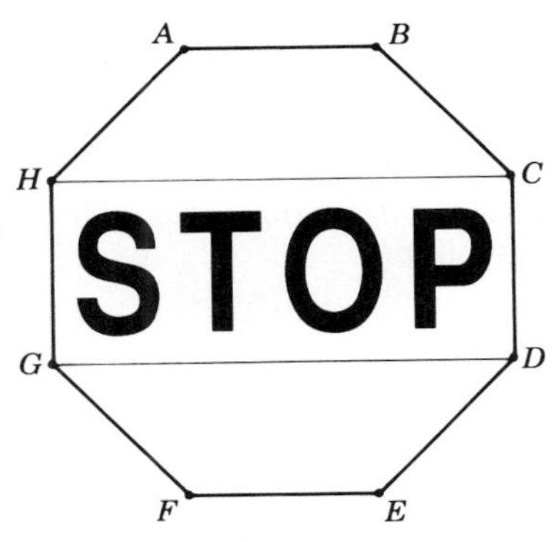	This polygon has eight sides. Points A, B, C, D, E, F, G, and H are its vertices. Each segment of the polygon is called a side. We write polygon $ABCDEFGH$.	A **polygon** is the union of segments meeting only at endpoints such that (1) at most, two segments meet at one point, and (2) each segment meets exactly two other segments.
	Points A, B, and C are on the circle. Point O is the center of the circle. $\overline{AB}$ is a *diameter* of the circle. $\overline{OB}$ is a *radius* of the circle. $\overline{BC}$ is a chord of the circle. We say circle O. We Write $\odot O$.	A **circle** is the set of all points in a plane that are a fixed distance from a given point in the plane. A **chord** of a circle is any segment with endpoints on the circle.

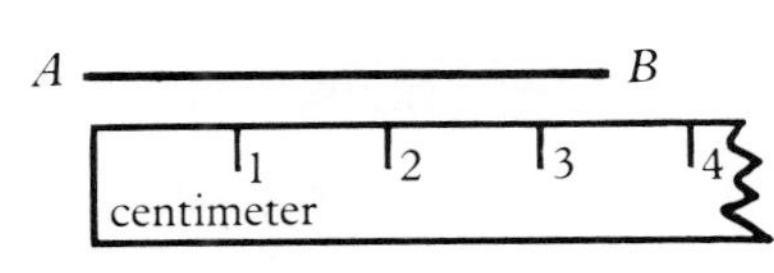

FIGURE 2.17

We look at a pair of segments and think "one is longer than the other" or "they are the same length." To describe the property of length, we assign a real number called the **length measure** to each segment. In Figure 2.17, the length of $\overline{AB}$ is 3.5 cm. We write $AB = 3.5$.

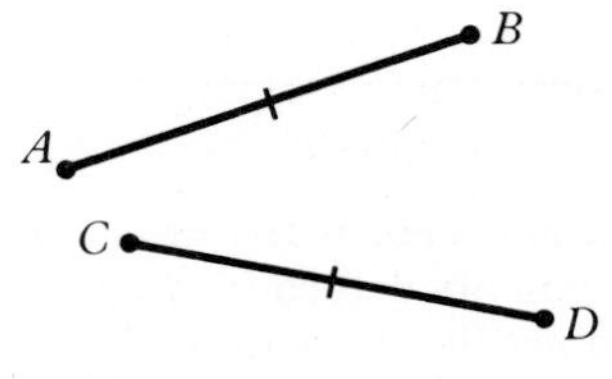

FIGURE 2.18

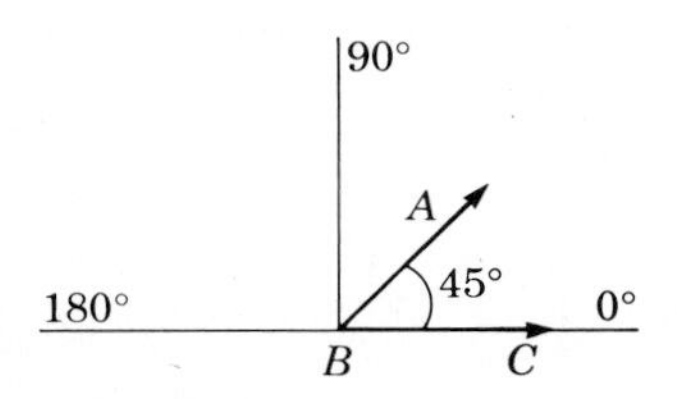

FIGURE 2.19

We have a special way to describe two segments that are the same length. Two segments are **congruent** if they have the same length. We say $\overline{AB}$ is congruent to $\overline{CD}$, (Fig. 2.18). We write $\overline{AB} \cong \overline{CD}$. We sometimes mark each segment to show that they are congruent.

An angle possesses a property that we might think of as "size" or "amount of rotation." To describe this property, we assign a real number between 0° and 180° to each angle. This number is called the **degree measure** (m) of the angle. In Figure 2.19, the degree measure of $\angle ABC$ is 45. We write $m\angle ABC = 45°$. We sometimes write that $\angle ABC$ has a measure of 45°.

EXPLORING WITH THE COMPUTER 2.2

1. Use LOGO commands to draw each of the following angles. (Draw only the portion shown by the heavy line.)

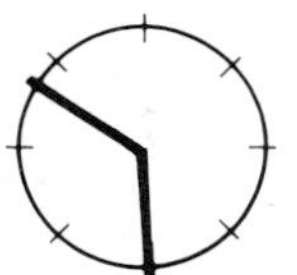
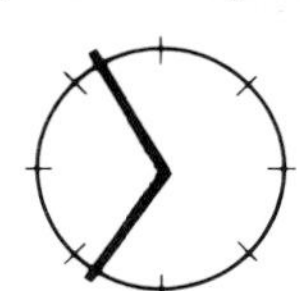

2. Use LOGO commands that draw figures like each of the following. What is the relationship between the numbered angles in each part?

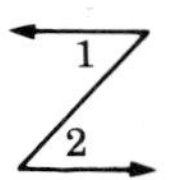

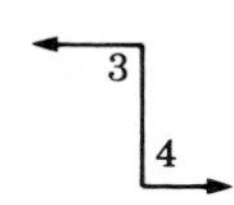

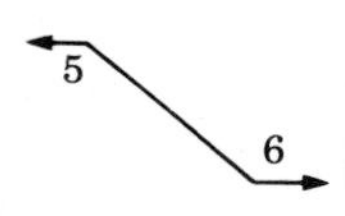

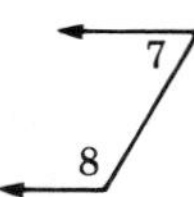

3. Use this LOGO procedure to answer the questions that follow:

```
TO ANGLE :D
FD 50 BK 50 RT:D FD 50 BK 50 LT:D
END
```

   a. What is the measure of the angle that this procedure draws? How does the turtle measure angles?
   b. Draw an angle that measures 60° in a clockwise direction. What angle would the turtle have to move in a counterclockwise direction to draw the same angle? Use the computer to check your answer.
   c. Repeat part b for clockwise angles of 10°, 85°, 150°, 180°, 210°, and 330°.

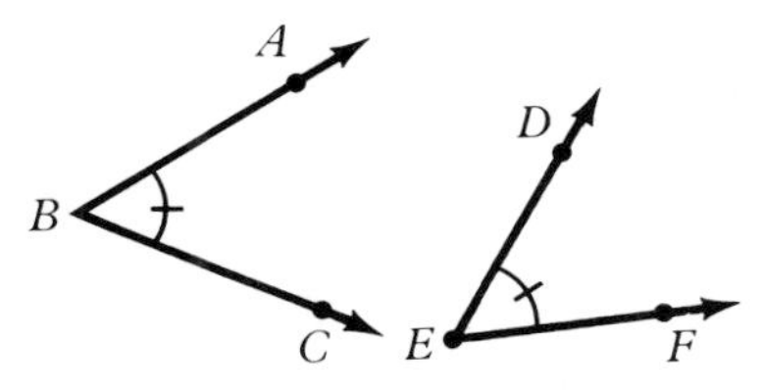

FIGURE 2.20

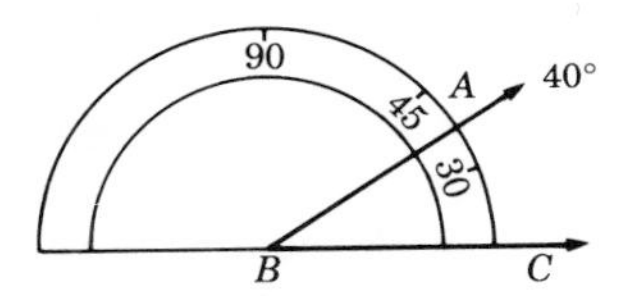

FIGURE 2.21

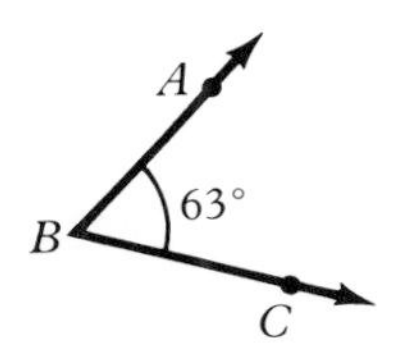

FIGURE 2.22

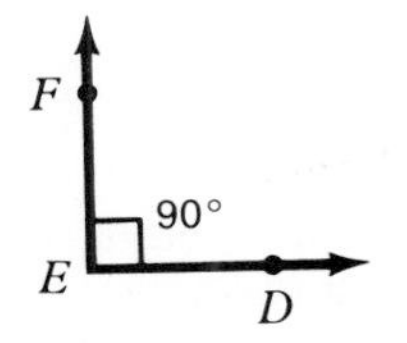

FIGURE 2.23

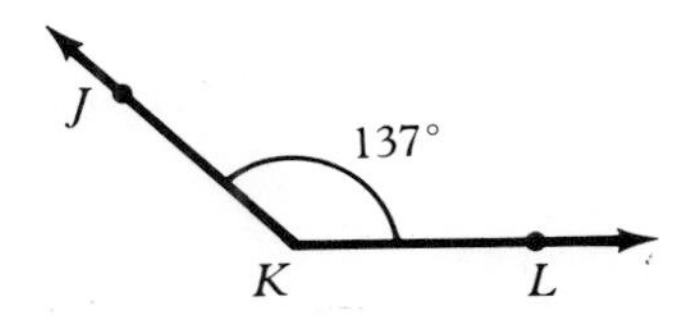

FIGURE 2.24

We have a special way to describe two angles that have the same measure. Two angles are **congruent** if they have the same measure. We say $\angle ABC$ is congruent to $\angle DEF$ (Fig. 2.20). We write $\angle ABC \cong \angle DEF$.

The measure of an angle is found using a protractor, as shown in Figure 2.21. Place the center point of the protractor on the vertex of the angle. Line up the mark labeled 0 on either scale with one side of the angle. Then, read the scale where it falls on the other side of the angle. The degree measure of $\angle ABC$ is 40, or $m\angle ABC = 40°$.

Angles can be classified according to their measures. An **acute angle** is an angle with measure less than 90° (Fig. 2.22). A **right angle** is an angle with measure of 90° (Fig. 2.23). An **obtuse angle** is an angle with measure greater than 90° (Fig. 2.24).

EXERCISE SET 2.2

1. Which geometric concepts are suggested by these physical situations?
   a. A taut string, held tightly at each end
   b. A beam of light from a strong searchlight
   c. A plastic hoop
   d. The blades of an open pair of scissors
   e. A fine wire bent in one place
   f. Railroad tracks
   g. A city intersection
2. Name a physical object that suggests these geometric figures:
   a. A line segment b. A ray c. An angle
   d. A triangle e. A quadrilateral f. A circle
   g. A polygon that is not a triangle or a quadrilateral
3. Draw and label the following geometric figures:
   a. $\overline{AB}$ b. $\overleftrightarrow{CD}$ c. $\angle JKL$
   d. $\overrightarrow{GH}$ e. $\overrightarrow{HG}$ f. $\triangle XYZ$
   g. Quadrilateral $RSTU$ h. Polygon $MNOPQ$
4. Describe how $\overline{RS}$ differs from $\overline{SR}$. How does $\overrightarrow{RS}$ differ from $\overrightarrow{SR}$?
5. Are Figures 2.25a through 2.25d angles? Why or why not?

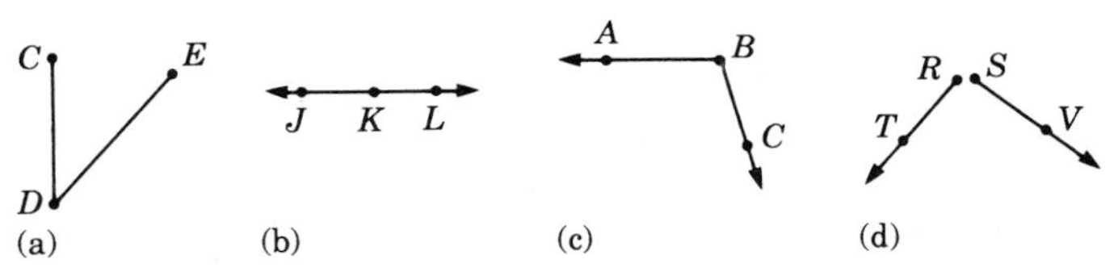

FIGURE 2.25

APPLICATION

Sprinkler Problem: Suppose you want to install permanent rotating water sprinklers that water a circular region with radius 30 ft. There is a light pole directly north of the back door of your house, 50 ft. from the door. You want to install two sprinklers as near the door as possible, but so that they will not hit the house or the light pole. You want to be able to water as much of the grass as possible. Where should you install the sprinklers? Explain your decision.

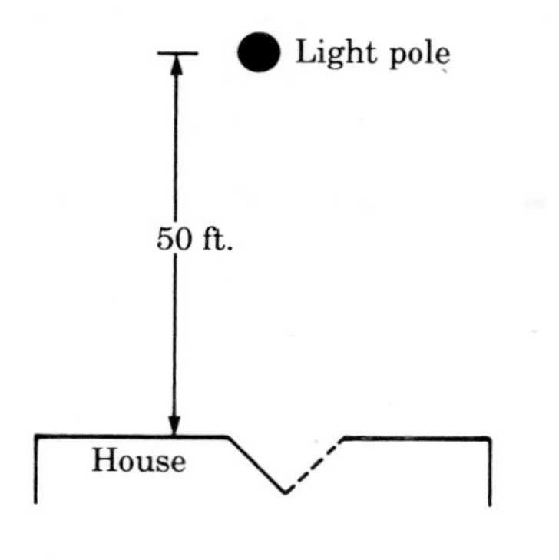

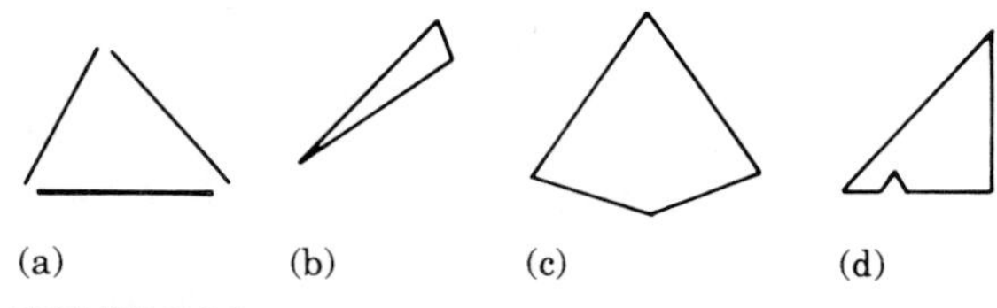

FIGURE 2.26

6. Are Figures 2.26a through 2.26d triangles? Explain why or why not.
7. Use only a straight edge to draw 5 different-length segments. Estimate the size of each segment. Then measure each segment. What was the average number of centimeters (to the nearest tenth centimeter) that your estimate differed from the actual measure?
8. Use only a ruler to draw 5 angles of different sizes. Estimate the size of each angle. Then measure each angle using a protractor. What was the average number of degrees (to the nearest whole number) that your estimate differed from the actual measure?
9. Decide whether these following statements are *sometimes true, always true,* or *never true.*
   a. $\overrightarrow{AB}$ is the same ray as $\overrightarrow{BA}$.
   b. The sum of the measures of 2 acute angles is more than 90°.
   c. The sum of the measure of 2 acute angles is more than 180°.
   d. The sum of the measures of 2 obtuse angles is less than or equal to 180°.
   e. A certain quadrilateral has 4 right angles.
   f. A certain polygon has 3 sides.
   g. A circle has radius greater than or equal to its diameter.
10. Form the longest possible segment on a geoboard with 5 nails on a side. If the vertical or horizontal distance between two adjacent nails is 1 unit, find the length of the longest segment without measuring.
11. Use a ruler and compass to construct $\triangle ABC$ with side AB and angles A and B (Fig. 2.27).
12. Are Figures 2.28a through 2.28d quadrilaterals? Explain why or why not.

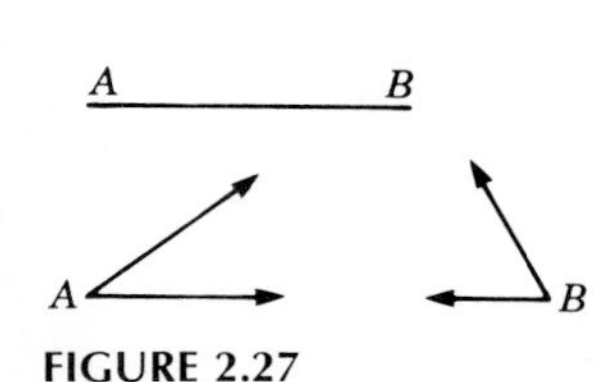

FIGURE 2.27

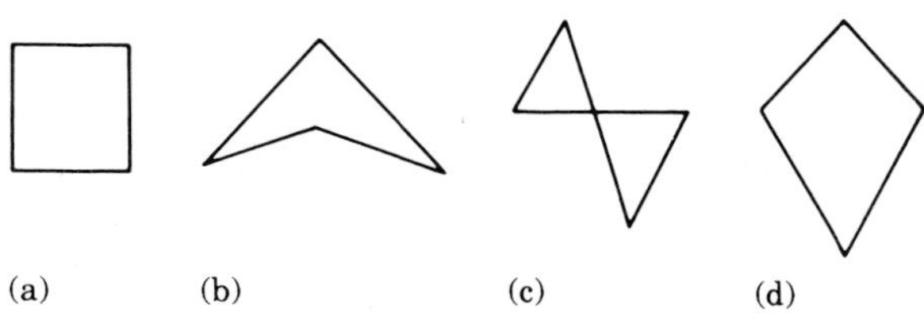

FIGURE 2.28

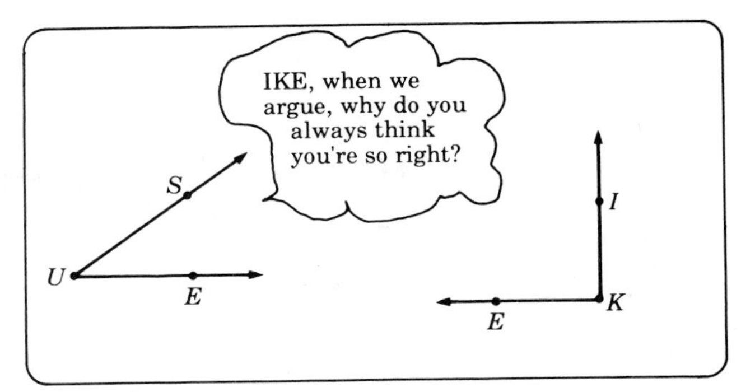

FIGURE 2.29

13. Figure 2.29 is a special cartoon called a "geomotoon." Create a geomotoon.
14. Study the ruler-and-compass construction procedures for copying and bisecting a segment (Appendix B) and, given segments with lengths a and b (Fig. 2.30), construct segments with the following lengths:
    a. Length $2a$
    b. Length $a + b$
    c. Length $a - b$
    d. $a/2$
    e. $b/4$
15. Refer to Figure 2.31 and answer the following:
    a. Name at least 8 triangles
    b. Name at least 8 quadrilaterals.
    c. Copy the figure. Then draw one more segment that will add exactly 3 more triangles.
16. Study the ruler-and-compass construction procedures for copying and bisecting an angle (Appendix B) and, given the angles with measures x and y (Fig. 2.32), construct angles with the following measures:
    a. $x + y$
    b. $x - y$
    c. $3y$
    d. $x/2$
    e. $x/4$

a

b

FIGURE 2.30

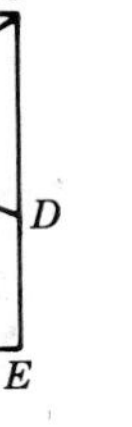

FIGURE 2.31

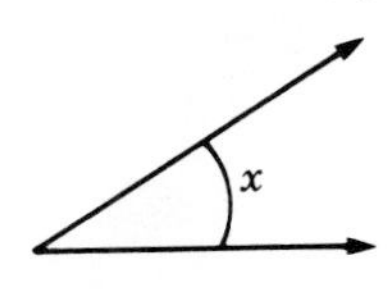

FIGURE 2.32

FIGURE 2.33

FIGURE 2.34

TEACHING NOTE

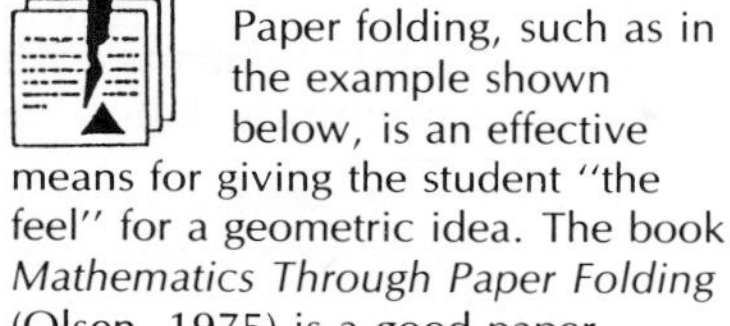

Paper folding, such as in the example shown below, is an effective means for giving the student "the feel" for a geometric idea. The book *Mathematics Through Paper Folding* (Olsen, 1975) is a good paper-folding resource book for grades 7 through 12.

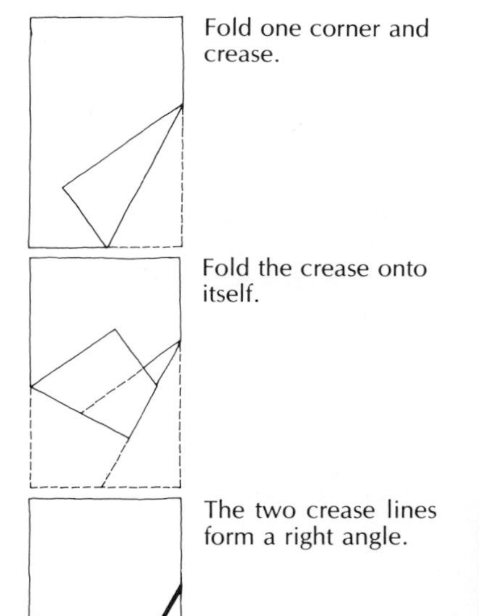

17. **Critical Thinking.** Each of the drawings in Figure 2.33 is a "quadtri." None of the drawings in Figure 2.34 is a "quadtri." Using this information, write a definition of a "quadtri." Compare it with definitions other students have written and work with them to form a group definition. Form some questions you would like to ask in order to be sure that your group definition is correct.

LINE AND ANGLE RELATIONSHIPS

Throughout the ages humans have used simple geometric shapes and relationships as the basis for the creation of interesting and useful objects. The architecture shown in Figure 2.35 makes creative use of relationships between lines and angles to provide an attractive, functional building. Investigation 2.3 provides an opportunity to discover some of these important relationships.

FIGURE 2.35

INVESTIGATION 2.3

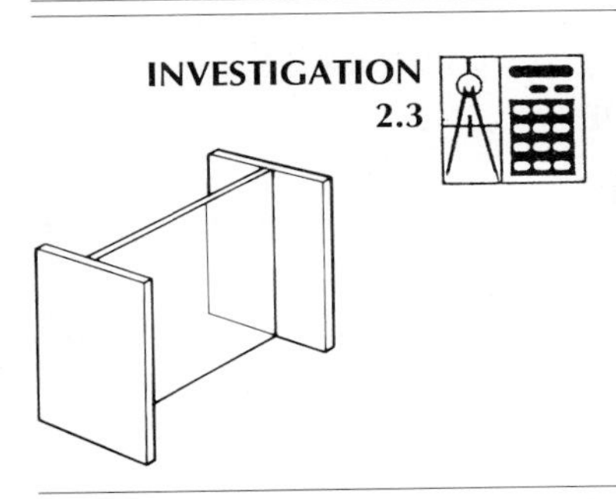

A MIRA, as shown here, is a device made from red plexiglass with the properties of a mirror, but that, at the same time, is transparent. Use a MIRA to do the following, if possible:

A. Find the middle point of a segment.
B. Draw a line that forms right angles with a given line.
C. Draw a pair of parallel lines.
D. Divide an angle into 2 congruent angles.

See Appendix B for additional activities with a MIRA.

The idea of perpendicular for lines, for a line and a plane, and for planes plays an important role in geometry. The definitions in Table 2.3 describe this idea.

TABLE 2.3

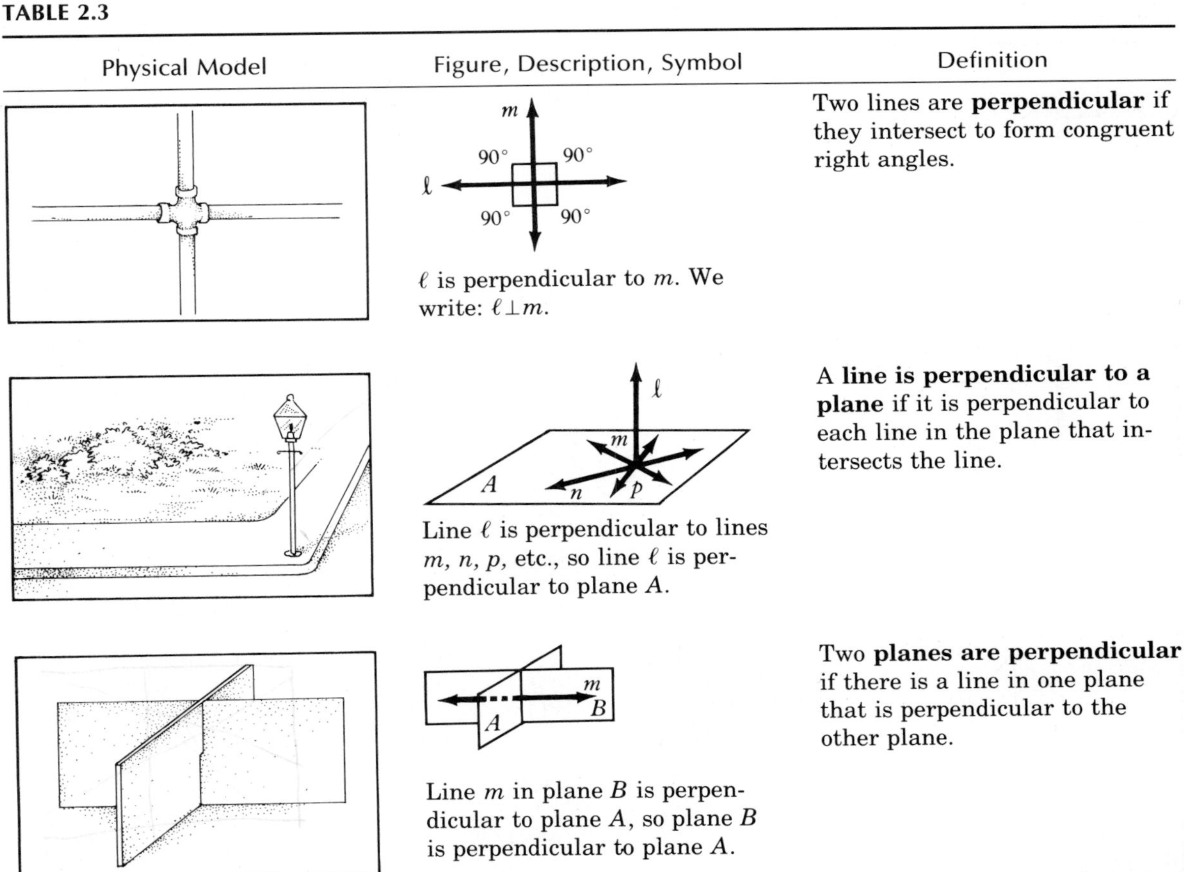

Physical Model	Figure, Description, Symbol	Definition
	ℓ is perpendicular to m. We write: $\ell \perp m$.	Two lines are **perpendicular** if they intersect to form congruent right angles.
	Line ℓ is perpendicular to lines m, n, p, etc., so line ℓ is perpendicular to plane A.	A **line is perpendicular to a plane** if it is perpendicular to each line in the plane that intersects the line.
	Line m in plane B is perpendicular to plane A, so plane B is perpendicular to plane A.	Two **planes are perpendicular** if there is a line in one plane that is perpendicular to the other plane.

Some other relationships between lines and angles are described in Table 2.4. The definitions in Table 2.5 concern relationships between pairs of angles.

TABLE 2.4

Physical Model	Figure, Description, Symbols	Definition
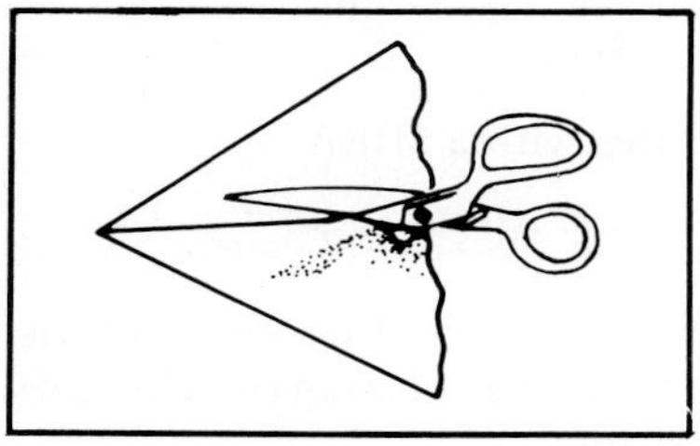	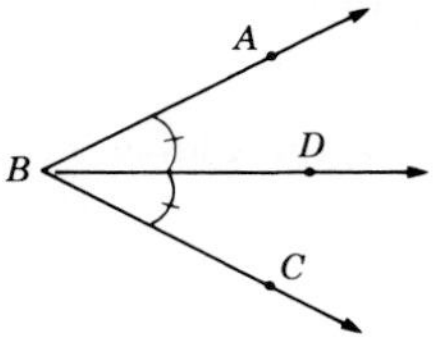 Ray BD is the *bisector* of $\angle ABC$. The points on $\overrightarrow{BD}$ are equal in distance from the sides of $\angle ABC$.	The **bisector of an angle** ABC is a ray BD in the interior of $\angle ABC$ such that $\angle ABD \cong \angle DBC$.
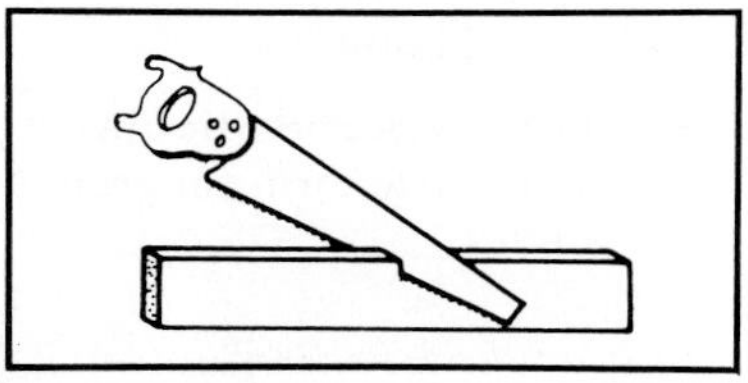	 Point C is the *midpoint* of $\overline{AB}$.	The **midpoint of a segment** is a point C between A and B such that $\overline{AC} \cong \overline{CB}$.
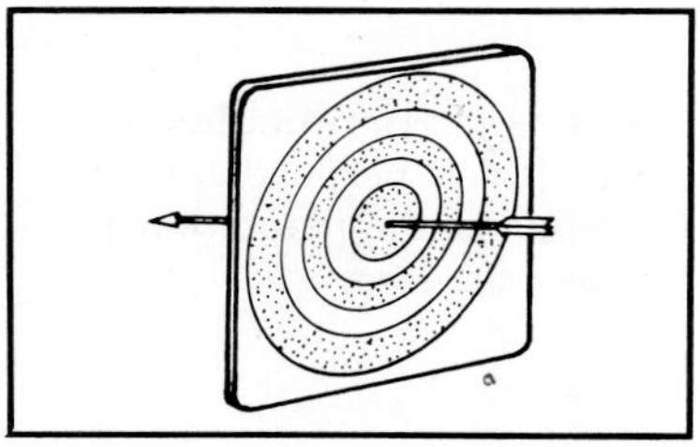	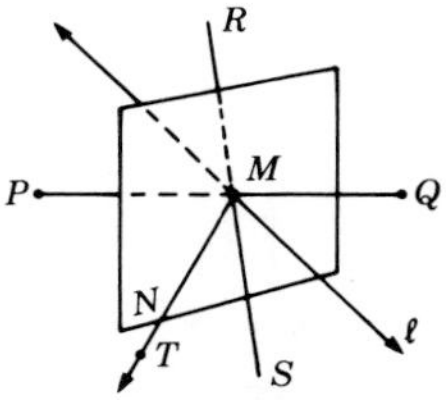 $\overline{RS}$, $\overrightarrow{MT}$, line ℓ, and plane N all intersect $\overline{PQ}$ at the midpoint M, and are bisectors of $\overline{PQ}$.	A **bisectors of a segment** is any point, segment, ray, line, or plane that contains the midpoint of the segment.
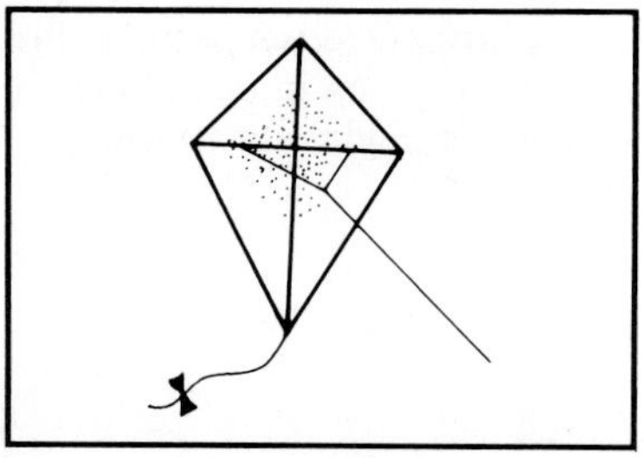	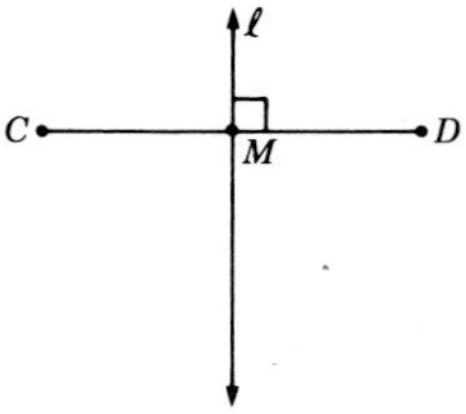 ℓ is the perpendicular bisector of $\overline{CD}$.	The **perpendicular bisector of a segment** is a line that is perpendicular to the segment and contains its midpoint.

TABLE 2.4 Continued

Physical Model	Figure, Description, Symbols	Definition
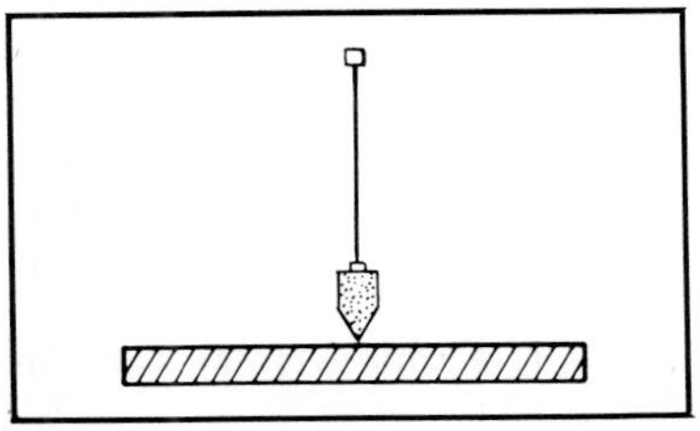	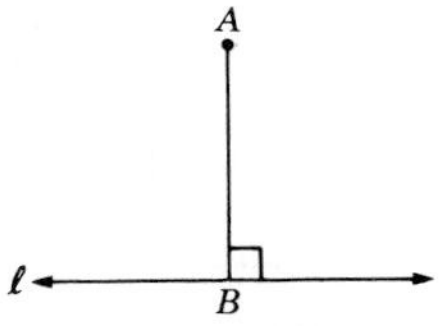 AB is the distance from point A to line ℓ.	The **distance from a point to a line** is the length of the segment drawn from the point perpendicular to the line.

TABLE 2.5

Figure	Description, Symbols	Definition
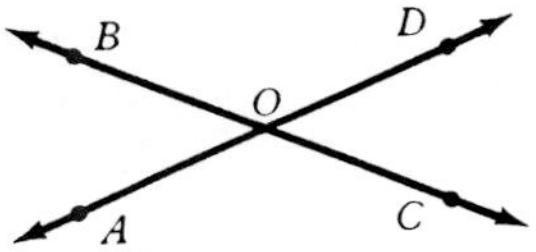 $\overleftrightarrow{BC}$ and $\overleftrightarrow{AD}$ intersect at O.	$\angle AOB$ and $\angle COD$ are vertical angles.	**Vertical angles** are two angles that are formed by two intersecting lines but which are not a linear pair of angles.
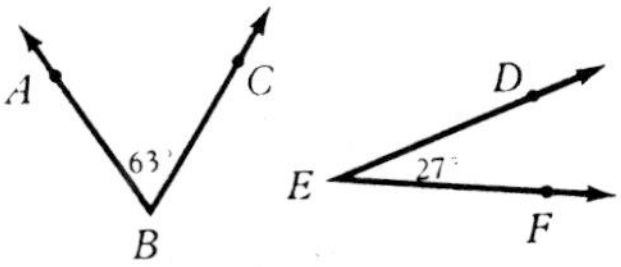 The sum of the measures of $\angle ABC$ and $\angle DEF$ is 90°.	$\angle ABC$ is complementary to $\angle DEF$. $\angle DEF$ is complementary to $\angle ABC$.	**Complementary angles** are two angles whose measures have a sum of 90°.
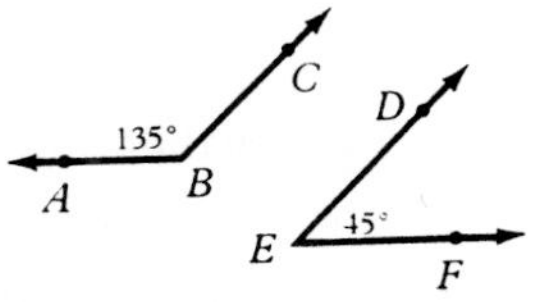 The sum of the measures of $\angle ABC$ and $\angle DEF$ is 180°.	$\angle ABC$ is supplementary to $\angle DEF$. $\angle DEF$ is supplementary to $\angle ABC$.	**Supplementary angles** are two angles whose measures have a sum of 180°.
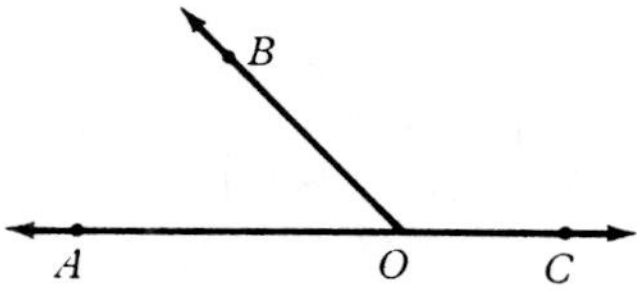 $\angle AOB$ and $\angle BOC$ have a common side, $\overrightarrow{OB}$. The union of the other two sides, $\overrightarrow{OA}$ and $\overrightarrow{OC}$, is a line.	$\angle AOB$ and $\angle BOC$ are a linear pair of angles.	A **linear pair of angles** is a pair of angles with a common side such that the union of the other two sides is a line.

EXPLORING WITH THE COMPUTER 2.3

Use geometry exploration software (see Appendix E) to make and test conjectures about the following questions. Use at least 3 different-shaped figures when testing each conjecture. Record your results.

1. What is the sum of the measures of the interior angles of a triangle? A quadrilateral?
2. What is the sum of the measures of 3 exterior angles of a triangle (one at each vertex)? Four exterior angles of a quadrilateral (one at each vertex).
3. How does an exterior angle of a triangle relate to the interior angles of the triangle?

Figure 2.36 is used to define relationships between certain pairs of angles when lines are cut by a transversal. Angles related as $\angle 1$ and $\angle 2$ are **alternate interior angles** (Fig. 2.36a). Angles related as $\angle 3$ and $\angle 4$ are **corresponding angles** (Fig. 2.36b). Angles related as $\angle 5$ and $\angle 6$ are **alternate exterior angles** (Fig. 2.36c).

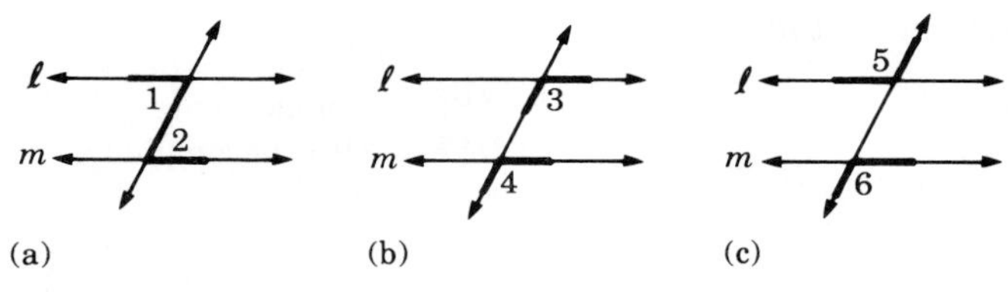

FIGURE 2.36

EXPLORING WITH THE COMPUTER 2.4

Use LOGO software and create separate instructions so that the turtle will draw each of the following:

1. A 50-unit segment that is perpendicular to a 100-unit segment
2. A 60° angle that is supplementary to a 120° angle
3. A pair of 45° vertical angles
4. A 60° angle and its bisector
5. A perpendicular bisector of a 100-unit segment
6. Two complementary angles with degree measures in the ratio 2:1
7. An angle that is supplementary to an angle one-third its measure in degrees.

Compare your instructions with others. Which instructions were most efficient?

APPLICATION

Old Wheel Problem: An archeologist found a part of an old wheel in a dig. She would like to determine how large the complete wheel was. Draw a diagram of such a wheel and decide how you could help the archeologist. Can you find more than one way to do it? (Note: The perpendicular bisector of a chord of a circle goes through the center of the circle.)

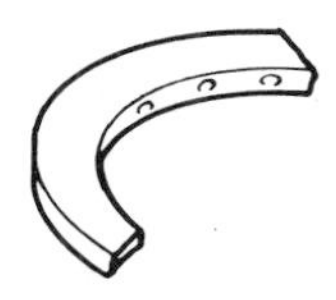

EXERCISE SET 2.3

1. Draw a line ℓ and construct a line r perpendicular to it. Call the point of intersection P. (The description of ruler-compass and MIRA constructions in Appendix B may be helpful.) Use this pair of perpendicular lines and answer the following questions. Explain your thinking.
   a. How many angles are formed? What are the measures of these angles?
   b. Do you think there are other lines perpendicular to line ℓ? How many?
   c. Do you think there is another line through point P that is perpendicular to line ℓ?
2. Draw a segment AB and construct its perpendicular bisector. Call the point of intersection M. Answer these questions and explain your reasoning.
   a. How do segments AM and MB compare.
   b. What can you say about the adjacent angles (angles with a common side) formed by the perpendicular lines?
3. Use a protractor to do the following:
   a. Draw a pair of angles that are complementary, but have no common vertex.
   b. Draw a pair of angles that are supplementary, but have no common vertex.
   c. Draw a pair of angles that are complementary and have a common vertex, but no common side.
   d. Draw a pair of angles that are supplementary and have a common vertex, but no common side.
   e. Draw a pair of angles that are complementary and have a common side.
   f. Draw a pair of supplementary, adjacent angles.
   g. Draw a linear pair of angles. What do you think is always true about the angles in a linear pair?
4. Use a ruler and compass to construct a pair of parallel lines (Fig. 2.37). Draw a transversal and label the angles as shown in the figure.
   a. Name 2 pairs of alternate interior angles.
   b. Name 4 pairs of corresponding angles.
   c. Name 2 pairs of alternate exterior angles.
   d. Name 2 pairs of interior angles on the same side of the transversal.
5. Use a ruler and compass to construct the figures in Exercise 4. Use a protractor to complete the following:
   a. Name 2 pairs of exterior angles on the same side of the transversal.
   b. Measure the angles in each pair of alternate interior angles. What do you think is true about pairs of alternate interior

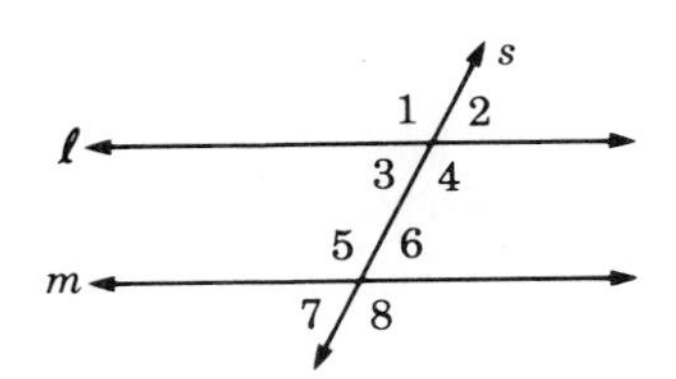

FIGURE 2.37

angles when parallel lines are cut by a transversal?

c. Measure to help you form a conjecture about pairs of corresponding angles when parallel lines are cut by a transversal.

d. What conjecture can you make about alternate exterior angles? What about exterior angles on the same side of the transversal? Finally, what about interior angles on the same side of the transversal?

6. Draw a transversal and construct 2 rays on the same side of the transversal that make congruent corresponding angles with the transversal. Extend the rays to form 2 lines cut by a transversal with a pair of corresponding angles congruent. What do you think is true about the 2 lines?

7. In a manner similar to that used for Exercise 6, construct 2 lines cut by a transversal with a pair of alternate interior angles congruent. What do you think is true about these 2 lines?

8. Use a MIRA to construct 2 lines that are perpendicular to the same line. What do you think is true about the 2 lines?

9. Draw an acute $\angle ABC$. Choose a point P not in the interior of $\angle ABC$ and construct lines $\overline{PD} \perp \overline{AB}$ and $\overline{PE} \perp \overline{BC}$. Measure $\angle ABC$ and $\angle DPE$. What do you think might be true about these angles? Try this exercise with different-size angles. Does it seem to be true for all angles?

10. Use the ruler-compass constructions described in Appendix B to construct the following:

    a. Construct a perpendicular to a line through a given point on the line.

    b. Construct a perpendicular to a line through a given point not on the line.

11. Study the constructions using a MIRA in Appendix B and use a MIRA to do parts a and b of Exercise 10.

12. Which of the construction techniques, ruler-compass or MIRA, do you think is more efficient for the constructions in Exercises 10 and 11. Which is more accurate? Explain.

13. Use ruler-compass to do the following constructions:

    a. Construct a 22 ½° angle.
    b. Construct a 135° angle.
    c. Construct a 112 ½° angle.
    d. Construct a rectangle with length 3 times the width.
    e. Construct a triangle with sides in a ratio 1:2:3.

14. Read Theorem 72 in Appendix F and explain how it justifies the procedure of using lined notebook paper to divide a wooden rod into equal parts (Fig. 2.38).

15. **Critical Thinking.** You have an adjustable compass in good working order (Fig. 2.39). You also own a rusty compass that is stuck and always has the same opening. (See Appendix B for a

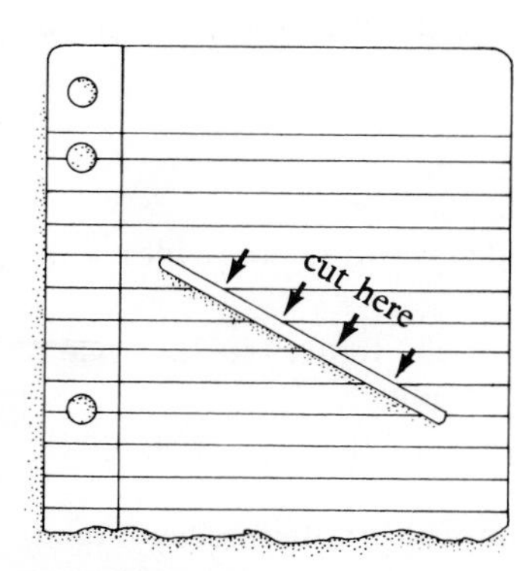

FIGURE 2.38

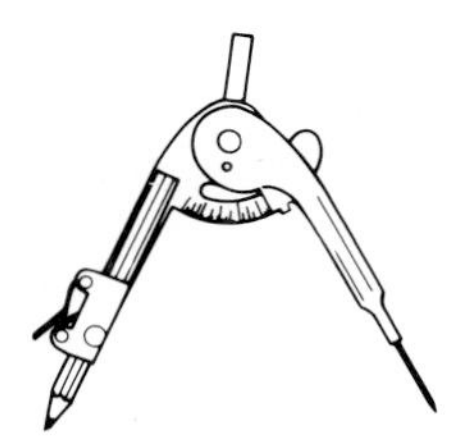

FIGURE 2.39

review of compass constructions.) Which of the following constructions must be done with your good adjustable compass? Which can be done with your rusty compass? Support your decisions.

a. Construct the perpendicular bisector of a segment (with your rusty compass, assume that the compass opening is less than one-half of the segment).
b. Construct the bisector of an angle.
c. Construct a perpendicular to a line from a point on the line.

THEOREMS ABOUT INTERSECTING LINES

In some of the exercises in Exercise Set 2.3, you may have arrived at some generalizations by constructing and measuring figures. A generalization that can be proven true by applying logic is called a **theorem**. In this section, four theorems about the measure of angles formed by intersecting lines that you may have discovered in earlier exercises are summarized without proof.

Theorem 1 Vertical angles are congruent (Fig. 2.40). That is,

$$m\angle 1 = m\angle 2$$
$$m\angle 3 = m\angle 4$$

Theorem 2 Angles that form a linear pair are supplementary. That is,

$$m\angle 1 + m\angle 3 = 180^\circ$$
$$m\angle 1 + m\angle 4 = 180^\circ$$

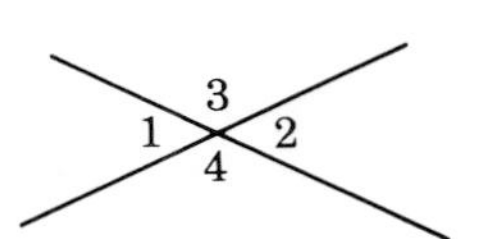

FIGURE 2.40

Theorem 3 Given parallel lines cut by a transversal, alternate interior angles are congruent (Fig. 2.41). That is,

$$m\angle 3 = m\angle 6$$
$$m\angle 4 = m\angle 5$$

Theorem 4 Given parallel lines cut by a transversal, corresponding angles are congruent. That is,

$$m\angle 1 = m\angle 5 \qquad m\angle 3 = m\angle 7$$
$$m\angle 2 = m\angle 6 \qquad m\angle 4 = m\angle 8$$

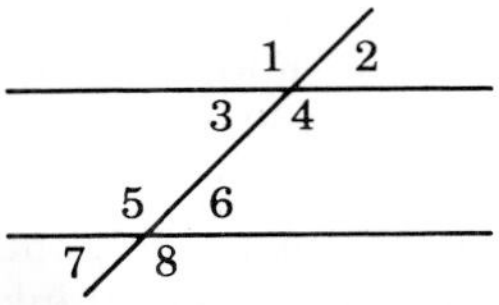

FIGURE 2.41

These theorems can be used to complete each of the exercises in the following set.

EXERCISE SET 2.4

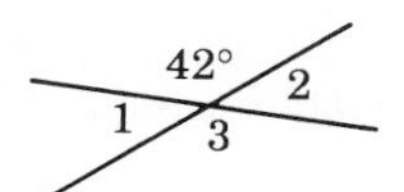

FIGURE 2.42

1. Refer to Figure 2.42. Complete the following statements and state the theorem that you used for each.
   a. $m\angle 1 =$ __
   b. $m\angle 2 =$ __
   c. $m\angle 3 =$ __

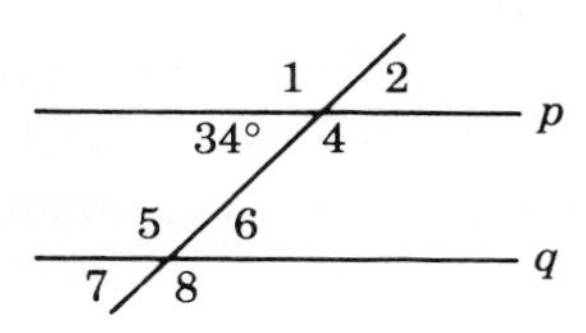

FIGURE 2.43

2. Refer to Figure 2.43. Complete the following statements and state the theorem you used for each.
   a. $m\angle 1 =$ __
   b. $m\angle 2 =$ __
   c. $m\angle 4 =$ __
   d. $m\angle 5 =$ __
   e. $m\angle 6 =$ __
   f. $m\angle 8 =$ __
   g. $m\angle 7 =$ __

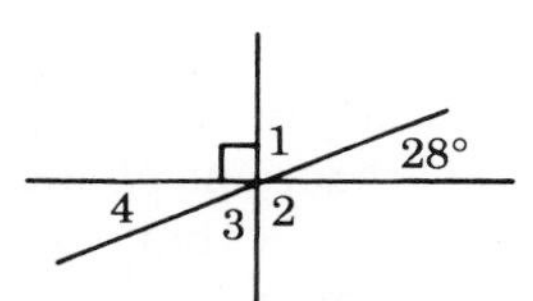

FIGURE 2.44

3. Refer to Figure 2.44. Complete the following statements and state the theorem you used for each.
   a. $m\angle 1 =$ __
   b. $m\angle 2 =$ __
   c. $m\angle 3 =$ __
   d. $m\angle 4 =$ __

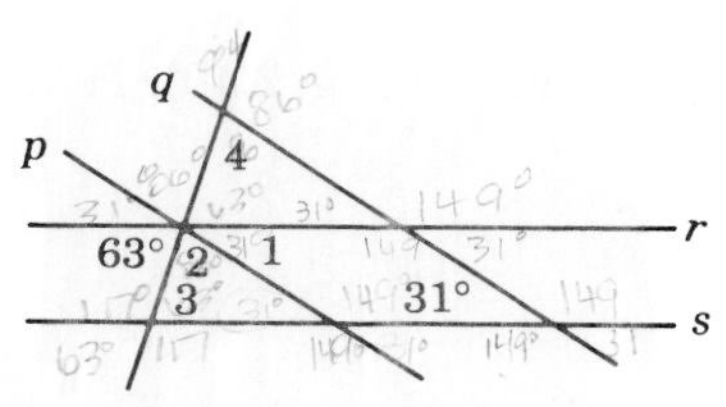

FIGURE 2.45

4. Refer to Figure 2.45. In this figure, $p \parallel q$ and $r \parallel s$. Complete the following statements and state the theorem you used for each.
   a. $m\angle 1 =$ __
   b. $m\angle 2 =$ __
   c. $m\angle 3 =$ __
   d. $m\angle 4 =$ __

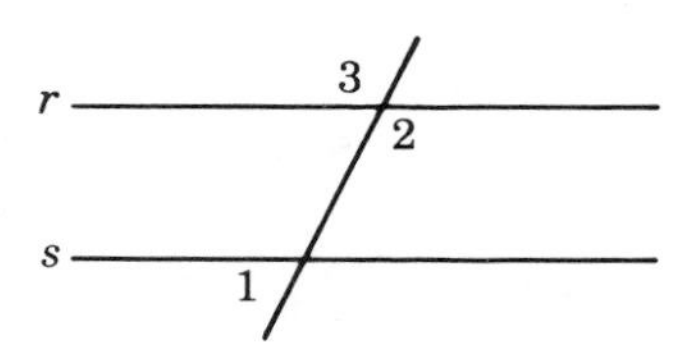

FIGURE 2.46

5. **Critical Thinking.** In Figure 2.46, if $r \parallel s$, give a convincing argument for each of the following.
   a. $\angle 1$ is supplementary to $\angle 3$.
   b. $\angle 1$ is supplementary to $\angle 2$.

Problem Solving: Developing Skills and Strategies

Some problems require special insight, and the only strategy seems to be to think about the problem long enough until the solution just "pops into your mind." However, in solving such a problem, you may use other strategies without even thinking about them. Try to solve the following problem.

Problem: A flower bed with opposite sides parallel and all sides equal is to be built in the center of circular patio. The plans, in Figure 2.47, show the flower bed inside a dotted rectangle. $AB = 4$ m and $BC = 6$ m. The contractor is puzzled as to how to calculate the length of the side of the flower bed. His daughter glances at the plan, says she knows the length, and hardly has to calculate at all. Do you think this is possible? Find the length.

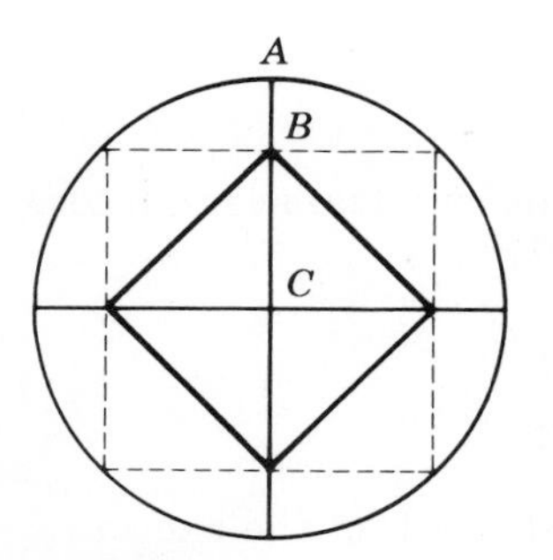

FIGURE 2.47

TEACHING NOTE

Teachers can integrate geometry with other subjects. To integrate geometry and art, have students fold and crease a paper several times in a way so the corners do not meet. Have them open the paper and color triangles, quadrilaterals, and other polygons, each a different color. Also have them color acute angles, right angles, and obtuse angles, each with a different color. Then have them select the color designs that are most pleasing.

MORE ABOUT TRIANGLES AND QUADRILATERALS

The ability to classify triangles and quadrilaterals is central to the development of important geometric concepts. Triangles, for example, can be classified in different ways (Fig. 2.48). An **isosceles triangle** is usually defined as a triangle with *at least* two sides congruent. Because of this, an equilateral triangle is considered to be a special type of isosceles triangle. Note that triangles can also be classified according to their symmetry properties. This classification will be described in a later section. For now, to begin to analyze these figures, try Investigation 2.4.

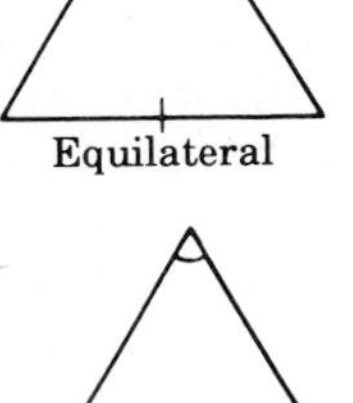

Equilateral

Isosceles

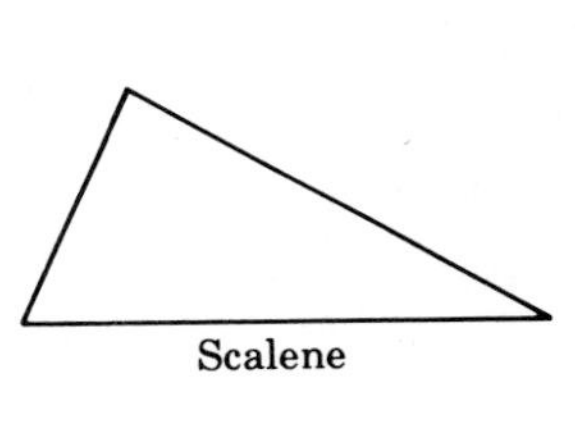

Scalene

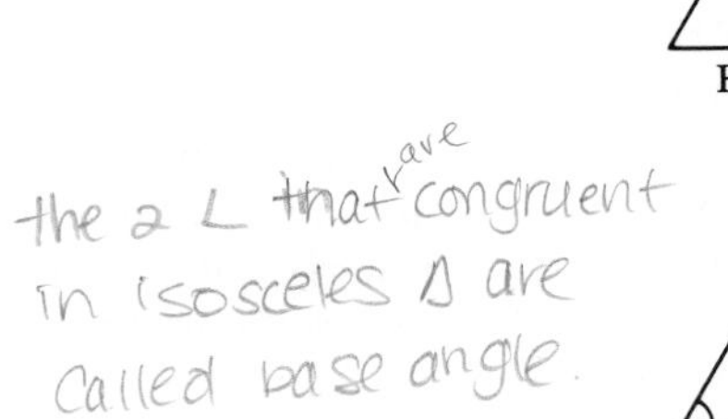

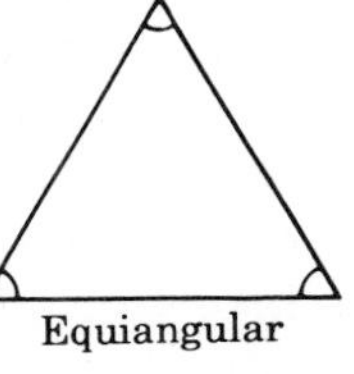

Equiangular

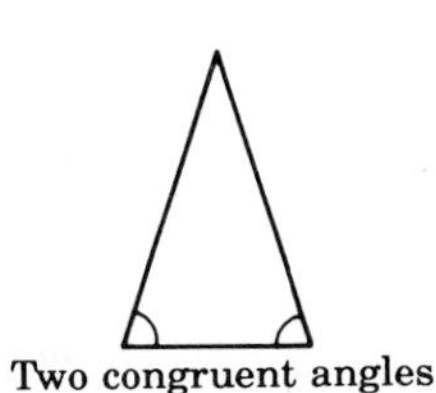

Two congruent angles

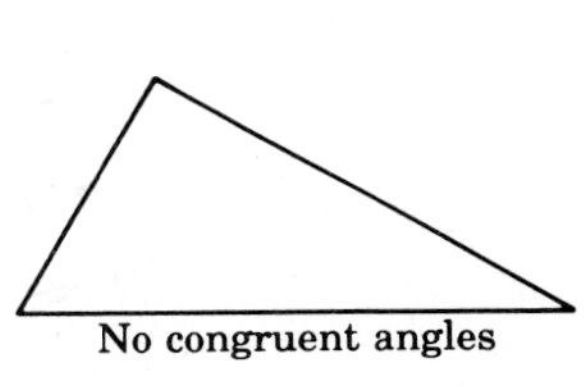

No congruent angles

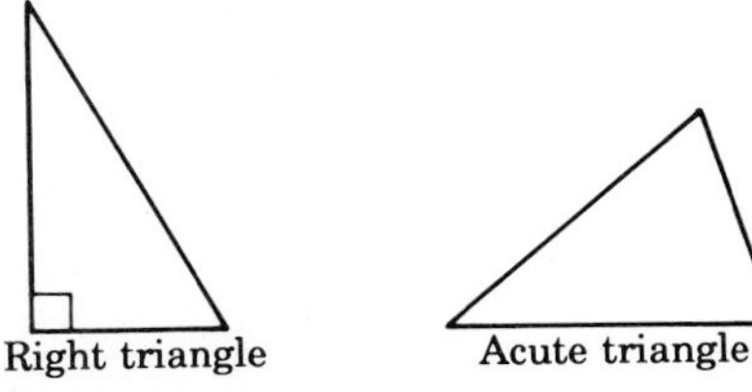

Right triangle

Acute triangle

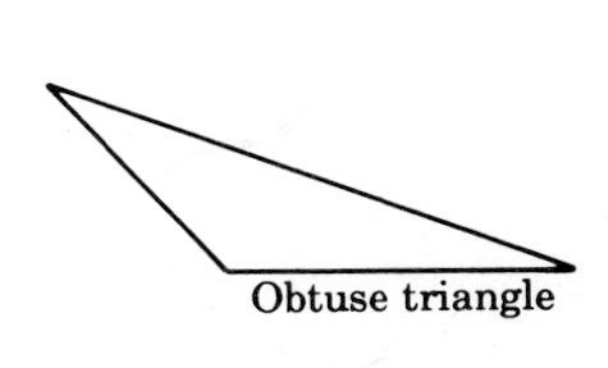

Obtuse triangle

FIGURE 2.48

INVESTIGATION 2.4

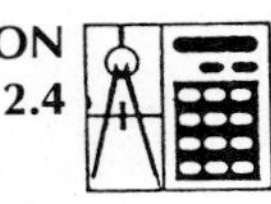

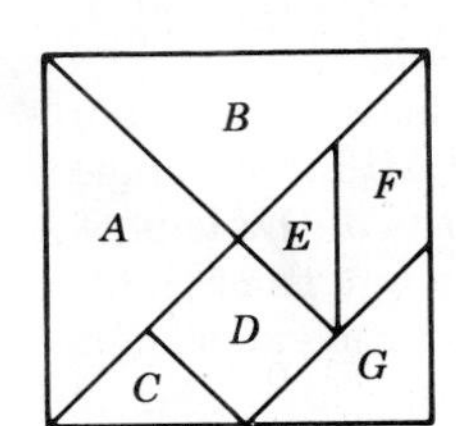

A. Look at the tangram puzzle.

1. Can you make a triangle with 2 tangram pieces? 3 pieces? 4 pieces? 5 pieces? 6 pieces? All 7 pieces? Draw and label pictures to show the triangles and the pieces used.
2. How are these triangles different? How are they alike?
3. How many of the triangles can you name?

B. Answer the questions in part A for a quadrilateral.

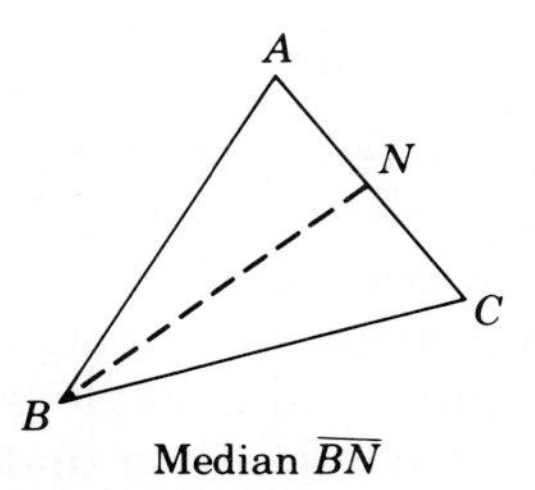

Median $\overline{BN}$

FIGURE 2.49

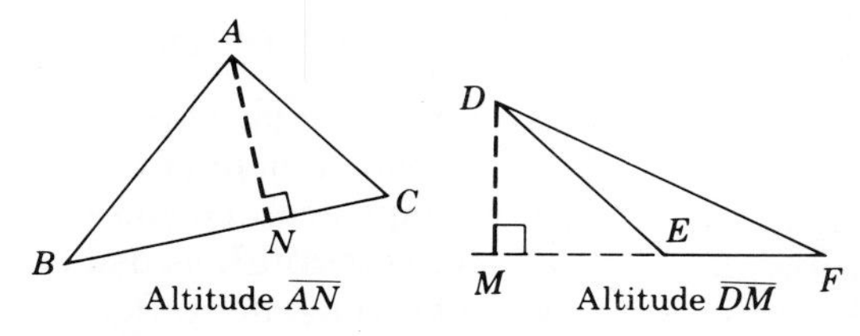

Altitude $\overline{AN}$ Altitude $\overline{DM}$

FIGURE 2.50

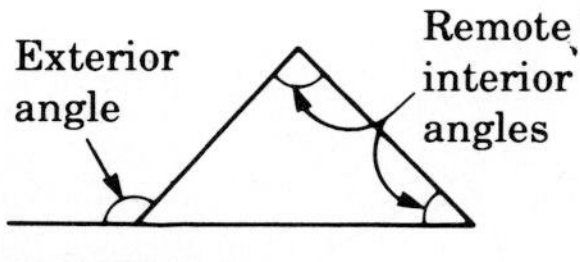

FIGURE 2.51

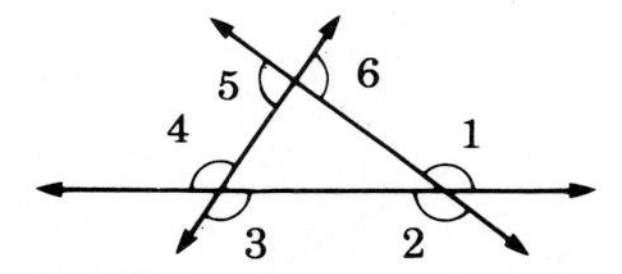

FIGURE 2.52

Some important segments, angles, and lines are associated with triangles. As shown in Figure 2.49, a **median** of a triangle is a segment from a vertex to the midpoint of the opposite side. An **altitude** of a triangle is a segment from a vertex perpendicular to the line containing the opposite side, as shown in Figure 2.50. The point of intersection of this line and the altitude is called the *foot* of the altitude.

As shown in Figures 2.51 and 2.52, an **exterior angle** of a triangle is an angle that forms a linear pair with one of the angles of the triangle. The **remote interior angles** with respect to an exterior angle are the two angles of the triangle that are not adjacent to the exterior angle. Note that every triangle has three interior and six exterior angles.

Properties involving sides and angles can also be used to classify quadrilaterals, as shown in Table 2.6. A special segment associated with a quadrilateral is called a diagonal. The **diagonal** of a quadrilateral is a segment joining any two nonconsecutive vertices of the quadrilateral.

EXPLORING WITH THE COMPUTER 2.5

Use geometry exploration software to make and test conjectures about the following questions. Use at least 3 different figures when testing each conjecture. Record your results.

1. How does the segment connecting the midpoints of two sides of a triangle compare to the third side of the triangle?
2. What can you discover about the relationship between the shortest leg and the hypotenuse of a right triangle with a 30° angle.
3. How does the sum of the squares of the lengths of the legs of a right triangle compare with the square of the length of the hypotenuse? What if the triangle is obtuse or acute instead of right?

TABLE 2.6

Quadrilateral	Name	Properties
	Trapezoid	Exactly one pair of opposite sides parallel
	Kite	At least two pairs of adjacent sides
	Parallelogram	Pairs of opposite sides parallel and the same length
	Rhombus	All sides the same length and opposite sides parallel
	Rectangle	Opposite sides parallel and the same length; all angles are right angles
	Square	All sides the same length; all angles are right angles

EXPLORING WITH THE COMPUTER 2.6

Use geometry exploration software to make and test conjectures about the following questions. Use at least 3 different figures when testing each conjecture. Record your results.

1. For which quadrilaterals are the diagonals perpendicular?
2. For which quadrilaterals do the diagonals bisect each other?
3. What type of figure is formed when the midpoints of the sides of a quadrilateral are connected?
4. What figure is formed when the midpoints of the sides of any isosceles trapezoid are connected?

APPLICATION

Factory Problem: A company owned an equilateral triangular shaped section of land bordered by three highways. The company executive officer wanted to build a factory on the land so that trucks would travel the least possible distance when going to the highways. Suppose you were the company surveyor. Where would you recommend that the highway be built? Support your choice.

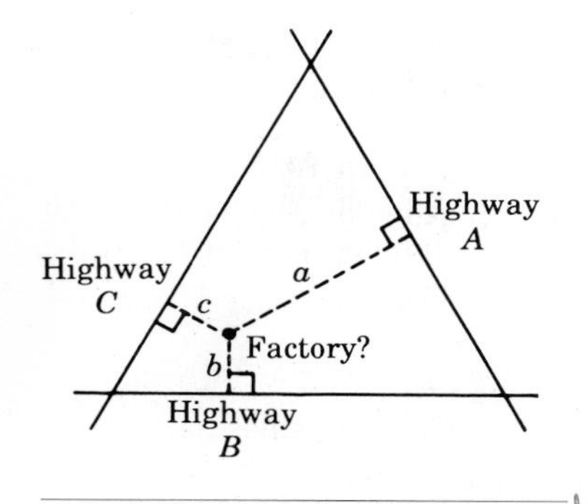

EXERCISE SET 2.5

1. Decide whether each statement about triangles is *always, sometimes,* or *never* true. Give reasons for your decision.
   a. A right triangle is an isosceles triangle.
   b. A right triangle is an acute triangle.
   c. An equilateral triangle is an isosceles triangle.
   d. An isosceles triangle is an equilateral triangle.
   e. An equilateral triangle is an acute triangle.
   f. An isosceles triangle is a scalene triangle.
   g. An isosceles triangle is a right triangle.
   h. An equiangular triangle is an equilateral triangle.
   i. An isosceles triangle is an acute triangle.
   j. A right triangle is an obtuse triangle.
   k. An obtuse triangle is an isosceles triangle.
   l. A right triangle is an equiangular triangle.
   m. A scalene triangle is an obtuse triangle.
2. Decide whether each statement about quadrilaterals is *always, sometimes,* or *never* true. Give reasons for your decision.
   a. A square is a rectangle.
   b. A rectangle is a square.
   c. A rectangle is a parallelogram.
   d. A parallelogram is a trapezoid.
   e. A square is a parallelogram.
   f. A rhombus is a kite.
   g. A rhombus is a square.
   h. A square is a rhombus.
   i. A rectangle is a kite.
   j. A parallelogram is neither a rhombus nor a rectangle.
   k. A quadrilateral is neither a parallelogram nor a trapezoid.
   l. A rhombus is a rectangle.
   m. A trapezoid is a kite.
3. Figure 2.53 is a Venn Diagram that shows the relationships between rectangles, parallelograms, and squares. Draw a Venn Diagram that shows the relationships between all triangles, scalene triangles, isosceles triangles, and equilateral triangles.
4. Draw a Venn Diagram that shows the relationship between all quadrilaterals, parallelograms, and trapezoids.
5. Draw a Venn Diagram that shows the relationship between rhombi, parallelograms, and rectangles.
6. Draw a Venn Diagram that shows the relationship between rhombi, squares, and kites.
7. Use a ruler and compass only to construct the following:
   a. An isosceles triangle
   b. An equilateral triangle
   c. A right triangle
   d. A square
   e. A parallelogram
8. Construct each figure in Exercise 7 with MIRA only.

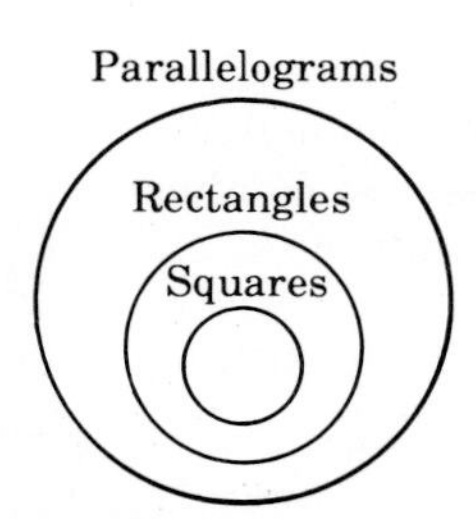

FIGURE 2.53

APPLICATION

Power Plant Problem: A new power plant is to be built to provide electricity for 3 major cities, Altitudnia, Bay City, and Centerville. Suppose you are the surveyor contracted to decide where the plant should be located in order that the sum of its distances from the three cities is the smallest, thus providing the most efficient system. Present and justify your decision. Could a model like the one shown be of help? How?

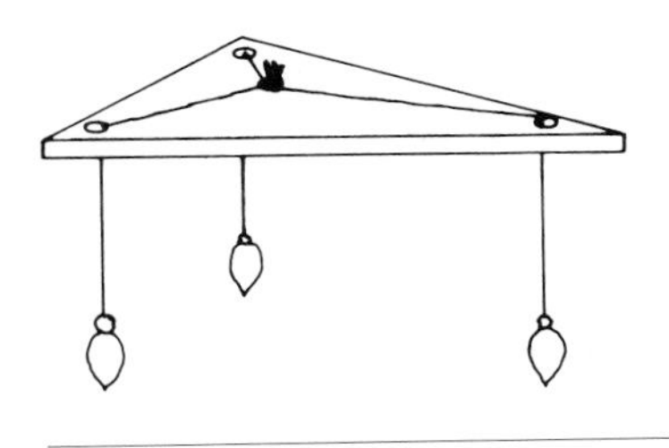

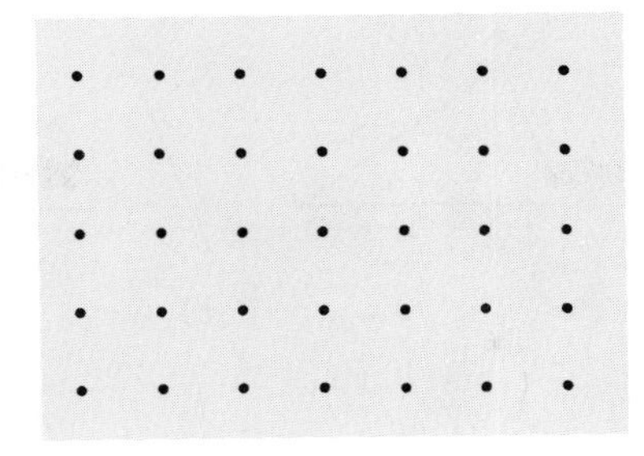

Square dot paper

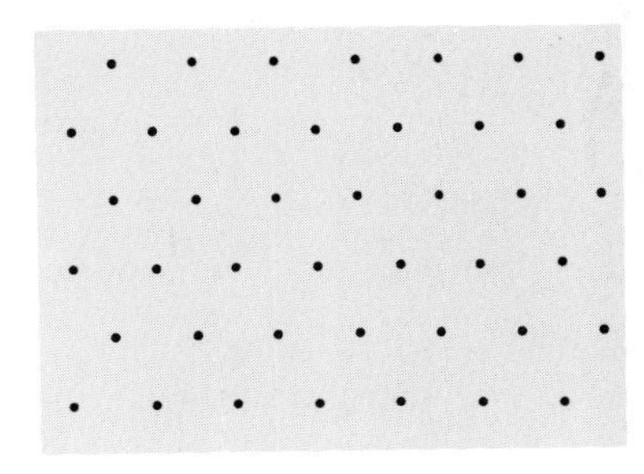

Isometric dot paper

FIGURE 2.54

9. Figure 2.54 illustrates square dot paper and equilateral triangular (isometric) dot paper. By drawing only straight lines from one point to another, which of the different types of triangles and quadrilaterals can you draw on each type of paper?
10. In Figures 2.55a through 2.55d, the segment suggested as the altitude of the triangle is given by the heavy line. For which of these figures is the altitude to side a not shown correctly? Explain your answer.
11. How many exterior angles does a triangle have? Draw a triangle and explain your reasoning.
12. How many medians does a triangle have? Do you think they are always inside the triangle? Explain your reasoning.

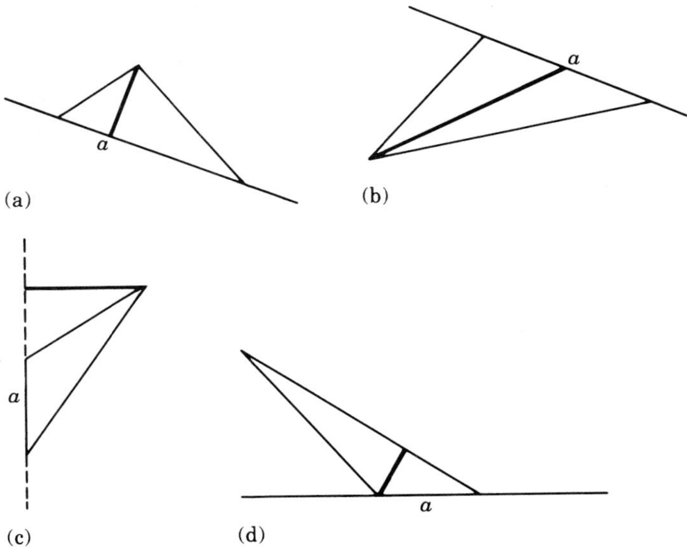

FIGURE 2.55

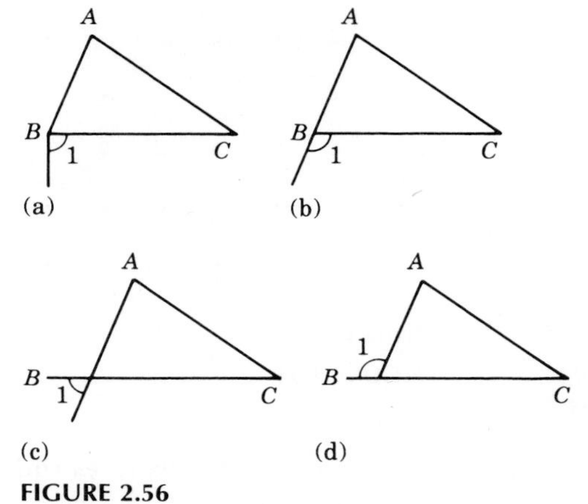

FIGURE 2.56

13. In Figures 2.56a through 2.56d, which of the angles labeled $\angle 1$ are not exterior angles of triangle ABC? Explain your answer.
14. Draw an acute and an obtuse triangle. Construct an altitude to a chosen side of each of these triangles.
15. How many altitudes does a given triangle have? Draw a triangle and sketch the altitudes to illustrate your answer.
16. How many diagonals does a triangle have? A quadrilateral? Draw figures to illustrate your answers.
17. **Critical Thinking.** Consider the four figures in Figure 2.57.
    a. Write a definition for a classification of quadrilateral that would include Figures 2.57a and 2.57b, but not 2.57c and 2.57d.
    b. Write a definition for a classification of quadrilateral that would include Figures 2.57a, 2.57b and 2.57c, but not 2.57d.
    c. Write a definition for a classification of quadrilateral that would include all the figures (Figs. 2.57a–d).

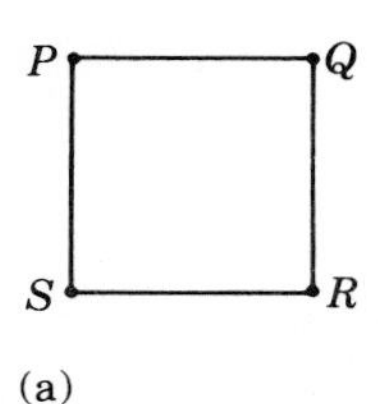

(a)

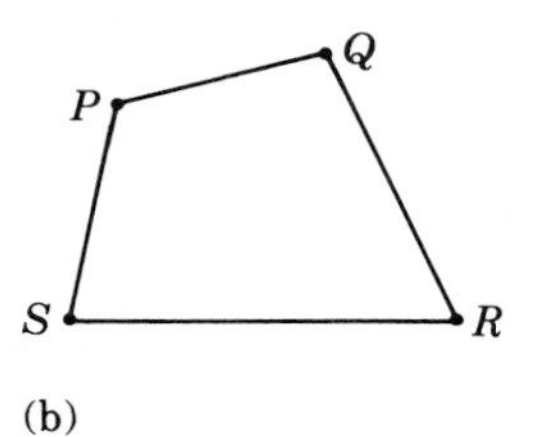

(b)

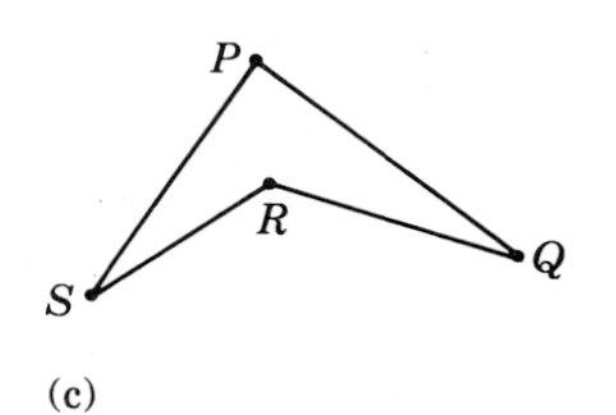

(c)

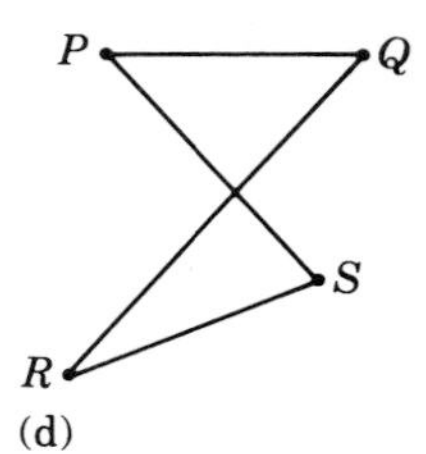

(d)

FIGURE 2.57

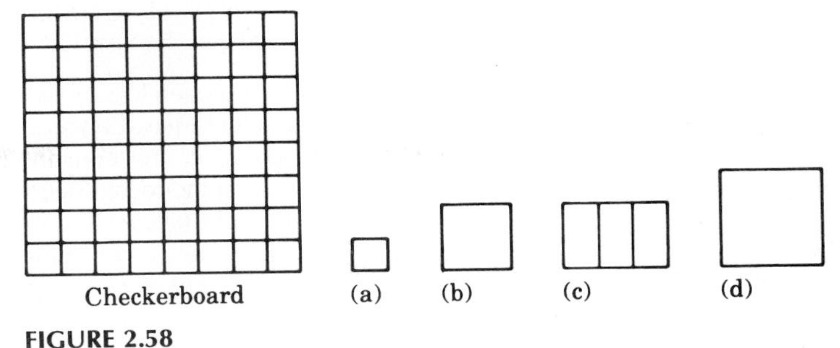

FIGURE 2.58

Problem Solving: Developing Strategies and Skills

The usefulness of the problem-solving strategy *draw a picture* was discussed earlier. Sometimes, however, a picture is *given* in the problem, and one's ability to solve the problem depends on knowing how to use the picture systematically and effectively. Problem 1 and Figure 2.58 illustrate this idea.

1. How many squares are on an ordinary checkerboard? To use Figure 2.58 effectively, let's look for a way to systematically count the squares on the checkerboard.
   a. How many small squares (Fig. 2.58a) in each row? How many rows? 8×8, or 64 in all.
   b. How many of the next-larger size square (Fig. 2.58b) in each row? Be sure to count overlapping squares. (Fig. 2.58c). How many rows? (Be sure to count overlapping rows too!) 7×7, or 49 in all.
   c. Answer the question for the next larger square (Fig. 2.58d). Be sure to count overlapping squares! Do you see a pattern?
   d. What is the solution to the problem?
2. How many different equilateral triangles can you find in the triangular patio of Figure 2.59? Try to use a systematic approach to count the different-size equilateral triangles.

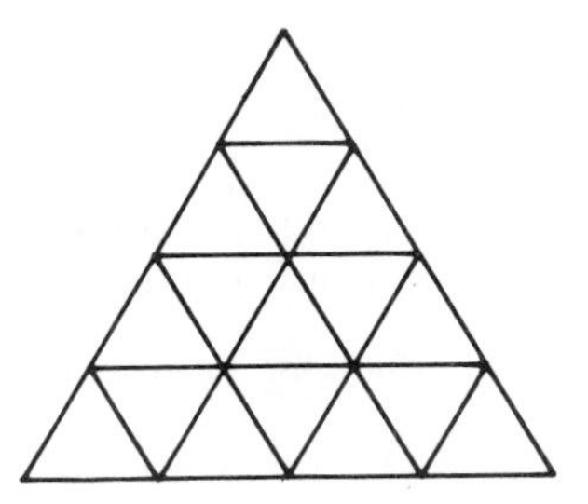

FIGURE 2.59

PEDAGOGY FOR TEACHERS

ACTIVITIES

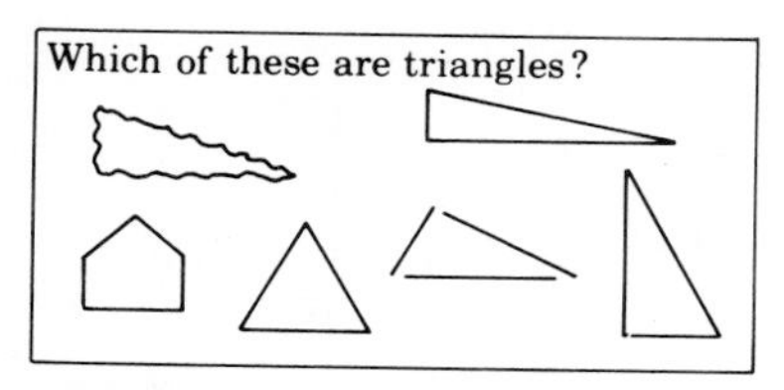

FIGURE 2.60

1. A concept has been learned when the student has a basic understanding of a mathematical idea. Concept cards are often used to help teach concepts. For example, the concept of a triangle may be taught to young children by using the ideas suggested by the concept card in Figure 2.60. Note that the card teaches the concept of triangle using examples along with the appropriate nonexamples. Choose at least one of the concepts presented in this chapter and develop a concept card that teaches the concept.

2. Study the activities in the book on pattern blocks available from Creative Publications (Roper, 1991). Describe at least one way these blocks could be used to help a student learn a geometric concept.

3. Articles on the use of the geoboard, appropriate for students at almost every level, are available in journals such as *The Arithmetic Teacher* and *The Mathematics Teacher*. Find an appropriate article for the level of your interest and evaluate the usefulness of the ideas presented.

4. Make a set of cards that can be used for a game of "Concentration" with students. The cards should include pictures of basic polygons, definitions, and so on.

COOPERATIVE LEARNING GROUP DISCUSSION

P.M. van Hiele, as a result of a study of students' mental development in geometry, identified the following five levels:

- —*Level 1: Visualization.* At this level, geometric concepts are visualized as total entities. They are not viewed as having attributes or components.
- —*Level 2: Analysis.* At this level, students begin to discern characteristics of figures through observation and experimentation.
- —*Level 3: Informal Deduction.* At this level, students establish the interrelationships of properties both within figures and between figures. At this level, classification is possible and definitions can be comprehended. Informal arguments can be followed and given.
- —*Level 4: Deduction.* At this level, logical reasoning and deduction are used within the context of an axiomatic system.
- —*Level 5: Rigor.* At this level, different systems of geometry can be compared, and the learner can work in a variety of axiomatic systems.

Read Chapter 1 of *Learning and Teaching Geometry* (NCTM, 1987) for more information and discuss answers to the following questions:

1. Do you think the levels really describe students' mental development in geometry.
2. At what grades in school do you think students usually operate at the levels given?

3. Do you think these levels provide a helpful framework for discussing the geometry curriculum?

Give the reasons for your answers.

CHAPTER REFERENCES

Goodnow, Judy. 1991. *Junior High Cooperative Problem Solving with Geoboards*. Sunnyvale, CA: Creative Publications.

Lindquist, Mary M., and Schulte, Albert P., Eds. 1987. *Learning and Teaching Geometry, K–12* (1987 Yearbook). Reston, VA: National Council of Teachers of Mathematics.

Olsen, Alton T. 1975. *Mathematics Through Paper Folding*. Palo Alto, CA: Dale Seymour Publications.

Roper, Ann. 1991. *Junior High Cooperative Problem Solving with Pattern Blocks*. Sunnyvale, CA: Creative Publications.

NCTM. 1987. *Geometry in Our World*. (Includes slides.) Reston, VA: National Council of Teachers of Mathematics.

SUGGESTED READINGS

Consult the following sources listed in the Bibliography for additional related readings: Hill (1987), Morrow (1991), Spikell (1990), and Teppo (1991).

CHAPTER THREE Discovering Polygon Relationships

INTRODUCTION

In the interesting book *Flatland* (Abbott, 1963), the author describes a "civilization" in which the inhabitants are restricted to the two-dimensional world of the plane. Our adaptation of his description is as follows:

> Our middle class consists of equilateral triangles. Our professional class is squares and pentagons. Next above these comes the nobility, of whom there are several degrees, beginning at hexagons and from thence rising in the number of their sides till they receive the honorable title of polygonal, or many-sided. When the number of the sides becomes so numerous, and the sides themselves so small that the figure cannot be distinguished from a circle, it is included in the circular or priestly order; and this is the highest class of all.

Although the "personalities" depicted in this chapter may not be quite like those in *Flatland,* Investigation 3.1 serves to introduce the idea of a polygon.

TEACHING NOTE

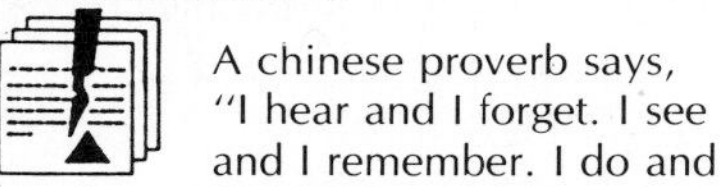

A chinese proverb says, "I hear and I forget. I see and I remember. I do and I understand." Piaget says, "motor activity is of enormous importance for the understanding of spatial thinking." Thus, it is important for students to have many physical experiences with geometric objects. The ideas of polygons can be enhanced by using cardboard cutouts or the sets of Pattern Blocks, available from Dale Seymour Publications.

MORE ABOUT POLYGONS

A definition of a simple polygon can now be given:

> Let $P_1, P_2, \ldots P_n$ be a set of n distinct points in a plane where $n > 2$. The union of the segments P_1P_2, P_2P_3, . . .

INVESTIGATION 3.1

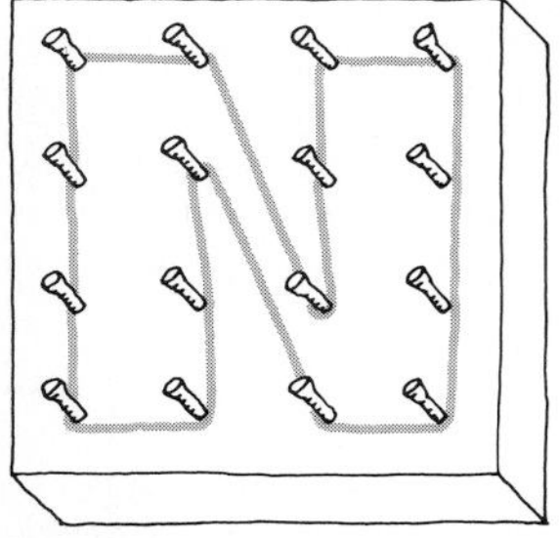

The figure shows a ten-sided figure on a 4×4 geoboard. (Note that no side crosses another side and no two vertices coincide.)

A. A student searched for such a figure with the most number of sides. On a 4×4 geoboard, she found a sixteen-sided figure. On a 6×6 geoboard, she found a 36-sided figure. Show these figures on dot paper.

B. Make a conjecture that might be true about the greatest number of sides for such figures on a 2×2 geoboard, a 3×3 geoboard, a 5×5 geoboard, and a 7×7 geoboard. Test your conjectures by showing figures with the largest possible number of sides on dot paper. What other conjectures can you make?

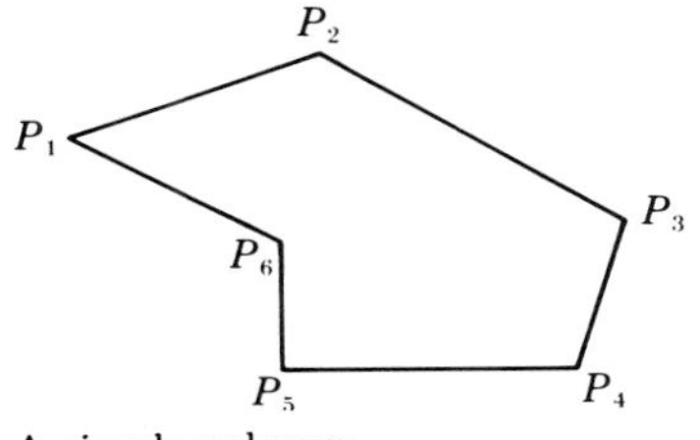

A simple polygon

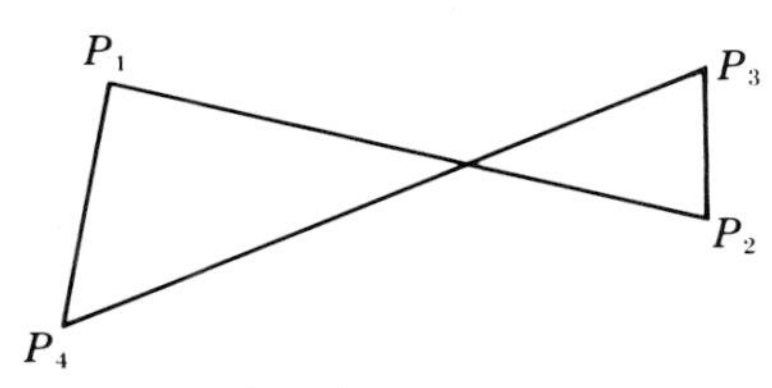

A nonsimple polygon

FIGURE 3.1

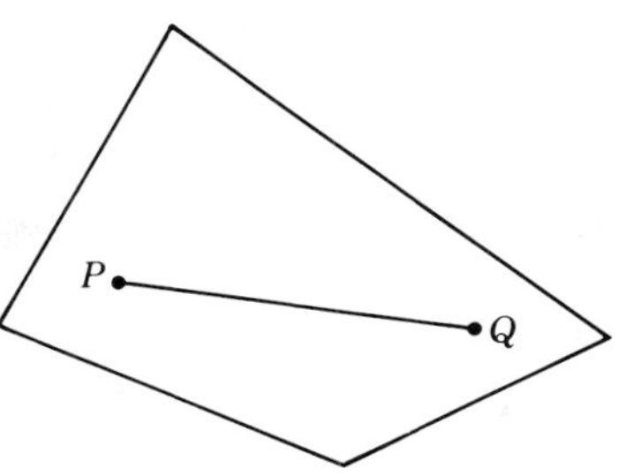

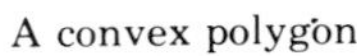
A convex polygon

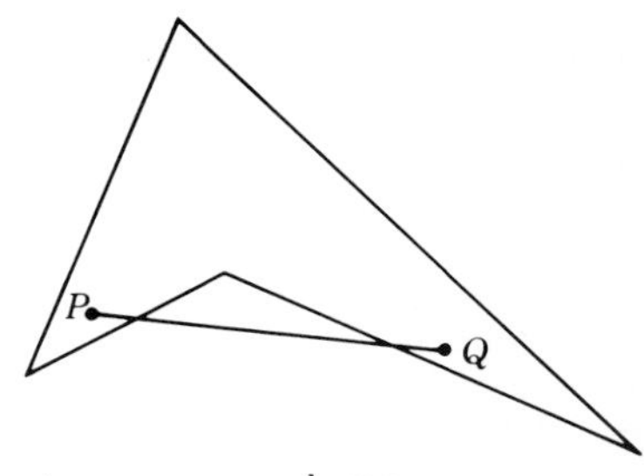

A nonconvex polygon

FIGURE 3.2

P_nP_1 is a **simple polygon** provided the segments intersect only at their endpoints and no two segments with a common endpoint are collinear.

The points $P_1, P_2, \ldots P_n$ are called the **vertices** of the polygon, and the segments $P_1P_2, P_2P_3, \ldots P_nP_1$ are called the **sides** of the polygon. A segment (other than a side) joining two vertices of a polygon is called a **diagonal** of the polygon. The union of two consecutive sides of the polygon is called an **angle of the polygon**.

If the segments described in the definition intersect at a point or points in addition to their endpoints, the polygon is called a **nonsimple polygon**. Simple and nonsimple polygons are illustrated in Figure 3.1.

Figure 3.2 illustrates a way of classifying simple polygons according to whether they are convex or nonconvex. A **convex** simple polygon satisfies the property that for each pair of points P and Q inside the polygon, the segment PQ lies inside the polygon. In the case where there exists at least one pair of points P and Q such that $\overline{PQ}$ does not lie inside the polygon, the polygon is called **nonconvex**. Investigation 3.2 provides practice with convex polygons.

INVESTIGATION 3.2

There are 13 convex polygons that can be formed each using all 7 pieces of the tangram puzzle.

How many of these polygons can you find?

TEACHING NOTE

Primary children can learn to recognize different polygons by feel. The teacher might have several different polygon shapes in a bag and ask the child to reach into the bag, pick one up, and describe it exclusively by feeling it inside the bag.

Polygons are named according to the number of sides they have, as indicated in Table 3.1. Polygons with 1 through 10 sides and 12 sides are named using Greek prefixes that indicate the number of sides. A polygon with 11 sides or more than 12

TABLE 3.1 Classification of Polygons by Number of Sides

Polygon	Number of Sides	Name
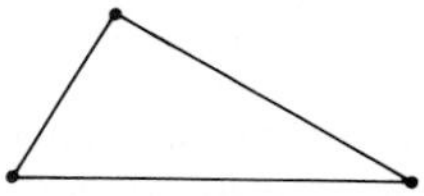	3	Triangle
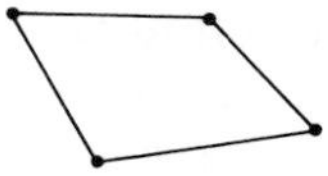	4	Quadrilateral
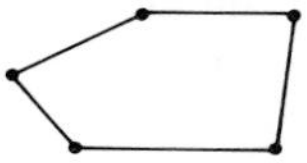	5	Pentagon
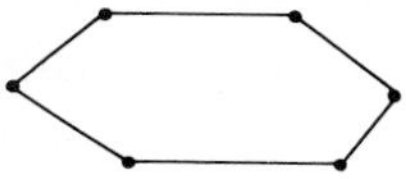	6	Hexagon
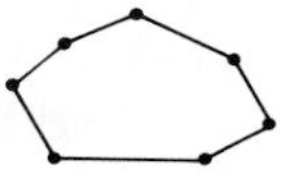	7	Heptagon
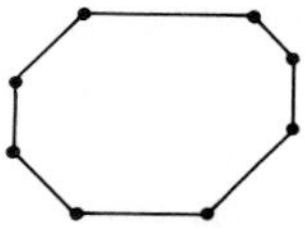	8	Octagon
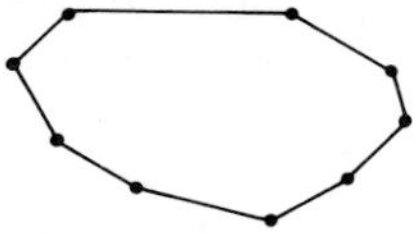	9	Nonagon
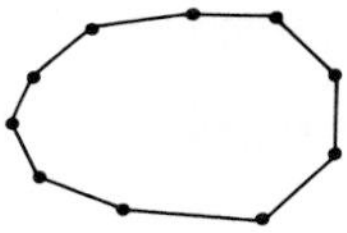	10	Decagon
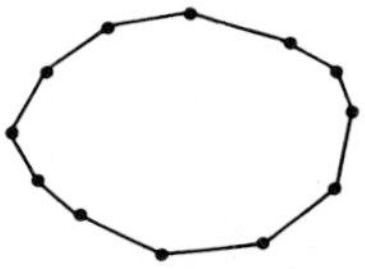	12	Dodecagon

sides is referred to as an ***n*-gon**, where n is the number of sides of the polygon. In the sections that follow, we will study ideas of symmetry of polygons, analyze some elementary polygons, introduce regular polygons, and explore the interesting world of the star polygons.

SYMMETRY OF POLYGONS

Each of the designs in Figure 3.3 appears to have a certain order, balance, or beauty that makes it interesting to observe. Several of the designs possess what is known as reflectional symmetry. Some possess rotational symmetry. While a more formal definition of symmetry will be given in a later chapter, a description for both reflectional and rotational symmetry that might be appropriate for helping students learn these ideas will be presented in this section.

FIGURE 3.3

TEACHING NOTE

Simple ideas of symmetry can be experienced by cutting valentines and snowflakes or by folding a paper with some finger paint on it. Activities such as those in *Symmetry with Pattern Blocks* (Wiebe, 1985) can also help children develop important symmetry ideas.

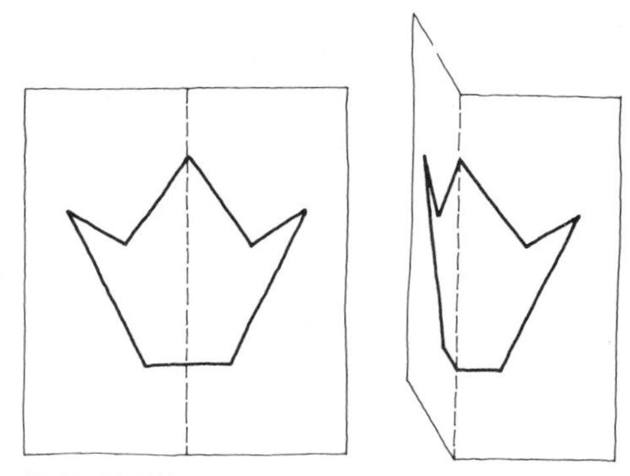

FIGURE 3.4

To test a geometric figure for reflectional symmetry, one might use what we will call (1) the "fold test" (Fig. 3.4) or (2) the "mirror test" (Fig. 3.5). To implement the fold test, trace the figure on a piece of paper. Check to see if the paper can be folded so that one-half of the figure *exactly* coincides with the other half. If such a fold line can be found, the figure is said to possess **reflectional symmetry**. The fold line is called the **line of symmetry**.

To implement the mirror test, try to place a mirror or piece of plexiglass on a figure in such a way that half of the figure and its reflection in the mirror appear to accurately form the entire figure. (The piece of plexiglass pictured in Figure 3.5 is

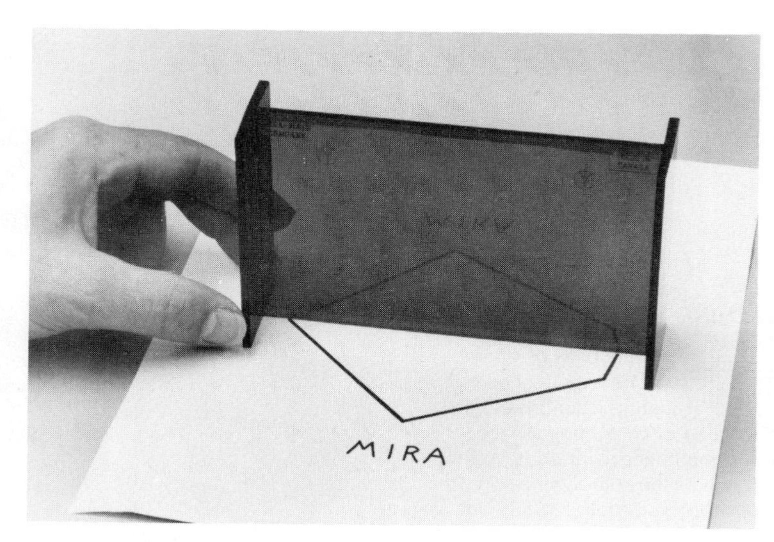

FIGURE 3.5

TEACHING NOTE

The *Mirror Puzzle Book* (Walter, 1985) can help students develop a concept of reflectional symmetry. The advanced puzzles are appropriate for secondary-level students. Some students enjoy making their own mirror puzzles.

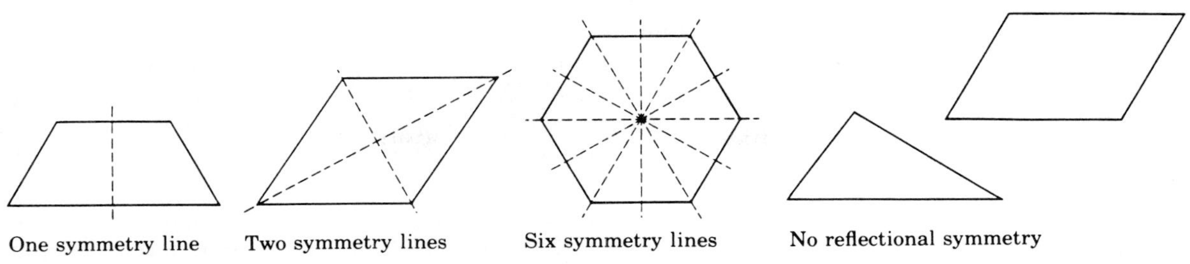

FIGURE 3.6

commercially produced and is called a MIRA.) If the mirror can be so placed, the figure is said to possess reflectional symmetry. The line of the mirror is the line of symmetry.

In Figure 3.6, we see examples of polygons with and without lines of symmetry. Use one of the tests described above to check each figure.

To test a figure for rotational symmetry, one might use what is called the "trace-and-turn" test. To implement this test, trace the figure on a piece of tracing paper or a piece of plastic and place the tracing directly on top of the original figure. Then, while holding one point fixed, turn the tracing until the tracing and the original again coincide, as shown in Figure 3.7. In the figure, we see that after the tracing is rotated counterclockwise about the center point through an angle of 45°, the figure and its tracing do not coincide. However, after a counterclockwise turn about the center point through 90°, the figure and its tracing do coincide (the tracing fits exactly upon the original figure). Thus, we say the figure has 90° **rotational symmetry.** The point that is held fixed is called the **center of rotational symmetry.**

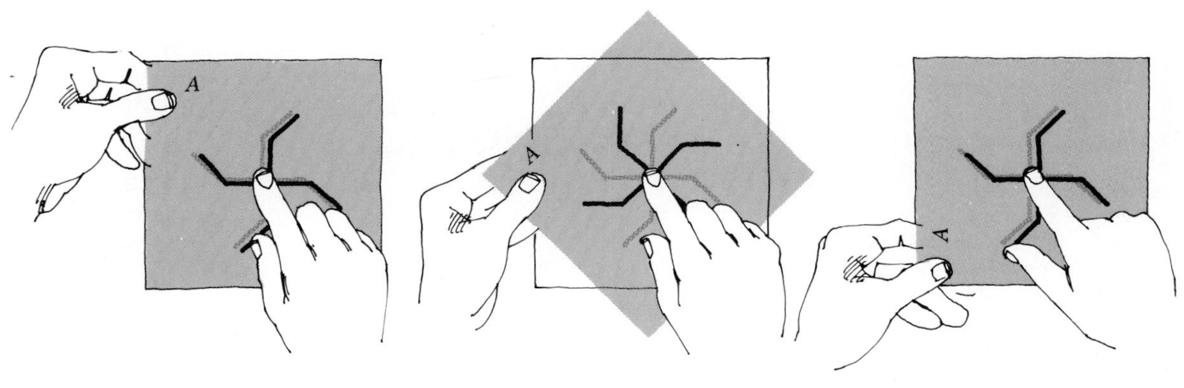

FIGURE 3.7

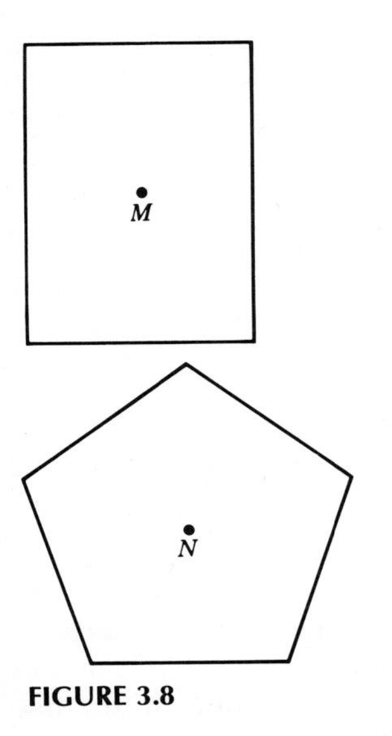

FIGURE 3.8

In Figure 3.8, we see a rectangle with 180° rotational symmetry about point M and a pentagon with 72° rotational symmetry about point N. Use the trace-and-turn test and your protractor to check the accuracy of these statements.

Since the tracing of the pentagon can also be made to fit exactly upon the original after rotations through 144°, 216°, and 288°, it is said to have 144° rotational symmetry, 216° rotational symmetry, and 288° rotational symmetry. Since a tracing of any figure fits upon the original figure after a rotation of 360°, we do not consider a complete rotation when describing rotational symmetry.

APPLICATION

The Box Pattern Problem: A manufacturing manager found that the machines in his factory could make boxes most efficiently if a box pattern with the following characteristics were used:

1. It has six squares, each touching other square(s) along complete sides only.
2. It has four squares in a row.
3. It has rotational symmetry.
4. It can be folded to make a box.

The manager asked an engineer to provide an analysis of all patterns satisfying characteristic 1 (he indicated that he had heard that there were 35 such patterns) and to identify those with characteristics 2, those with characteristics 3, and finally those with all of the listed characteristics. He also asked the engineer to give reasons to assure him that all possibilities had been considered. Cooperate with others in a group to prepare the engineer's report.

EXERCISE SET 3.1

1. For each of the following conditions, sketch a triangle, if it is possible, that satisfies the conditions:
   a. No symmetry lines
   b. Exactly 1 symmetry line
   c. Exactly 2 symmetry lines
   d. Exactly 3 symmetry lines
   e. More than 3 symmetry lines
   f. Reflectional symmetry, but no rotational symmetry
   g. Rotational symmetry, but no reflectional symmetry
2. For each of the figures in Figure 3.9, sketch all lines of symmetry and describe all rotational symmetries.
3. Sketch each of the following quadrilaterals: a square, a rhombus, a rectangle, a parallelogram, a trapezoid, and a kite.
   a. Describe all the lines of symmetry for each.
   b. Describe all rotational symmetries for each.
4. Sketch, if possible, a quadrilateral other than those mentioned in Exercise 3 that satisfies the following conditions. You need not restrict your consideration to convex quadrilaterals.
   a. It has a line of symmetry, but no rotational symmetry.
   b. It has rotational symmetry, but no line of symmetry.

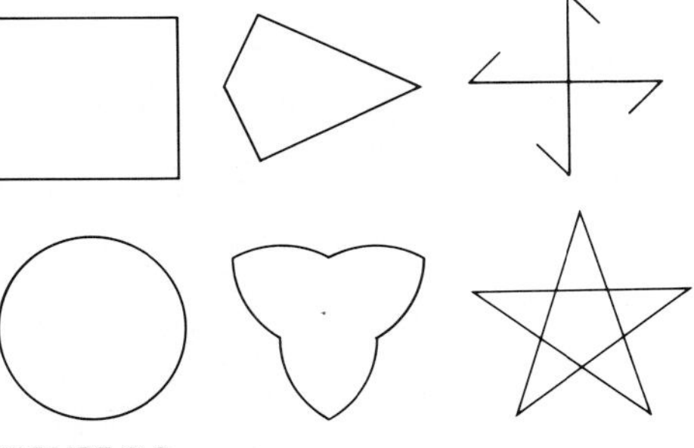

FIGURE 3.9

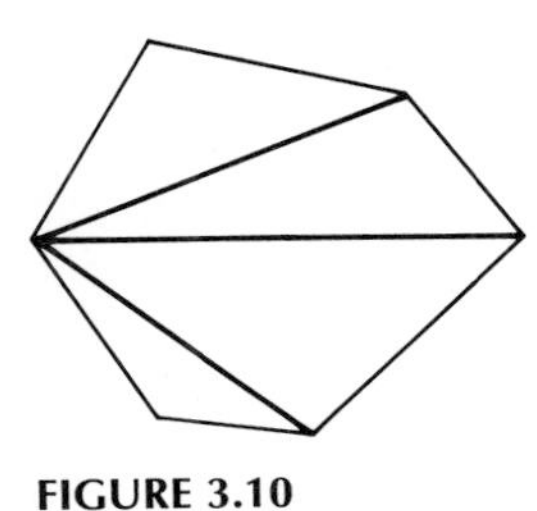

FIGURE 3.10

TABLE 3.2

Number of Sides of Polygon	Maximum Number of Nonintersecting Diagonals	Number of Resulting Triangles
3	0	1
4	1	
5		
6		
7		
.		
.		
.		
n		

5. When the maximum number of nonintersecting diagonals is drawn, a polygon is divided up into triangles (Fig. 3.10). Complete Table 3.2 and find a formula that enables you to find the number of triangles when you know the number of sides.
6. Use the results of Exercise 5 to complete Table 3.3. Search for a formula relating the number of sides of a polygon and the sum of the angle measures for the vertex angles of a polygon.

TABLE 3.3

Number of Sides	Sum of Angle Measures
3	180°
4	
5	
6	
7	
.	
.	
.	
n	

7. A *pentominoe* is a polygon made from 5 connected squares (Fig. 3.11). Each square can touch another only on a complete side.
   a. Find all 12 pentominoes, and show them on a geoboard or on dot paper.
   b. Which pentominoes have 90° rotational symmetry about some point? 180° rotational symmetry?
   c. Which pentominoes have exactly one line of reflectional symmetry?
   d. Which pentominoes have more than one line of reflectional symmetry?

*8. Sketch a Venn diagram showing the interrelationships among the types of quadrilaterals mentioned in Exercise 3.

*9. A *parpolygon* is a polygon with an even number of sides in which opposite sides are equal and parallel. Every parpolygon with $2n$ sides can be dissected into $[n(n - 1)]/2$ parallelograms.
   a. Show (by drawing pictures) examples of this theorem for $n = 2$, $n = 3$, and $n = 4$.
   b. Describe the dissection when the parpolygon is equilateral.

10. Four of the eight designs in Figure 3.3 possess symmetry. One of these four has only reflectional symmetry, two possess only rotational symmetry, and the fourth has both rotational and reflectional symmetry. Find these four designs. [*Hint*: Note that regions of different colors must be considered as different in determining the symmetry properties.]

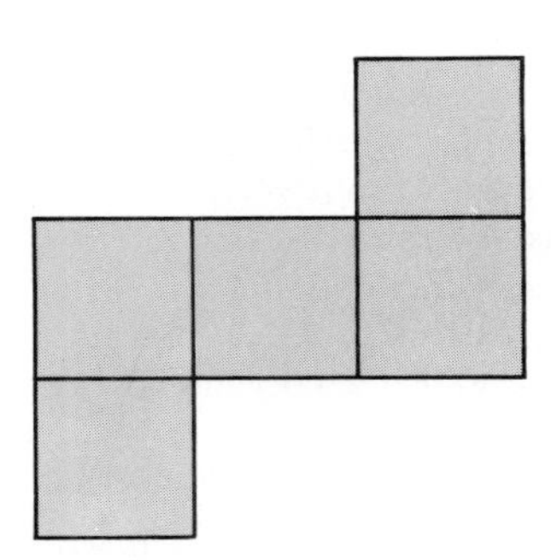

FIGURE 3.11

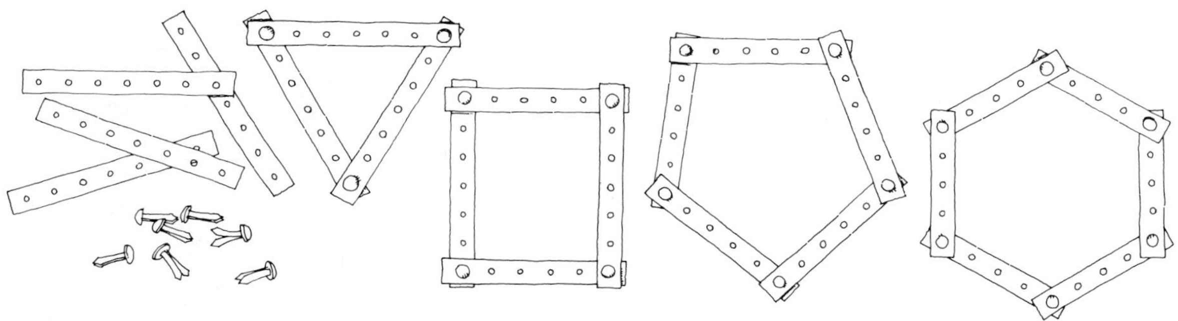

FIGURE 3.12

*11. Refer to Figure 3.12 and use some cardboard strips (with punched holes and brass fasteners) to explore the following questions:

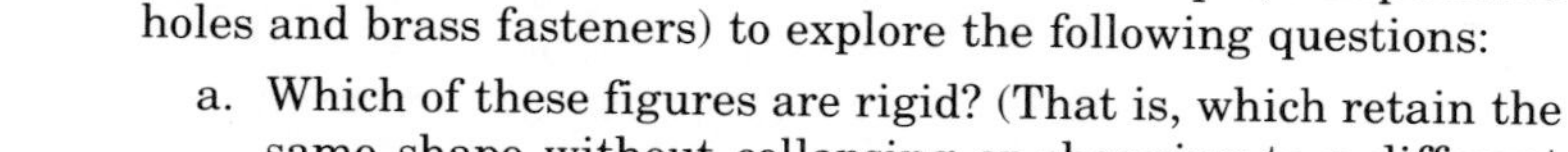

a. Which of these figures are rigid? (That is, which retain the same shape without collapsing or changing to a different shape when pressure is applied?)
b. For each figure that is not rigid, what is the fewest number of strips that is required to make it rigid?
c. Can you formulate a general rule for the number of strips required to make an n-gon rigid?

*12. A *pierced polygon* is the union of two disjoint polygons, one in the interior of the other. We use the notation ${}_nP_m$ to denote a pierced polygon with an m-gon interior to an n-gon. Suppose a pierced polygonal region is divided into triangular regions (Fig. 3.13). Find the relationship describing the number of triangles resulting from a pierced polygon ${}_nP_m$.

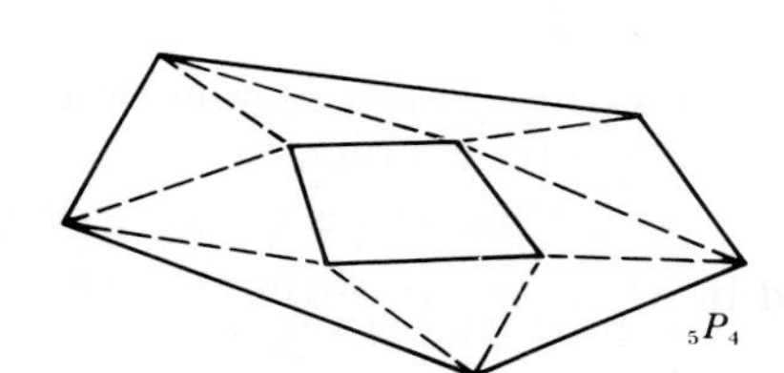

FIGURE 3.13

*13. Let ${}_nP_{p,q}$ represent an n-gon pierced by a p-gon and a q-gon. If the polygon ${}_nP_{p,q}$ is divided into triangles, what is the relationship between the numbers n, p, and q (Fig. 3.14)?

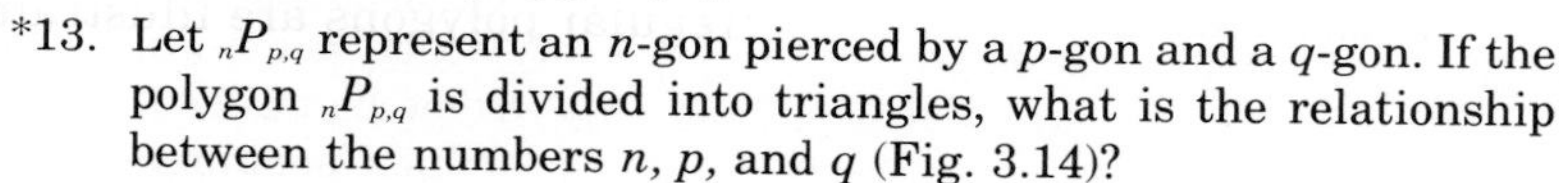

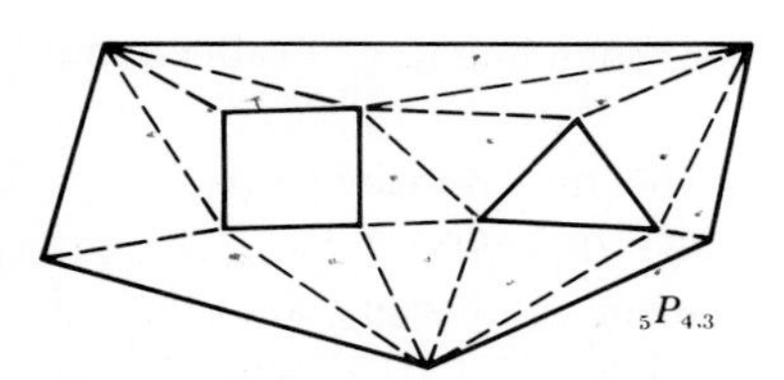

FIGURE 3.14

14. A triangle has either 3 lines of symmetry, 1 line of symmetry, or 0 lines of symmetry. What are the possible number of lines of symmetry for the following:

a. Pentagons? b. Hexagons? c. Octagons?

15. **Critical Thinking.** When asked to find how may diagonals an n-gon has, a student reasoned as follows (Fig. 3.15):

> First, if I choose a vertex V of an n-sided polygon, I can't draw diagonals to V itself or to the two vertices adjacent to V. This means I can't draw to three of the n vertices of the n-sided polygon. So, I can draw from V to $n-3$ other vertices. Second, since there are n vertices in all and I can draw diagonals from each of these n vertices to $n-3$ vertices, I can draw $n(n-3)$ diagonals. Therefore, a polygon always has $n(n-3)$ diagonals.

a. Is this reasoning correct.
b. Do you agree with the student's conclusion? Explain why or why not.

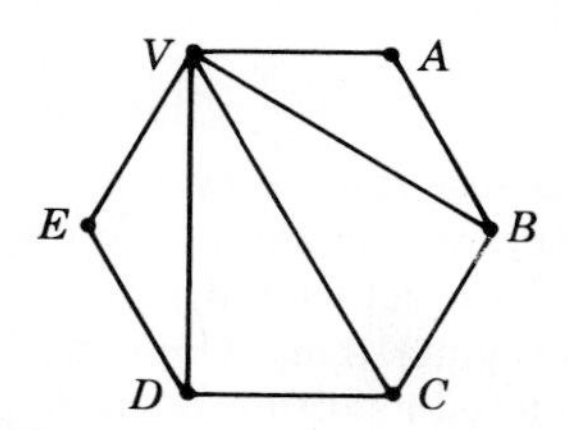

FIGURE 3.15

Square regular 4-gon

FIGURE 3.16

Problem Solving: Developing Skills and Strategies

The strategy called **make an organized list** can often be used to solve certain geometry problems. It may be surprising that sometimes these problems can also be solved by drawing a picture and counting. Consider the following problem.

Problem: A research center had 10 separate buildings, A–J. The director decided to lay cables between buildings so that the computers in each building could be connected to the computers in every other building. How many cables must be layed to complete the project?

A solution might be found by completing an organized list like the following:

AB, AC, AD, AE, AF, AG, AH, AI, AJ
BC, BD, BE, BF, BG, BH, BI, BJ
CD, CE, . . .

1. Use a list like the one shown to find the solution.
2. Can you find the solution in a different way by drawing a picture and using ideas about diagonals of polygons?

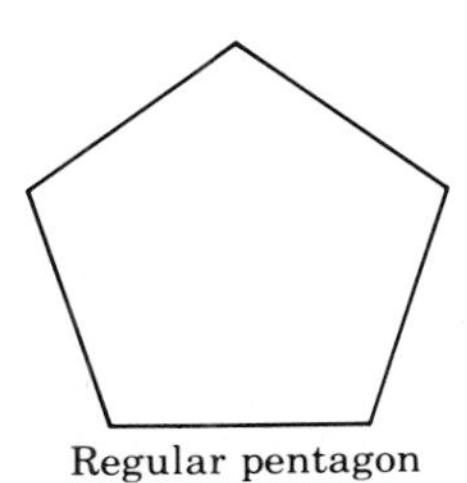

Regular pentagon

FIGURE 3.17

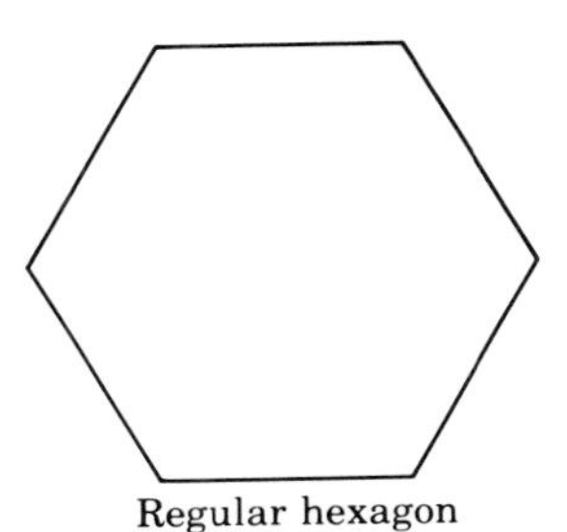

Regular hexagon

FIGURE 3.18

REGULAR POLYGONS

A **regular polygon** is a simple polygon with all sides the same length and all vertex angles the same measure. A regular polygon with n sides is called a *regular n-gon.* Three examples of regular polygons are illustrated in Figures 3.16 through 3.18. Investigation 3.3 explores the world of regular polygons.

INVESTIGATION 3.3

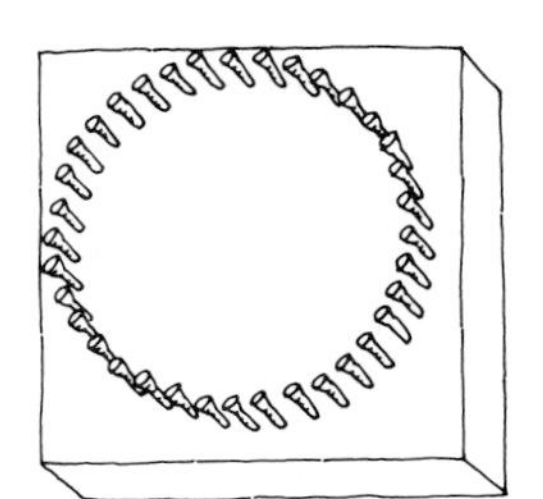

Regular polygons with interesting shapes can be constructed with yarn or string on a circle geoboard. These shapes cannot be constructed on a square or triangular geoboard.

A. On a circle geoboard (or dot paper) with 36 nails, how many different polygons with sides of equal length can you construct?
B. Can you find any relationship between the number of nails (36) and the number of sides of polygons constructed in part A?

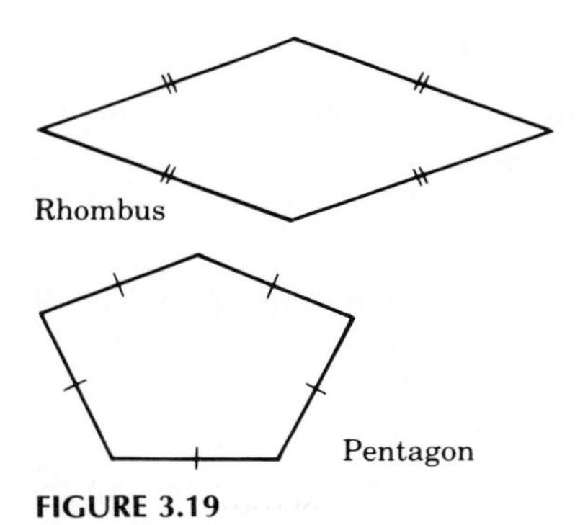

FIGURE 3.19

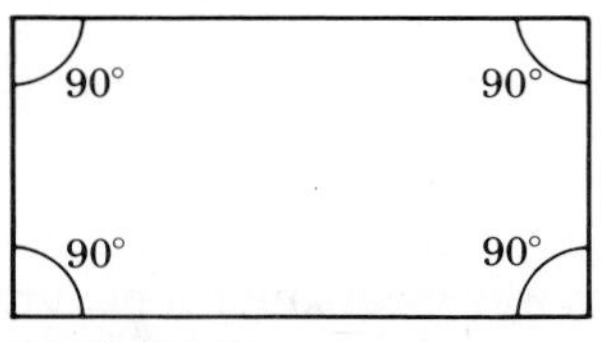

FIGURE 3.20

It can be proven that a polygon with equal sides when inscribed in a circle (like those constructed in Investigation 3.3) will also have equal vertex angles. To further illustrate the importance of equal sides *and* equal angles in the definition of regular polygons, we observe that there are polygons with all sides the same length that do not have vertex angles of equal measure and, consequently, are not regular polygons. The rhombus and pentagon shown in Figure 3.19 have all sides the same length. Neither, however, is a regular polygon. Why?

Similarly, there are polygons that have all vertex angles of equal measure, but that are not regular polygons. The rectangle is a simple example of such a polygon (Fig. 3.20). Why is the rectangle not a regular polygon?

An important question concerning regular polygons is, *How many degrees are there in each vertex of a regular n-gon?* Table 3.4 provides the answer to this question. We conclude from the table that there are $[180(n - 2)]/n$ degrees in each vertex angle of a regular n-gon. Sometimes we may find it convenient to write this number in the equivalent form:

$$180\left(1 - \frac{2}{n}\right)$$

TABLE 3.4 Classification of Regular Polygons by Vertex Angles

Regular Polygons	Number of Sides n	Diagonals Divide into How Many Triangles?	Total Number of Degrees in All the Triangles (i.e., The Sum of Measures of the Vertex Angles)	Number of Degrees in *One* Vertex Angle
Equilateral triangle	3	1	1(180)	$\frac{1(180)}{3} = 60°$
Square	4	2	2(180)	$\frac{2(180)}{4} = 90°$

TABLE 3.4 Continued

Regular Polygons	Number of Sides n	Diagonals Divide into How Many Triangles?	Total Number of Degrees in All the Triangles (i.e., The Sum of Measures of the Vertex Angles)	Number of Degrees in *One* Vertex Angle
Pentagon	5	3	3(180)	$\frac{3(180)}{5} = 108°$
Hexagon	6	4	4(180)	$\frac{4(180)}{6} = 120°$
⋮	⋮	⋮	⋮	⋮
	n	$n - 2$	$(n - 2)(180)$	$\frac{(n - 2)180}{n}$

TEACHING NOTE

The concept of a regular polygon can be introduced nicely using homemade cardboard strips. A pentagon, for example, can be made by fastening 5 strips together. Since a variety of different pentagons can be demonstrated with this model, it becomes clear to the student that not only side length, but also angle size should be considered when describing a regular polygon.

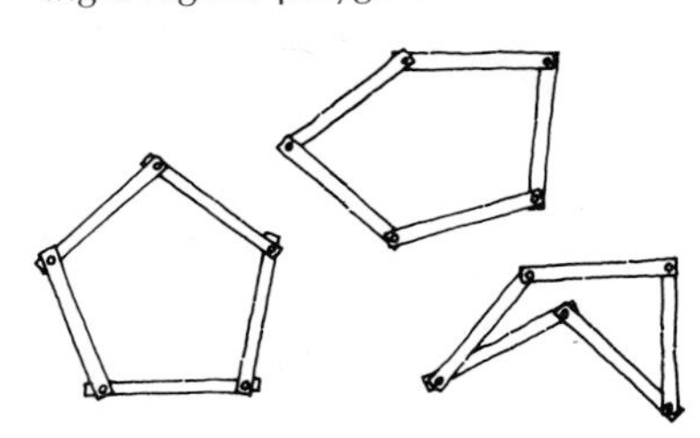

An angle with vertex at the center of a regular polygon and sides that contain adjacent vertices of the polygon is called a **central angle** of a polygon (Fig. 3.21). Thus, $\angle AOB$ is a central angle of pentagon $ABCDE$. A central angle of a regular n-gon has measure $360°/n$.

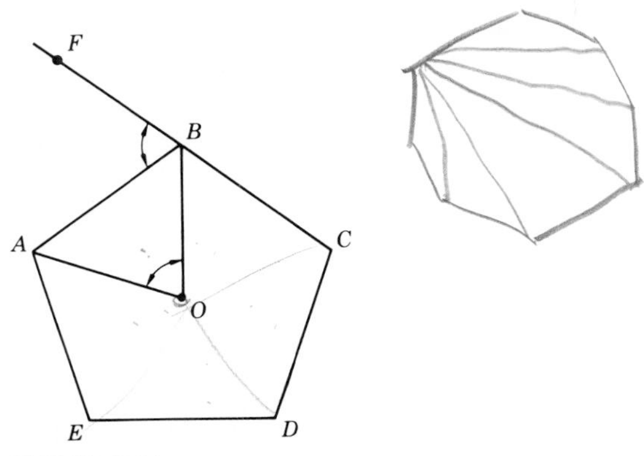

FIGURE 3.21

EXPLORING WITH THE COMPUTER 3.1

Use geometry exploration software to make and text conjectures about the following questions. Use at least 3 different figures when testing each conjecture. Record your results.

1. How does the measure of a central angle of any regular polygon compare to the measure of an exterior angle of the polygon?
2. How does the measure of an interior angle of a regular polygon compare to the measure of the central angle? Can you explain why?

An angle formed by one side of a polygon and the extension of an adjacent side is called an **exterior angle** of the polygon. Thus, $\angle ABF$ is an exterior angle of pentagon $ABCDE$. Investigation 3.4 will help you discover how to find the measure of an exterior angle of a regular n-gon.

CONSTRUCTION OF REGULAR POLYGONS

To draw a regular n-gon, it is convenient to begin with a circle and mark n equally spaced points on the circle. Two vertices of the n-gon are found by using a protractor to measure a

INVESTIGATION 3.4

Draw an equilateral triangle with exterior angles. Make a cardboard arrow and lay it on the triangle in the starting position shown. Slide and turn the arrow through each of the exterior angles of the triangle.

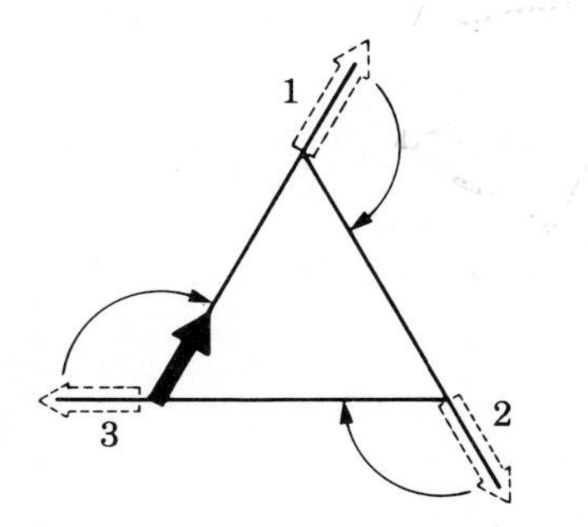

A. What is the ending position of the arrow? Through how many degrees did it turn?
B. What conjecture can you make about the sum of the exterior angles of an equilateral triangle?
C. Try this procedure for other regular polygons (square, regular pentagon, regular hexagon, etc.). How would you complete these conjectures?
   1. The sum of the exterior angles of a regular polygon is ____?
   2. An exterior angle of a regular polygon has measure ____?

(If a computer with LOGO is available, you may want to do the investigation with the LOGO turtle, rather than the cardboard arrow.)

TEACHING NOTE

While some activities using the compass and the straight edge have been utilized with second-grade children, it is generally believed that the motor skills of the second grader are not well enough developed for extensive handling of the compass. Upper-grade and junior high school students enjoy some compass constructions and the making of designs using the ruler and compass. The problems of the constructability of the polygons are sometimes motivational for secondary school students.

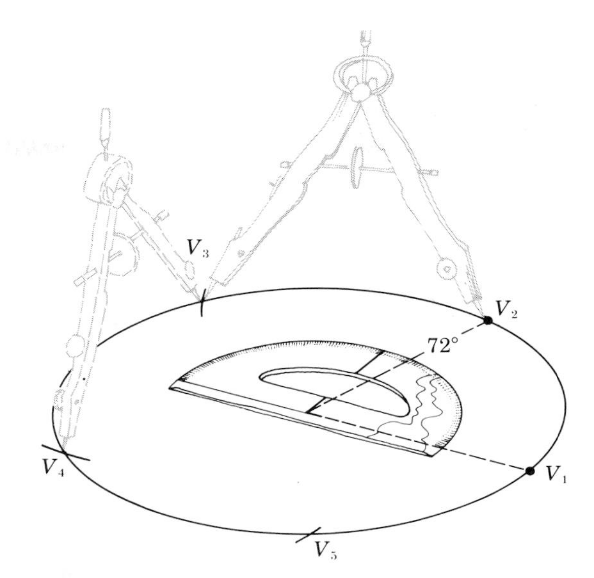

FIGURE 3.22

central angle of the n-gon. A compass is then used to mark the remaining vertices on the circle. These vertices are connected to form the n-gon. This procedure is illustrated for the pentagon in Figure 3.22.

Another method that is appropriate and usually accurate enough for most practical purposes is the process of constructing regular polygons using *only* a compass and an unmarked straight edge. This method has long been of historical and theoretical interest. Investigation 3.5 will help answer the question, *Can each regular n-gon be constructed using* only *a compass and a straight edge?*

INVESTIGATION 3.5

If necessary, see Appendix B for instruction on use of basic ruler-and-compass constructions.

A. How many regular polygons can you construct using only a compass and an unmarked straight edge? Start with a circle.

B. When you have constructed a certain n-gon, you may list the other n-gons that are easily constructible using that n-gon.

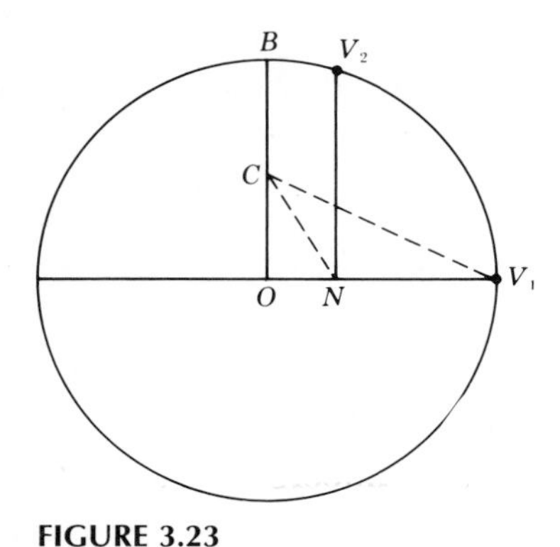

FIGURE 3.23

As you may have discovered in Investigation 3.5, the regular pentagon is not so easily constructed using a straight edge and compass. To construct a regular 5-gon, first draw a circle with center O (Fig. 3.23). Then proceed to find on the circle the vertices V_1, V_2, V_3, V_4, and V_5 of the regular pentagon as follows:

1. Label any point on the circle V_1 and draw $\overline{OB}$ perpendicular to $\overline{OV_1}$.
2. Join V_1 to C, the midpoint of $\overline{OB}$.
3. Bisect angle OCV_1 to obtain the point N, $\overline{OV_1}$.
4. Construct the perpendicular to $\overline{OV_1}$ at N and obtain the point V_2.

The segment V_1V_2 is one side of a regular pentagon and from it, points V_3, V_4, and V_5 can be found. These points are then connected to complete the construction.

Once a method of constructing a regular 3-gon, a regular 4-gon, and a regular 5-gon is available, other n-gons can be constructed by successively bisecting the sides of these basic regular polygons. The following sequences describe the regular polygons that can be constructed in this way.

$$3, 6, 12, 24 \ldots 3\cdot 2^{k-1} \ldots k = 1, 2, 3, \ldots$$

$$4, 8, 16, 32 \ldots 4\cdot 2^{k-1} \ldots k = 1, 2, 3, \ldots$$

$$5, 10, 20, 40 \ldots 5\cdot 2^{k-1} \ldots k = 1, 2, 3, \ldots$$

Beyond this, the n-gons that can be constructed depend upon the number n. For example, a regular 5-gon can be con-

---

EXPLORING WITH THE COMPUTER 3.2

Read the section on Procedures and on Variables in Appendix A, Using LOGO in Geometry. Study this procedure, try it a few times, and then predict what will be produced for each command given below. Check your prediction.

```
TO POLYGON :N :L
REPEAT :N[FD :L RT 360/:N]
END
```

1. POLYGON 4 10
2. POLYGON 5 8
3. POLYGON 10 5
4. POLYGON 300 1

---

structed, but a regular 7-gon cannot be constructed using these tools. The nineteenth-century mathematician Carl Friedrich Gauss, using the notion of Fermat Primes, completely answered this question by proving the following theorem.

> **Theorem:** A regular n-gon can be constructed with compass and straight edge if and only if the odd prime factors of n are distinct Fermat Primes.

A **Fermat prime** is a prime number that can be produced by the formula $F_k = 2^{2^k} + 1$. The only numbers of this form known to be prime are listed below.

$$F_0 = 2^{2^0} + 1 = 2^1 + 1 = 3$$

$$F_1 = 2^{2^1} + 1 = 2^2 + 1 = 5$$

$$F_2 = 2^{2^2} + 1 = 2^4 + 1 = 17$$

$$F_3 = 2^{2^3} + 1 = 2^8 + 1 = 257$$

$$F_4 = 2^{2^4} + 1 = 2^{16} + 1 = 65537$$

Much to the surprise of Pierre de Fermat (1601–1665), after whom these primes are named, the next number generated by the formula,

$$F_5 = 2^{2^5} + 1 = 2^{32} + 1 = (641) \cdot (6{,}700{,}417)$$

has more than two factors and, hence, is not a prime number. Thus, according to Gauss' theorem, a 7-gon cannot be constructed because 7 is not a Fermat Prime, a 55-gon cannot be constructed because 11 is not a Fermat Prime, and a 9-gon cannot be constructed because the odd factors of 9 are not *distinct* Fermat Primes (i.e., 3 occurs *twice* as an odd factor of 9).

EXERCISE SET 3.2

1. How many degrees are there in an exterior angle of an equilateral triangle? A square? A regular pentagon? A regular hexagon? A regular octagon?
2. Use the information in Exercise 1 to give the number of degrees in a vertex angle of the polygons mentioned in Exercise 1.
3. How many different regular polygons can you form using a circular geoboard with 24 nails?

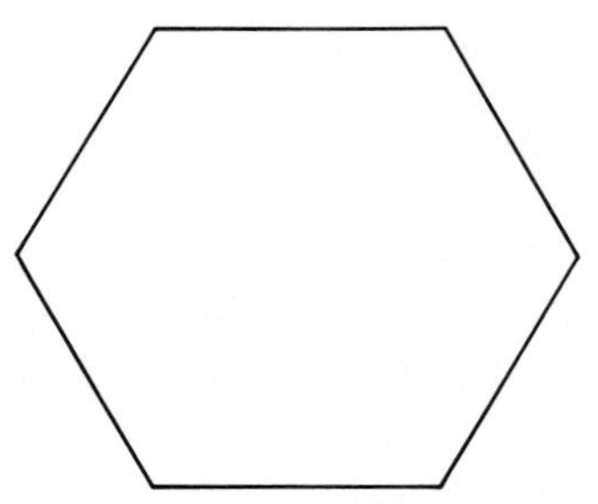

FIGURE 3.24

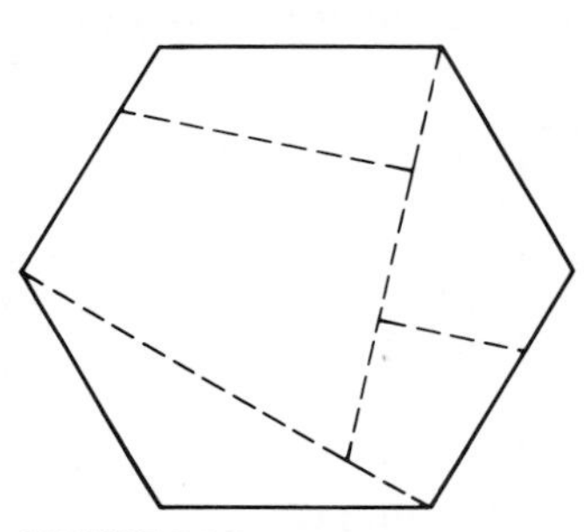

FIGURE 3.25

4. Find the number of degrees in the vertex angle of a regular
   a. Heptagon b. Octagon c. Nonagon
   d. Decagon e. Dodecagon f. 100-gon
5. a. Find the number of degrees in the central angle of each polygon in Exercise 4.
   b. If a regular polygon with a very small central angle is chosen, what can you say about the measure of a vertex angle of this polygon?
6. Use a ruler, compass, and protractor to construct a regular heptagon.
7. Use only a compass and straight edge to construct a regular
   a. 3-gon b. 4-gon c. 6-gon d. 8-gon
8. Use only a straight edge and a compass to construct a regular pentagon.
9. a. How many lines of symmetry do the regular pentagon, hexagon, and octagon possess?
   b. Generalize the findings in part a and describe the lines of symmetry for any regular n-gons.
10. a. Copy a regular hexagon as in Figure 3.24. Can you cut the hexagon to produce 6 equilateral triangles?
    b. Can you cut the hexagon to produce 2 isosceles trapezoids?
    c. Make 2 copies of the regular hexagon. Can you cut the 2 hexagons to form 3 equilateral triangles?
11. Copy the regular hexagon in Figure 3.25, and cut it apart along the dotted lines. Can you place the pieces back together to form a square?
12. Use the theorem proved by Gauss and list all regular polygons with fewer than 100 sides that can be constructed with compass and straight edge.

*13. Cut a regular hexagon into 18 kites all the same size and shape. [*Hint:* Draw all the diagonals of the hexagon. Then draw 6 more short segments, strategically located.]

*14. Cut a regular hexagon into 12 rhombi all the same size and shape.

*15. Copy the 3 small hexagons in Figure 3.26, and cut as indicated by the dotted lines. Can you place the 13 pieces together to form the large hexagon? [*Hint:* A six-pointed star is involved.]

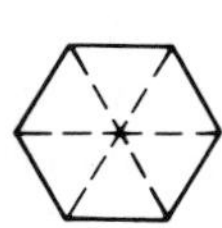

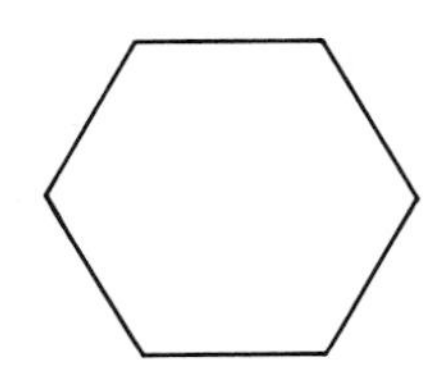

FIGURE 3.26

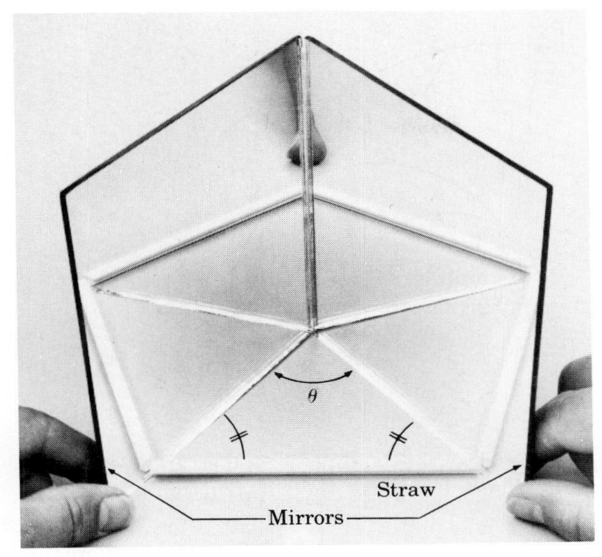

FIGURE 3.27

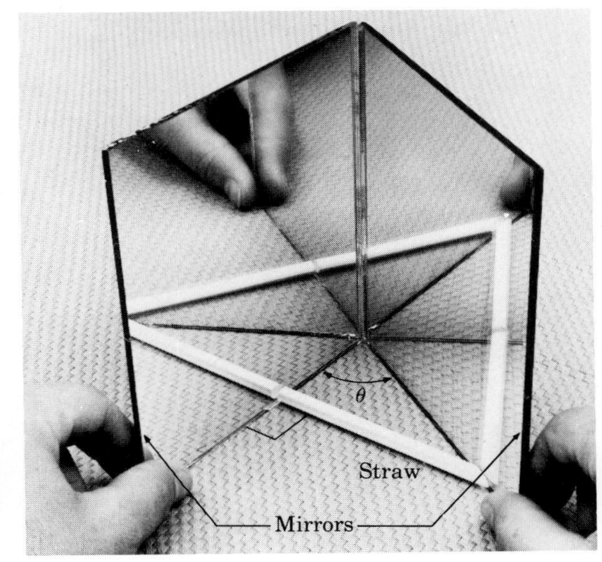

FIGURE 3.28

TABLE 3.5

Angle Between the Mirrors (**θ**)	Number of Sides of Regular Polygon Formed
	3
	4
	5
	.
	.
	.
	n

*16. If a pair of mirrors are hinged and a straw is laid between them as shown in Figure 3.27, a regular polygon can be seen for certain values of θ (the measure of the angle formed by the two mirrors).

a. Experiment with mirrors and a straw to determine the relationship between the angle θ and the number of sides for the regular polygon formed. Complete Table 3.4.

b. Repeat the experiment in part a with the straw positioned as shown in Figure 3.28.

17. Draw a picture of a nonconvex pentagon with 3 collinear vertices.

18. Draw a picture of a nonconvex pentagon with 2 diagonals, one with 3 diagonals, and one with 4 diagonals.

19. A regular hexagon can be partitioned into triangles by drawing diagonals in 14 different ways. Draw a picture of each way.

*20. Copy the 2 small dodecagons in Figure 3.29, and cut as indicated by the dotted lines. Can you place the 13 pieces together to form the large dodecagon?

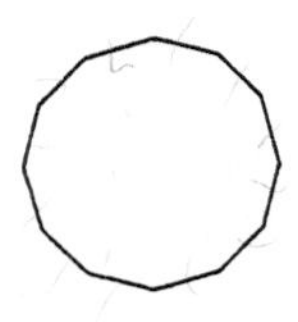
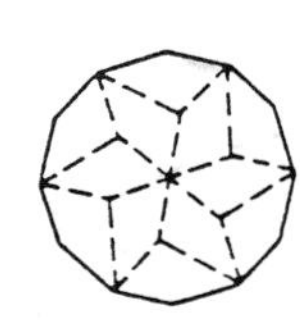
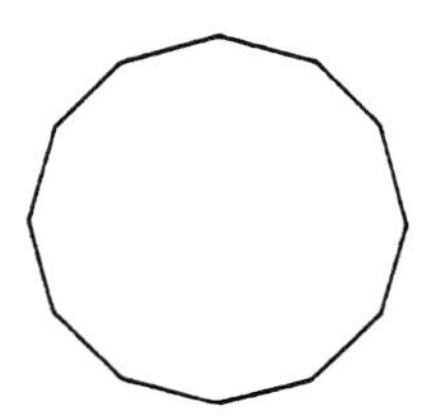

FIGURE 3.29

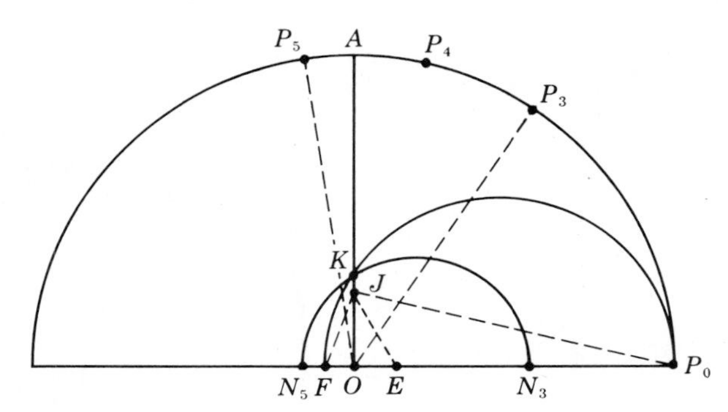

FIGURE 3.30

*21. The following outline describes a compass and straight-edge construction for the 17-gon. Complete the construction of a 17-gon (Fig. 3.30).

1. Construct a circle with center O and draw a diameter through O and P_0.
2. Construct $\overline{OA}$ perpendicular to $\overleftrightarrow{OP_0}$ and find point J one-quarter of the way from O to A. Draw $\overline{P_0J}$.
3. Label E and F on diameter $\overline{OP_0}$ so that $m(\angle OJE)$ is 1/4 $m(\angle OJP_0)$ and $m(\angle FJE)$ is 45°.
4. Draw a circle with diameter $\overline{FP_0}$ and label the point in which that circle intersects $\overline{OA}$ as K.
5. Draw a circle with center E and radius EK. Label as N_3 and N_5 the points of intersection of this circle with the diameter $\overline{OP_0}$.
6. Draw perpendiculars $\overline{N_5P_5}$ and $\overline{N_3P_3}$.
7. Bisect $\angle P_3OP_5$ to obtain point P_4. Then, arc P_3P_4 is 1/17 of the circumference of the original circle.

22. **Critical Thinking.** Write a paragraph explaining how you can use the generalization that "the sum of the exterior angles of a regular n-gon is 360°" to produce a formula for finding the measure of a vertex angle of a regular n-gon. Which formula, the one you created in this exercise or the formula $[180(n-2)]/n$ discovered in Table 3.4, would be easier to recreate if forgotten. Explain the reasons for your choice.

STAR POLYGONS

The five-pointed star, often called a *pentagram,* was chosen as the sacred symbol of the Pythagorean Society of ancient Greeks because of its special beauty and because of the appearance of the Golden Ratio when certain of its segments were compared. This symbol was worn on their clothing, and around the five

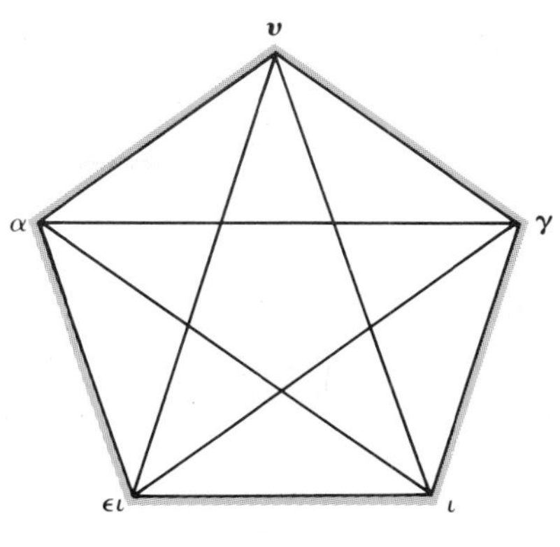

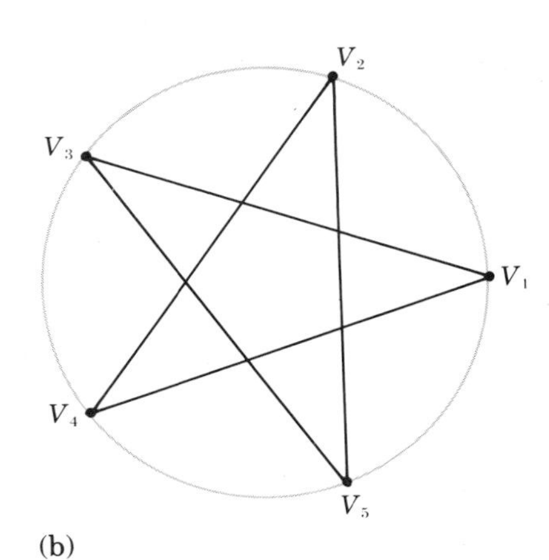

FIGURE 3.31 (a) (b)

points were placed the letters of the Greek word for health, "υγιεια," or hygeia (Fig. 3.21a).

This star, like the regular pentagon, can be constructed from five equally spaced points on a circle. Beginning at V_1 and going around the circle in one direction, every second point is joined by a segment. The star is a nonsimple polygon and is called a **star polygon**. Like the regular pentagon, it has five vertices and five sides. Since all five segments are of the same length and all vertex angles are of the same measure, it is called a **regular star polygon.** Since it is constructed from *five* points with every *second* point joined (i.e., $V_1\ V_3\ V_5\ V_2\ V_4\ V_1$ in Fig. 3.31b), this regular star polygon is denoted by $\{\frac{5}{2}\}$.

regular usually refers to equal sides and/or angles

In Figure 3.32, we analyze all the possibilities using 5 equally spaced points on a circle. Note that there is exactly one star polygon with 5 sides, since $\{\frac{5}{2}\}$ and $\{\frac{5}{3}\}$ represent the same set of points and, hence, the same star polygon.

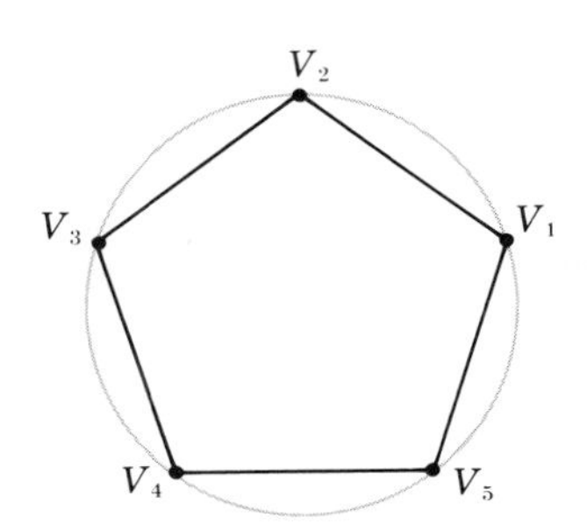

5 points, with consecutive points joined ($V_1V_2V_3V_4V_5V_1$) or with every fourth point joined ($V_1V_5V_4V_3V_2V_1$), is not a star polygon.

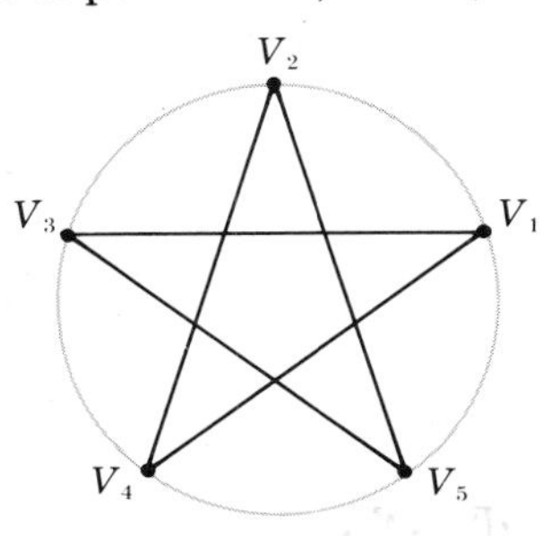

5 points, with every second point joined ($V_1V_3V_5V_2V_4V_1$), is a star polygon $\{\frac{5}{2}\}$.

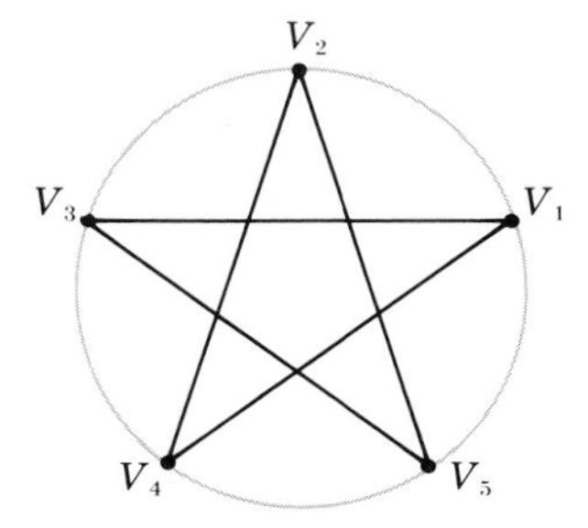

5 points, with every third point joined ($V_1V_4V_2V_5V_3V_1$), is a star polygon $\{\frac{5}{3}\}$.

FIGURE 3.32

The process for constructing the star polygon $\{\frac{5}{2}\}$ can be generalized, although we need to proceed with a degree of caution. Suppose we begin with 8 equally spaced points on a circle and join every second point (Fig. 3.33a). A closed path is obtained before all 8 points have been reached. We do not obtain a star polygon. On the other hand, if we join every third point (Fig. 3.33b), a nonsimple closed polygon with 8 sides is obtained that may also be called a regular star polygon. Since we begin with 8 equally spaced points and joined every third point, this star polygon is denoted $\{\frac{8}{3}\}$.

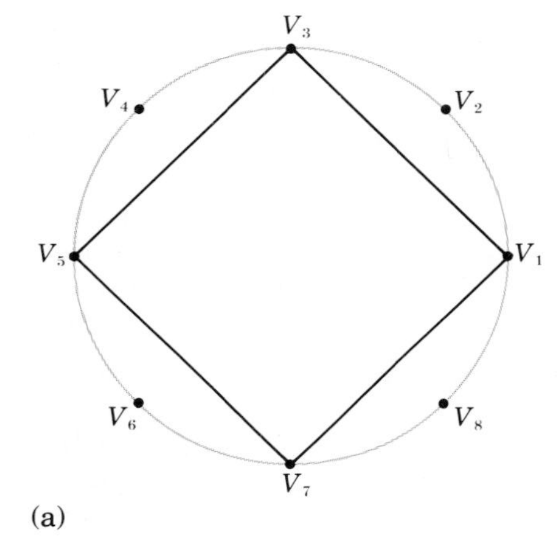

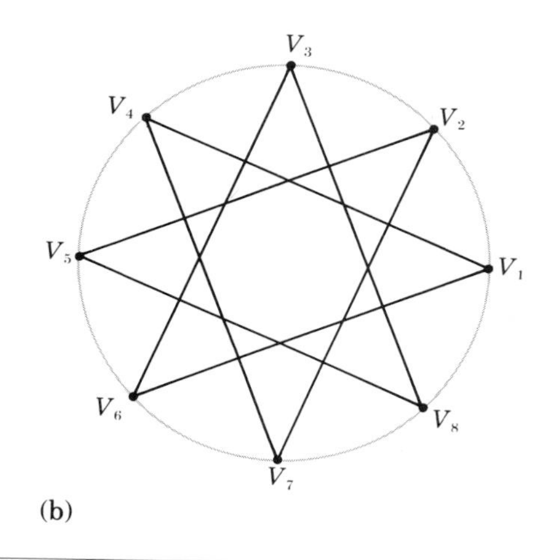

FIGURE 3.33 (a) (b)

EXPLORING WITH THE COMPUTER 3.3

Use LOGO software and the following LOGO procedure to draw star polygons. Imagine a circle on which n equally spaced points have been drawn. Begin drawing at one point and draw a path by joining every Dth point.

```
TO STAR:N:D
MAKE "A 360/:N
REPEAT:N[FD 50 RT:D*:A]
END
```

:N the number of equally spaced points
:D the path joins every Dth point

1. For what pairs of numbers :N,:D does STAR:N:D complete a star polygon?
2. Sometimes a star polygon STAR:N:X is the same as the star polygon STAR:N:Y. What must be true about X and Y for this to occur?
3. What other conjectures can you make about the polygons that are drawn by STAR:N:D?

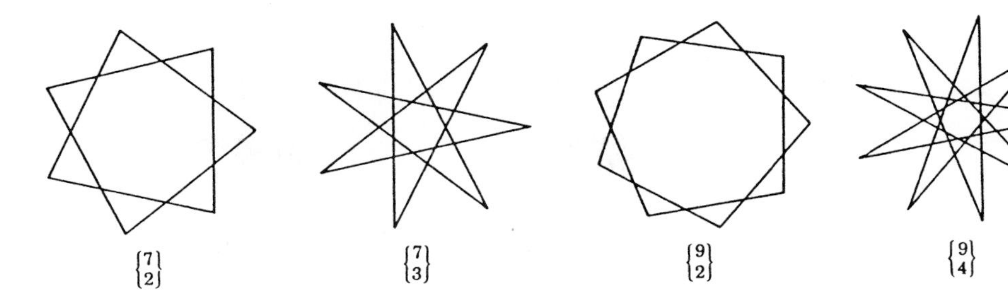

FIGURE 3.34

Usually, a regular star polygon that has been constructed from n equally spaced points on a circle by joining every d^{th} point is denoted $\left\{ {n \atop d} \right\}$. We have already seen that there is no star polygon for the pair $\left\{ {8 \atop 2} \right\}$, which raises the question, *Which pairs of numbers* $\left\{ {n \atop d} \right\}$ *represent a regular star polygon?* The following exercises will help answer this question. Readers should convince themselves that the star polygons $\left\{ {5 \atop 2} \right\}$ and $\left\{ {8 \atop 3} \right\}$, together with the star polygons in Figure 3.34, are the only star polygons with fewer than 10 sides.

EXERCISE SET 3.3

1. Answer these questions for each star polygon in Figure 3.34.
   a. How many sides does each polygon possess?
   b. If the vertices are labeled V_1, V_2, V_3, V_4, etc., name all the sides of each polygon.
   c. How many lines of symmetry does each possess?
   d. Describe the rotational symmetry of each polygon.
2. Using circle dot paper with 8 equally spaced points, show by drawings that $\left\{ {8 \atop 3} \right\}$ and $\left\{ {8 \atop 5} \right\}$ are the only eight-sided star polygons and that, in fact, they are the identically same star polygon.
3. Use circle dot paper with ten equally spaced dots to determine how many distinct ten-sided regular star polygons exist.
4. Use circle dot paper with 12 equally spaced dots to determine how many distinct twelve-sided regular star polygons exist.
5. Using the experience gained in Exercises 2, 3, and 4, can you make any guesses as to which star polygons exist? In other words, there is a numerical relationship between n and d if the star polygon $\left\{ {n \atop d} \right\}$ exists. What is it?
6. $\left\{ {5 \atop 2} \right\}$ and $\left\{ {5 \atop 3} \right\}$ represent the same star polygon.
   $\left\{ {7 \atop 3} \right\}$ and $\left\{ {7 \atop 4} \right\}$ represent the same star polygon.
   $\left\{ {8 \atop 3} \right\}$ and $\left\{ {8 \atop 5} \right\}$ represent the same star polygon.
   a. Do the following represent the star polygon: $\left\{ {10 \atop 3} \right\}$? $\left\{ {11 \atop 5} \right\}$? $\left\{ {12 \atop 7} \right\}$?
   b. In general, the star polygon $\left\{ {n \atop d} \right\}$ is the same as the star polygon $\left\{ {n \atop ?} \right\}$. Explain why this is true.

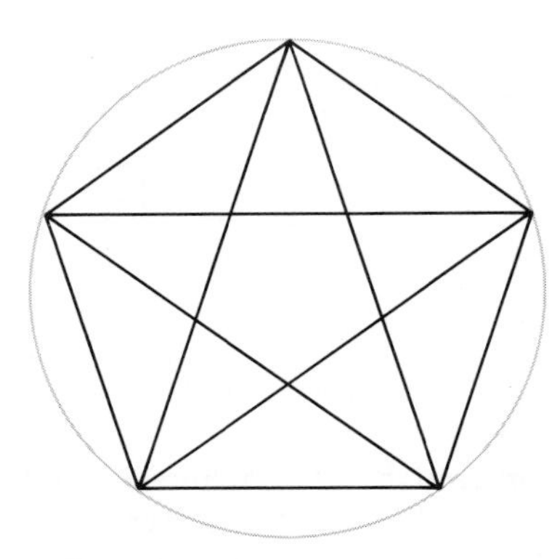

FIGURE 3.35

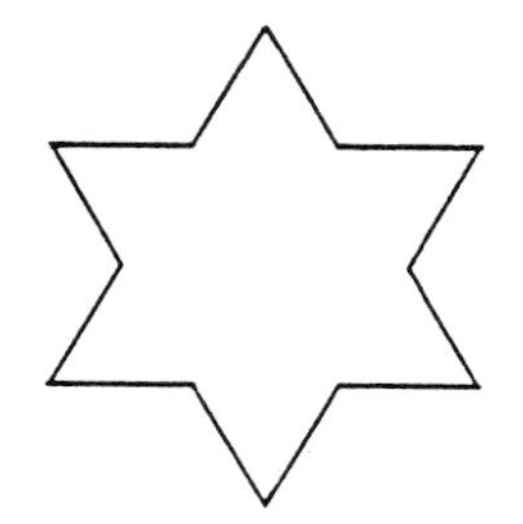

FIGURE 3.36

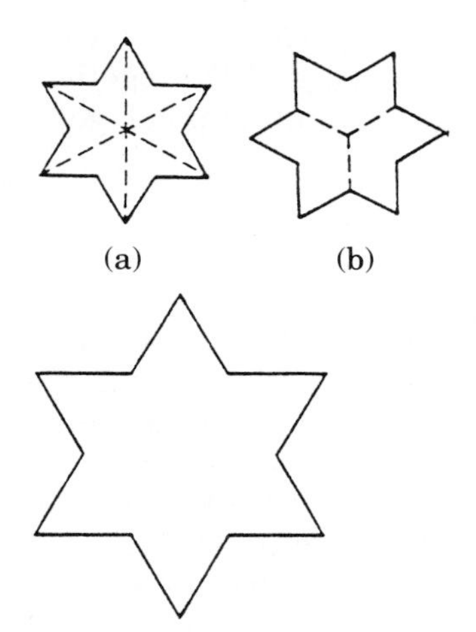

FIGURE 3.37

7. **Theorem:** The n-sided star polygon $\{{n \atop d}\}$ exists if and only if $d \neq 1$, $d \neq (n - 1)$, and n and d are relatively prime. Use this theorem to determine how many eleven-sided star polygons exist.

*8. **Theorem:** The number of numbers less than n and relatively prime to n is $n[1-(1/p_1)][1-(1/p_2)] \ldots$, where $p_1, p_2, \ldots$ are prime factors of n.
   a. How many numbers are less than 36 and relatively prime to 36?
   b. How many star polygons are there with 36 sides?

9. Figure 3.35 is sometimes called the five-point *mystic rose*.
   a. How many different geometric figures can you find in this five-point mystic rose?
   b. How many different geometric figures can you find in a six-point mystic rose?

10. Copy and cut up 3 copies of the small star in Figure 3.36, and fit the resulting 12 pieces together to form the large six-pointed star.

11. Make one copy of star A and two copies of star B, and cut them up as indicated (Fig. 3.37). Fit the resulting 12 pieces together to form the large six-pointed star.

12. **Critical Thinking.** Do you believe that the following generalization is probably true? "The measure of a 'point angle' of the star polygon $\{n/d\}$ is $|n - 2d|180°/n$." What evidence can you find to support your belief?

TEACHING NOTE

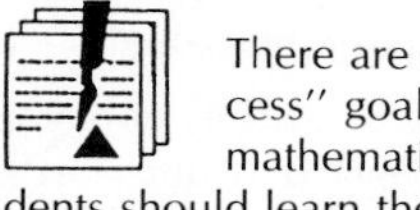

There are certain "process" goals for learning mathematics. That is, students should learn the process of active inquiry, of discovery of relationships, of formulating and testing conjectures, and of critical and analytical thinking. In many cases, the process is more important than the content. Thus, students of all ages should have as many opportunities as possible to discover geometric relationships.

DISCOVERING THEOREMS ABOUT POLYGONS

Many interesting and unexpected relationships for polygons have been discovered by mathematicians over the course of many centuries. Relationships so regular and so surprising as to be considered beautiful have emerged from figures as simple as a triangle or as irregular as a general quadrilateral. In this section we shall explore a few such relationships involving triangles and quadrilaterals. Many of these beautiful relationships can be discovered with a computer and geometry exploration software.

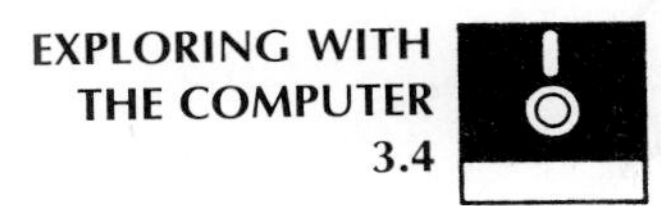

Use geometry exploration software to make and test conjectures about the following questions. Use at least three different figures when testing each conjecture. Record your results.

1. What relationships can you discover when all 3 medians of a triangle are drawn? Compare lengths of segments on the medians in your figure.
2. Answer question 1 when "medians" is replaced with "altitudes." With "perpendicular bisectors of the sides."
3. Draw all 3 medians, all 3 altitudes, and all 3 perpendicular bisectors of the sides on one general triangle. What do you discover?

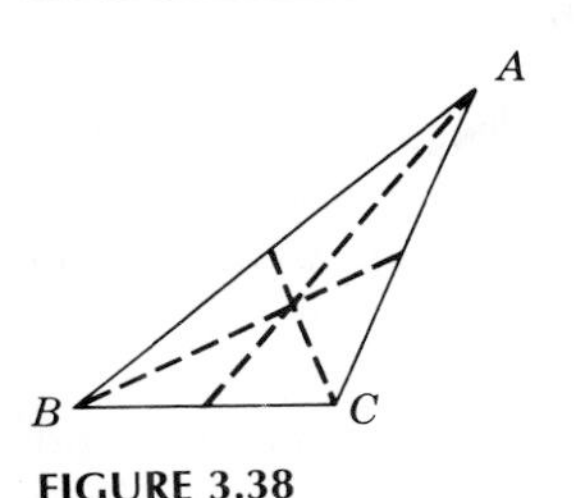

FIGURE 3.38

Accurate construction of the 3 medians in a triangle, as in Figure 3.38, may lead to the observation that the medians are concurrent. Is this concurrence a coincidence, or are the medians concurrent for all triangles? The answer is emphatically stated in the following theorem, which is stated here without proof.

Theorem 1: For any triangle, the medians are concurrent.

At first glance this theorem seems almost expected. Perhaps the reader has learned of the theorem in a previous geometry course and is not surprised to rediscover it. However, on second glance this concurrence relation is almost astounding. Three randomly chosen lines are certainly not expected to be concurrent. Although medians are not randomly selected segments, the triangle itself was randomly chosen; the triangle was determined from three randomly selected nonconcurrent lines.

For similar reasons the next two theorems, which the reader may have discovered, are often unexpected.

Theorem 2: For any triangle, the lines containing the altitudes are concurrent.

Theorem 3: For any triangle, the perpendicular bisectors of the sides are concurrent.

These three theorems describe three points associated with each triangle. The point of intersection of the medians is the **centroid** (G); the point of intersection of the altitudes is the **orthocenter** (H); and the point of intersection of the perpendicular bisectors is the **circumcenter** (O). The circumcenter is

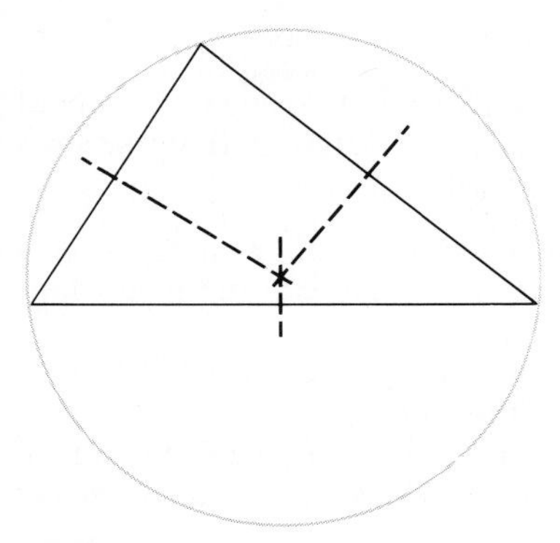

FIGURE 3.39

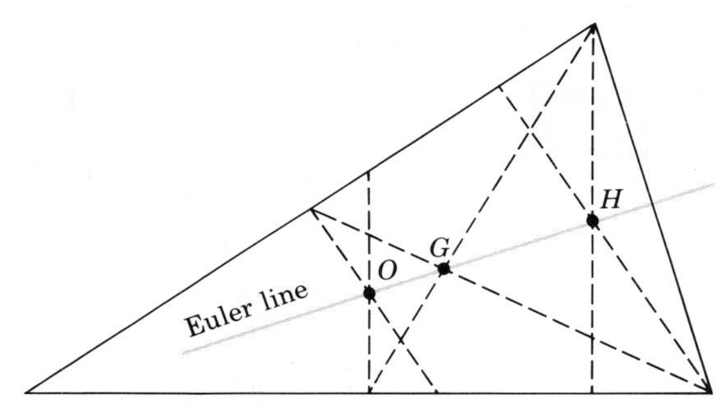

FIGURE 3.40

the center of a circle called the **circumscribed circle** that contains all three vertices of a triangle (Fig. 3.39).

A theorem perhaps more stunning and beautiful than the first three is one that relates the centroid, the orthocenter, and the circumcenter. Who would expect these three points to be collinear?

> **Theorem 4:** In any triangle the centroid, orthocenter, and circumcenter are collinear (Fig. 3.40).

The line containing these three points is called the **Euler line**, named after the Swiss mathematician Leonard Euler who discovered this theorem. Some questions that the reader might explore are as follows:

1. Is point G always between points O and H?
2. Is there any numerical relationship relating the distances between these three points?
3. What happens when the original triangle is isosceles? Equilateral?

In each exercise of Exercise Set 3.4, a setting is provided in which at least one interesting potential theorem can be discovered. The reader may find it helpful to work on these exercises with another person and together seek interesting relationships or patterns of regularity (i.e., regular polygons, concurrence, collinearity, etc.).

The following suggestions might help the reader in attempting to discover interesting relationships in a given situation:

1. Use large figures since they make it easier to recognize relationships.

2. Use at least three different types of figures for each situation to verify that the relationship discovered does not depend upon consideration of a figure of a particular size or shape.
3. Use geometry exploration software whenever possible. It reduces the time required and increases the accuracy of measurement.
4. When actual measurement must be used, a ruler marked in millimeters is best.
5. A MIRA is often helpful in making constructions since fewer construction lines take less time and don't obscure the relationship.
6. Varying the conditions of the problem may suggest a new, perhaps interesting, relationship. Often a "what if" question sparks an idea. For example, if the situation you have been exploring deals with quadrilaterals, you might ask, *What if I began with a triangle or a pentagon*?

EXERCISE SET 3.4

1. What is the sum of the acute angles of a right triangle? How would you convince someone that this is true?
2. Draw a triangle and connect the midpoints of its sides. Measure perimeters of the small and large triangles in this situation. Make a conjecture and test it with other examples. If your conjecture is true, how could you use it to convince someone your conjecture for this exercise is true?
3. Draw the medians of a triangle and measure the distance from the point of their intersection to a vertex of the triangle and to the midpoint of the side opposite that vertex. How do these distances relate to each other and to the median?
4. Begin with any triangle. Draw the 3 angle bisectors of the triangle.
   a. What do you discover about these 3 bisectors?
   b. What do these 3 lines have to do with the inscribed circle of a triangle, the circle that touches each side of the triangle at exactly one point?
5. Draw a rhombus and its diagonals. Measure the angles formed by the diagonals. Make a conjecture and test it with other examples.
6. Draw a trapezoid with a pair of nonparallel sides and find the midpoints of these sides. Measure the segment formed. What relationship can you discover between the measure of this segment and the measures of the 2 bases? Make a conjecture and test it with other examples.

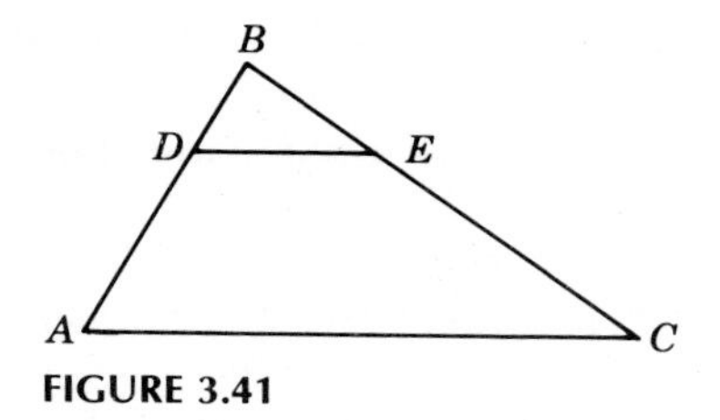

FIGURE 3.41

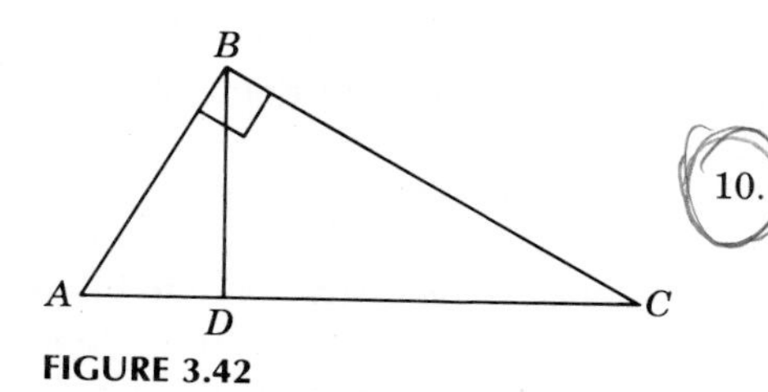

FIGURE 3.42

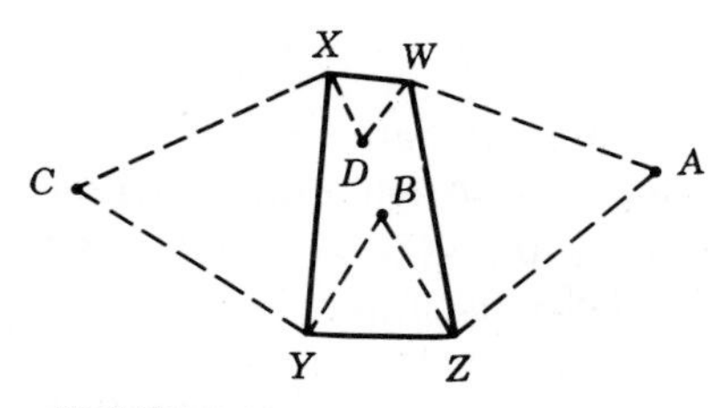

FIGURE 3.43

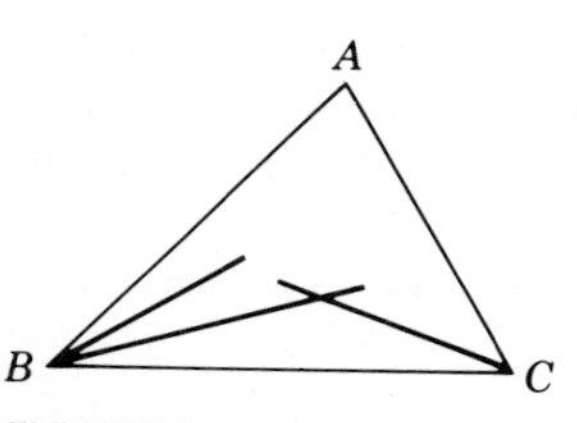

FIGURE 3.44

7. Draw a triangle ABC and any segment DE parallel to AC (Fig. 3.41). Find DB/AD and BE/EC. Make a conjecture and test it with other examples.
8. Construct a right triangle ABC with BD perpendicular to the hypotenuse AC (Fig. 3.43). Measure and calculate to find $BD \cdot BD$, $AD \cdot DC$, and $AB \cdot BC$. Make a conjecture and test it using other examples.
9. Draw with a straight edge any quadrilateral $WXYZ$ (Fig. 3.43). Construct equilateral triangles on each side of the quadrilateral alternately inside and outside the quadrilateral and label the new vertices A, B, C, and D. What is true about $\overleftrightarrow{AD}$ and $\overleftrightarrow{BC}$? About $\overleftrightarrow{AB}$ and $\overleftrightarrow{CD}$? State a generalization.
10. a. Begin with any triangle, $\triangle ABC$ (Fig. 3.44). Use a protractor to trisect each of the angles ($<B$ in Fig. 3.44). Extend adjacent trisectors until they intersect (as for $<B$ and $<C$ in Fig. 3.44). Connect these 3 points of intersection. What do you discover?
    b. Repeat part a by trisecting the exterior angles of $\triangle ABC$.
11. Begin with any triangle, $\triangle ABC$. On each side of this triangle, construct equilateral triangles external to the triangle (Fig. 3.45). Find the centroid of each of these 3 triangles and join them by segments. What do you discover?
12. Begin with any triangle, $\triangle ABC$. Find the midpoints of the sides and label them L, M, and N (Fig. 3.46).
    a. How does $\triangle LMN$ compare with each of the triangles $\triangle ALM$, $\triangle CLN$, and $\triangle BNM$?
    b. For each of the 3 triangles containing a vertex of $\triangle ABC$ in part a, construct the circumcenters and the orthocenters. Connect these 3 circumcenters. Similarly, connect these 3 orthocenters. What do you discover?
13. Begin with any convex quadrilateral. On each side construct a square lying external to the quadrilateral (Fig. 3.47). Find the centers of these squares and label them X_1, X_2, X_3, and X_4. If X_1 and X_3 are centers of opposite squares, compare segments $\overline{X_1X_3}$ and $\overline{X_2X_4}$. What do you discover?
14. Begin with any parallelogram $ABCD$. On each side of the parallelogram construct squares lying external to the parallelogram

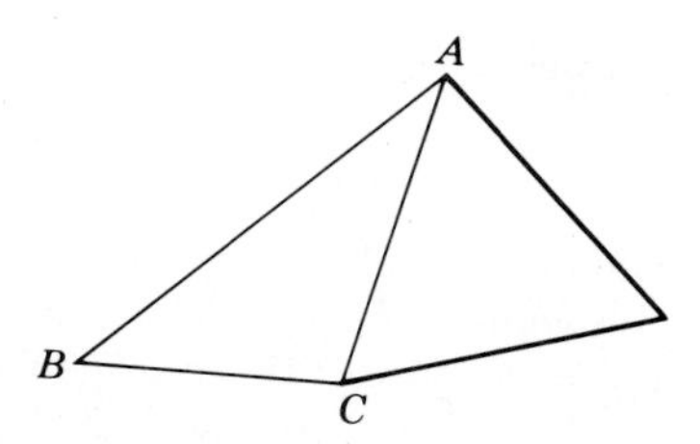

FIGURE 3.45

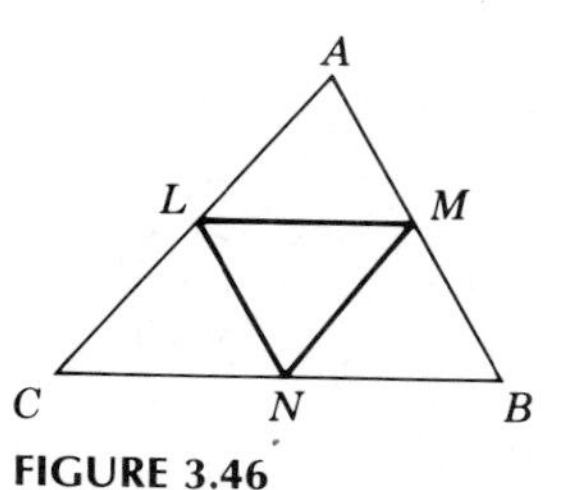

FIGURE 3.46

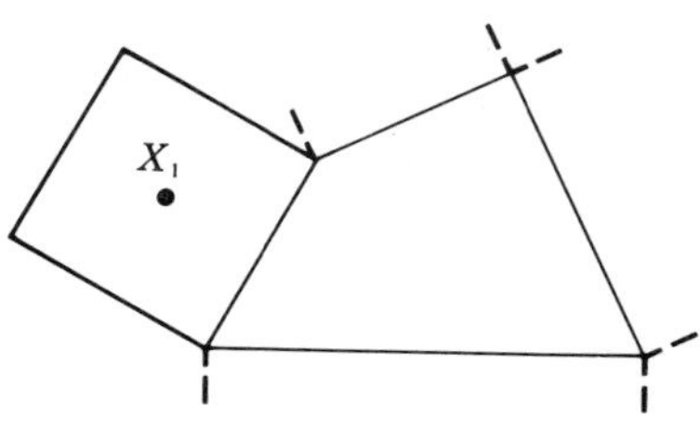

FIGURE 3.47

(Fig. 3.48). The centers of these squares, X_1, X_2, X_3, and X_4 are the vertices of what type of figure?

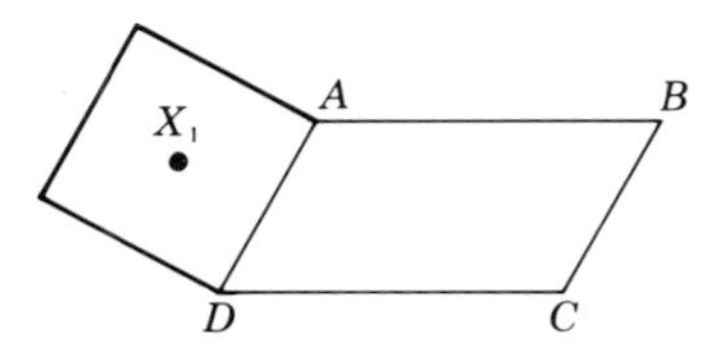

FIGURE 3.48

15. Begin with any convex quadrilateral $ABCD$. Construct equilateral triangles ABM_1, BCM_2, CDM_3, and ADM_4 where they are alternately exterior and interior to the quadrilateral (Fig. 3.49). The points M_1, M_2, M_3, and M_4 are vertices of what kind of polygon?

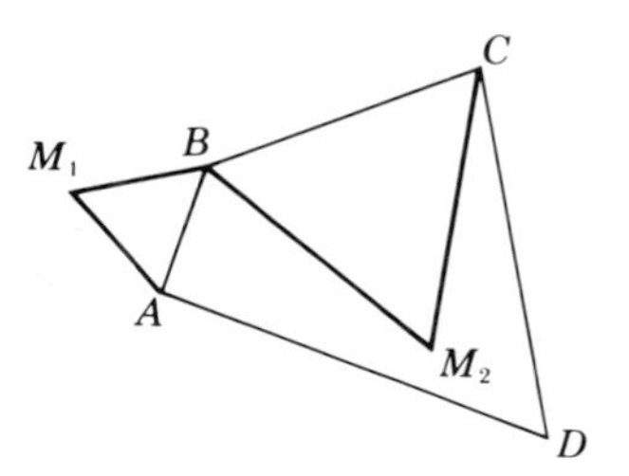

FIGURE 3.49

16. Begin with any triangle, $\triangle ABC$. Construct squares on sides $\overline{AB}$ and $\overline{BC}$ external to the triangle (Fig. 3.50). If M is the midpiont of $\overline{AC}$, and X and Y are the centers of the constructed squares, how are these 3 points related?

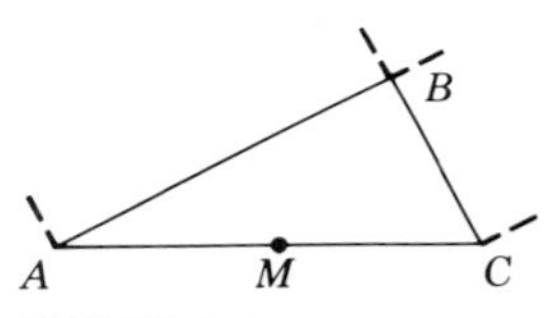

FIGURE 3.50

17. Begin with any triangle. Construct the following 9 points: the midpoints of the 3 sides, the feet of the 3 altitudes, and the midpoints of the segments from the 3 vertices to the orthocenter.
    a. How are these 9 points related?
    b. Is there any relationship between the discovery found in part a and the Euler line?

*18. Begin with $\triangle ABC$. Bisect $\angle A$ (Fig. 3.51). Measure to the nearest millimeter) to find the ratio BP/PC.
    a. Can you find another ratio in the triangle that is approximately equal to this ratio?
    b. Try this with other angles in $\triangle ABC$ and with other triangles. Does it seem to hold true? Can you state a theorem?

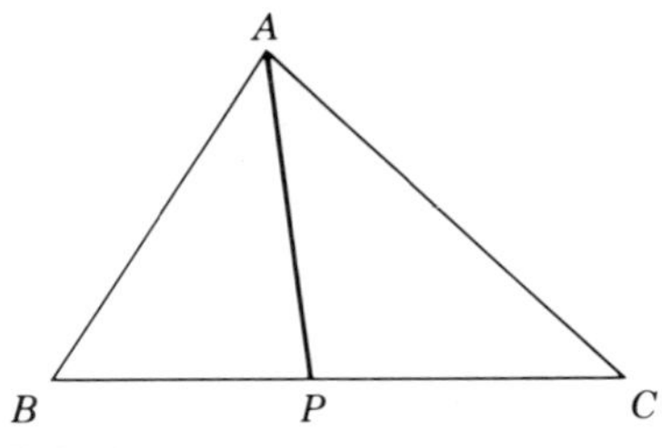

FIGURE 3.51

*19. Begin with any $\triangle ABC$. Bisect the external angles at A and C and the internal angle at B (Fig. 3.52).
    a. What do you discover?
    b. Try this with other angles in the triangle and with other triangles. Does you discovery hold true? Can you state a theorem?
    c. Consult a reference and find out what an *excenter* of a triangle is. How does it apply to this situation?

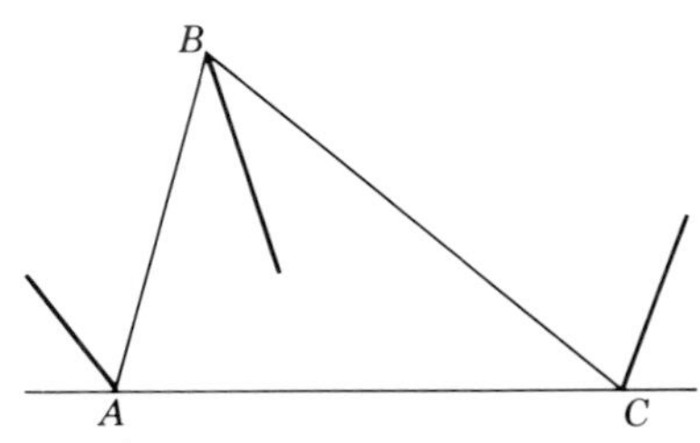

FIGURE 3.52

*20. Start with $\triangle ABC$. Bisect $\angle A$ and $\angle B$. Bisect the external angle at C. If the sides of $\triangle ABC$ are necessarily extended, what can you discover about the points where the 3 bisectors intersect the sides of the triangle opposite the bisected angles? Can you state a theorem?

*21. The *incenter* of a triangle is the point of intersection of the angle bisectors. Begin with $\triangle ABC$. Find the incenter and draw the inscribed circle. What can you say about the lines determined by connecting vertices to the points of tangency of the incircle? Can you state a theorem?

*22. Begin with $\triangle ABC$. Find the circumcenter and draw the circumscribed circle. Now choose *any* point on this circle and construct the perpendiculars from this point to each of the 3 sides of the triangle (sides may be extended if necessary). What do you discover? Do this again with a different point on the circle. With a different triangle. What do you discover?

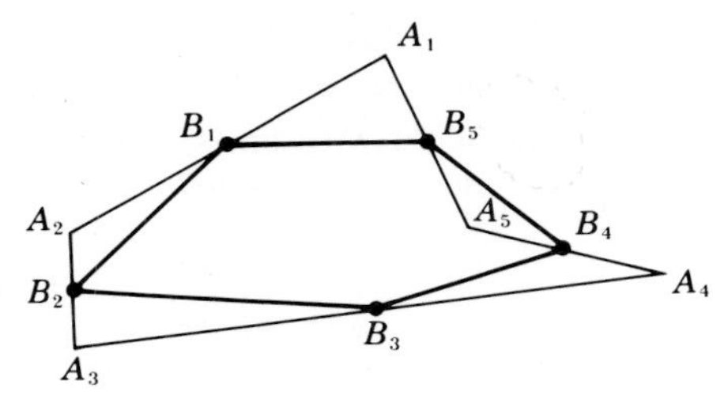

FIGURE 3.53

*23. Begin with $\triangle ABC$. Choose one point, anywhere you like, on each side of the triangle. Construct 3 circles, each of which is determined by a vertex point and the 2 points nearest it. Do this again with different points and with a different $\triangle$. What do you discover?

*24. Begin with $\triangle ABC$. Find an accurate way to locate by construction points P_1 (halfway around the $\triangle$ from vertex A), P_2 (halfway around the $\triangle$ from vertex B), and P_3 (halfway around the $\triangle$ from vertex C). Consider lines AP_1, BP_2, and CP_3. What do you discover?

25. **Critical Thinking.** If someone said, "All frogs are green," you could prove that this statement was untrue by showing the person just one frog that was some color other than green! In doing this, you would have produced a counterexample to the generalization. A *counterexample* is a single example that shows a generalization to be false.

    a. Can you find a counterexample to the following statement: A polygon with congruent sides is a regular polygon?
    b. Now refer to Figure 3.53, and try to find a counterexample to prove that the following generalization is false:

    Generalization: The n-gon $B_1 \ldots B_n$ is always convex even if $A_1 \ldots A_n$ is not.

PEDAGOGY FOR TEACHERS

ACTIVITIES

1. Plan a sequence of activities on symmetry for students. You may want to include paper folding and cutting, ink blot designs, symmetry in nature, repeating geometric figures on graph paper, and mirror activities. Include objectives for each activity and one test question with which you could ascertain whether or not the objective had been met.

2. Do posters in the classroom have value? Make a poster that would motivate students to draw star polygons and investigate their properties. Design the poster in such a way that students are encouraged to become actively involved and leave the results of their work in an envelope attached to the poster. Place the poster in a classroom and devise a means for evaluating the effectiveness and value of the poster.

3. Look in references such as *Recreational Problems in Geometric Dissections and How to Solve Them* (Lindgren, 1972) or at Exercises 2.6, p.94, in *Tilings and Patterns* (Grunbaum and Shepard, 1989), and find at least five dissection problems involving polygons that you feel would be interesting to students at a given level. Try these puzzles with selected students, and evaluate their usefulness.

4. Select a grade level, and work with a group of students using activities from *The Mirror Puzzle Book* (Walter, 1985). Ask specific questions that you have previously prepared to stimulate their interest in the activities. Observe students, and try to answer the following questions.

   a. What is the most difficult mirror puzzle that these students can complete correctly.
   b. Can these students create mirror puzzles themselves?
   c. Are the mirror puzzles interesting and appropriate for these students?
   d. Decide what goals from the NCTM Standards appear to be met by these activities.

COOPERATIVE LEARNING GROUP DISCUSSION

Discuss the value of having students at a selected grade level use geometry exploration software to make and test conjectures in geometry. Arrive at group responses to the following questions.

1. At what levels, if any, can students profit from using geometry exploration software?
2. What is the value of students using such software?
3. How could the use of such software help implement the NCTM Standards?

After formulating responses to these questions, read the article by Chazan (Chazan, 1990), and discuss how the ideas in the article compare with your responses.

CHAPTER REFERENCES

Abbott, Edwin A. 1963. *Flatland—A Romance of Many Dimensions* (5th revised edition). New York: Barnes and Noble Books.

Chazan, Daniel. 1990. "Students' Micro-Aided Exploration in Geometry" (Implementing the Standards). *The Mathematics Teacher* (NCTM), 83(8).

Grunbaum, Branko, and Shephard, G.C. 1989. *Tillings and Patterns—An Introduction*. New York: W. H. Freeman and Company.

Lindgren, Harry. 1972. *Recreational Problems in Geometric Dissections and How to Solve Them*. New York: Dover.

Walter, Marion. 1985. *The Mirror Puzzle Book*. New York: Parkwest Publications.

Weibe, Arthur. 1985. *Symmetry with Pattern Blocks*. Fresno, CA: Creative Teaching Associates.

SUGGESTED READINGS

Consult the following sources listed in the Bibliography for additional related readings: Bright (1989), O'Daffer (1990), Posamentier (1984), and Zaslavsky (1990).

CHAPTER FOUR Patterns of Polygons: Tessellations

INTRODUCTION

In Chapter 3 we studied properties of polygons. In this chapter we explore patterns formed by combinations of polygons in the plane. These patterns are intimately related to what might be called the "art of ornamentation." Many examples of floor, wall, and fabric patterns from early Egyptian, Greek, and Chinese civilizations survive to this day. Figure 4.1 shows several early Greek frieze patterns; Figure 4.2 shows Egyptian wall patterns; and Figure 4.3 shows patterns of Chinese origin.

Weyl (1952) writes:

> One can hardly overestimate the depth of geometric imagination and inventiveness reflected in these patterns. Their construction is far from being mathematically trivial. The art of ornament contains in implicit form the oldest piece of higher mathematics known to us.

The mathematics referred to in this quotation is the theorem that there are exactly 17 two-dimensional crystallographic groups. (This theorem provides the basis for the popular statement that "there are exactly 17 essentially different wallpaper patterns.") The first direct mathematical treatment of this theorem was given by Fedorov in 1891, yet the ornaments decorating the walls of the Alhambra in Granada indicate that all 17 types of patterns were known to the Moors.* In fact, the Moors may have provided a creative stimulus for Maurits Escher (1898–1972), the Dutch painter whose art has been a special source of enjoyment for persons interested in mathematics. In his book Escher (1960) says:

> This is the richest source of inspiration I have ever struck; nor has it yet dried up . . . a surface can be regularly divided into, or filled up with, similar-shaped figures (congruent) which are contiguous to one another, without leaving any open spaces. The Moors were past masters of this.

*The Moors occupied Spain from 711 to 1492. They were forbidden by their religious credo to draw living objects.

They decorated walls and floors, particularly in the Alhambra in Spain, by placing congruent multi-colored pieces of majolica (tiles) together without leaving any spaces between.

FIGURE 4.1

FIGURE 4.2

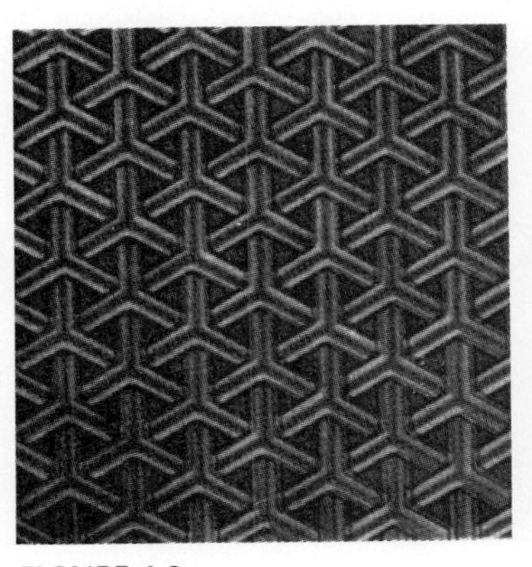

FIGURE 4.3

Alhambra drawings by M. C. Escher

FIGURE 4.4

Figure 4.4 shows some of the works of the Moors (as sketched by Escher). Figure 4.5 shows this type of art coming to life in one of Escher's creations.

Today, also, we find many interesting ornamental patterns in our architecture (Fig. 4.6). While a thorough analysis of patterns would take us beyond the scope of this book, we shall explore a particular type of pattern called a tiling or a tesselation.

Day and Night, M. C. Escher

FIGURE 4.5

FIGURE 4.6

This chapter is divided into three sections. In the first section we study tessellations that can be generated by repeated use of one polygon shape. In the second section we consider tessellations formed by using more than one polygon shape, but with the requirement that all the polygons be regular polygons. Finally the third section illustrates how mirrors can be used to generate many interesting tessellations and how it is possible to construct tessellations made of curved figures, rather than polygon shapes.

TEACHING NOTE

One of the purposes for studying geometry is to discover order and patterns in the universe. Students should be encouraged to look for examples of tessellations in their environment. A few of the situations in which they occur are in floor tilings, in wallpaper designs, in brick walls, in architectural patterns, and in various aspects of nature.

TESSELLATIONS OF POLYGONS

In this section we shall study patterns in the plane constructed using polygonal figures that completely cover the plane with no "holes" and no "overlapping." Such a pattern is called a **tessellation**, or a **tiling**, of the plane.

Before reading further, complete a thorough search for answers to the question asked in Investigation 4.1.

INVESTIGATION 4.1

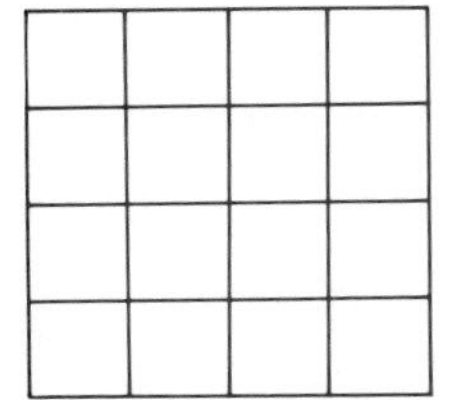

Square regions can be arranged into a repeating pattern that completely covers the plane, as shown in the figure on the left. There are no "holes" and no "overlapping" areas. We say that the squares "tile" or "tessellate" the plane.

A. Use cardboard or plastic polygons to search for other quadrilaterals, triangles, or regular polygons that will tessellate the plane. Then copy those polygons in the figure below that will tessellate the plane. Use tracing paper to show the tessellations.
B. Can you state some general conclusions about which polygons will tessellate the plane?

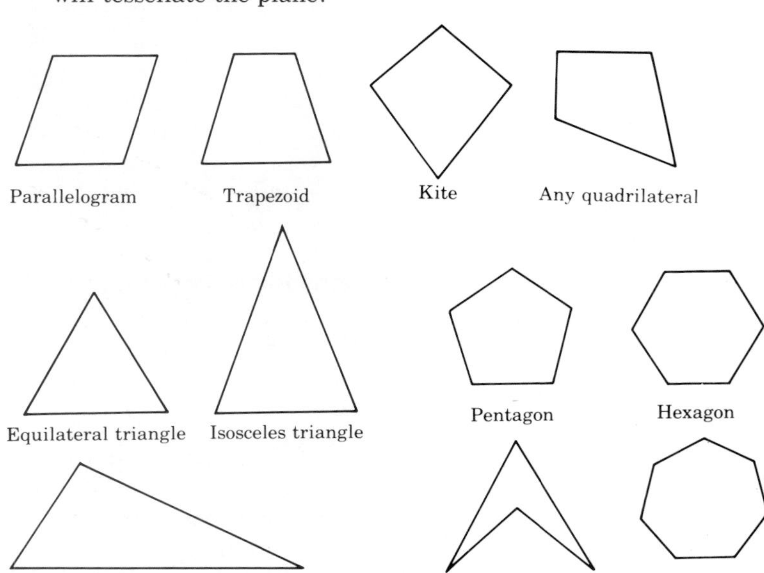

TABLE 4.1 Regular Tessellations

A Portion of the Tessellation	Description of the Tessellation	Notation
	Made up of regular 4-gons with 4 of them surrounding each vertex	4,4,4,4 or 4^4
	Made up of regular 3-gons with 6 of them surrounding each vertex	3,3,3,3,3,3 or 3^6
	Made up of regular 6-gons with 3 of them surrounding each vertex	6,6,6 or 6^3

One of the questions posed in Investigation 4.1 is, *Which regular polygons tessellate the plane?* Table 4.1 describes three basic tessellations of the plane with regular polygons.

The tessellations of equilateral triangles and squares in Figure 4.7 are different from the tessellations in Table 4.1 in that the vertices of the triangles and squares in the figure coincide with the midpoints of the sides of other squares and triangles.

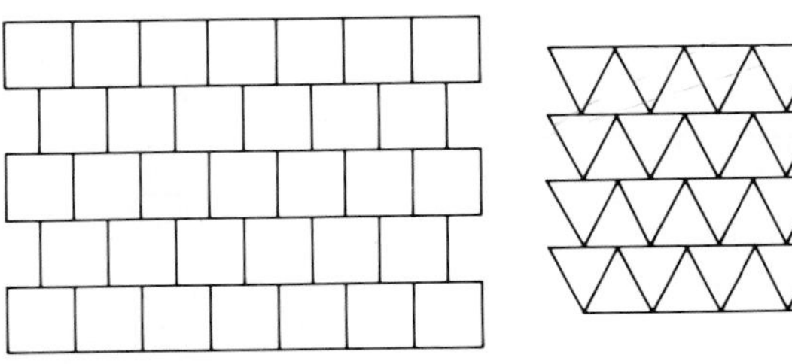

FIGURE 4.7

EXPLORING WITH THE COMPUTER 4.1

Use LOGO software to complete the following:

1. Write a LOGO procedure SQUARE that draws a square.
2. Use the procedure SQUARE to write a procedure ROW. SQUARE that draws a row of squares.
3. Use the procedure ROW.SQUARE to write a procedure that is called SQUARE.TESS that will draw the regular tessellation of squares.
4. Select one of the tessellation patterns in Table 4.1 or Figure 4.7. Write a procedure to draw the tessellation.

To explain the difference more formally, we refer to what are called vertex figures. In Figure 4.8, the polygons drawn in gray are polygons whose vertices are the midpoints of the polygonal edges emanating from a vertex of the tessellation and are known as **vertex figures**. A tessellation is a **regular tessellation** if it is constructed from regular convex polygons of one size and shape such that each vertex figure is a regular polygon. It can be proven that in regular tessellations all vertex figures are congruent to each other.

Note (Fig. 4.8) that the vertex figure for the regular tessellation 3^6 is a regular hexagon; for the regular tessellation 6^3, it is an equilateral triangle; and for the regular tessellation 4^4, the vertex figure is a square. On the other hand, we see that the vertex figures for the tessellations in Figure 4.7 are isosceles triangles and trapezoids, respectively, and are not regular polygons (Fig. 4.9). Consequently, the tessellations in Figure 4.9 are not regular tessellations.

The question remains, *How many regular tessellations are there in addition to those in Table 4.1?* Considering the regular

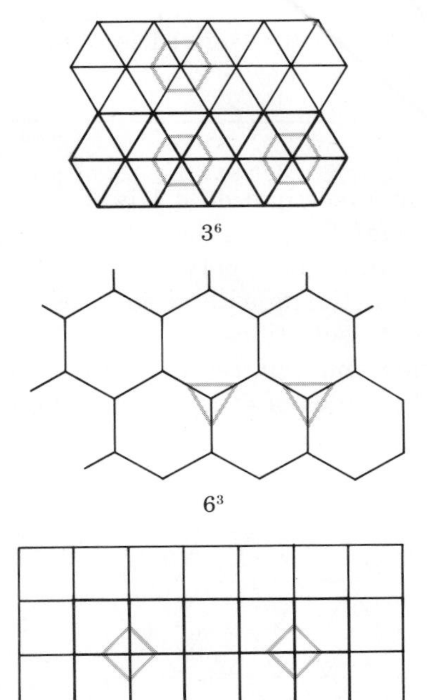

FIGURE 4.8

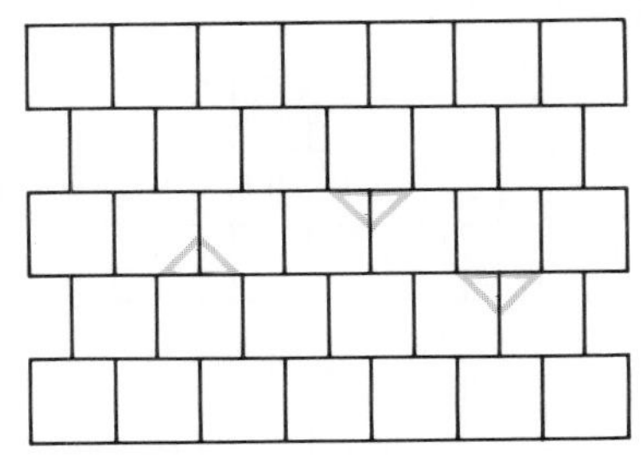

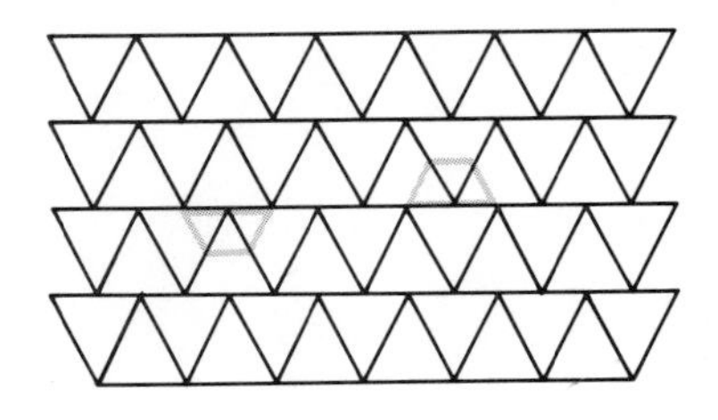

FIGURE 4.9

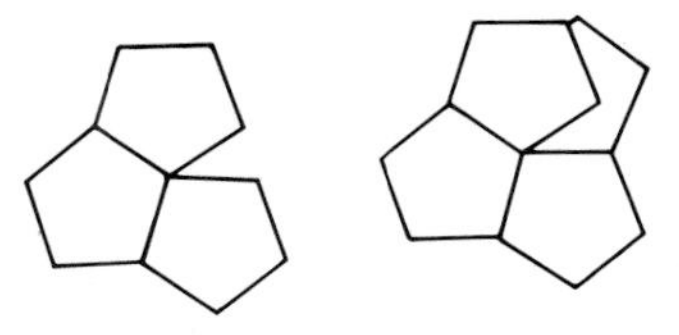

FIGURE 4.10

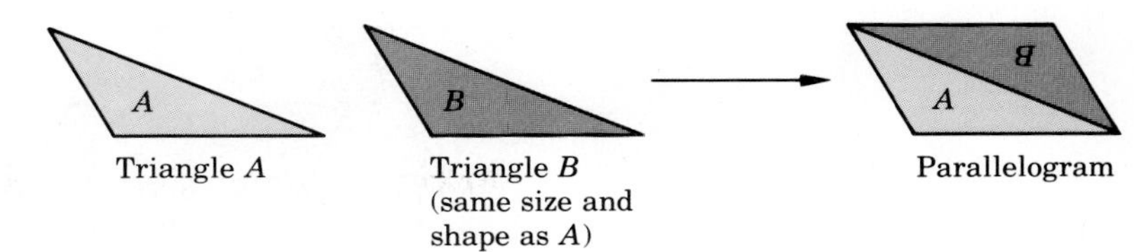

FIGURE 4.11

pentagon, we see that three regular pentagons surrounding a vertex leave a gap and four regular pentagons surrounding a vertex overlap (Fig. 4.10). So, there can be no regular tessellation of pentagons.

Attempts to tessellate with other regular polygons will show that if there exists a regular tessellation of p-gons, the vertex angle of the p-gons must divide 360°. Furthermore, there must be at least three polygons around each vertex in a regular tessellation so the vertex angle cannot be greater than 120°. In Table 4.2, we see that the only vertex angles that are less than or equal to 120°, and that also divide 360°, are the vertex angles of equilateral triangles, squares, and regular hexagons. Consequently, the tessellations 3^6, 4^4, and 6^3 pictured in Table 4.1 are the only regular tessellations.

Another question suggested by Investigation 4.1 is, *Which triangles will tessellate the plane?* To answer this question, consider any triangle, and note that this triangle can be paired with another triangle of the same size and shape to form a parallelogram (Fig. 4.11). Since all parallelograms tessellate

TABLE 4.2

Number of Sides of a Polygon	Number of Degrees in One Vertex Angle of the Polygon
3	60°
4	90°
5	108°
6	120°
7	128 4/7°
8	135°
.	.
.	.
.	.
n	$\frac{(n-2)180}{n}$

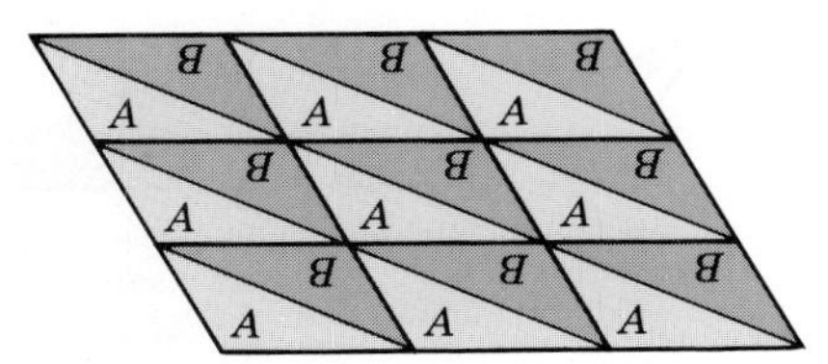

FIGURE 4.12

the plane (the reader should use cutouts or drawings to see that this is plausible), a tessellation of the arbitrarily chosen triangle is effected by the tessellation of the parallelograms (Fig. 4.12). Thus, *any triangle will tessellate the plane.*

A third question suggested by the investigation is, *Which quadrilaterals will tessellate the plane?* To answer this question, consider *any* quadrilateral and note that

1. Successive 180° rotations about the midpoints (A, B, and C) of the sides of the quadrilateral will generate four copies of the quadrilateral around a vertex (V) (Fig. 4.13). (The reader should check this with tracing paper.) In particular, each of the angles around V is a distinct angle of the quadrilateral, and these four angles have a sum of 360°.
2. Since this process can be carried out at each vertex, it is possible to generate a tessellation of the original quadrilateral.

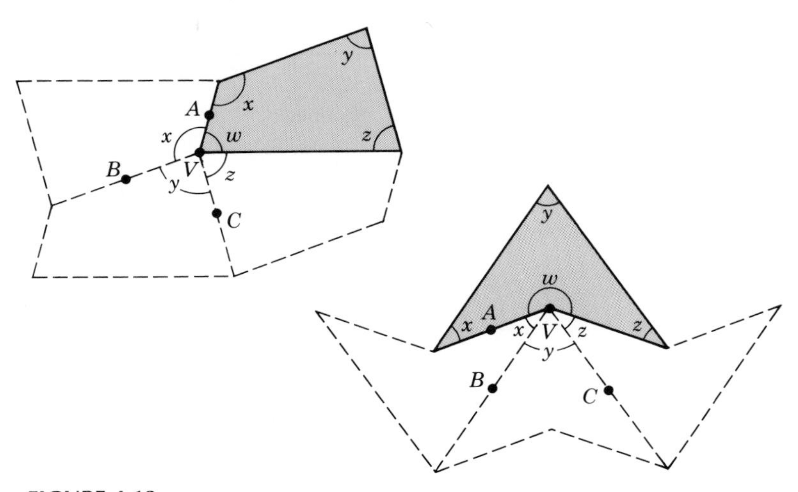

FIGURE 4.13

TEACHING NOTE

Tessellation Jobcards and *Exploration Pack*, available from Creative Publications, provide useful materials for exploring tessellations. Students should be encouraged to think about the sides and angles of the various polygons as they investigate tessellation properties.

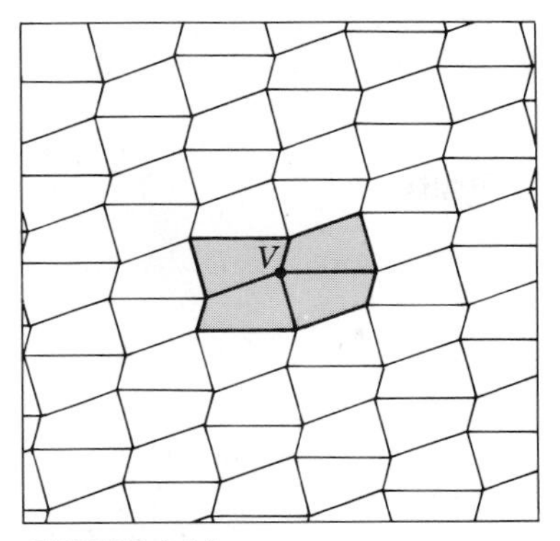

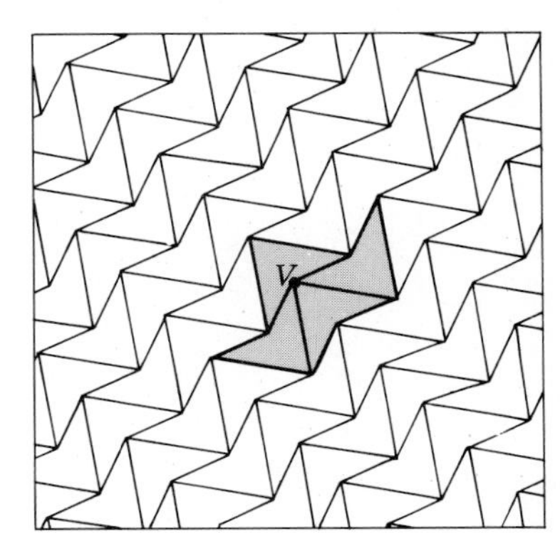

FIGURE 4.14

Note that the preceding conclusions can also be made about a nonconvex quadrilateral (the reader should check this with tracing paper). Pictures of the two tessellations generated in this manner are shown in Figure 4.14. These observations lead to the following generalization: *Any quadrilateral will tessellate the plane.*

In summary, we have learned that (a) equilateral triangles, squares, and regular hexagons are the only regular polygons that tessellate the plane, (b) all triangles tessellate the plane, and (c) all quadrilaterals tessellate the plane.

Since all triangles and quadrilaterals tessellate the plane, it is natural to ask the question, *What types of pentagons and hexagons will tessellate the plane*? While a regular pentagon will not tessellate, it is interesting to note that there is a pentagon (see region A in Fig. 4.15) with all sides congruent (but

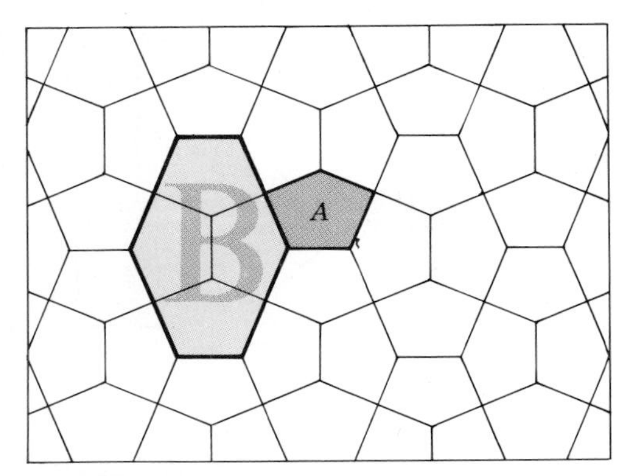

FIGURE 4.15

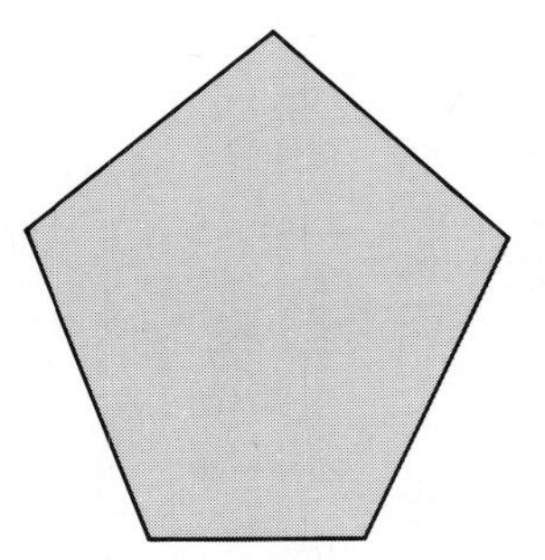

Nontessellating pentagon

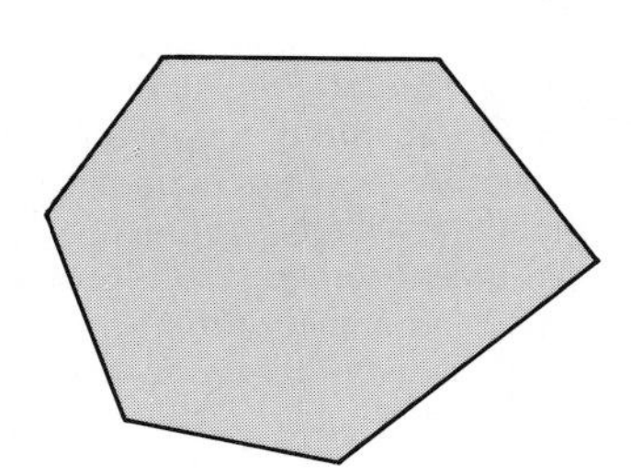

Nontessellating hexagon

FIGURE 4.16

with different-size angles) that will tessellate the plane. A portion of this tessellation is shown in Figure 4.15. If four of these pentagonal regions are considered together (see region B), an interesting hexagonal shape results that will tessellate the plane.

Thus, we have seen a nonregular pentagon and a nonregular hexagon that *will* tessellate the plane. The pentagon and the hexagon shown in Figure 4.16, however, are examples of these types of figures that *will not* tessellate the plane. Grunbaum and Shephard (1987) give a complete discussion of tilings of pentagons and hexagons.

It can be proven that no *convex* polygon with more than six sides will tile the plane. However, other irregular polygons will tessellate the plane, and some interesting methods can be used to construct these tessellations. Figure 4.17 illustrates

FIGURE 4.17

(a)

(b)

(c)

FIGURE 4.18

how a tessellation of irregular shapes can be constructed from two tessellations, one superimposed upon the other. Figure 4.18 shows other interesting types of tessellations. It is possible to tessellate the plane with polygons shaped like certain letters of the alphabet (Fig. 4.18a). A hexaminoe is a figure made of six squares. Certain hexaminoes will tessellate the plane (Fig. 4.18b). Certain odd-shaped figures that one wouldn't expect to tessellate the plane provide interesting tessellations (Fig. 4.18c).

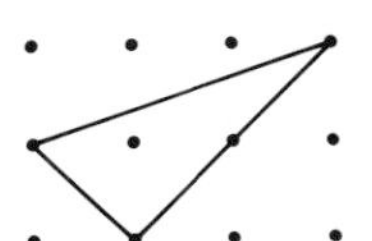

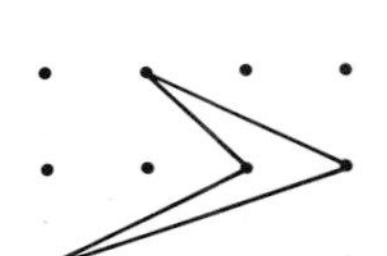

FIGURE 4.19

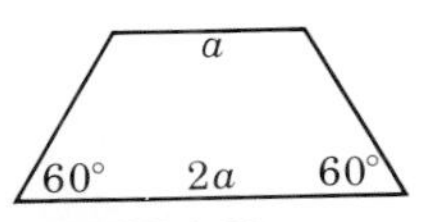

FIGURE 4.20

EXERCISE SET 4.1

1. Refer to Figure 4.19 and complete the following:
   a. On square dot paper draw a tessellation using this triangle as the basic figure.
   b. Draw in red the vertex figures for the tessellation you drew in part a.
   c. Are these vertex figures on your tessellation of one single shape? If not, redraw your tessellation so that there is only one shape of vertex figure.
   d. On square dot paper draw a tessellation using this quadrilateral. Make sure that your tessellation possesses only one type of vertex figure.
2. Using a trapezoid with the shape shown in Figure 4.20, construct the following:
   a. A tessellation with this trapezoid so that each vertex figure is a rhombus
   b. A second tessellation in which the vertex figures are not kites

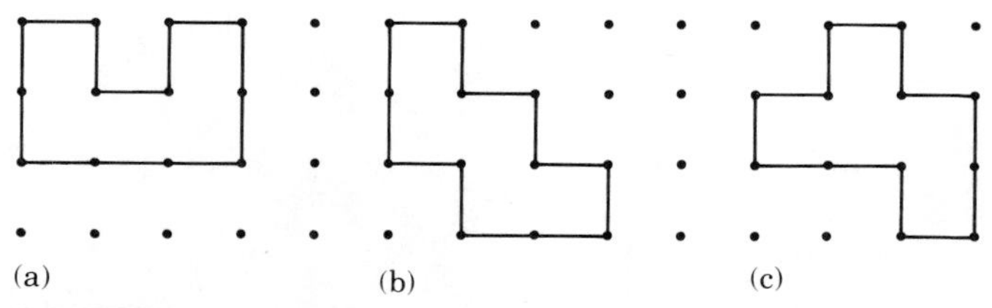

FIGURE 4.21

3. Using dot paper draw, if possible, tessellations of each of the pentominoes given in Figure 4.21.
4. The shaded portion of the tessellation in Figure 4.22 suggests what famous theorem from plane geometry?
5. An equilateral triangle is called a *reptile* (an abbreviation for "repeating tile") because 4 equilateral triangles can be arranged to form a larger equilateral triangle (Fig. 4.23). Which of the figures in Figure 4.24 are reptiles? Use appropriate dot paper and show the larger figure in each case.

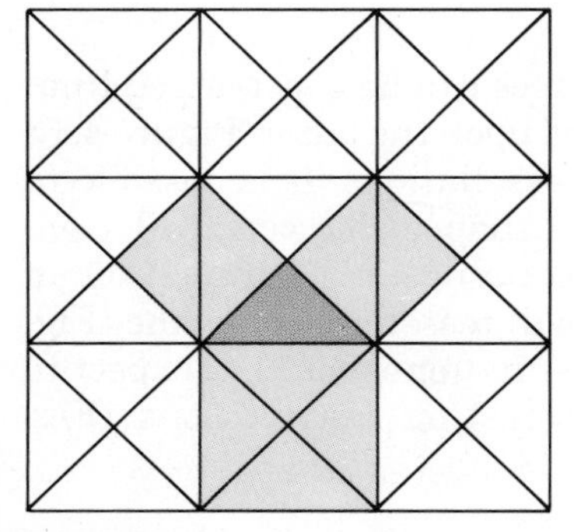

FIGURE 4.22

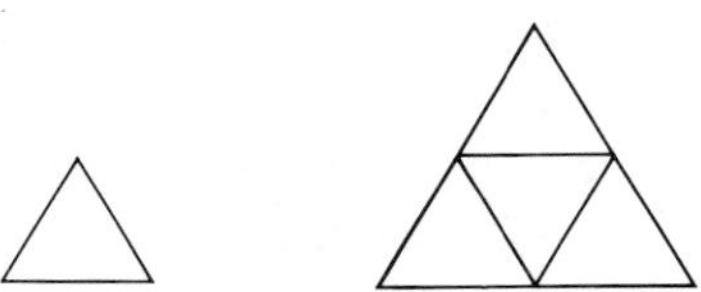

A "reptile"

FIGURE 4.23

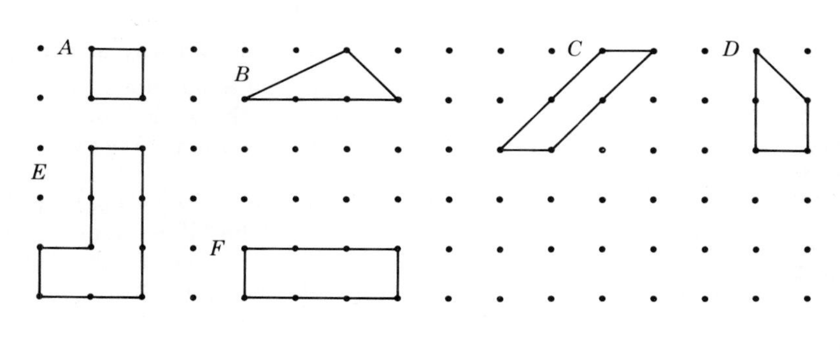

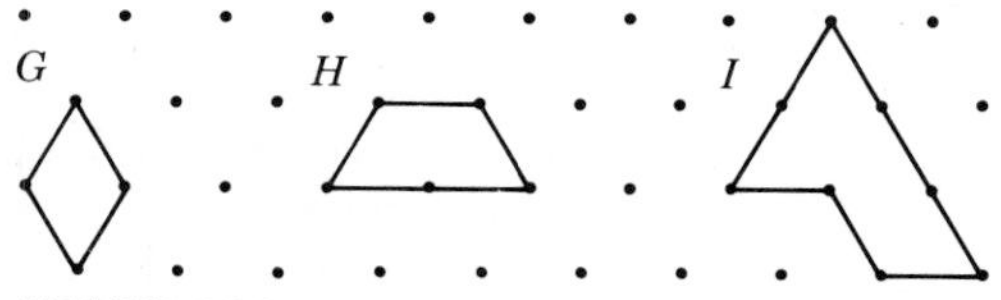

FIGURE 4.24

APPLICATION

Tiling Problem: A floor restoration specialist was called upon to restore the floor tile in an ancient temple. It was known that the floor was tiled using only equilateral triangles and regular hexagons. The only remaining pieces of original tile were as shown here. Can you help the specialist and draw a diagram of what the original tile looked like? (Tracing paper or cutout polygons might be useful.) Is there more than one possibility? Explain.

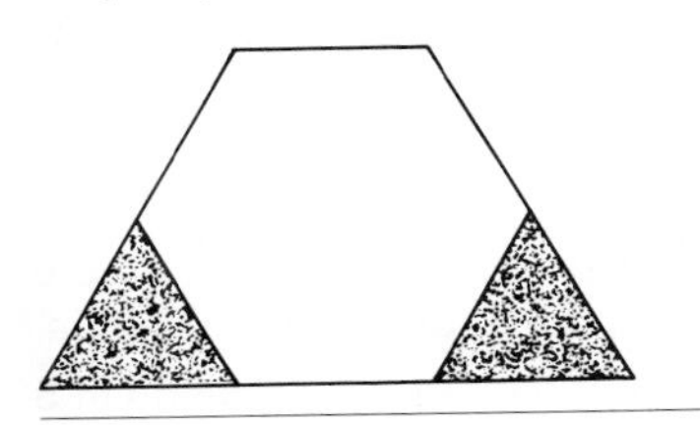

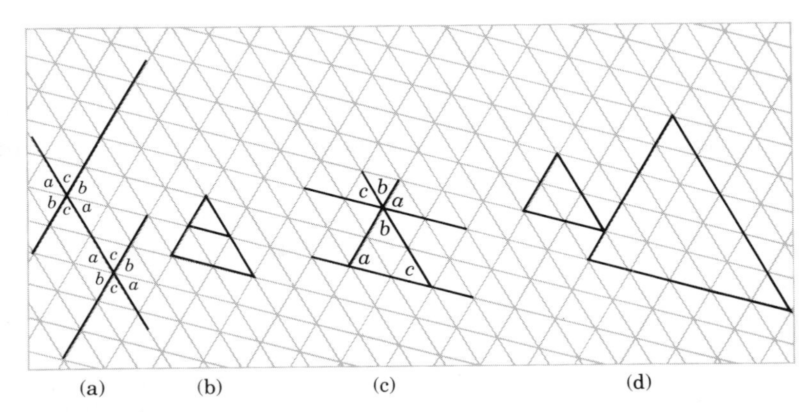

FIGURE 4.25

6. Each figure on the tessellation of the triangles in Figure 4.25 suggests at least one theorem from plane geometry. State at least one theorem for each part.
7. Figure 4.18a shows a tessellation using the letter T as the basic figure. Which of these letters of the alphabet in Figure 4.26 will tessellate the plane? Show a portion of the tessellation(s).
8. Refer to Figure 4.27 and complete the following:
   a. Use tracing paper and trace 4 copies of this tessellation. Draw around 2 adjacent quadrilaterals to form a different-shaped hexagon on each copy.
   b. What does the experience in part a suggest about finding hexagons that tessellate the plane?
9. The *dual* of a tessellation is a new tessellation obtained by connecting the centers (centroids) of polygons that share a common side. The dual (shown in gray in Fig. 4.28) of the tessellation of regular hexagons is the tessellation 3^6. Describe the dual of the regular tessellation of
   a. Squares
   b. Equilateral triangles

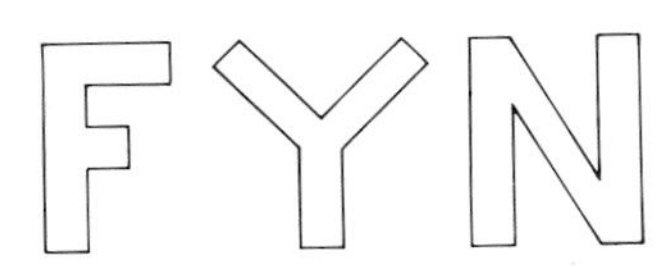

FIGURE 4.26

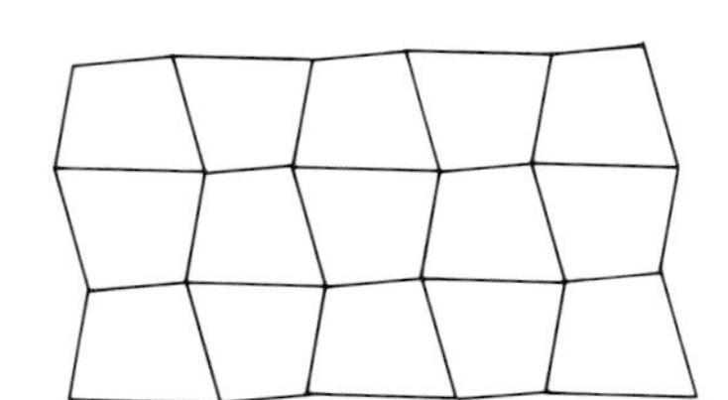

FIGURE 4.27

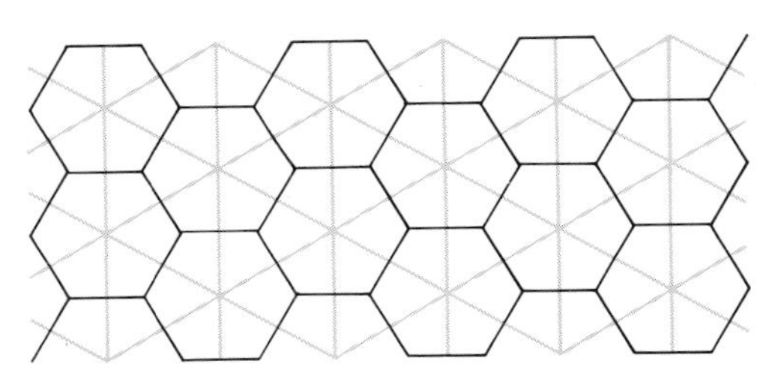

FIGURE 4.28

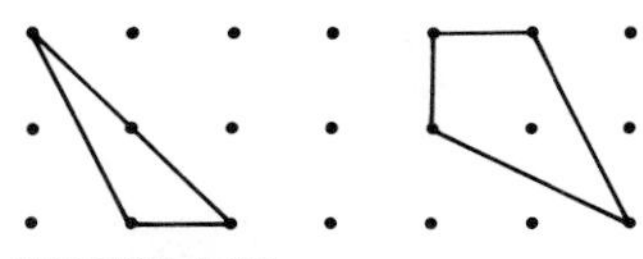
FIGURE 4.29

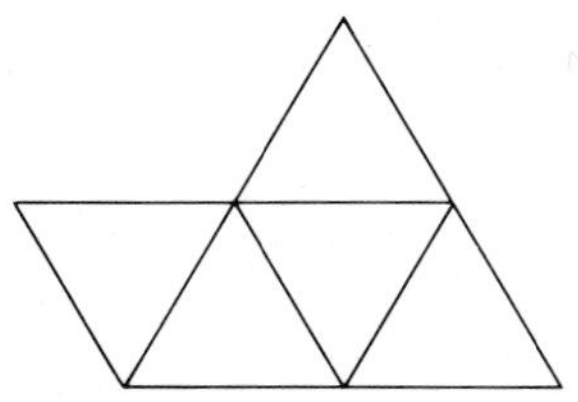
FIGURE 4.30

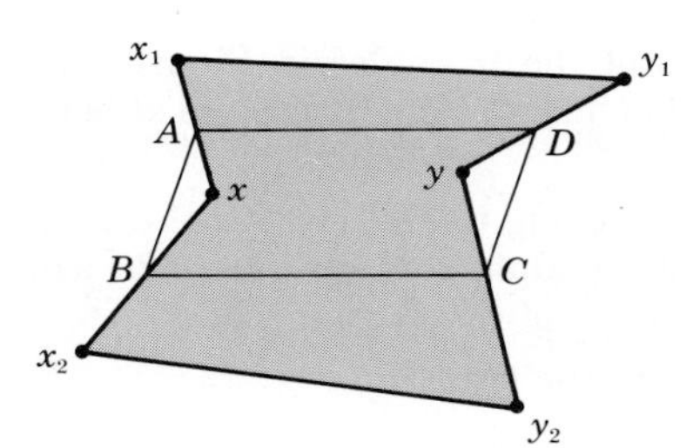

FIGURE 4.31

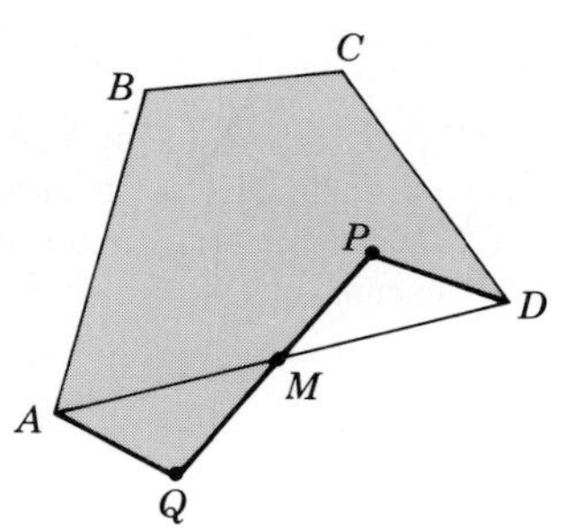

FIGURE 4.32

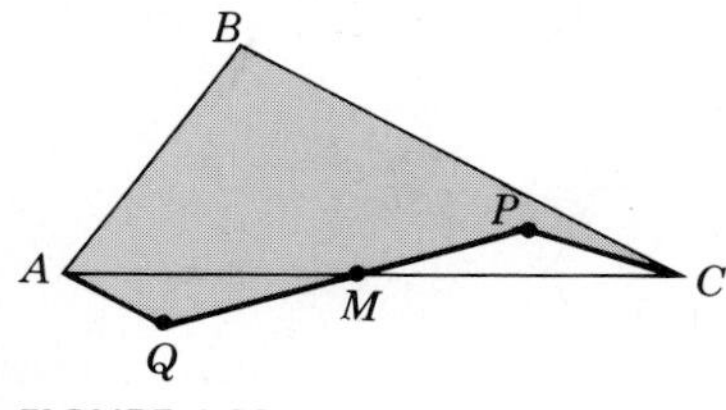

FIGURE 4.33

10. Given the triangle and quadrilateral in Figure 4.29:
    a. What is the maximum number of these triangles that can be drawn without overlap on square dot paper with 7 dots on a side?
    b. What is the maximum number of these quadrilaterals that can be drawn without overlap on square dot paper with 7 dots on a side.

*11. There are 5 tetrominoes (pentominoe-like figures made up of 4 squares). Draw pictures of these tetrominoes and decide which of these will tessellate the plane.

*12. There are 12 pentominoes.
    a. Draw pictures of these pentominoes and decide which of them will tessellate the plane.
    b. Use one copy of each of the pentominoes to "tile" a rectangle region with length 10 and width 6 (the unit of length is the length of one side of one of the squares that form the pentominoes).

*13. There are 35 hexominoes (pentominoe-like figures made up of 6 squares). Find at least 3 hexominoes that will tessellate the plane.

*14. There are 4 pentiamonds. A *pentiamond* is a figure made up of 5 connected equilateral triangles in such a way that each triangle ways has at least one side in common with an adjacent triangle (Fig. 4.30). Draw a picture of each pentiamond and decide which of these pentiamonds will tessellate the plane.

*15. There are 12 hexiamonds (figures made from 6 equilateral triangles in the manner described in Exercise 14). Find at least 2 hexiamonds that will tessellate the plane.

Exercises 16 through 19 describe a collection of pentagons and hexagons that tessellate.

*16. Consider a parallelogram $ABCD$ (Fig. 4.31) and let x and y be two interior points. Construct points x_1 and x_2 so that A is the midpoint of xx_1 and B is the midpoint of xx_2. Construct y_1 and y_2 so that D is the midpoint of yy_1 and C is the midpoint of yy_2. The hexagon $x_1xx_2y_2yy_1$ will always tessellate the plane. Use tracing paper and accurate drawings to convince yourself that this is true.

*17. Consider any quadrilateral $ABCD$ (Fig. 4.32). Find the midpoint M of any side (i.e., AD) of the quadrilateral and draw any segment PQ such that P is inside $ABCD$ and M is the midpoint of PQ. Draw segments PD and AQ. The hexagon $ABCDPQ$ will always tessellate the plane. Try this with a quadrilateral of your choice and use tracing paper to draw a portion of the tessellation to convince yourself that this is true.

*18. Draw any triangle ABC (Fig. 4.33). Find the midpoint M of any side (i.e., AC) of the triangle and draw any segment PQ such that P is inside the $\triangle$ and M is the midpoint of PQ. Draw segments PC and QA. Pentagon $ABCPQ$ will always tessellate the plane.

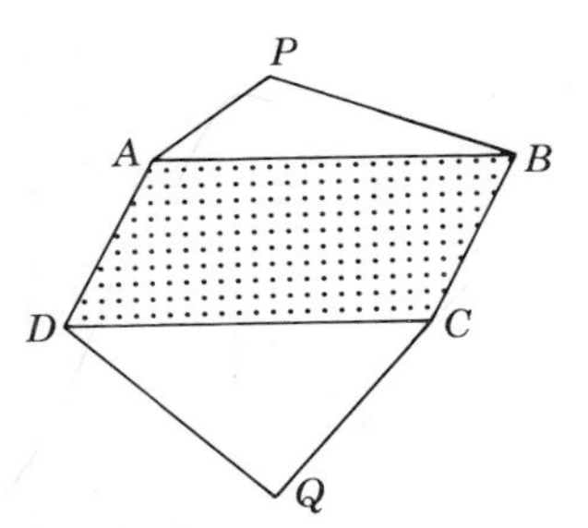

FIGURE 4.34

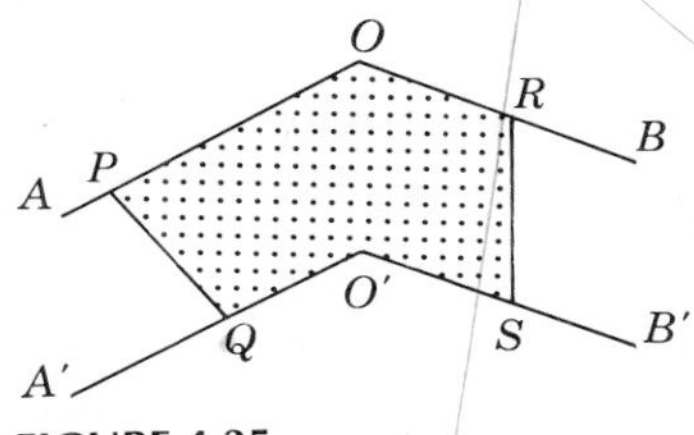

FIGURE 4.35

Try this with a triangle of your choice, using tracing paper to draw a portion of the tessellation.

19. **Critical Thinking.** An important aspect of being skilled in critical thinking is being aware of your problem-solving thought processes so that you can communicate your solution methods to others. For the following exercises, write out a solution that explains to someone else how you thought through the problem.
    a. Choose points P and Q outside a parallelogram $ABCD$ so that triangles APB and CDQ are both acute triangles (Fig. 4.34). Show that hexagon $APBCQD$ tessellates the plane.
    b. Draw parallel rays OA and $O'A'$ and parallel rays OB and $O'B'$ (Fig. 4.35). Choose points P and Q on the first rays and R and S on the second rays. Explain that the hexagon $PORSO'Q$ tessellates the plane.

Problem Solving: Developing Skills and Strategies

When considering a complex, difficult problem, it is often advisable to break down the problem into related, but simpler subproblems. This strategy is called **consider a simpler related problem**.

Consider this complex problem: Find all types of polygons that tessellate the plane. What are some of the simpler related problems that you could solve first that are suggested by the discussion in the previous section.

TESSELLATIONS WITH REGULAR POLYGONS

Combinations of several polygon shapes often yield tessellations with interesting and beautiful patterns, patterns with artistic appeal (Fig. 4.36). However, in order to concentrate on the mathematics of tessellations and not on the art in tessel-

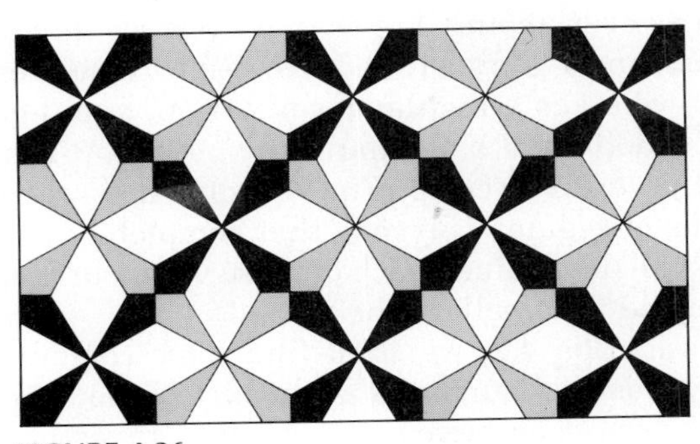

FIGURE 4.36

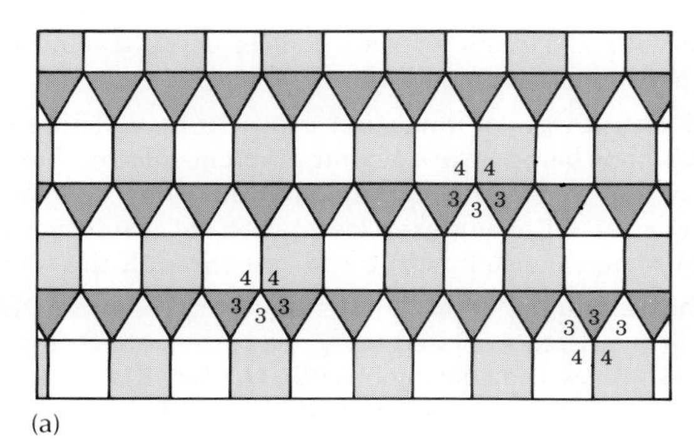

(a) (b)

FIGURE 4.37

lations, the discussion in the remainder of this chapter is restricted to tessellations formed by combinations of *regular* polygons. Two examples are shown in Figure 4.37.

There are many tessellations that are formed by a combination of several regular polygons, and these tessellations can be classified according to certain properties. Note that the tessellation in Figure 4.37a possesses only one type of vertex figure, while the tessellation in Figure 4.37b possesses two types of vertex figures.

The tessellation in Figure 4.37a is an example of a type of tessellation called a semiregular tessellation. A tessellation is a **semiregular tessellation** if it is composed of regular polygons of two or more types in such a way that all vertex figures are identical; that is, the arrangement of polygons at each vertex is the same. Why is the tessellation in Figure 4.37b not semiregular?

In the preceding section of this chapter, we have seen that there are only three regular tessellations. In this section we answer the question, *How many semiregular tessellations are there?* As a first step in answering this question, we search for combinations of regular polygons that completely surround a vertex with no overlapping. For example, we see that a regular 12-gon, 6-gon, and 4-gon surround a vertex with no overlapping (Fig. 4.38a), whereas a regular 8-gon, 6-gon, and 3-gon leave a gap (Fig. 4.38b), and a regular 8-gon, 4-gon, 3-gon, and 4-gon around a vertex overlap (Fig. 4.38c). Once we find some combinations of regular polygons that completely surround a vertex, we shall determine which of these vertex arrangements can be extended to tessellate the plane.

In Investigation 4.2, we shall find some combinations of regular polygons that surround a vertex with no overlapping. The following question arises as a result of Investigation 4.2:

TEACHING NOTE

Activities with tessellations can sometimes catch the interest of students who haven't been very successful in other aspects of mathematics. The book, *Introduction to Tesselations* (Seymour and Britton, 1988) suggests good ideas for student activities. Also, see the *Arithmetic Teacher* articles in the suggested readings for additional things your students can do with tessellations.

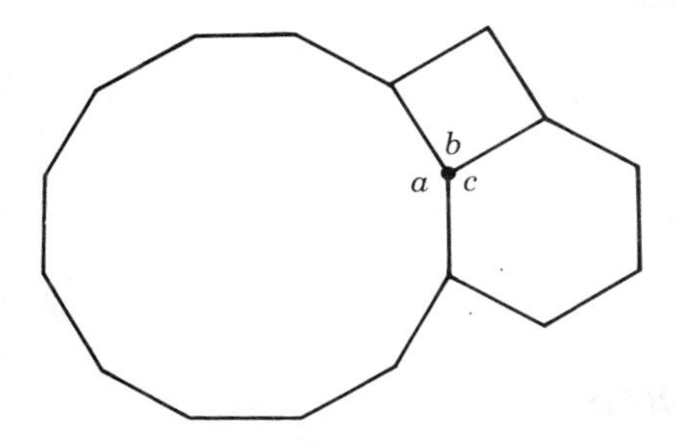

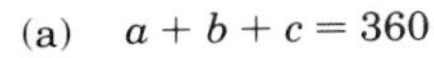
(a) $a + b + c = 360$

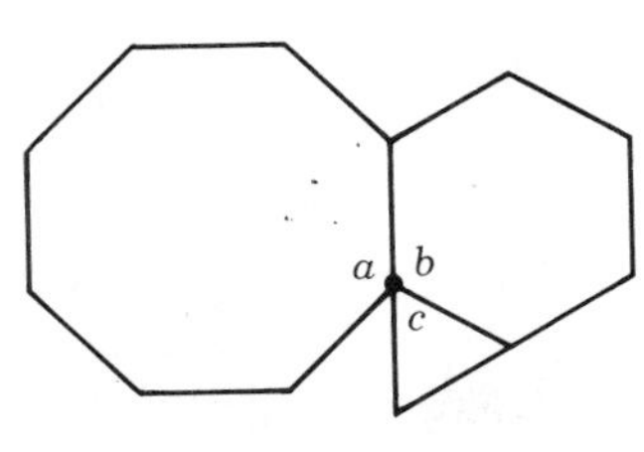

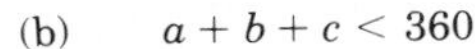
(b) $a + b + c < 360$

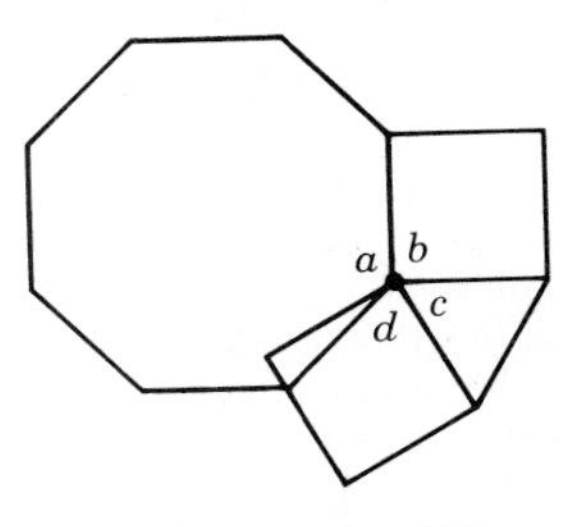

(c) $a + b + c + d > 360$

FIGURE 4.38

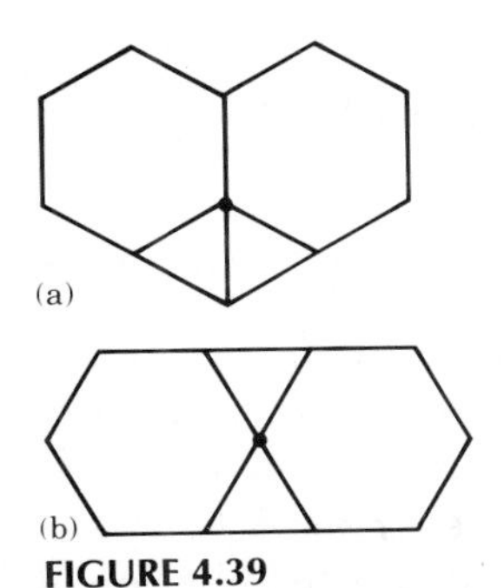

FIGURE 4.39

Should different orderings of the same set of polygons be considered different ways of surrounding a vertex? For example, are the arrangements in Figures 4.39a and 4.39b the same or are they different? It is natural to consider them to be different since one of the figures has one line of symmetry, while the other has two lines of symmetry. In summary, when counting different orderings as different arrangements, the reader should have found the following while completing Investigation 4.2.

INVESTIGATION 4.2

Cut out several copies of an equilateral triangle, a square, a regular hexagon, a regular octagon, and a regular dodecagon with sides all the same length.

A. In how many different ways can you fit 3 of these polygons around a point? (A particular type of polygon may be repeated in the arrangement.)
B. Answer the question in part A for 4, 5, 6, and more than 6 polygons surrounding a point.
C. Use tracing paper to record each of the ways of arranging the figures that you found. Devise a notation to describe each arrangement.

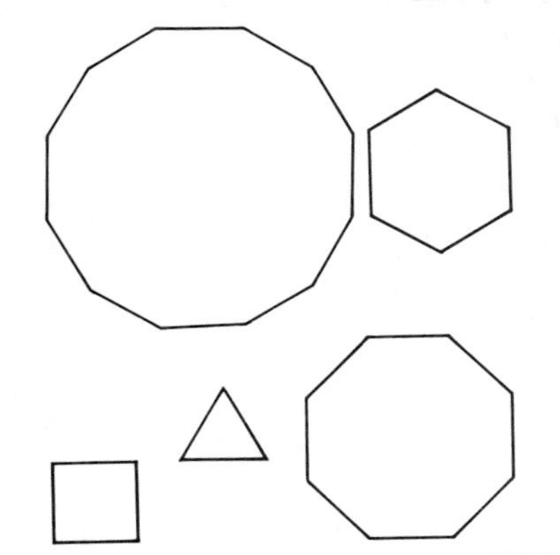

TEACHING NOTE

Make copies of 8 semiregular tessellations such as those shown in Figure 4.41 (p. 106) to give to students to color in an interesting way. Students at different ages and with varying abilities in mathematics enjoy the artistic aspects of this activity, and the ability to recognize geometric patterns is enhanced. Older students can try to create tessellations with different reflectional symmetry properties when color is considered. The Tessellation Teaching Masters available from Dale Seymour Publications are also useful.

Using only an equilateral triangle, a square, a regular hexagon, a regular octagon, and a regular dodecagon, we find there are

— Four ways of surrounding a vertex with three of these polygons
— Seven ways of surrounding a vertex with four of these polygons
— Three ways of surrounding a vertex with five of these polygons
— One way of surrounding a vertex with six of these polygons
— Zero ways of surrounding a vertex with more than six of these polygons

Later in this chapter we will verify that a vertex can also be surrounded by

— Two 5-gons and a 10-gon
— A square, a 5-gon, and a 20-gon
— An equilateral triangle, a 7-gon, and a 42-gon
— An equilateral triangle, an octagon, and a 24-gon
— An equilateral triangle, a 9-gon, and an 18-con
— An equilateral triangle, a 10-gon, and a 15-gon

(The reader may verify these statements by calculating the number of degrees in the vertex angle of each polygon and checking to see if the angles around a vertex sum to 360°.)

We shall also show that *the 21 arrangements described above are the only possible arrangements of regular polygons around a vertex.*

Since three of these arrangements produce the familiar regular tessellations (4,4,4,4; 3,3,3,3,3,3; and 6,6,6), we are left with 18 possibilities to consider. A crucial question immediately arises: *Can every one of these 18 arrangements of regular polygons around a single vertex be extended to form a semiregular tessellation?* The answer is no. For example, consider an arrangement of regular polygons around vertex V that we shall denote by 6,6,3,3 (Fig. 4.40). If we attempt to extend this configuration so that the same arrangement will occur around vertex V_1, we now find that it will be impossible to effect this same arrangement around V_2. Thus, the arrangement 6,6,3,3 cannot be extended to produce a semiregular tessellation.

Which of the vertex arrangements found in Investigation 4.2 cannot be extended to form a semiregular tessellation? This question motivates Investigation 4.3.

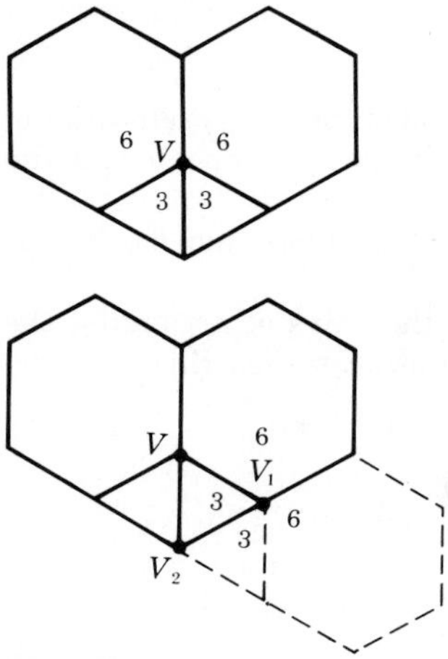

FIGURE 4.40

INVESTIGATION 4.3

If we exclude the regular tessellations, there are four of the arrangements you found in Investigation 4.2 that cannot be extended to form a semiregular tessellation.

Can you find them and explain why they cannot be extended? (Tracing paper and the figures in Investigation 4.2 may be helpful.)

In Investigation 4.3, after four of the arrangements found in Investigation 4.2 are eliminated, the remaining arrangements of 3-gons, 4-gons, 6-gons, 8-gons, and 12-gons produce eight semiregular tessellations (Fig. 4.41). These eight are the only semiregular tessellations that can be formed by using the regular polygons from Investigation 4.2.

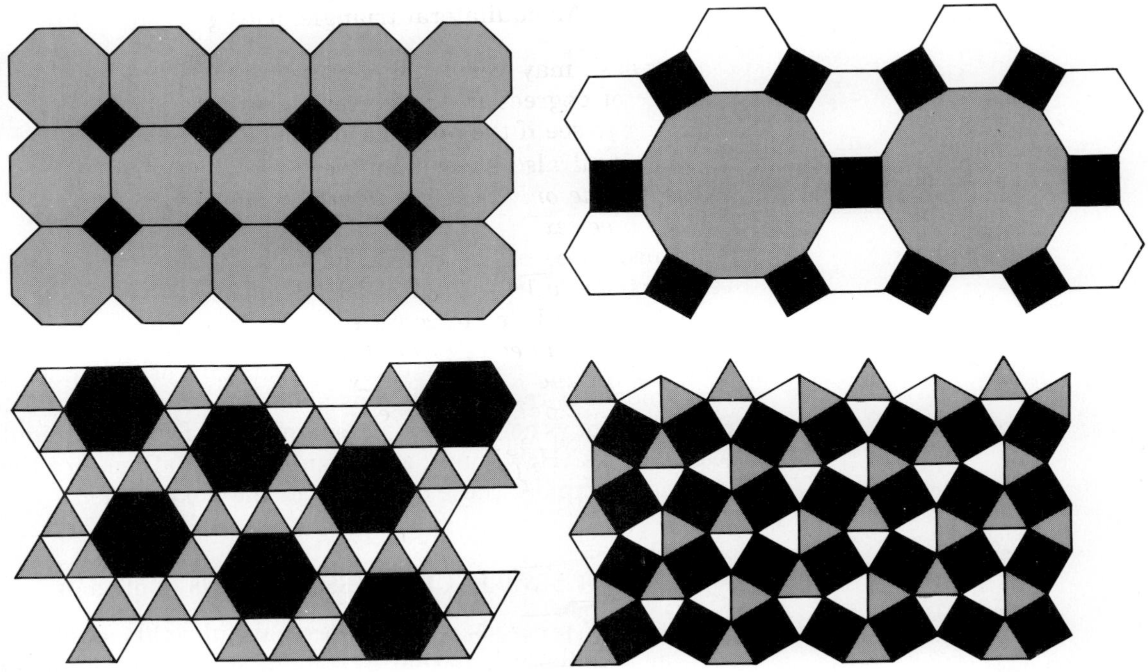

FIGURE 4.41

(Figure continues)

EXPLORING WITH THE COMPUTER 4.2

Use LOGO software to complete the following:

1. Explore ways to write a procedure that produces one of the semiregular tessellations pictured in Figure 4.41.
2. Present your successful procedure or write a paragraph describing what you attempted to do and the difficulties you encountered.

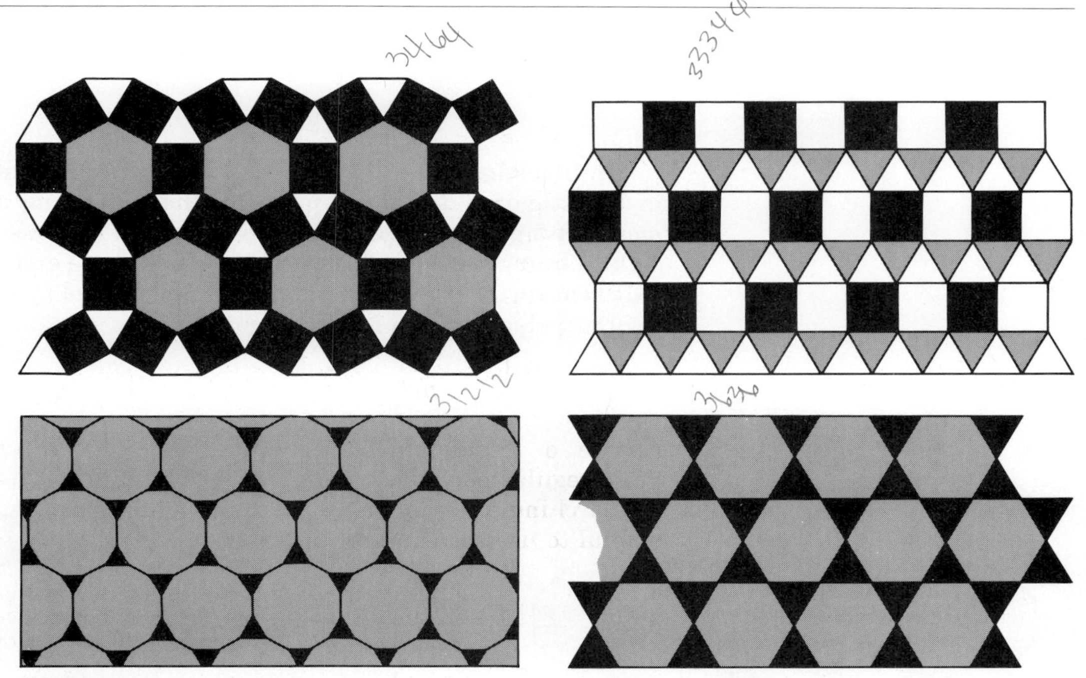

FIGURE 4.41 Continued

There are questions concerning semiregular polygons that remain unanswered. For example, can any of the other arrangements of regular polygons about a point, which include pentagons or other n-gons, be extended to form semiregular tessellations? How many semiregular tessellations are there in all? We explore these questions in the next section.

ANALYSIS OF SEMIREGULAR TESSELLATIONS (Optional)

This section proceeds with a more careful analysis of semiregular tessellations so that at the end of the section, we can declare that all semiregular tessellations have been found.

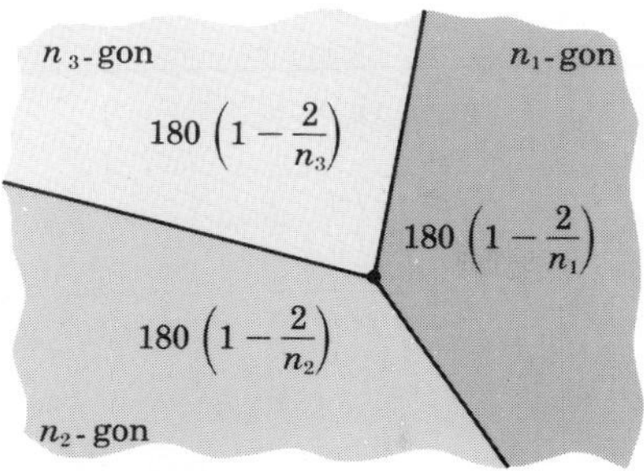

FIGURE 4.42

We have seen that a vertex angle of a regular n-gon measures $180[1 - (2/n)]$ degrees. If there are three regular polygons, an n_1-gon, an n_2-gon, and an n_3-gon that completely surrounds a vertex with no overlapping (Fig. 4.42), then the three numbers (n_1, n_2, n_3) must satisfy the equation

$$\text{(i)} \qquad 180\left(1 - \frac{2}{n_1}\right) + 180\left(1 - \frac{2}{n_2}\right) + 180\left(1 - \frac{2}{n_3}\right) = 360$$

Any set of whole-number solutions to equation (i) represents three regular polygons surrounding a vertex with no overlap. In searching for whole-number solutions to equation (i), it is helpful to use the equivalent equation (ii), which we derive.

$$\text{(i)} \qquad 180\left(1 - \frac{2}{n_1}\right) + 180\left(1 - \frac{2}{n_2}\right) + 180\left(1 - \frac{2}{n_3}\right) = 360$$

$$\Leftrightarrow \left(1 - \frac{2}{n_1}\right) + \left(1 - \frac{2}{n_2}\right) + \left(1 - \frac{2}{n_3}\right) = 2$$

$$\Leftrightarrow 3 - \left(\frac{2}{n_1} + \frac{2}{n_2} + \frac{2}{n_3}\right) = 2$$

$$\text{(ii)} \qquad \Leftrightarrow \frac{2}{n_1} + \frac{2}{n_2} + \frac{2}{n_3} = 1$$

$$\Leftrightarrow \frac{1}{n_1} + \frac{1}{n_2} + \frac{1}{n_3} = \frac{1}{2}$$

Since there can also be four, five, or six regular polygons surrounding a vertex with no overlapping, we also need to consider solutions to the three additional equations recorded in Table 4.3 that are similar to equation (i).

TABLE 4.3

Number of Polygons Surrounding Vertex	Resulting Equation	Equivalent Equation
3	$180\left(1-\frac{2}{n_1}\right)+180\left(1-\frac{2}{n_2}\right)+180\left(1-\frac{2}{n_3}\right)=360$	$\frac{1}{n_1}+\frac{1}{n_2}+\frac{1}{n_3}=\frac{1}{2}$
4	$180\left(1-\frac{2}{n_1}\right)+180\left(1-\frac{2}{n_2}\right)+180\left(1-\frac{2}{n_3}\right)+180\left(1-\frac{2}{n_4}\right)=360$	$\frac{1}{n_1}+\frac{1}{n_2}+\frac{1}{n_3}+\frac{1}{n_4}=1$
5	$180\left(1-\frac{2}{n_1}\right)+180\left(1-\frac{2}{n_2}\right)+180\left(1-\frac{2}{n_3}\right)+180\left(1-\frac{2}{n_4}\right)+180\left(1-\frac{2}{n_5}\right)=360$	$\frac{1}{n_1}+\frac{1}{n_2}+\frac{1}{n_3}+\frac{1}{n_4}+\frac{1}{n_5}=\frac{3}{2}$
6	$180\left(1-\frac{2}{n_1}\right)+180\left(1-\frac{2}{n_2}\right)+180\left(1-\frac{2}{n_3}\right)+180\left(1-\frac{2}{n_4}\right)+180\left(1-\frac{2}{n_5}\right)+180\left(1-\frac{2}{n_6}\right)=360$	$\frac{1}{n_1}+\frac{1}{n_2}+\frac{1}{n_3}+\frac{1}{n_4}+\frac{1}{n_5}+\frac{1}{n_6}=2$

We seek all whole-number solutions to the equations in Table 4.3. Each solution describes a set of regular polygons that completely surround a vertex with no overlapping, and hence, each solution represents a potential vertex arrangement in a semiregular tessellation. There is a total of 17 solutions to the four equations in the right column of Table 4.3, all of which are listed in Table 4.4. By considering the vertex arrangements represented by each solution, we can find all semiregular tessellations. Note that the solutions given in lines 12, 13, 14, and 16 of Table 4.4 can be considered in different orders to produce different vertex arrangements.

A careful study of the solutions shown in Table 4.4 and the pictures of the eight semiregular tessellations found so far (Fig. 4.41) provides the following information.

1. The solutions in lines 1, 11, and 17 represent the vertex arrangements of three regular tessellations.
2. The solutions in lines 4, 5, 10, 13, 14, 15, and 16 provide the vertex arrangements that can be extended to form the

TABLE 4.4

Equations	Solutions n_1	n_2	n_3	n_4	n_5	n_6	Line Number
$\frac{1}{n_1}+\frac{1}{n_2}+\frac{1}{n_3}=\frac{1}{2}$	6	6	6				1
	5	5	10				2
	4	5	20				3
	4	**6**	**12**				**4**
	4	**8**	**8**				5
	3	7	42				**6**
	3	8	24				7
	3	9	18				8
	3	10	15				9
	3	**12**	**12**				**10**
$\frac{1}{n_1}+\frac{1}{n_2}+\frac{1}{n_3}+\frac{1}{n_4}=1$	**4**	**4**	**4**	**4**			**11**
	3	3	4	12			12
	3	**3**	**6**	**6**			**13**
	3	**4**	**4**	**6**			**14**
$\frac{1}{n_1}+\frac{1}{n_2}+\frac{1}{n_3}+\frac{1}{n_4}+\frac{1}{n_5}=\frac{3}{2}$	**3**	**3**	**3**	**3**	**6**		15
	3	**3**	**3**	**4**	**4**		16
$\frac{1}{n_1}+\frac{1}{n_2}+\frac{1}{n_3}+\frac{1}{n_4}+\frac{1}{n_5}+\frac{1}{n_6}=2$	3	3	3	3	3	3	17

semiregular tessellations in Figure 4.41. The eight particular arrangements that form the tessellations are given in boldface type in Table 4.5.

3. The solution in line 12, which is 3,3,4,12, provides no vertex arrangement that can be extended to form a semiregular tessellation. (Recall your work in Investigation 4.3.) That is, neither 3,3,4,12 nor 3,4,3,12 describe a semiregular tessellation (Fig. 4.43).
4. It remains to consider the solutions in lines 2, 3, 6, 7, 8, and 9 of Table 4.4.

TABLE 4.5

Equation Solutions	Possible Vertex Arrangements	
Line 4	**4,6,12**	
Line 5	**4,8,8**	
Line 10	**3,12,12**	
Line 13	**3,6,3,6**	3,3,6,6
Line 14	**3,4,6,4**	3,4,4,6
Line 15	**3,3,3,3,6**	
Line 16	**3,3,4,3,4**	**3,3,3,4,4**

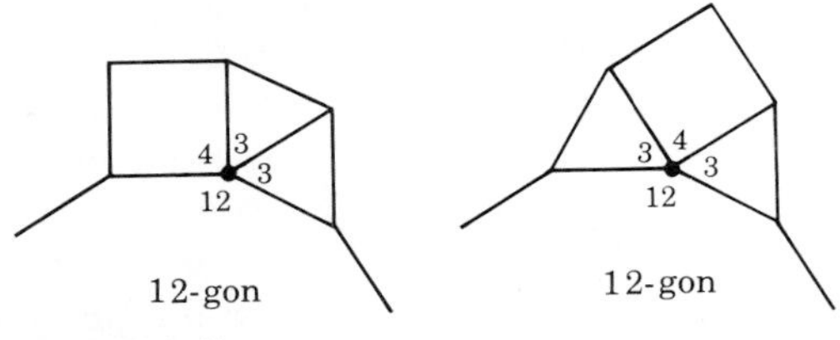

FIGURE 4.43

EXPLORING WITH THE COMPUTER 4.3

1. For each of the vertex arrangements discussed in Table 4.4, use the LOGO procedure POLYGON to write a procedure that will draw the vertex arrangement.

```
TO POLYGON :N
REPEAT :N[FD 40 RT 360/:N]
END
```

2. Use this BASIC program to find all the solutions to the first equation in Table 4.3.

```
.10 FOR H = 3 TO 8
20 FOR I = H TO 15
30 FOR J = ITO 50
70 IF 1/H + 1/I + 1/J = .0.5 THEN PRINT H,I,J,:PRINT
80 NEXT J:NEXT I:NEXT H
90 END
```

3. Use this BASIC program to find all the solutions to the second equation in Table 4.3.

```
10 FOR H = 3 TO 8
20 FOR I = H TO 15
30 FOR J = I TO 50
40 FOR K = J TO 15
70 IF 1/H + 1/I + 1/J + 1/K = 1 THEN PRINT H,I,J,K,:PRINT
80 NEXT K:NEXT J:NEXT I:NEXT H
90 END
```

4. Modify the programs in parts 2 and 3 to write a program that finds all solutions to the third and fourth equations in Table 4.3.

Let us consider lines 6 through 9 in Table 4.4. Each of these solutions suggests a vertex arrangement in which an equilateral triangle and two other regular polygons surround a vertex. Figure 4.44 suggests that this type of arrangement cannot be extended to form a semiregular tessellation unless the other two regular polygons are the same. Note that if the "lower" polygon has vertex angles α, the arrangements at each vertex won't be the same. Similarly, if it has vertex angles β, the arrangements at each vertex won't be the same. Thus, it is impossible to extend this arrangement to form a semiregular tessellation unless $\alpha = \beta$. Because of this, none of the solutions in lines 6 through 9 represent semiregular tessellations since no two of the polygons are the same.

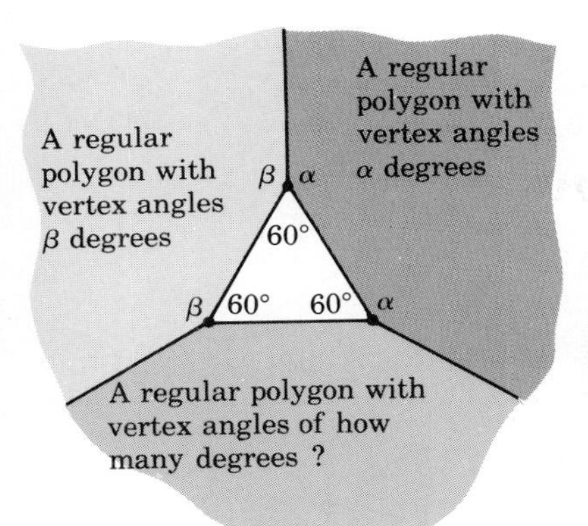

FIGURE 4.44

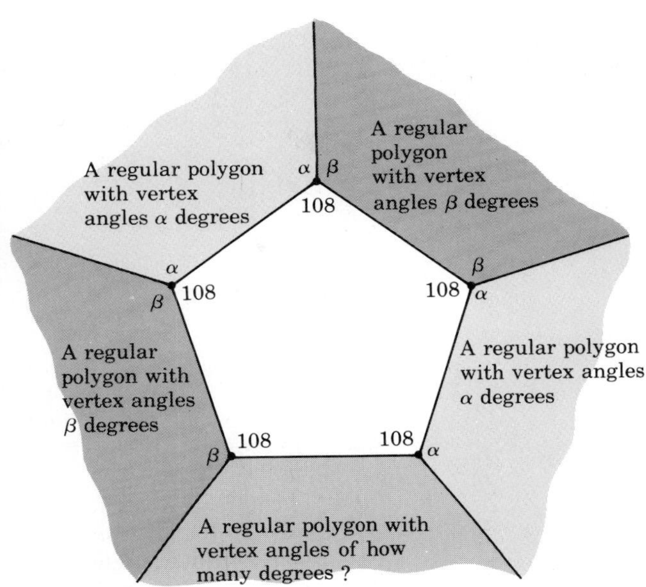

FIGURE 4.45

In a similar manner, Figure 4.45 suggests that an arrangement with a regular pentagon and two other regular polygons about each vertex cannot be extended to form a semiregular tessellation unless the other two regular polygons are the same. Thus, it is impossible to extend this arrangement to form a semiregular tessellation unless $\alpha = \beta$. Since the two polygons other than the pentagon in lines 2 and 3 are not the same, these solutions do not represent semiregular tessellations. We conclude that the eight semiregular tessellations in Figure 4.41 are the *only* semiregular tessellations.

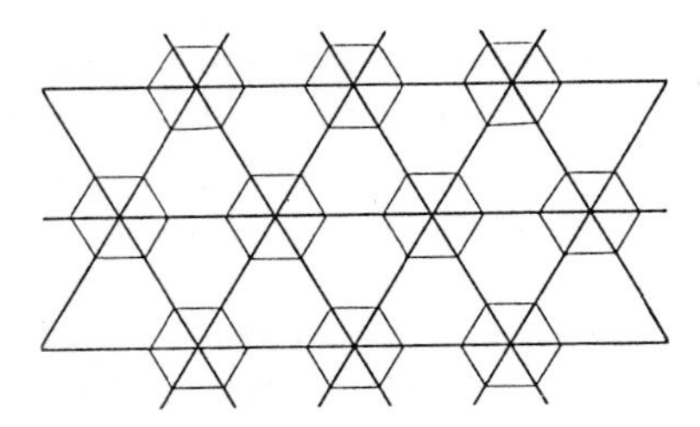

FIGURE 4.46

EXERCISE SET 4.2

1. The top left semiregular tessellation in Figure 4.41 is sometimes denoted 4,8,8 because each vertex of the tessellations is surrounded by a square and 2 octagons. What would be the comparable notation for the other 7 tessellations in the figure?
2. Figure 4.46 shows a tiling with two vertex types, a 3,3,6,6 type and a 3,3,3,3,3,3 type. Draw another tiling of regular polygons that has a regular pattern of 2 vertex types.

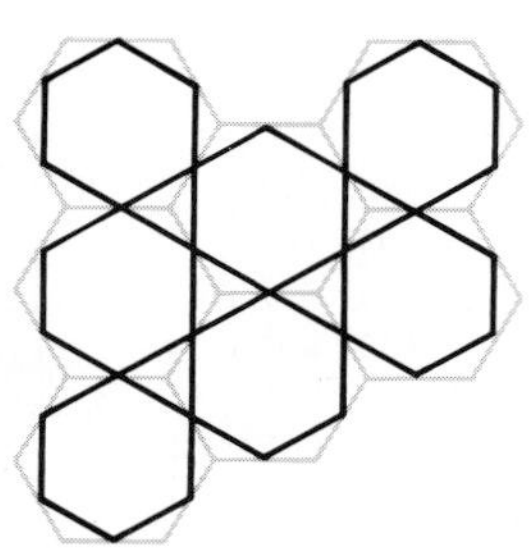

FIGURE 4.47

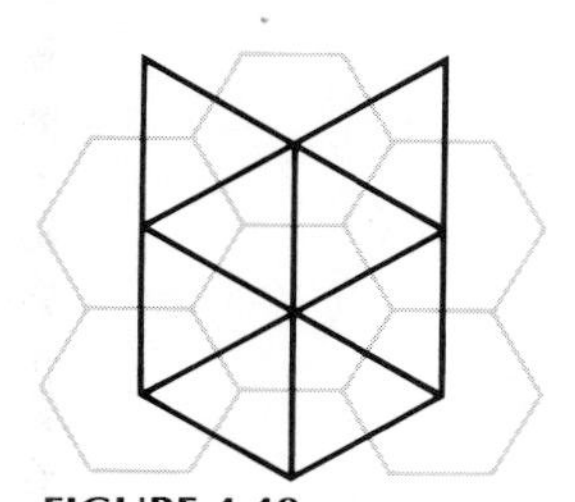

FIGURE 4.48

3. Show that by replacing each hexagon with 6 equilateral triangles in one of the semiregular tessellations, you can create a tessellation with vertex types 3,3,4,12 and 3,3,3,3,3,3. Draw a portion of that tessellation.
4. Show that by pasting together strips that have been cut from the regular tessellation of squares and the regular tessellation of equilateral triangles, you can create a tessellation with vertex types 3,3,3,4,4 and 3,3,3,3,3,3.
5. Figure 4.47 illustrates the *vertex figure tessellation* that is created by drawing all the vertex figures of the tessellation of regular hexagons. Choose one of the semiregular tessellations of Figure 4.41, trace it, and draw its vertex figure tessellation.
6. Figure 4.48 shows the *dual tessellation* that is created by connecting the centers of neighboring polygons. Choose one of the semiregular tessellations and draw its dual tessellation.
7. **Critical Thinking.** The tiling shown in Figure 4.49 appears to be a tiling of squares, regular pentagons, regular hexagons, regular heptagons, and regular octagons. Explain how you know that some of these polygons are apparently not regular, even though they appear to be.

Problem Solving: Developing Skills and Strategies

The previous sections have provided the solution to the problem of determining how many semiregular tessellations there are in a given figure. To solve this problem we considered several simpler problems. Develop a presentation that you could make to your class that explains how we have used the strategy "consider a simpler related problem."

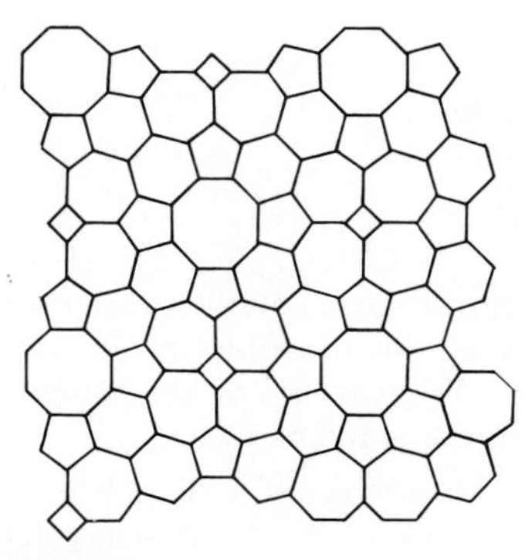

FIGURE 4.49

THREE-MIRROR KALEIDOSCOPE

It is both exciting and instructive to generate tessellations using mirrors. Fasten three mirrors together to form the vertical faces of an equilateral triangular prism. Place construction paper or

TEACHING NOTE

Students of all ages find the 3-mirror kaleidoscope fascinating. The creation of patterns for the kaleidoscope provides an experimental situation that encourages a creative approach to design. This activity also provides a situation rich in opportunities for a deeper look into properties of polygons, including angles, sides, diagonals, and lines of symmetry. The *Kaleidoscope Math Kit*, available from Creative Publications, is useful for creating and experimenting with kaleidoscopes.

FIGURE 4.50

colored acetate patterns in the base of the kaleidoscope. Viewers (Fig. 4.50) behold an interesting and sometimes beautiful tessellation as they peer over the edge. Try it!

The tessellation in Figure 4.51 was generated by the indicated pattern being placed in the base of a kaleidoscope. Investigation 4.4 provides an experience that should help build the reader's intuition about symmetry while generating some interesting tessellations. Figure 4.52 may provide a hint for parts B and C of Investigation 4.4.

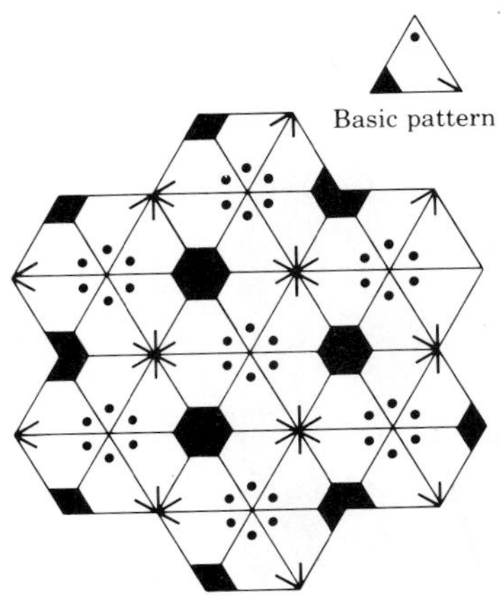

FIGURE 4.51

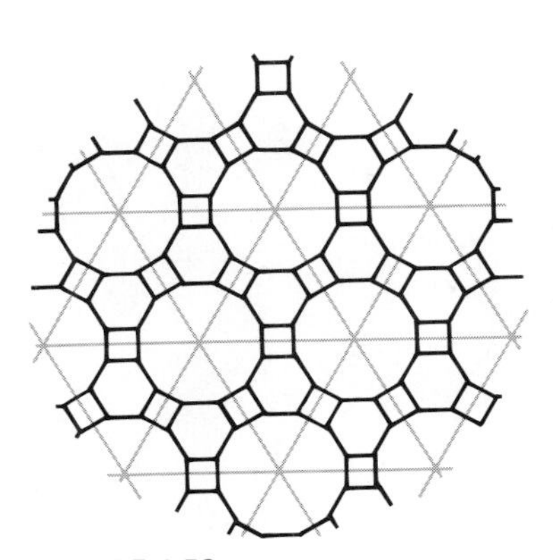

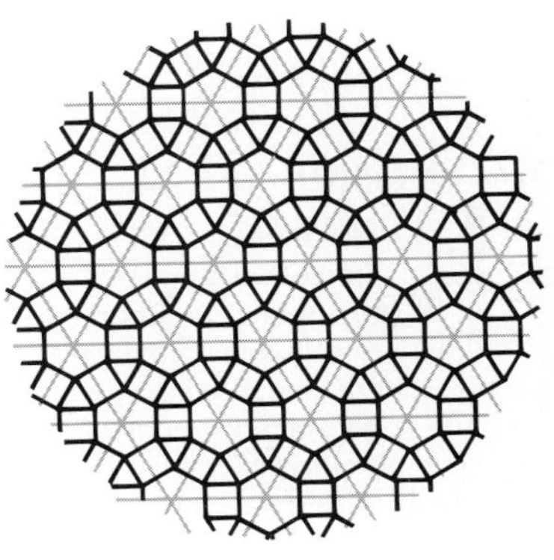

FIGURE 4.52

INVESTIGATION 4.4

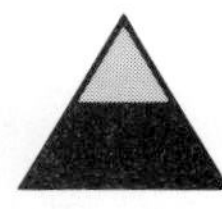

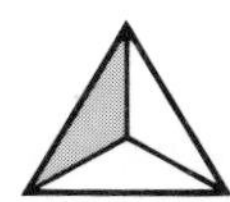

A. Each of the patterns in the figure shown, when placed in the base of a 3-mirror kaleidoscope, generates a tessellation. For each pattern, guess what the tessellation will look like. Check your guess with a kaleidoscope.
B. Can the regular tessellations be generated by placing patterns in a 3-mirror kaleidoscope? If so, show the patterns.
C. How many of the semiregular tessellations can be generated by placing patterns in a 3-mirror kaleidoscope? Show the patterns.

TESSELLATIONS OF CURVED PATTERNS

It is possible to create elaborate and interesting tessellations formed by patterns of curved figures. In Figure 4.53, we see some examples of this created by the artist Maurits Escher.

Many of these tessellations of curved figures can be formed by modifying a tessellation of polygons. One approach involves

Drawing by M. C. Escher

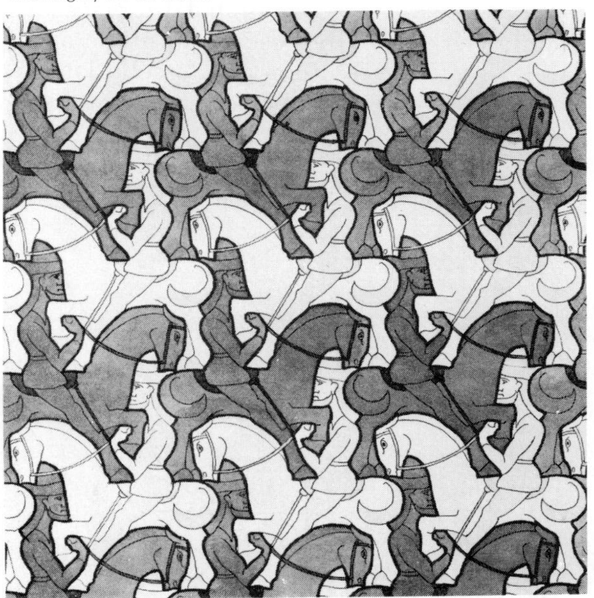

FIGURE 4.53

TEACHING NOTE

Teachers of mathematics are continually involved in attitude development. Work with tessellations provides an avenue for creative interest in geometry and helps students see mathematics as an enjoyable activity. Also, work with tessellations provides an avenue for integrating geometry and art.

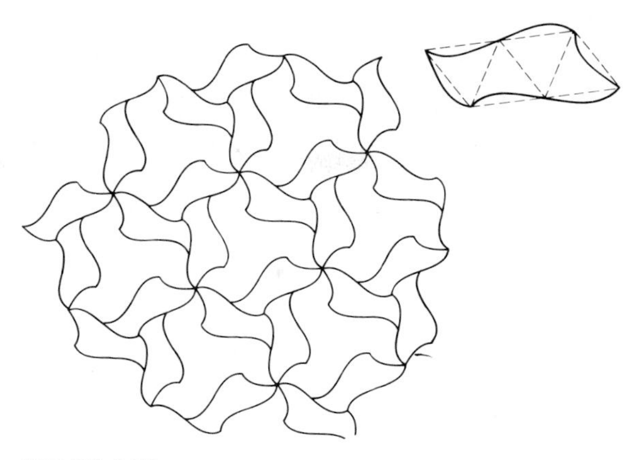

FIGURE 4.54

replacing one or more edges of each polygon with a curved line. In Figure 4.54, we see how the regular tessellation of equilateral triangles can be transformed into a tessellation figure.

A second approach involves providing the interior of each polygon with some pattern or design. For example, a square designed as in Figure 4.55 generates the tessellation of Figure 4.56. In Figure 4.57, we see how a tessellation of quadrilaterals can be converted into two shoals of fish swimming in opposite directions. Interested readers might wish to create some of their own tessellations by employing these techniques.

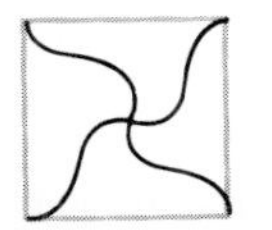

FIGURE 4.55

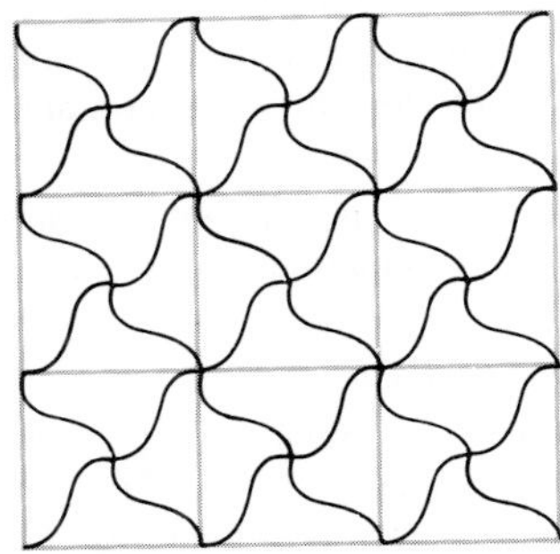

FIGURE 4.56

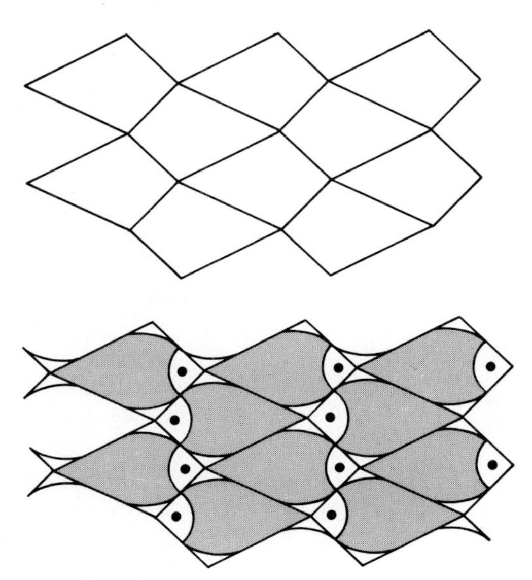

FIGURE 4.57

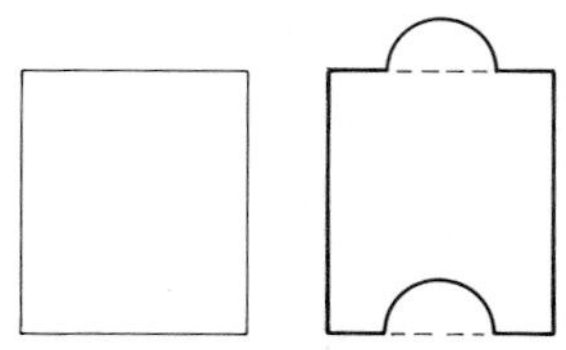

FIGURE 4.58

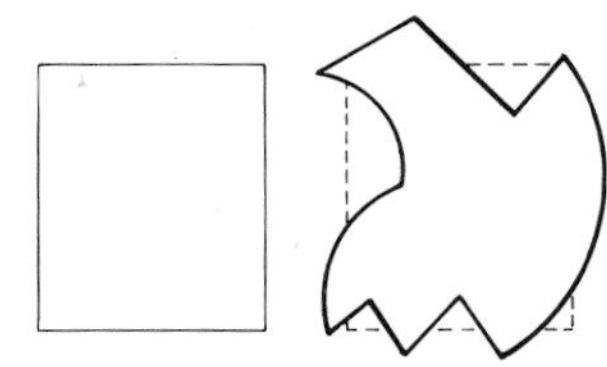

FIGURE 4.59

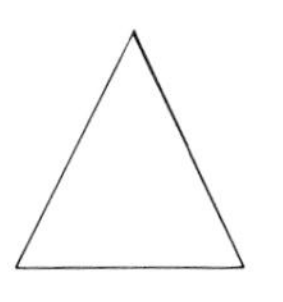

FIGURE 4.60

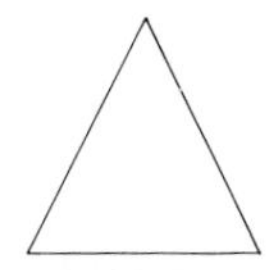
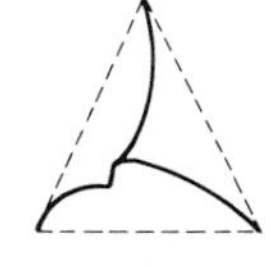

FIGURE 4.61

EXERCISE SET 4.3

1. Suppose that each square of a regular tessellation of squares has been altered as shown in Figure 4.58. Draw a portion of the resulting tessellation.
2. Suppose that each square of a regular tessellation of squares has been altered as shown in Figure 4.59. Draw a portion of the resulting tessellation.
3. Suppose that each triangle of a regular tessellation of equilateral triangles has been altered as shown in Figure 4.60. Draw a portion of the resulting tessellation.
4. Draw in each triangle of a regular tessellation of equilateral triangles the pattern shown in Figure 4.61. Then, erase the lines of the original triangle. Draw a portion of the resulting tessellation.
5. Be creative. Create a tessellation of curved-line patterns by combining several techniques.

PEDAGOGY FOR TEACHERS

ACTIVITIES

1. Procure some laboratory materials, such as *Tessellation Jobcards: Pattern Blocks* produced by Creative Publications. Use these materials with a student at an appropriate level. Summarize the student's reaction to the materials. How would you improve this experience for students?

2. Devise an investigation card that has the potential to get upper-grade or junior high school students involved in a search for all the semiregular tessellations.

3. Discuss tessellations with someone from the art department in your school and list some ideas for integrating these ideas of mathematics with an art project.

4. Devise a way to capture a student's interest and get him or her involved in some work with tessellations. The work might include coloring tessellations in interesting ways, making curved-line tessellations, searching for semiregular tessellations, and so on. Your means of motivating the student might range from a photography project to an art contest.

5. Analyze a representative elementary school or upper-level mathematics text series in the following way: (a) Check each grade level and see if any activities that involve tessellations are included; (b) if the text series does not include these activities, list the page number of specific places in the series in which you feel such activities might be included. Be specific about what activities you would include.

6. Make a set of worksheets that include some or all of the semiregular tessellations and present them to primary school children. Ask the children to color them in interesting ways. Display the results for your fellow teachers to observe and analyze.

7. Write a sequence of discovery questions that would lead a student to use tessellations to discover the theorems suggested in Figure 4.23.

COOPERATIVE LEARNING GROUP DISCUSSION

Tessellations and the study of patterns are topics that cross disciplines.

1. What concepts from geometry does a student study when studying a unit on tessellations?
2. What concepts from art could be taught through a unit of tessellations?
3. How would you adapt a unit on tessellations and the study of patterns for various grade levels? Be as complete as possible in discussing this question.

CHAPTER REFERENCES

Escher, Maurits C. 1960. *The Graphic Work of M.C. Escher.* New York: Hawthorn Books, Inc.

Grunbaum, Branko, and Shephard, G.C. 1989. *Tilings and Patterns: An Introduction.* New York: W. H. Freeman and Company.

Holiday, E. 1970. *Altair Design.* London: Pantheon.

Weyl, Hermann. 1952. *Symmetry.* Princeton, N.J.: Princeton University Press.

Seymour, Dale, and Britton, Jill. 1988. *Introduction to Tessellations.* Palo Alto, CA: Dale Seymour Publications.

SUGGESTED READINGS

Consult the following sources listed in the Bibliography for additional readings: Clauss (1991), Giganti and Cittadino (1990), Hill (1987), Kaiser (1988), and Woods (1988).

CHAPTER FIVE Geometry in Three Dimensions

INTRODUCTION

Plato (427–348 B.C.), who required all his pupils to obtain a thorough knowledge of mathematics before being initiated into his philosophy, was asked, "What does God do?" He is said to have responded, "God eternally geometrizes." It is in the natural context of our three-dimensional environment that we consider the impact of Plato's statement. Whether or not we agree, we must surely sense profound wonderment when we consider the myriad of exciting shapes that occur in the microscopic world of matter; in the natural world of rocks, plants, and animals; in the created world of the architect and artist; and in the vast world of outer space that we can barely see (Figs. 5.1–5.3). If we but pay attention to the crystals under a microscope, to the skeletons of tiny sea creatures, to shells, sunflowers, and honeybees, and to the human body and the cathedrals humankind has built, we might find that a course in geometry would be an afterthought.

Yet, we often do not have the inclination nor the opportunity to be so observant. In fact, many very simple aspects of geometry in familiar everyday situations often escape us. To begin our study of three-dimensional space, we consider one of

FIGURE 5.1

FIGURE 5.2

FIGURE 5.3

INVESTIGATION 5.1

A wall (theoretically of unlimited length and height) can be built by repeating a basic pattern. For this investigation, use either children's building blocks or rectangular cardboard cutouts that simulate blocks.

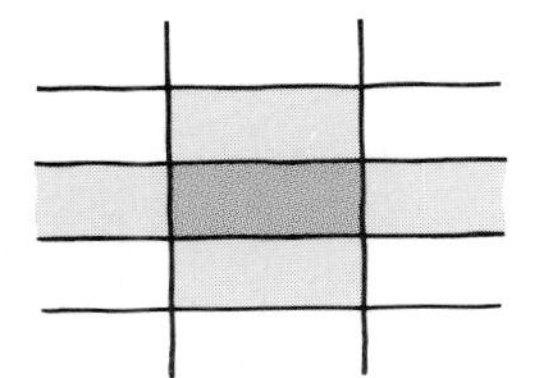

A. Build a wall with layers that can shift; that is, the wall can be sliced with a horizontal plane without cutting any bricks.
B. Build a wall that can be sliced with both horizontal and vertical planes without cutting any bricks.
C. Build a wall that cannot be sliced with a plane without cutting bricks (except the plane of the table on which you are working).
D. Build a wall in such a way that each brick in the center of the wall touches 5 neighboring bricks. Two bricks are called "neighboring" only if they touch along a flat surface. For example, the gray brick shown here has 4 neighbors.
E. Build a wall in such a way that each brick in the center of the wall touches 6 neighboring bricks.
F. Sketch a picture of each of your walls.

those situations. The reader who has observed a brick wall being built or a child playing with blocks may not have thought of this simple situation as a source of some rather interesting geometrical questions. Investigation 5.1 provides an opportunity to explore the spatial relationships involved in wall building.

Extending Investigation 5.1 by considering walls of at least three bricks thick (with every other layer shifted slightly), we see in Figure 5.4 that it is possible for a brick to have 14 neighbors. This fact is related to a space tessellation that is described later in this chapter.

TEACHING NOTE

Every school classroom should have a set of geometry solids made from wood, styrofoam, plexiglass, or cardboard. These forms provide the very small child with a rich environment in which spatial perceptions of polyhedra are developed by comparing and contrasting the models, using touch and sight.

The older child uses the models to explore quantitative questions, such as finding volumes and counting faces, edges, and vertices.

FIGURE 5.4

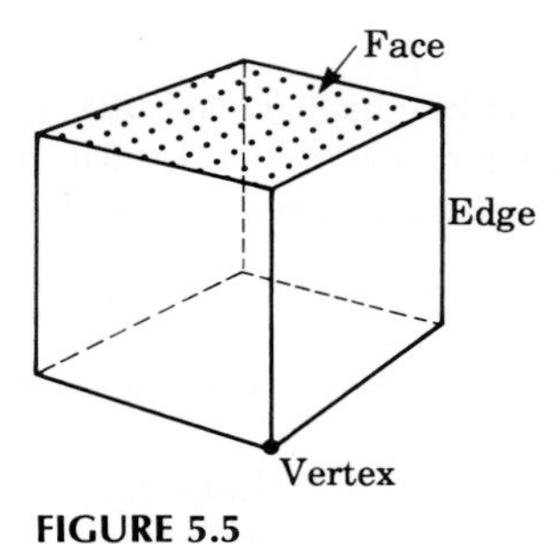

FIGURE 5.5

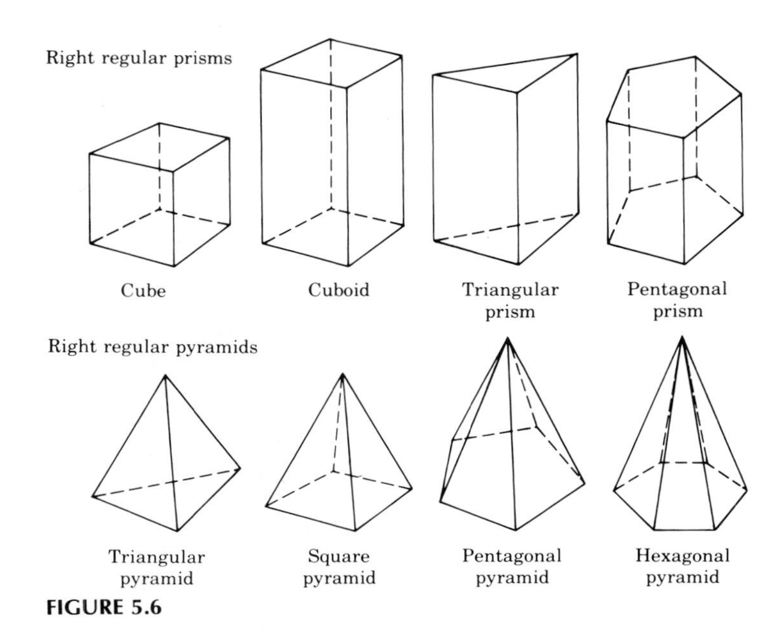

FIGURE 5.6

The "bricks" in Investigation 5.1 exemplify one of the simplest of an important class of solid figures, called *polyhedra.* A **polyhedron** is a finite set of joined polygons that enclose a finite region of space. If this space region is a convex region, the polyhedron is called a **convex polyhedron**. (A convex space region is defined in a manner similar to that used in defining a convex polygon.)

Each of the polygonal regions of a polyhedron is called a **face** (Fig. 5.5). Each side of a face, called an **edge**, is the edge of exactly one other face. **Vertices** of a polyhedron are the vertices of the faces, each of which may have a number of faces in common.

Figure 5.6 illustrates selected examples of complex polyhedra. The **prism** is a polyhedron with a pair of identical faces that lie in parallel planes. These two faces are called the **bases** of the prism. Corresponding vertices of these bases are joined by edges to form faces that are parallelogram regions. Prisms

TEACHING NOTE

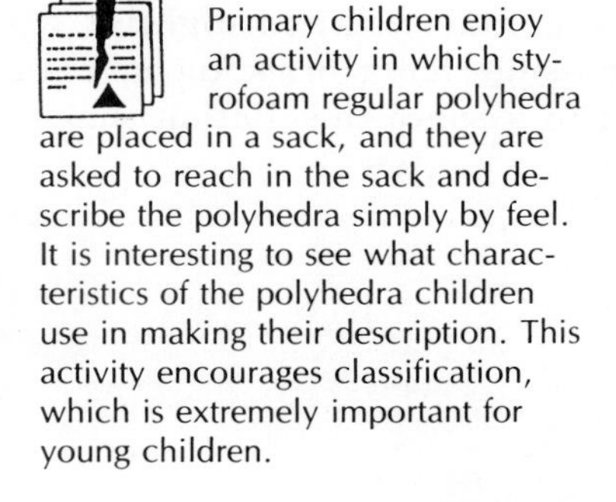

Primary children enjoy an activity in which styrofoam regular polyhedra are placed in a sack, and they are asked to reach in the sack and describe the polyhedra simply by feel. It is interesting to see what characteristics of the polyhedra children use in making their description. This activity encourages classification, which is extremely important for young children.

EXPLORING WITH THE COMPUTER 5.1

Use 3-D computer software, such as *Building Perspective* and *Super Factory*, (Sunburst Publications) to create solid figures and answer questions about them. Write an evaluation of the software that includes what it allowed you to do, and what value it had, and whether or not you enjoyed it and why.

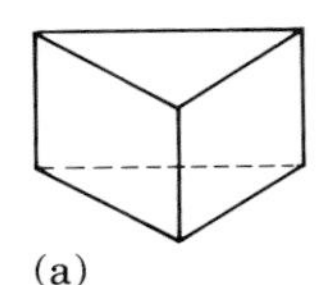
(a)

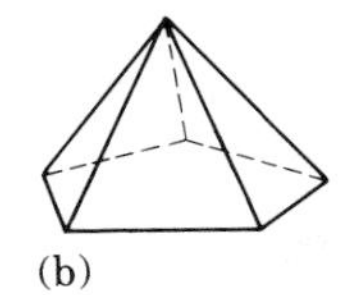
(b)

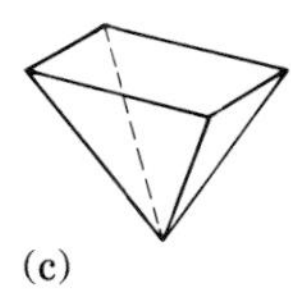
(c)

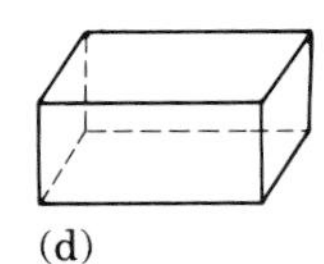
(d)

FIGURE 5.7

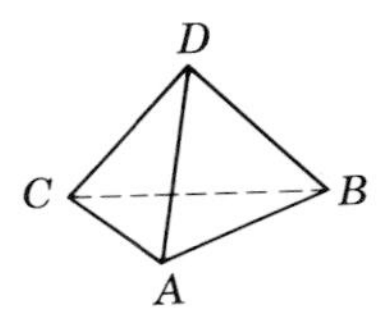

FIGURE 5.8

are named according to their polygon bases. For example, if the bases are pentagons, the prism is called a *pentagonal prism*.

A **pyramid** is a polyhedron formed when each vertex of a polygonal region in one plane is joined to a point not in that plane by a segment. The point is called the **apex**, and the polygonal region is called the **base** of the polyhedron. Pyramids are named according to their bases. For example, a pyramid with a hexagonal base is called a *hexagonal pyramid*.

EXERCISE SET 5.1

1. Name each of the polyhedra shown in Figure 5.7. If a polyhedron can be named in more than one way, give as many names as possible for it.
2. For the polyhedron shown in Figure 5.8, name the following:
   a. All the faces b. All the edges 3. All the vertices
3. Fill out Table 5.1 for each of the polyhedra shown in Figure 5.9.
4. How many faces, edges, and vertices does a pyramid with an n-gon base have?
5. How many faces, edges, and vertices does a prism with an n-gon base have?
6. For the polyhedra in Exercises 3, 4, and 5, compare F, V, and E. Write an equation that shows the relationship between these variables.
7. Draw a pyramid with a triangular base.
8. Draw a prism with pentagon bases.
9. Draw a polyhedra that is neither a prism nor a pyramid.

TABLE 5.1

	Number of Faces (F)	Number of Edges (E)	Number of Vertices (V)
a.			
b.			
c.			

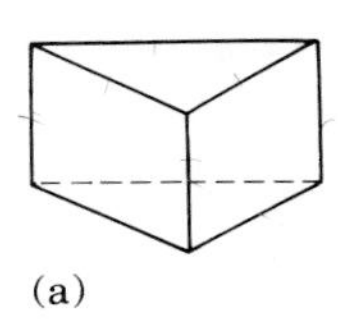
(a)

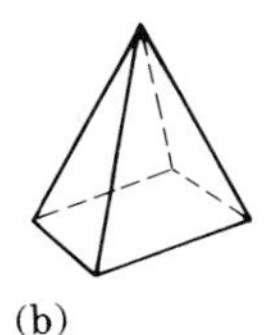
(b)

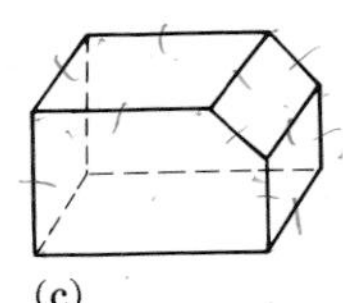
(c)

FIGURE 5.9

APPLICATION

Wiring Problem: An electrician wants to use the least amount of wire through the walls or ceiling of a 10 ft. by 10 ft. cubical room to connect switch box A to exit box B. Can you help her decide where to run the wire? Draw a diagram and show the calculations to support your decision. It may be helpful to utilize a net (pattern) for the cube.

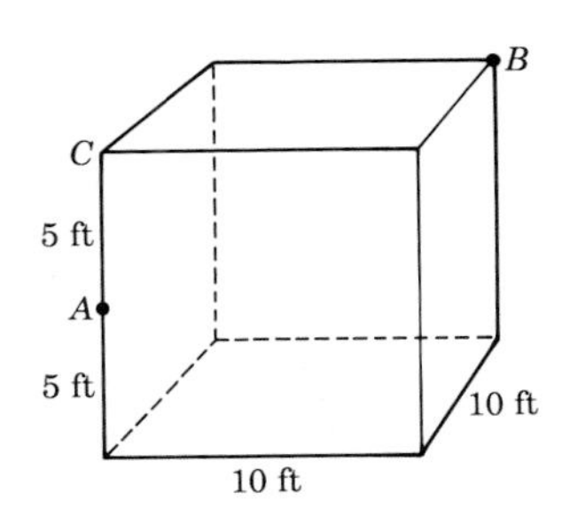

TEACHING NOTE

Many students at various ages need hands-on experiences in order to develop the concepts of solid geometry. Building and investigating polyhedra made from straws and pipe cleaners can provide this experience. *Geo-D-Stix*, a kit available from Dale Seymour Publications, provides useful materials for this type of activity.

10. **Critical Thinking**. Decide whether each of the following statements is *always true*, *sometimes true*, or *never true*:
    a. A pyramid has a square base.
    b. A prism has a pair of bases that lie in parallel planes.
    c. The building blocks that you used in Investigation 5.1 are prism shapes.
    d. A pyramid has triangular faces.
    e. A prism has rectangular faces.

    Write three additional statements about polyhedra: One that is *always true*, one that is *sometimes true*, and one that is *never true*.

REGULAR POLYHEDRA

In the book *The Hunting of the Snark* by Lewis Carroll (an alias for C. L. Dodgson, the mathematician-writer who wrote *Alice in Wonderland*), the following appears:

> You boil it in sawdust: you salt it in glue:
> You condense it with locusts and tape:
> Still keeping one principal object in view—
> To preserve its symmetrical shape.

Although Investigation 5.2 does not require such a drastic operation, it does provide an opportunity for some simple model building. It has been said that the best way to learn about polyhedra is to make them; the next best way is to handle them. This investigation provides an opportunity for both. Of course, in addition to the experience of manipulating physical

INVESTIGATION 5.2

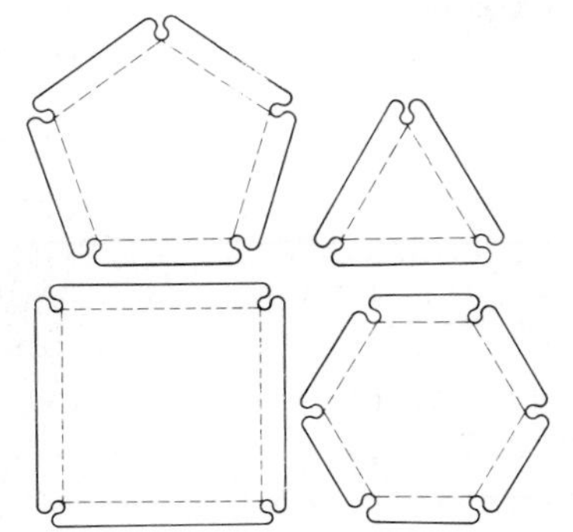

One class of polyhedra includes those in which all faces are the same size and shape.

A. Use commercially produced materials, or cut from posterboard as many square patterns shaped like the one in the figure as you need and put them together with rubber bands to from a cube. Can you construct any other solid figure with square faces?
B. The figure shows three other regular polygon patterns. Can you use larger patterns like these and make other polyhedra with each face the same shaped polygon? How many different such polyhedra can you make?
C. Can you make polyhedra from any other regular polygon patterns?

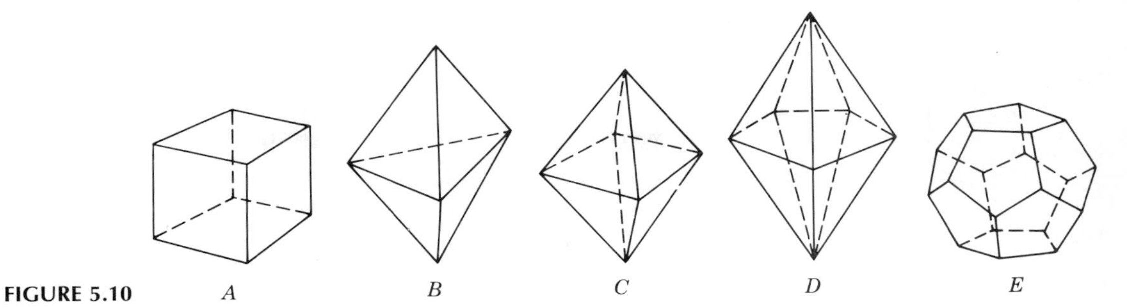

FIGURE 5.10

objects, the reader should be seeking conjectures about any relationships observed. For example, the relationship $V + F = E + 2$ you discovered in Exercise 6, Exercise Set 5.1, is called **Euler's Formula**. You can test the formula using the models in Investigation 5.2.

Some of the polyhedra you may have made in Investigation 5.2 are pictured in Figure 5.10. In these polyhedra, each face is a regular polygon and all faces are congruent (same size and same shape). Figures A, C, and E have an important characteristic, however, that is not found in Figures B and D.

In order to focus on this characteristic, recall that in Chapter 3, we discussed plane tessellations of regular polygons. The tessellation shown in Figure 5.11, consisting of pentagons with all sides equal, is often mistaken for an example of a tessellation of the plane with regular pentagons. However, to have a regular polygon, all vertex angles must be of equal measure and all sides must be of equal length.

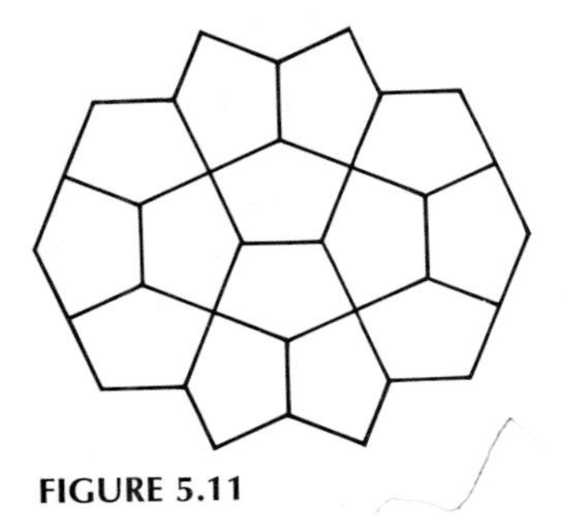

FIGURE 5.11

Similarly, in space a *regular polyhedron* is characterized not only by the condition that the faces are congruent regular polygons, but also by the further condition that the edges and vertices are all "alike." This can be described more precisely by saying that all vertex figures must be regular polygons of the same type. (A vertex figure of a polyhedron is a polygon whose vertices are midpoints of the edges emanating from a vertex of the polyhedron.) In the cube in Figure 5.12, the vertex figure is an equilateral triangle. With this condition in mind, the reader should decide which polyhedra in Figure 5.10 would be classified as regular polyhedra.

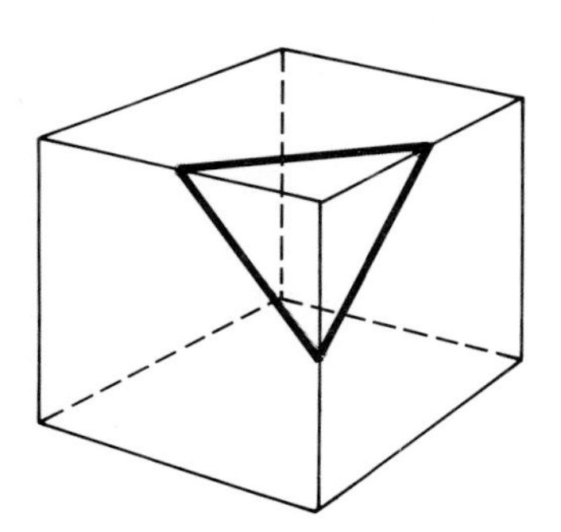

FIGURE 5.12

Now that the characteristics of regular polyhedra have been specified, a natural question arises: *How many regular polyhedra are there*? Perhaps the experiences from Investigation 5.2 will help. First we shall consider regular polyhedra with equilateral triangular faces. We discovered when studying

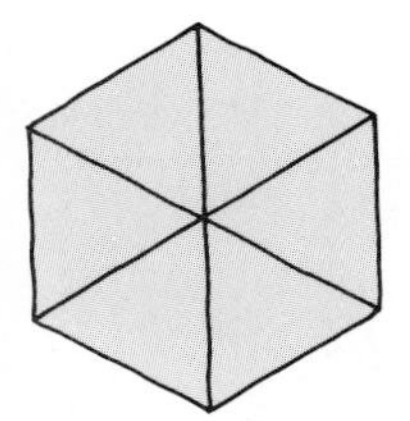

FIGURE 5.13

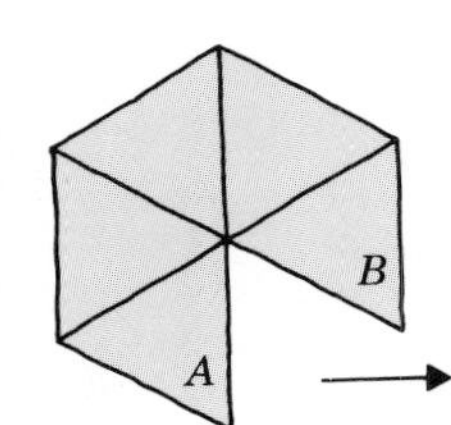

FIGURE 5.14

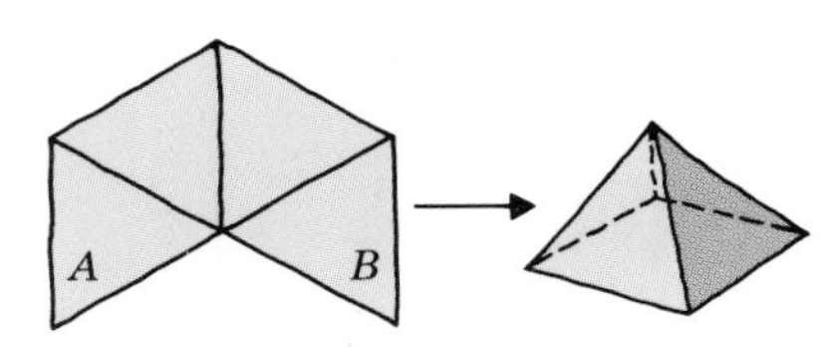

FIGURE 5.15

tessellations that six equilateral triangles completely surrounded a vertex in the plane, as shown in Figure 5.13. So any regular polyhedron can have at most five equilateral triangular faces surrounding a vertex.

Beginning with five equilateral triangles, as shown in Figure 5.14, and joining points A and B, we produce a three-dimensional "roof," which is a vertex arrangement for a regular polyhedron of twenty faces.

Beginning with four equilateral triangles, as shown in Figure 5.15, and joining corners A and B, we form a three-dimensional "roof" and find that two of these can be placed together to form an eight-faced regular polyhedron.

Beginning with three equilateral triangles, as in Figure 5.16, and joining corners A and B, we form a three-dimensional "roof" and find that one more equilateral triangle may be placed with this "roof" to form a four-faced regular polyhedron.

Next, we consider polyhedra with square faces. Since a vertex in the plane can be surrounded by four squares, as in Figure 5.17, we see that a regular polyhedron with square faces can have at most three squares at each vertex.

Suppose we remove one square from the four in Figure 5.17 and join corners A and B (Fig. 5.18). We form a three-dimensional "roof" and find that two of these "roofs" may be placed together to form a regular polyhedron with six square faces—the *cube*. This is the only regular polyhedron with square faces.

TEACHING NOTE

While younger children develop concepts of point, line, and plane by feeling vertices, edges, and faces of polyhedra models, older students enjoy building the models themselves. Books by Wenninger (1970) and Hilton and Pederson (1988) provide some very good activities for students of all ages.

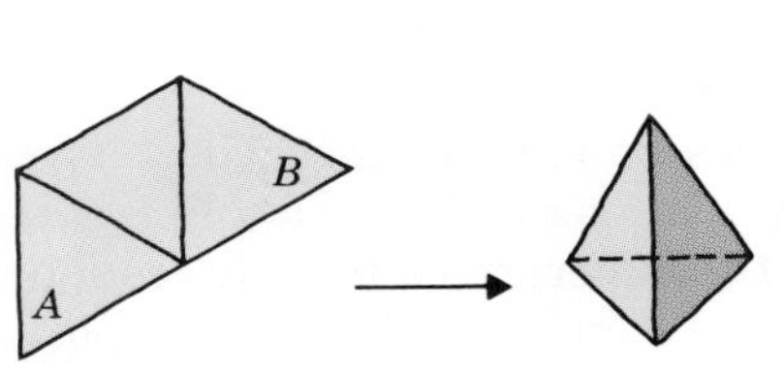

FIGURE 5.16

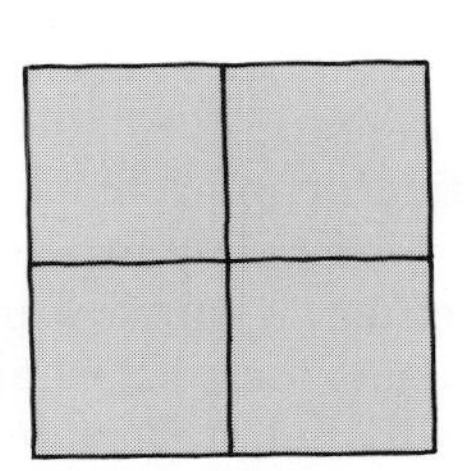

FIGURE 5.17

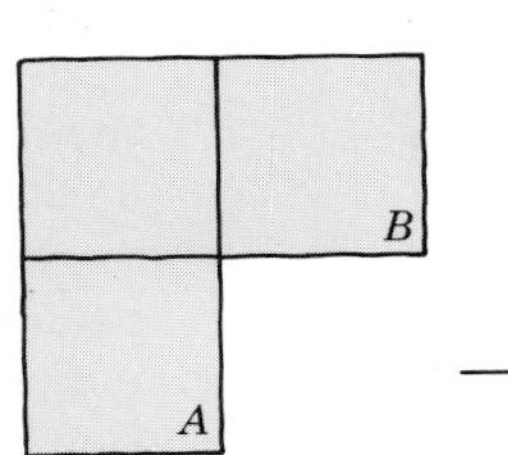
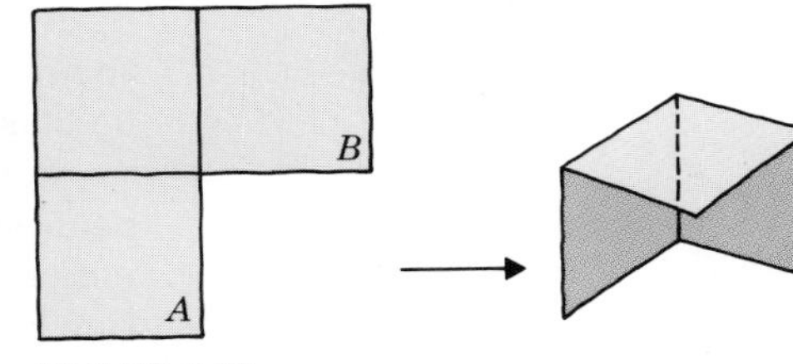

FIGURE 5.18

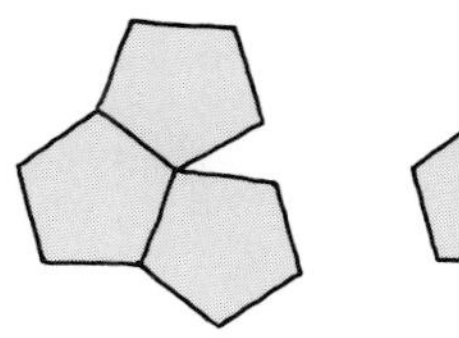

FIGURE 5.19

The plane cannot be tessellated with regular pentagons because three about a point leave a gap, and four about a point overlap, as shown in Figure 5.19. With three pentagons, however, a three-dimensional "roof" can be formed (Fig. 5.20), and it is found that four of these "roofs" will fit together to form a regular polyhedron with 12 regular pentagons as faces.

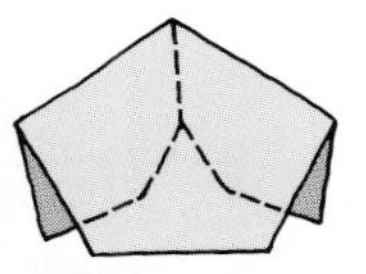

FIGURE 5.20

When we try to place three hexagons about a point, as in Figure 5.21, we again find that no three-dimensional "roof" is formed, and, hence, no polyhedra can be made. For regular polygons with more than six sides, we find that three or more such polygons placed together at a point always overlap.

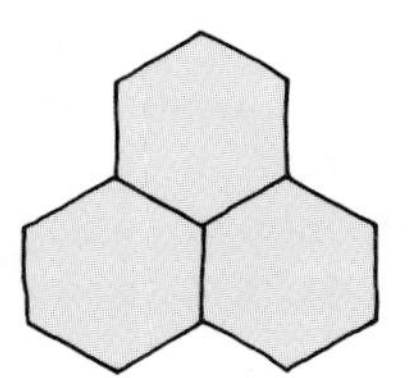

FIGURE 5.21

This intuitive consideration of "roofs," first using equilateral triangles, then squares, regular pentagons, regular hexagons, and larger polygons, strongly suggests that there are exactly five convex regular polyhedra. A more formal explanation that there can be no more than five convex regular polyhedra is given next.

Suppose there exists a regular convex polyhedron with each face a regular p-gon, as pictured in Figure 5.22. We have seen in Chapter 3 that each vertex angle has a degree measurement of $180((p - 2)/p)$. If on the convex polyhedron there are q p-gons surrounding each vertex, the sum of the angle measures around each vertex of the polyhedron would be $q[180((p - 2)/p)]$. Note, however, that $q[180((p - 2)/p)]$ must be less than 360, for if it is equal to 360, the faces all lie in a plane, and we would not have a solid figure. Hence, we have

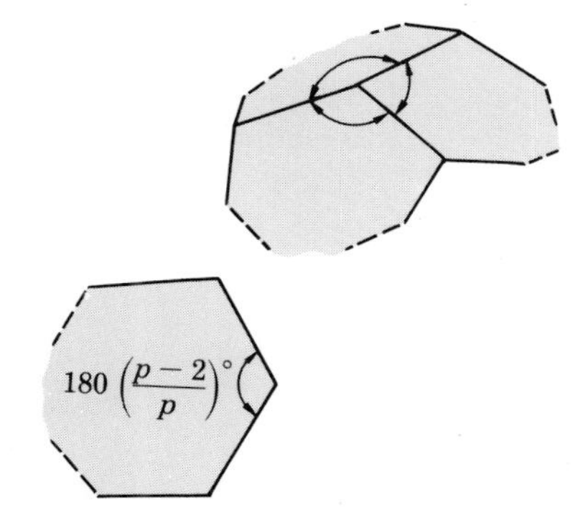

FIGURE 5.22

$$q\left[180\,\frac{(p - 2)}{p}\right] < 360$$

$$\Leftrightarrow q\,\frac{(p - 2)}{p} < 2$$

$$\Leftrightarrow q(p - 2) < 2p$$

$$\Leftrightarrow qp - 2q < 2p$$

$$\Leftrightarrow qp - 2q - 2p < 0$$

$$\Leftrightarrow qp - 2q - 2p + 4 < 4$$

$$\Leftrightarrow (p - 2)(q - 2) < 4$$

Since a face of the convex regular polyhedron must have more than two sides, and more than two faces are required around each vertex, we see that both p and q must be greater than 2. Table 5.2 shows the possible values for p and q. No other integer values of p and q satisfy the inequality $(p - 2)(q - 2) < 4$.

TEACHING NOTE

Students at the middle grades and above might be stimulated to look at figures and construct nets, rather than vice versa. Also, spatial perception might be tested by presenting nets, such as those shown with 6 squares, and asking students to decide from the picture whether or not they are nets for a particular polyhedron—in this case, a cube.

TABLE 5.2

p	q	$p - 2$	$q - 2$	$(p - 2)(q - 2)$
3	3	1	1	1
3	4	1	2	2
3	5	1	3	3
4	3	2	1	2
5	3	3	1	3

For if one of p or q is 6 or greater, and the other is at least 3, then one factor in the inequality is 4 or greater, and the other is at least 1. In any case, the product $(p - 2)(q - 2)$ is not less than 4.

Hence, we have only five possible convex regular polyhedra. Three of these have equilateral triangular faces—the **tetrahedron, octahedron**, and **icosahedron**; one has square faces—the cube; and one has pentagonal faces—the **dodecahedron**. These polyhedra do exist and are shown with patterns for making them—called **nets**—in Figure 5.23.

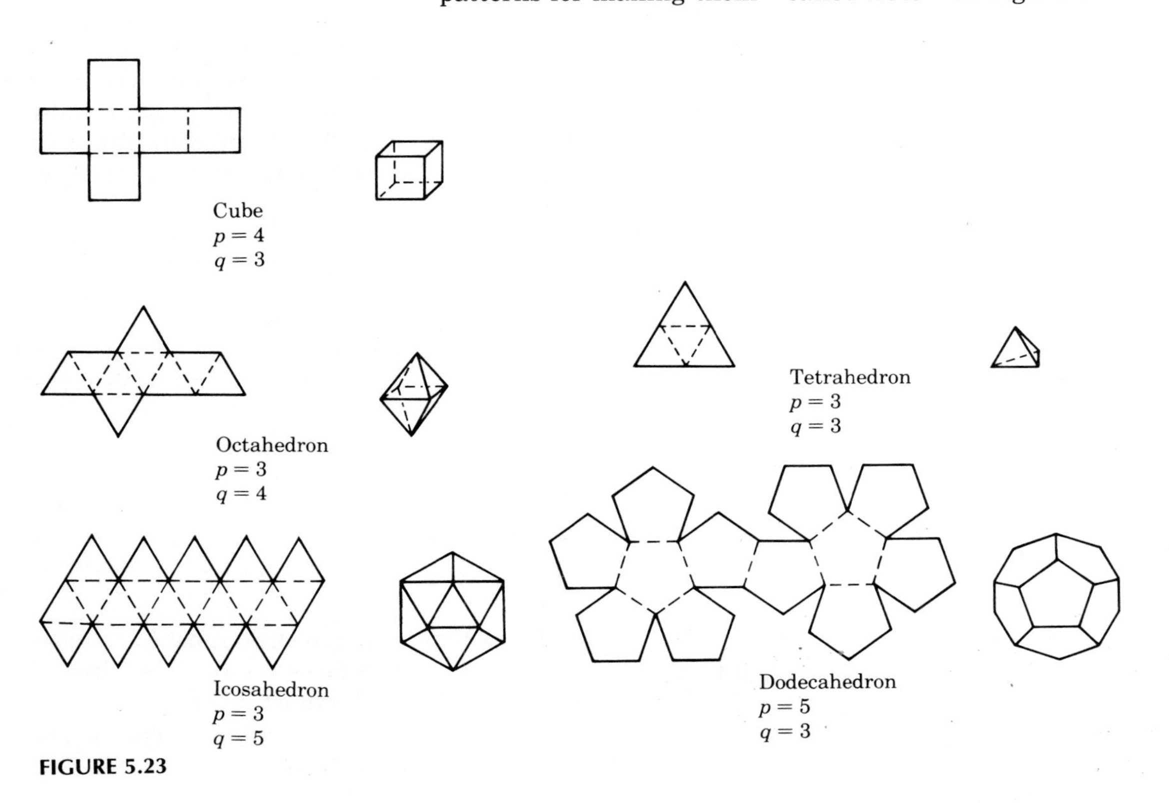

FIGURE 5.23

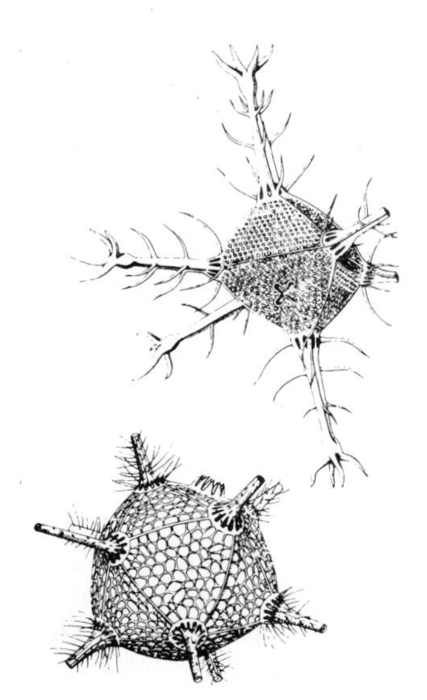

FIGURE 5.24

FIGURE 5.25

It is interesting to note that nature exhibits the regular convex polyhedra in a variety of exciting ways (Figs. 5.24–5.25). Skeletons of tiny sea creatures, made of silica and measuring only a fraction of a millimeter in diameter, have the form of the octahedron, icosahedron, and the dodecahedron. These creatures, called radiolarians, provide amazingly accurate models of these regular polyhedra. Cubes are realized in the shape of salt crystals, while a crystal of pyrite bears a close resemblance to the regular dodecahedron. Crystalline structures as simple as that of carbon are tetrahedral and octahedral in form. Models of the regular polyhedra occur in modified form in other crystals. For example, the crystals of argentite, the fluorite crystal, resemble octahedra in which vertices have been removed in certain ways.

It is not known who first discovered the regular polyhedra. Perhaps nature was observed and, as is so often the case, the mathematical abstractions followed. It is these abstractions of nature's polyhedra, first investigated carefully by the Greek Theaetetus in 398 B.C., that we consider more carefully in the exercises that follow.

EXERCISE SET 5.2

1. An old puzzle asks us to make 4 triangles using 6 match sticks. How is a tetrahedron related to this problem?

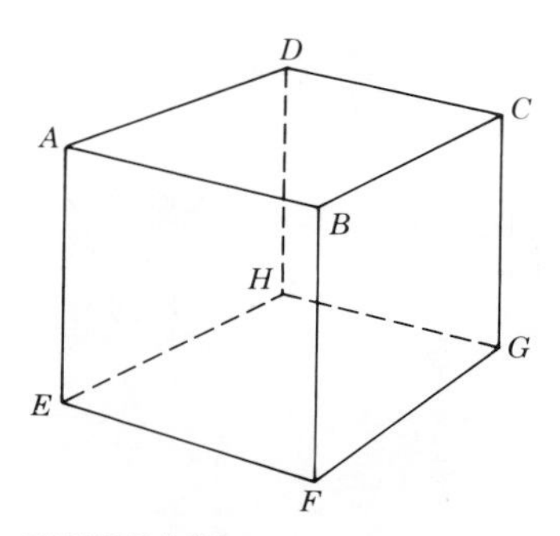

FIGURE 5.26

TABLE 5.3

	Number of Edges (E)	Number of Vertices (V)	Number of Faces (F)
Tetrahedron			
Cube			
Octahedron			
Dodecahedron			
Icosahedron			

TABLE 5.4

Regular Polyhedron	Schläfli Symbol
Cube	(4, 3)
Tetrahedron	
Octahedron	
Dodecahedron	
Icosahedron	

2. Use a model (if available) or a photograph to count the number of edges for each of the regular convex polyhedra. Count the number of vertices for each and the number of faces. Copy and complete Table 5.3. Is Euler's Formula satisfied for the regular polyhedra?
3. Consider the cube with vertices *ABCDEFGH* shown in Figure 5.26.
   a. Two edges of the cube are called opposite if they are parallel and are not the edges of a single face. List pairs of opposite edges. How many are there?
   b. Two vertices of a cube are called opposite if they are not joined by an edge and are not the vertices of a single face. List pairs of opposite vertices. How many are there?
4. In the cube pictured in Figure 5.26, vertices *B*, *D*, *E*, and *G* are the vertices of a tetrahedron inscribed in the cube.
   a. Sketch a cube (in one color) and draw in the edges of an inscribed tetrahedron (in another color).
   b. Can you find additional inscribed tetrahedra? How many are there?
5. Since each face of the cube contains 4 edges, and 3 edges meet at each vertex, we use a symbolism suggested by Schläfli to describe the cube. Specifically, the cube is described by (4,3). Complete Table 5.4 to describe the other polyhedra.
6. When 2-dimensional pictures are made of 3-dimensional solids (as in Exercise 2), dashed lines are used to show edges that are hidden from view. Practice making accurate drawings of a cube, a tetrahedron, and an octahedron. Be sure to include the dashed lines.
7. A pattern that can be cut out and fastened together to form a polyhedron is called a net for the polyhedron. Figure 5.23 shows a net for each of the regular polyhedra. Make an appropriate-sized net and construct a model of one of the regular polyhedra.
8. Which of the figures in Figure 5.27 are nets for a cube?

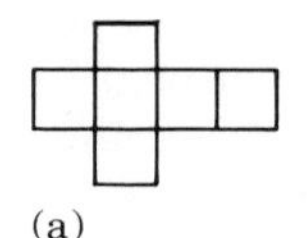
(a)

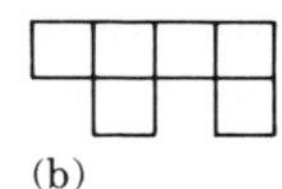
(b)

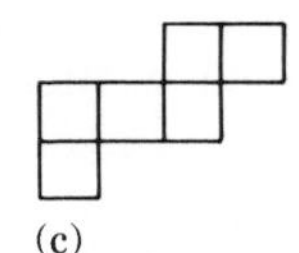
(c)

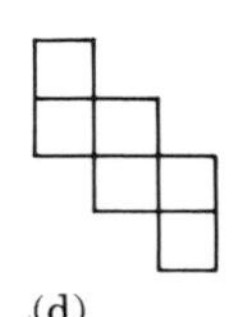
(d)

FIGURE 5.27

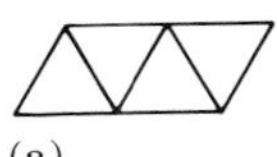
(a)

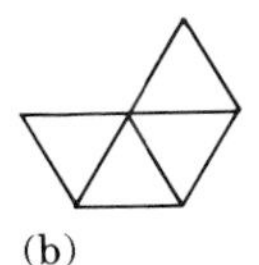
(b)

(c)

FIGURE 5.28

FIGURE 5.29

9. Which of the figures in Figure 5.28 are nets for a tetrahedron?
10. If a cube is drawn and a point is marked at the center of each face of the cube, and these points are connected, an octahedron is formed. We say that the *dual* of a cube is an octahedron. More generally, the dual of any convex regular polyhedron is another convex regular polyhedron formed by connecting the centers of the faces of the first polyhedron (Fig. 5.29).
    a. List each of the 5 regular polyhedra and find the dual of each.
    b. Compare your answers in part a with the table you made in Exercise 5 (Table 5.4).
11. Suppose a wire model of a polyhedron is made showing just the edges of the polyhedron. If the model is viewed in perspective from a position just outside the center of one face, this face appears as a frame and all the remaining edges are seen in its interior (Fig. 5.30). If this view is drawn for a solid, a diagram called a Schlegel diagram is formed. Draw a Schlegel diagram for a tetrahedron and an octahedron.

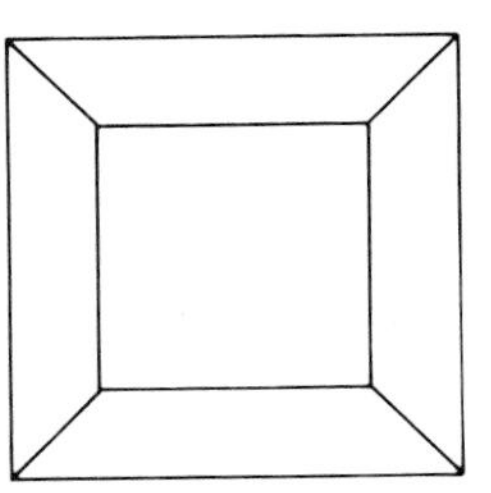

A Schlegel diagram for a cube

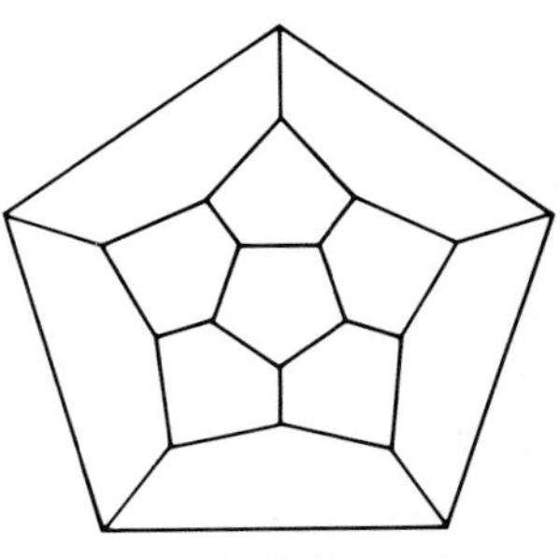

A Schlegel diagram for a regular dodecahedron

FIGURE 5.30

FIGURE 5.31

12. A *deltahedron* is a solid with equilateral triangles as faces (Fig. 5.31). The tetrahedron, with 4 equilateral triangular faces, is the smallest deltahedron. The icosahedron, with twenty equilateral triangular faces, is the largest deltahedron.
    a. Which other regular polyhedron is also a deltahedron? How many faces does it have?
    b. The deltahedra listed so far suggest that each deltahedron has an even number of faces. Use the method of Investigation 5.2 to construct deltahedra of 12 and 14 faces.
13. This exercise outlines a method of constructing a tetrahedron from a sealed envelope. Refer to Figure 5.32.
    a. Construct point C so that $\triangle ABC$ is an equilateral triangle.
    b. Cut along $\overline{DE}$, through C, and parallel to AB.
    c. Fold along $\overline{AC}$ and $\overline{BC}$ back and forth in both directions.
    d. Let C' be the point on the reverse side corresponding to point C.
    e. Pinch the envelope so that points D and E are joined and C and C' are separated. Tape along segment CC' and the tetrahedron is complete.

    Construct a tetrahedron using this method.
14. This exercise outlines a quick way of constructing a dodecahedron.
    a. Cut from poster board two nets like the one shown at the left in Figure 5.33.
    b. Crease sharply along edges of $ABCDE$.
    c. Place one net upon the other, rotate one 36°, and weave an elastic band alternately above and below the corners.
    d. Raise your hand and watch the dodecahedron pop up.

    Construct a dodecahedron using this method.

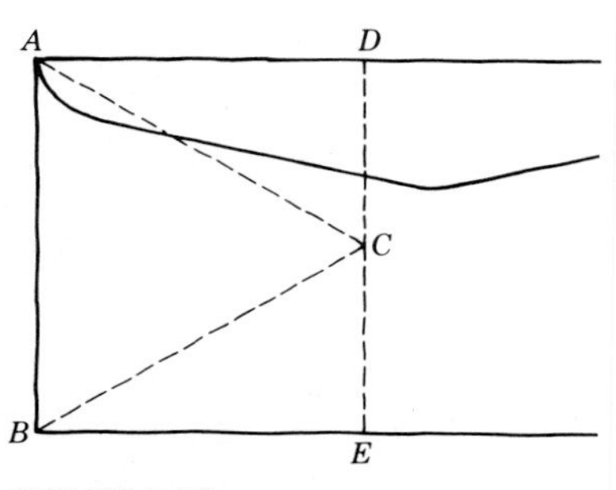

FIGURE 5.32

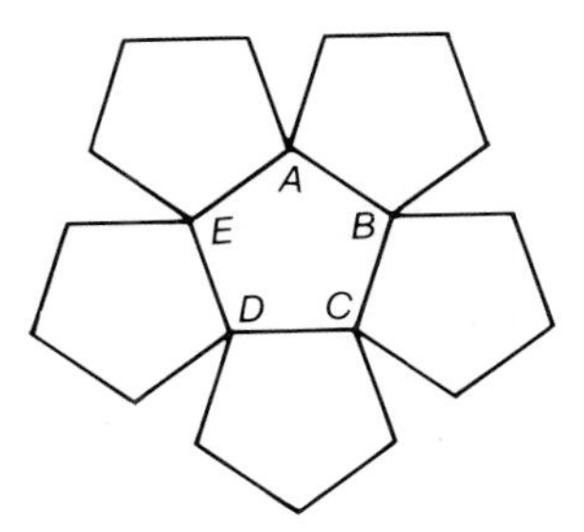

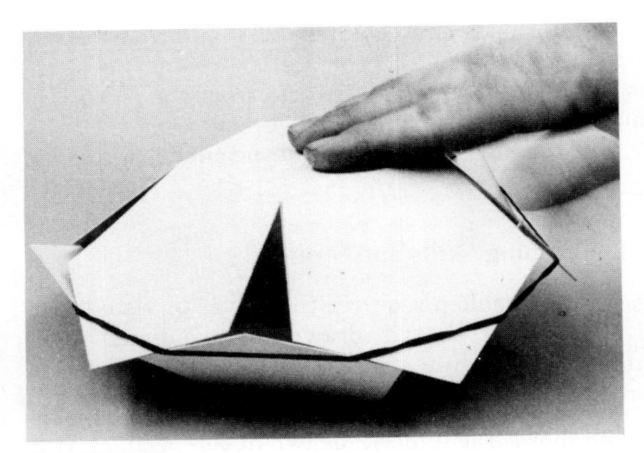

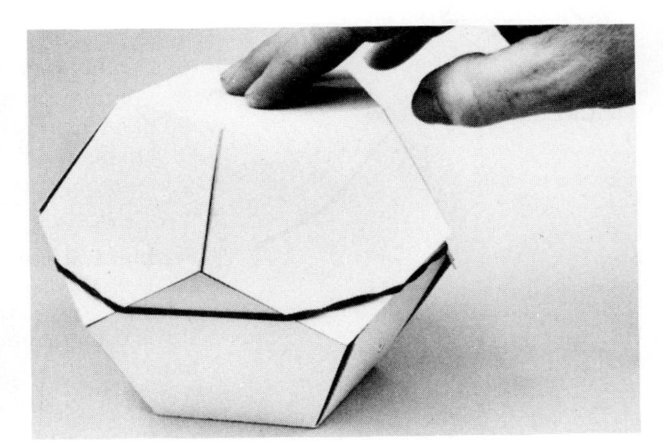

FIGURE 5.33

15. Puzzles
    a. Can you make 2 solids from these nets (Fig. 5.34) and place them together to form a tetrahedron?
    b. Can you make 3 solids from these nets (Fig. 5.35) and place them together to form a cube?
    c. Can you make 2 solids from these nets (Fig. 5.36) and put them together to form a cube?

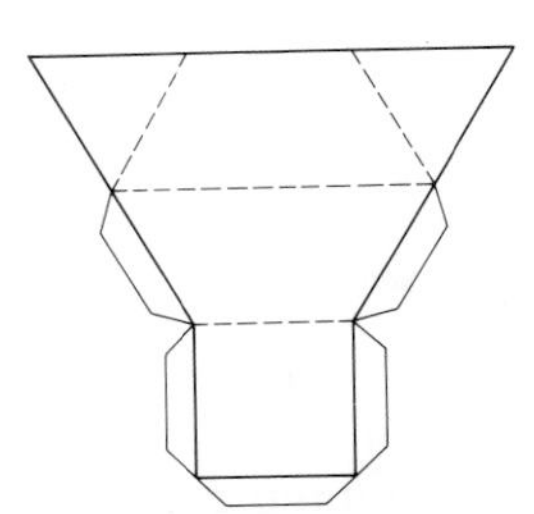

FIGURE 5.34

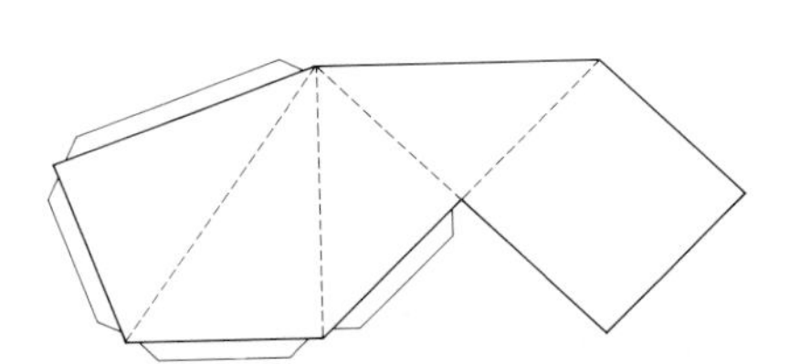

FIGURE 5.35

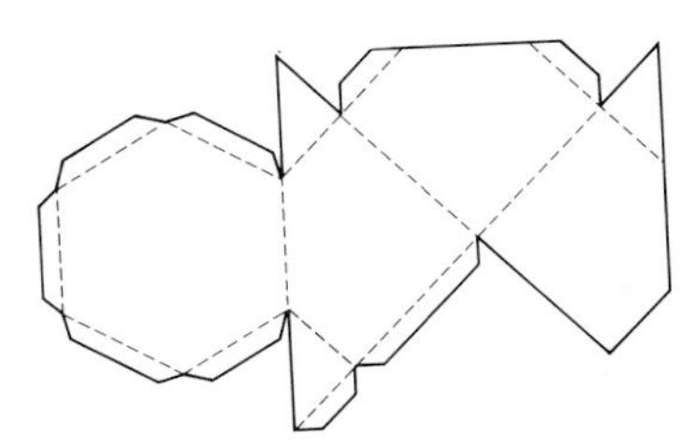

FIGURE 5.36

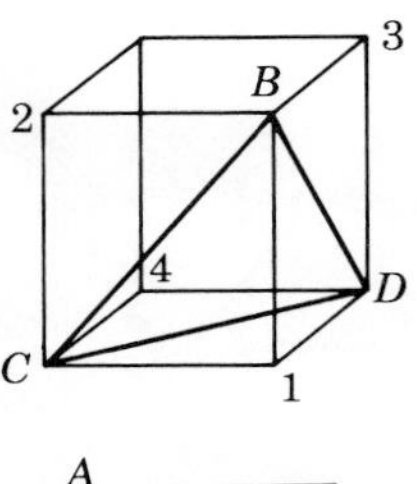

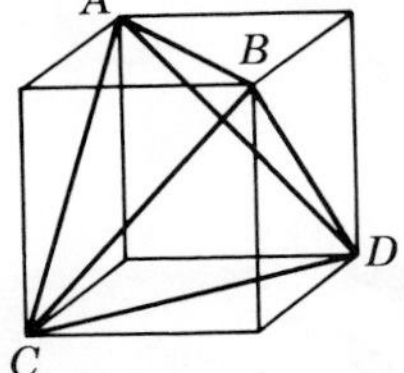

FIGURE 5.37

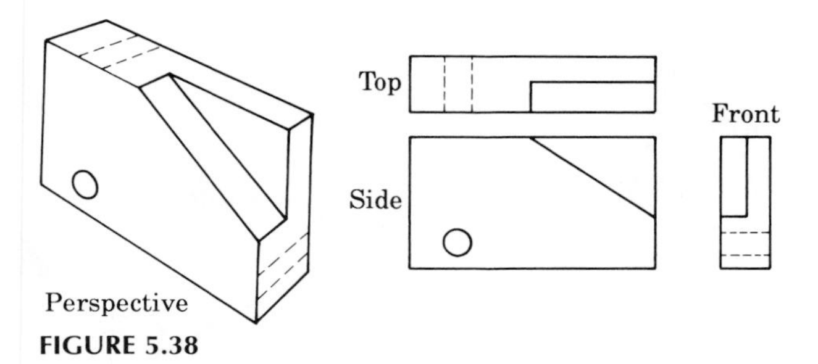

FIGURE 5.38

16. **Critical Thinking.** Refer to Figure 5.37. Slice a plane through points *B*, *C*, and *D* of this cube to obtain a triangular-based pyramid with apex 1 and base *BCD*. In a similar manner, suppose that pyramids with apex 2, 3, and 4 are cut from the cube. Use logical reasoning to write a convincing argument that the polyhedron *ABCD* that remains in the center of the cube is a regular tetrahedron.

Problem Solving: Developing Skills and Strategies

Often, solving a geometry problem requires the solver to visualize a three-dimensional solid object that is drawn on two-dimensional paper. We call this strategy **using space visualization**. Use this strategy for this problem.

Problem: Figure 5.38 shows a "cut block" drawn in perspective, with a top view, side view, and front view shown. Notice the use of dashed lines to show cuts hidden from view. Draw in perspective 2 cut blocks, each different from the one shown here, that have the same top and front view as the block shown in the figure.

SYMMETRY IN SPACE

From pre-Egyptian times, through the age of Theaetetus and Euclid, and into the twentieth century, the desire to find order and regularity in nature and in one's creations has stimulated tremendous activity and accomplishment. Herman Weyl, in his beautiful book *Symmetry* (1952), asserts:

> Symmetry, as wide and narrow as you may define its meaning, is one idea by which man through the ages has tried to comprehend and create order, beauty, and perfection.

For example, symmetry is an important theme in the study of molecular structure. Diamonds and sodium chloride are but a few of the crystalline structures possessing symmetry.

Even in the simple task of getting better acquainted with polyhedra, it is instructive to use the idea of symmetry. In

TEACHING NOTE

One important goal in teaching geometry is to help students develop their space perception. If your school art department has a hot wire for cutting styrofoam, the styrofoam polyhedra available from Cuisenaire Company can be sliced along a plane of symmetry, and a mirror can be used to illustrate the idea of a plane of symmetry.

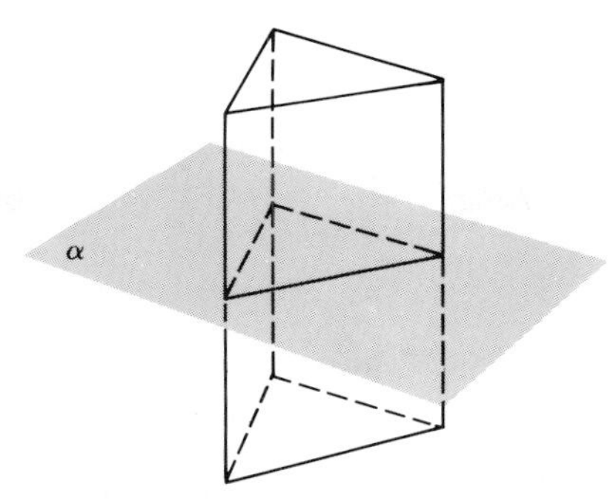

FIGURE 5.39

Figure 5.39, α is called a **plane of symmetry** of the equilateral triangular prism, while in Figure 5.40, β is not a plane of symmetry for the triangular pyramid.

To test whether a plane is a plane of symmetry for a polyhedra, you can visualize a mirror. If the plane in question, say α in Figure 5.39, were a mirror, and the part of the prism behind this mirror were removed, observers could look in the mirror and see the original, complete prism. That is, one-half of the prism is the mirror image of the other half. Hence, the plane α would qualify as a plane of symmetry of the prism. On the other hand, if the plane in question, say β in Figure 5.40, were a mirror, and the part of the pyramid on the lower side of this mirror were removed, observers could not see the original, complete pyramid when they looked in the mirror. Instead they would see a six-faced deltahedra, as pictured in Figure 5.10b. Hence, the plane β is not a plane of symmetry for the pyramid.

If a plane of symmetry exists for a polyhedron, we say the polyhedron has **reflectional symmetry**. Investigation 5.3 explores planes of symmetry for selected polyhedra.

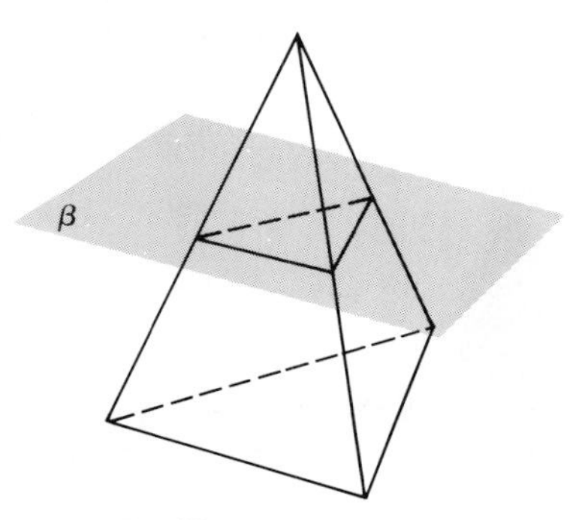

FIGURE 5.40

INVESTIGATION 5.3

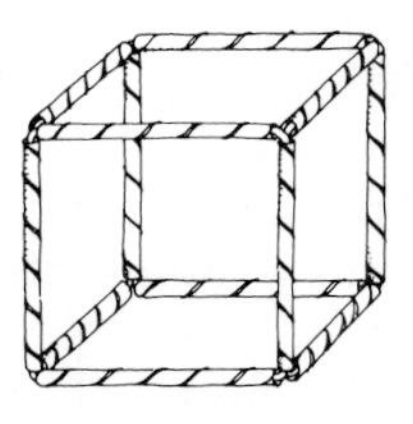

A. Use models to count the total number of planes of symmetry for the tetrahedron, cube, and octahedron by using *one* of the following techniques.

1. Construct the models from drinking straws and pipe cleaners and identify symmetry planes with 4 x 6 file cards glued into place.
2. Construct the models from nets and visualize the symmetry planes.
3. Use commercially produced models and attempt to visualize the symmetry planes. (Styrofoam models are helpful since they can be cut into pieces to illustrate planes of symmetry.)

B. In part A, did you devise a systematic method of counting the symmetry planes? Compare your methods with others in your class.

C. Use the method your class has decided is most efficient and count the number of symmetry planes for the dodecahedron and the icosahedron.

In addition to reflectional symmetry, a consideration of cyclic, or rotational, symmetry often supplies one with a fresh, new view of a solid. For example, if the flower in Figure 5.41

FIGURE 5.41

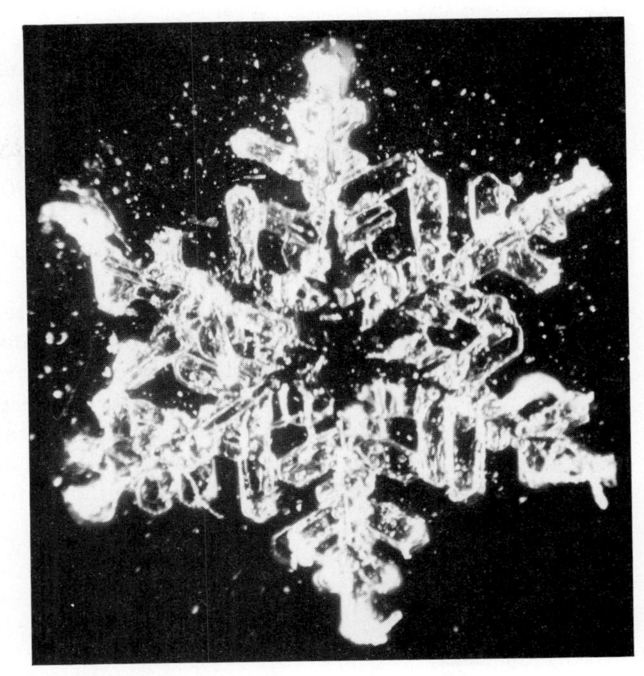

FIGURE 5.42

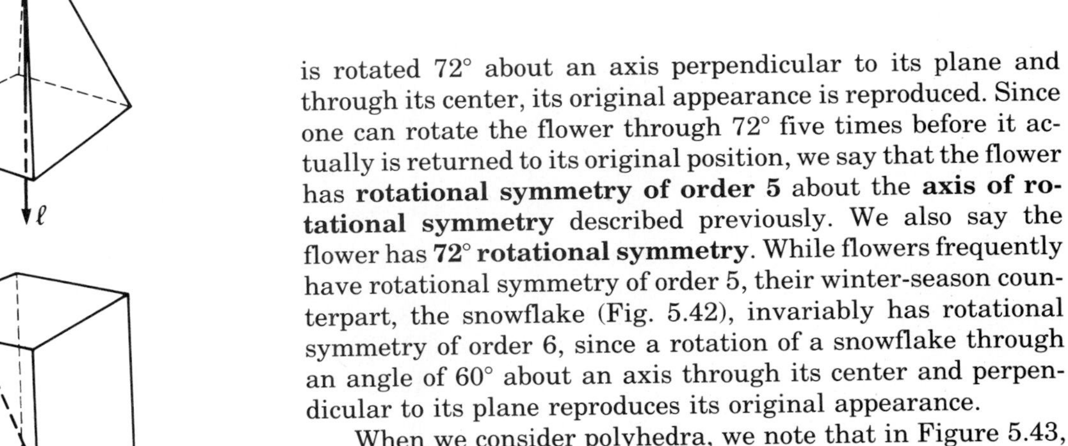

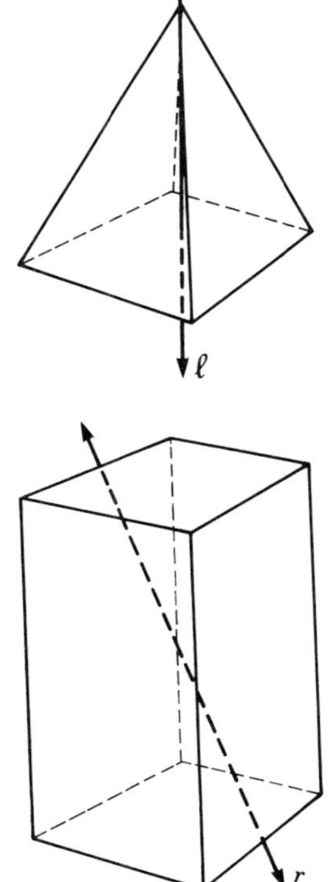

FIGURE 5.43

is rotated 72° about an axis perpendicular to its plane and through its center, its original appearance is reproduced. Since one can rotate the flower through 72° five times before it actually is returned to its original position, we say that the flower has **rotational symmetry of order 5** about the **axis of rotational symmetry** described previously. We also say the flower has **72° rotational symmetry**. While flowers frequently have rotational symmetry of order 5, their winter-season counterpart, the snowflake (Fig. 5.42), invariably has rotational symmetry of order 6, since a rotation of a snowflake through an angle of 60° about an axis through its center and perpendicular to its plane reproduces its original appearance.

When we consider polyhedra, we note that in Figure 5.43, line ℓ is an axis of rotational symmetry of the square pyramid, while line r is not an axis of rotational symmetry of the cuboid. As with reflectional symmetry, a careful definition of rotational symmetry is deferred to a later chapter; at this stage, an intuitive test is presented for deciding whether or not a polyhedron has an axis of rotational symmetry.

If one thinks of a line in question, say line ℓ in Figure 5.43, as a rod or a straw and of the pyramid as a model, the person could rotate the model (through an angle other than 360°) so that the pyramid appears to be in the same position as before the rotation. Hence, the line ℓ would qualify as an axis of rotational symmetry of the pyramid. Since the pyramid can be rotated through an angle of 90° four times before it returns to its actual original position, we conclude that the pyramid has rotational symmetry of order 4, or 90° rotational symmetry about this axis.

On the other hand, if line r is considered and one uses models, the cuboid (rectangular parallelepiped) model cannot be rotated about the rod r (through an angle other than 360°) so as to appear in the same position as before the rotation. Hence, r is not an axis of rotational symmetry of the cuboid. Investigation 5.4 provides an opportunity for you to search for axes of rotational symmetry for selected polyhedra.

INVESTIGATION 5.4

A. Use a paper net or an envelope to make a model of a tetrahedron. Cut off a vertex, cut a small slot in the center of one face, and insert a drinking straw as shown in the figure. The straw represents an axis of rotational symmetry of the tetrahedron.

1. The tetrahedron has rotational symmetry of what order about this axis?
2. Can you cut slots and use a straw to show each of the axes of rotational symmetry of the tetrahedron?

B. Use a paper net, make a model of a cube, cut slots, and use a straw to show each of the axes of rotational symmetry of the cube. How many axes of rotational symmetry does a cube have?

C. Use a paper model and drinking straws to determine the number of axes of rotational symmetry of an octahedron.

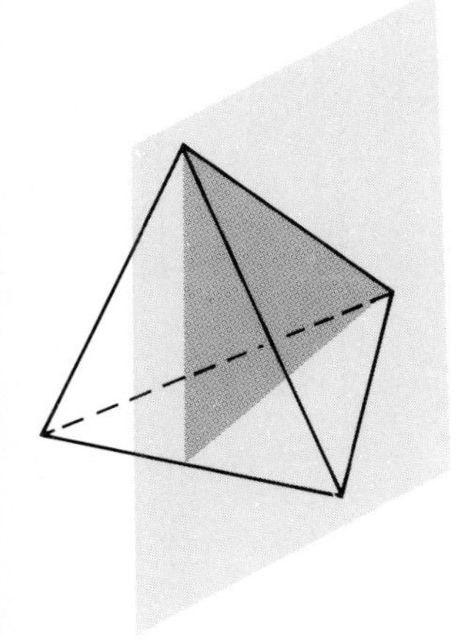

FIGURE 5.44

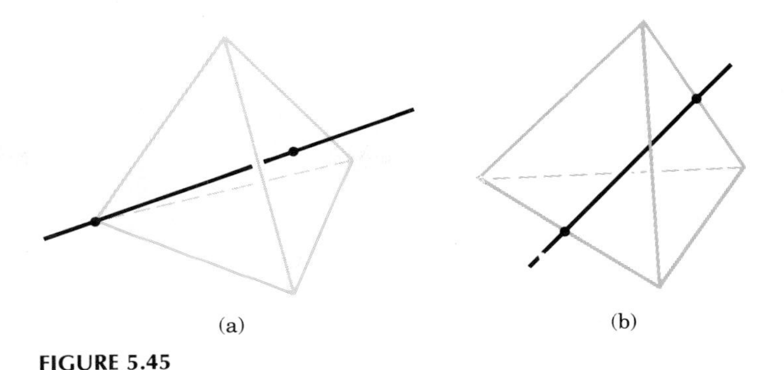

FIGURE 5.45

In Investigation 5.3, the analysis of the planes of symmetry of the tetrahedron should have established that such a plane contains an edge of a tetrahedron and the midpoint of an opposite edge. (See Fig. 5.44.) Since a tetrahedron has six edges, it has six different planes of symmetry.

In Investigation 5.4, the analysis of the axes of rotational symmetry of the tetrahedron should have revealed that there are several types of axes of symmetry. One type passes through one vertex and the center of a face opposite this vertex, as shown in Figure 5.45a. Since the regular tetrahedron has four vertices, it has four such axes. Since the face is an equilateral triangle, the order of rotational symmetry about each of these axes is three.

Another type of axis of symmetry for the tetrahedron passes through the midpoints of two opposite edges, as in Figure 5.45b. Since the tetrahedron has three different pairs of opposite edges, it has three such axes of rotational symmetry. Since a rotation of the tetrahedron through 180° about one of these axes returns it to a position that appears the same as before the rotation, the order of rotational symmetry about each of these axes is 2.

EXERCISE SET 5.3

1. Draw pictures of the following 3-dimensional figures and show one plane of symmetry.

a. Cube	b. Tetrahedron
c. Octahedron	d. Square pyramid
e. Cuboid	f. Pentagonal prism
g. Cylinder	h. Hexagonal prism

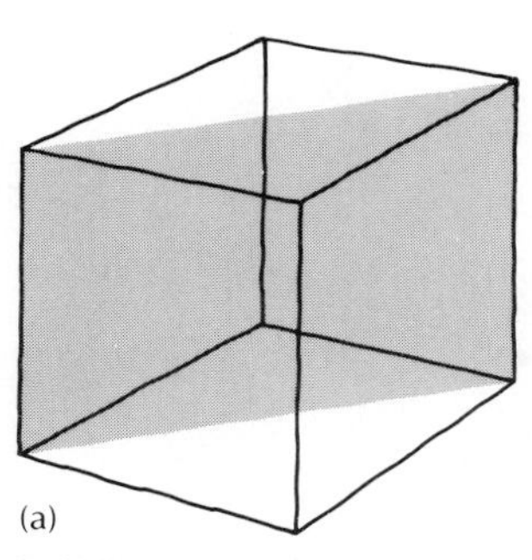
(a)

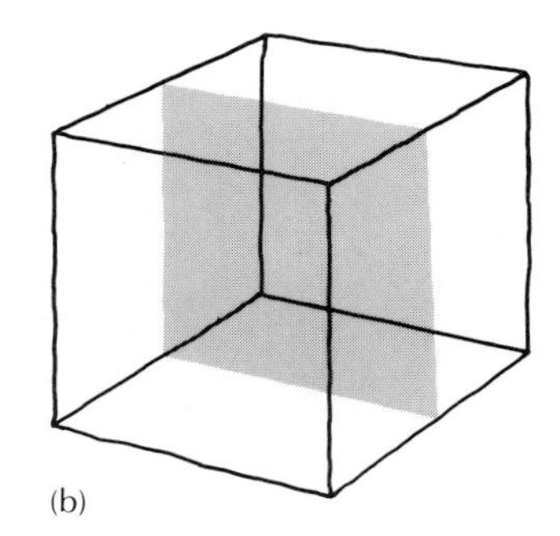
(b)

FIGURE 5.46

2. Refer to Figure 5.46 to complete the following exercises.
   a. A plane passing through opposite edges of a cube is a plane of symmetry for the cube (Fig. 5.46a). How many planes of symmetry of this type are there for a cube?
   b. A plane midway between a pair of opposite faces of a cube is a plane of symmetry for the cube (Fig. 5.46b). How many planes of symmetry of this type are there for a cube?
3. There are 3 types of axes of symmetry for a cube (Fig. 5.47).
   a. How many axes of symmetry of order 3 are there for a cube?
   b. How many axes of symmetry of order 2 are there for a cube?
   c. How many axes of symmetry of order 4 are there for a cube?
4. How many planes of symmetry does each of the following polyhedra have?
   a. A square pyramid
   b. A pentagonal prism
   c. A cuboid
   d. A cone
   e. An octagonal prism
   f. An isosceles triangular prism

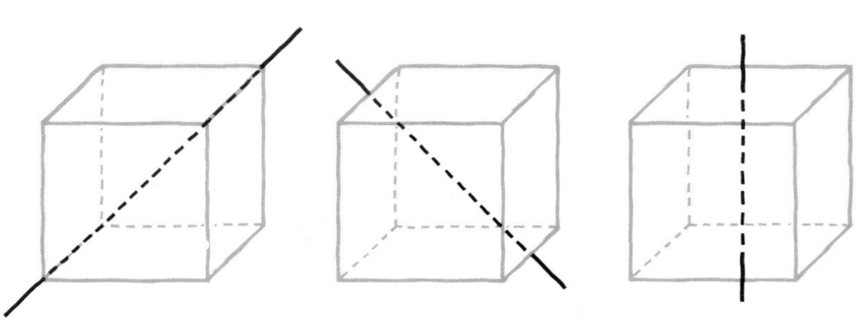

FIGURE 5.47

TABLE 5.5

Polyhedron	Number of Axes of Rotational Symmetry: Order 2	Order 3	Order 4	Order 5	Order 6	Total	Number of Planes of Symmetry
Regular tetrahedron							
Equilateral triangular prism							
Square pyramid							
Cube							
Regular octahedron							

5. Complete Table 5.5. The strategy outlined in Exercises 2 and 3 may help.
6. When an index card with a hole cut in it (Fig. 5.48a) is placed over a cube so that it "just fits" (Fig. 5.48b), a model of a plane intersecting a cube is produced. The intersection of a cube and a plane is called a *cross section* of a cube. Use index cards and a model of a cube. Cut a hole in an index card that fits on your cube to show each of the following cross sections of the cube:
   a. Square
   b. Isosceles triangle
   c. Equilateral triangle
   d. Trapezoid
   e. Parallelogram
   f. Pentagon
   g. Regular hexagon
7. How many axes of rotational symmetry does a cuboid have?
8. Which of the following objects possess neither reflectional nor rotational symmetry?
   a. Coffee cup
   b. Bowling ball
   c. Hand saw
   d. French horn
   e. Scissors
   f. Table knife

(a)

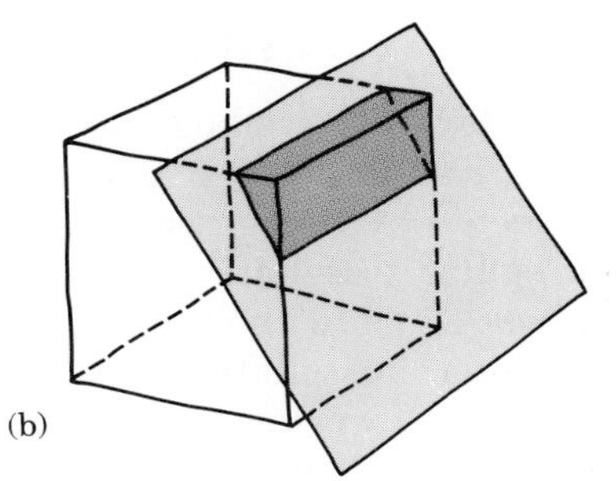
(b)

FIGURE 5.48

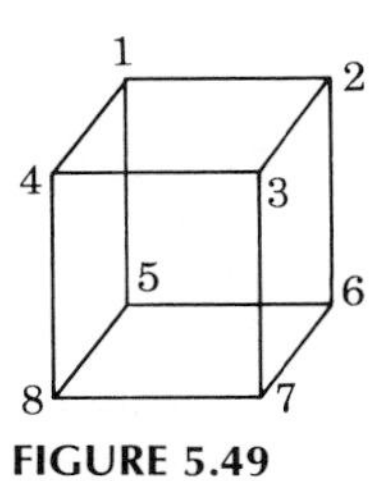

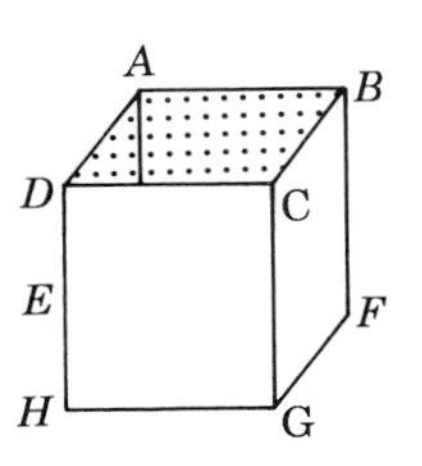

FIGURE 5.49

9. **Critical Thinking.** Proficiency at gathering and organizing data is often an important part of effective critical thinking. Use the problem posed as a data-collecting experiment. You may conduct this experiment mentally or make models.
   a. The cube in Figure 5.49 can be placed in the box in many different ways. For example, one way would match the vertices of the cube and corners of the box like this:

      2–A 6–E
      3–B 7–F
      4–C 8–G
      1–D 5–H

      Record as many ways different ways as you can for this cube to fit into the box.
   b. How have you used the axes of symmetry of the cube in your counting?

FIGURE 5.50

Problem Solving: Developing Skills and Strategies

Use the strategy **using space visualization** to determine whether the plane through $\triangle ABC$ in Figure 5.50 is a plane of symmetry of the cube. How is $\triangle ABC$ related to Exercise 3 in Exercise Set 5.2?

TEACHING NOTE

Many opportunities exist for older students to become involved with building models of semiregular polyhedra. An analysis of the Euler characteristic and of the lines and planes of symmetry of these polyhedra can often be quite challenging. *Shapes, Spaces, and Symmetry* (Holden, 1971) has an excellent description of a method for making polyhedra models.

SEMIREGULAR POLYHEDRA (Optional)

In Chapter 4 we studied both regular tessellations and semiregular tessellations. Recall that regular tessellations were constructed by using regular polygons of one type, whereas semiregular tessellations were constructed by using regular polygons of more than one type. In addition, all vertices of semiregular tessellations are surrounded in the same way. More simply, we say that all vertex figures of a semiregular tessellation are alike.

INVESTIGATION 5.5

Use materials like those described in Investigation 5.2.

A. What polyhedra can you make using the following?

1. 4 equilateral triangles and 4 regular hexagons
2. 8 equilateral triangles and 6 regular octagons
3. 8 equilateral triangles and 6 squares
4. 6 squares and 8 regular hexagons
5. 12 squares, 8 regular hexagons, and 6 regular octagons

B. Study the descriptions in this section and decide upon the names of the polyhedra you have made.

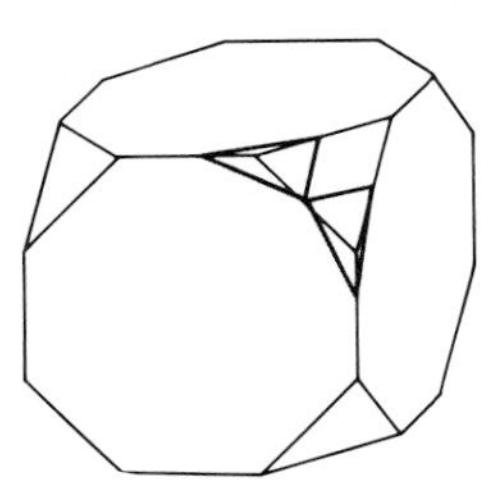

FIGURE 5.51

There are similar differences between the regular and the semiregular polyhedra. A **semiregular polyhedron** is a polyhedron constructed from regular polygons of more than one type such that all vertex figures are alike. For example, in Figure 5.51, we see a polyhedron with regular octagon and equilateral triangular faces with isosceles triangles for vertex figures. This polyhedron is one of the 14 convex semiregular polyhedra. Some of these polyhedra are considered in Investigation 5.5.

Several of the semiregular polyhedra can be obtained from the regular polyhedra through a process known as **truncating**, or "slicing off the vertices." In the case of a cube, if the plane of this slice is perpendicular to an axis of symmetry through the vertex, the new face resulting from the truncation is an equilateral triangle. If all eight vertices are truncated so that the original square faces of the cube become regular octagons, the new polyhedron is called a **truncated cube**. Figure 5.52 illustrates that the polyhedron of Figure 5.51 is a truncated cube.

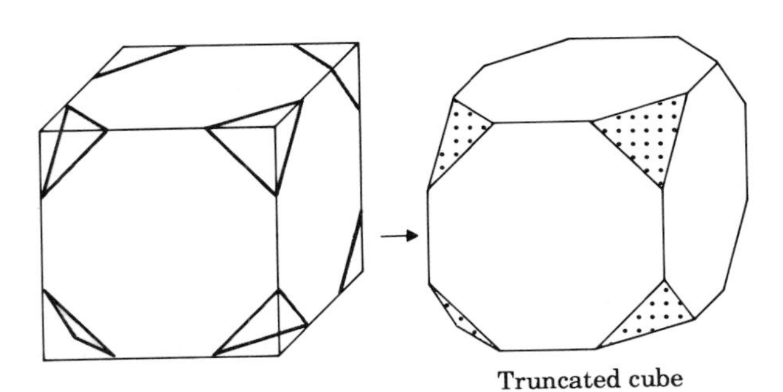

FIGURE 5.52

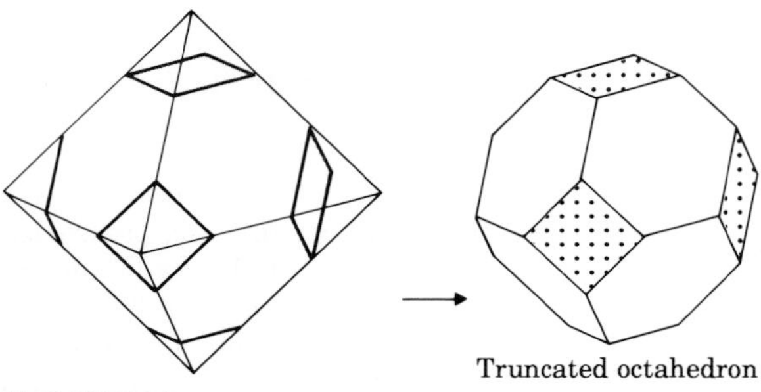

FIGURE 5.53

Similarly, as in Figure 5.53, if we truncate each vertex of the octahedron (the dual of the cube) so that the original equilateral faces become regular hexagons after the truncation, we produce a **truncated octahedron**.

A third solid, which we might think of as lying between the truncated octahedron and the truncated cube, is obtained by dividing each edge of *either the cube or the octahedron* into halves, connecting the points, as shown in Figure 5.54, and removing the corners. It is called a **cube octahedron**. We can also view this as a truncation of the vertices by planes that pass through the midpoints of the edges. So, from the cube and octahedron we obtain three semiregular polyhedra, pictured in Figure 5.55.

There are two semiregular polyhedra that we can imagine as being obtained from the cube octahedron through a truncation followed by a slight distortion. We see in Figure 5.56 that each vertex of a cube octahedron can be truncated so that the equilateral triangular faces become regular hexagons and the square faces become regular octagons. This truncation transforms the vertices into rectangles that we imagine are

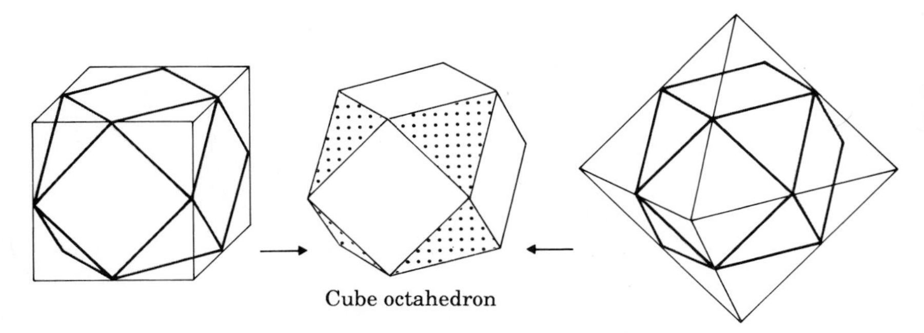

FIGURE 5.54

FIGURE 5.55

distorted into squares. This new polyhedron carries two names—the **truncated cube octahedron** or **great rhombicuboctahedron**.

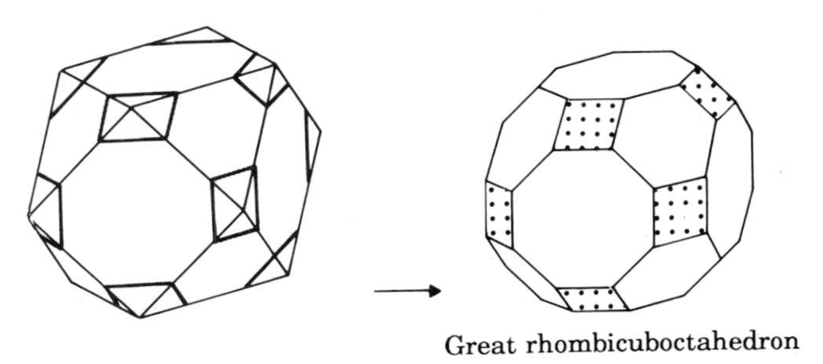

FIGURE 5.56

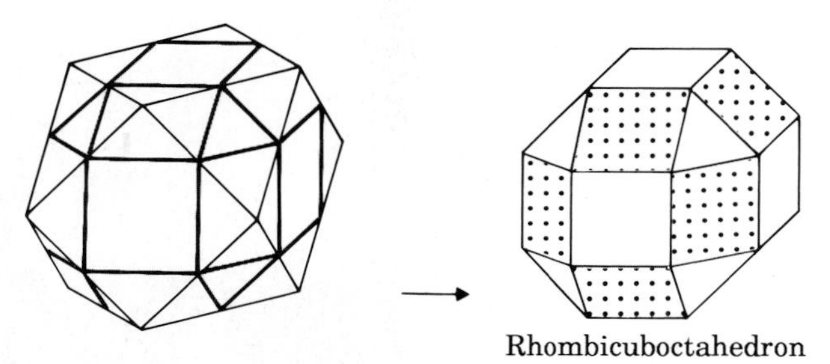

FIGURE 5.57

The second polyhedron obtained from the cube octahedron is called the **rhombicuboctahedron**. We find the midpoints of the edges of the cube octahedron and connect them as shown in Figure 5.57. Cutting off the vertices along the shaded lines yields a polyhedron with square faces, equilateral triangular faces, and rectangular faces. Again, imagine these rectangular faces are distorted into squares. The resulting polyhedron is the rhombicuboctahedron.

So we see that through the truncation process, we obtain—beginning with the cube and the octahedron—five semiregular polyhedra, as pictured in Figure 5.58 and listed in Table 5.6.

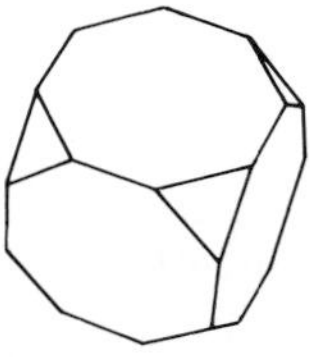

Truncated cube

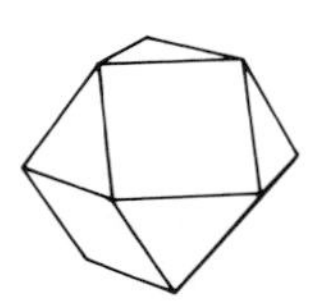

Cube octahedron

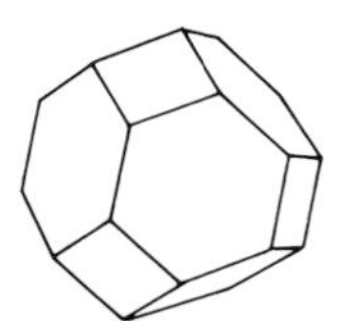

Truncated octahedron

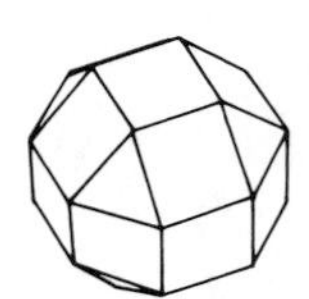

Rhombicuboctahedron

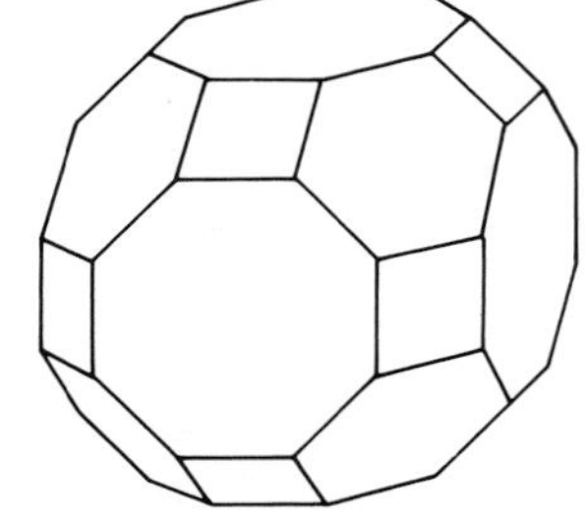

Great rhombicuboctahedron

FIGURE 5.58

TABLE 5.6

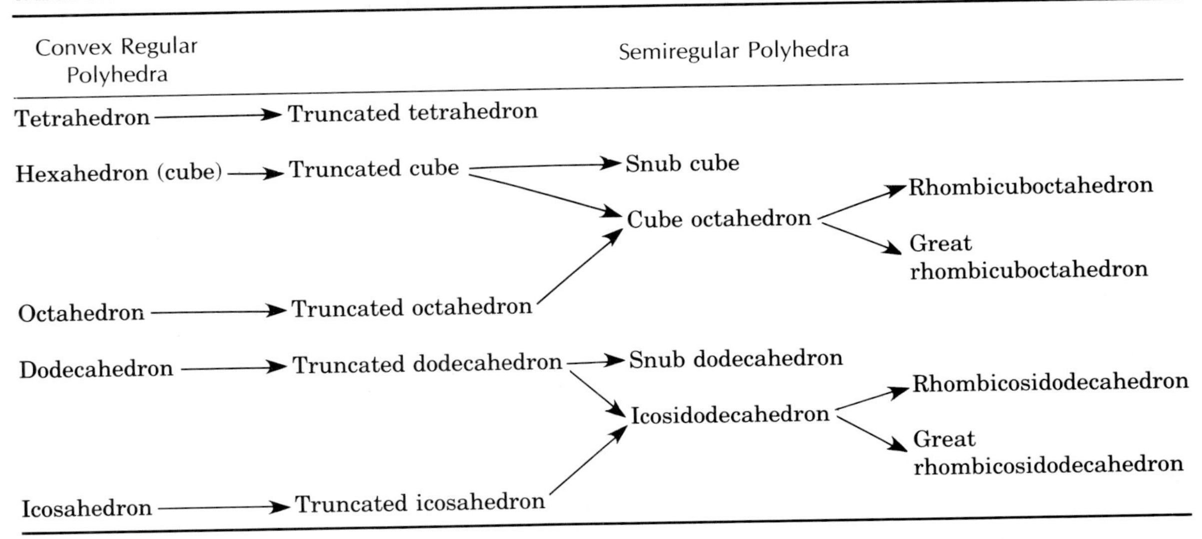

Convex Regular Polyhedra	Semiregular Polyhedra		
Tetrahedron →	Truncated tetrahedron		
Hexahedron (cube) →	Truncated cube →	Snub cube	
	(Truncated cube) →	Cube octahedron →	Rhombicuboctahedron
		(Cube octahedron) →	Great rhombicuboctahedron
Octahedron →	Truncated octahedron →	(Cube octahedron)	
Dodecahedron →	Truncated dodecahedron →	Snub dodecahedron	
	(Truncated dodecahedron) →	Icosidodecahedron →	Rhombicosidodecahedron
		(Icosidodecahedron) →	Great rhombicosidodecahedron
Icosahedron →	Truncated icosahedron →	(Icosidodecahedron)	

Truncating the icosahedron and the dodecahedron, we obtain the truncated icosahedron and truncated dodecahedron, with the icosidodecahedron being midway between these two and playing a role analogous to the cube octahedron (Fig. 5.59).

FIGURE 5.59

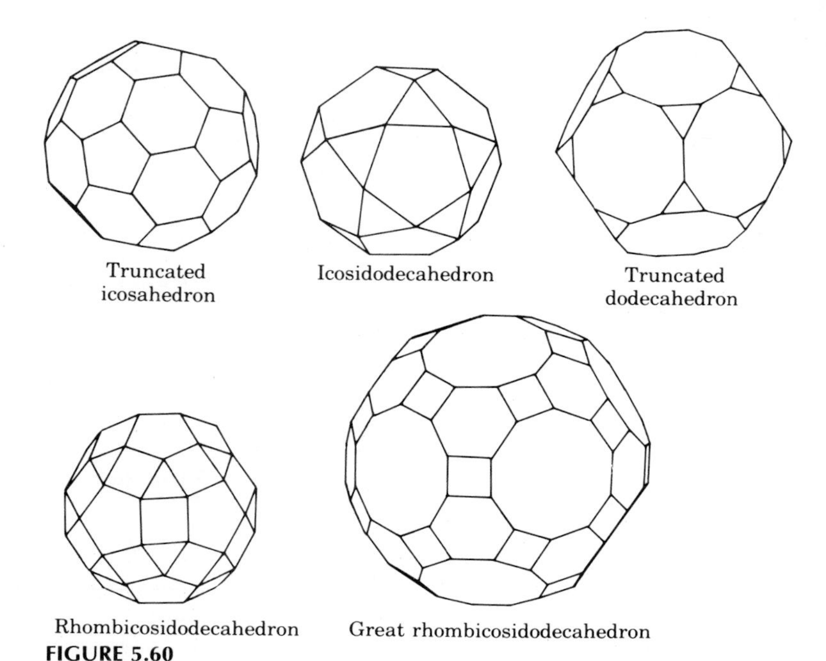

FIGURE 5.60

Truncation followed by distortion applied to the icosidodecahedron yields the **great rhombicosidodecahedron** and the **rhombicosidodecahedron**. So the dodecahedron and the icosahedron yield five more semiregular polyhedra, as pictured in Figure 5.60.

The three remaining polyhedra listed in Table 5.6 are the truncated tetrahedron, the snub cube, and the snub dodecahedron that are pictured in Figure 5.61. The **truncated tetrahedron** is formed by dividing edges of the tetrahedron into 3 parts, connecting the points, and removing the corners. The

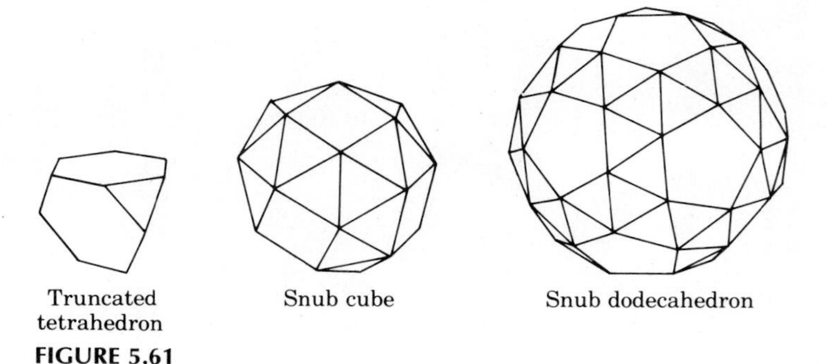

FIGURE 5.61

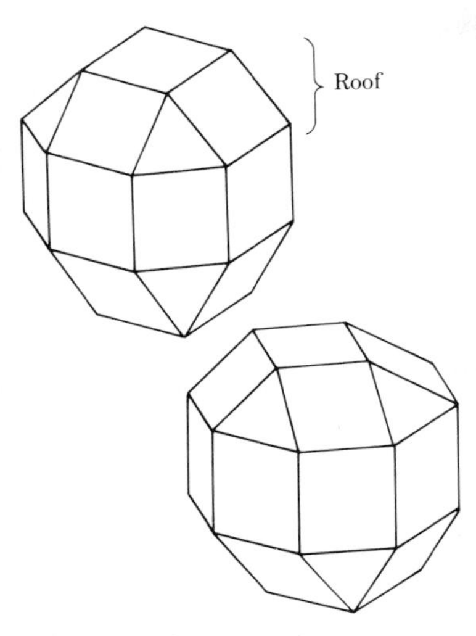

FIGURE 5.62

snub cube can be produced by drawing one diagonal in each of the 12 squares of the rhombicuboctahedron that correspond to the edges of the cube and then distorting each of the right triangles thus produced into an equilateral triangle. The diagonals are chosen so that just one diagonal passes through each of the 24 vertices. The **snub dodecahedron** can be reproduced in a manner similar to that of the snub cube by drawing appropriate diagonals on each of the 30 squares of the rhombicosidodecahedron. These 13 semiregular polyhedron were known to Archimedes and, hence, are often referred to as the Archimedean polyhedra.

Since the time of Archimedes, it has been proved that there is one and only one additional semiregular polyhedron (except for the prisms and antiprisms referred to in Exercise 6 in Exercise Set 5.4)—the so-called **pseudorhombicuboctahedron**. This polyhedra can be obtained from the rhombicuboctahedron by cutting off the "roof" (Fig. 5.62) and giving it a one-eighth turn. The chart in Table 5.7 gives some "vital statistics" for each of the 14 semiregular polyhedra. The number of each type of face and the number of vertices, edges, and faces are listed.

TABLE 5.7

Polyhedron	Triangles	Squares	Pentagons	Hexagons	Octagons	Decagons	V	E	F
Tetrahedron	4	—	—	—	—	—	4	6	4
Hexahedron	—	6	—	—	—	—	8	12	6
Octahedron	8	—	—	—	—	—	6	12	8
Dodecahedron	—	—	12	—	—	—	20	30	12
Icosahedron	20	—	—	—	—	—	12	30	20
Tr. tetrahedron	4	—	—	4	—	—	12	18	8
Tr. cube	8	—	—	—	6	—	24	36	14
Tr. octahedron	—	6	—	8	—	—			
Tr. dodecahedron	20	—	—	—	—	12			
Tr. icosahedron	—	—	12	20	—	—	60	90	32
Cube octahedron	8	6	—	—	—	—	12	24	14
Rhombicuboctahedron	8	18	—	—	—	—	24	48	26
Pseudorhombicubocta	8	18	—	—	—	—	24	48	26
Great rhombicuboctahedron	—	12	—	8	6	—	48	72	26
Snub cube	32	6	—	—	—	—	24	60	38
Icosidodecahedron	20	—	12	—	—	—	30	60	32
Rhombicosidodecahedron	20	30	12	—	—	—	60	120	62
Great rhombicosidodecahedron	—	30	—	20	—	12	120	180	62
Snub dodecahedron	80	—	12	—	—	—	60	150	92

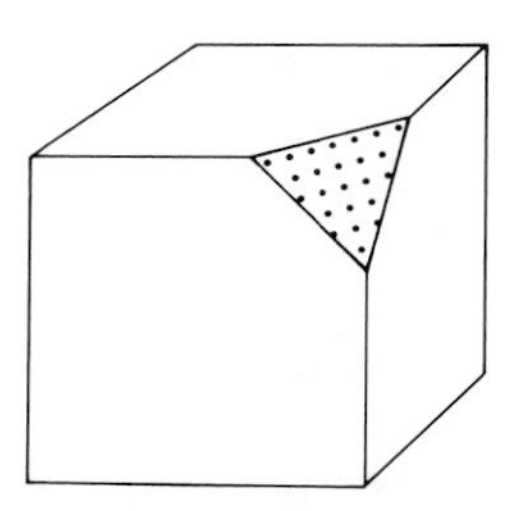

FIGURE 5.63

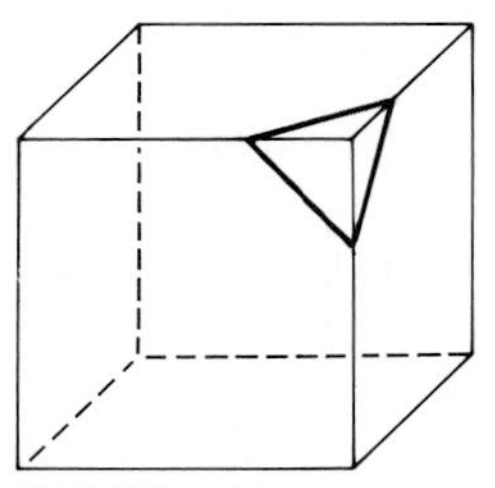

FIGURE 5.64

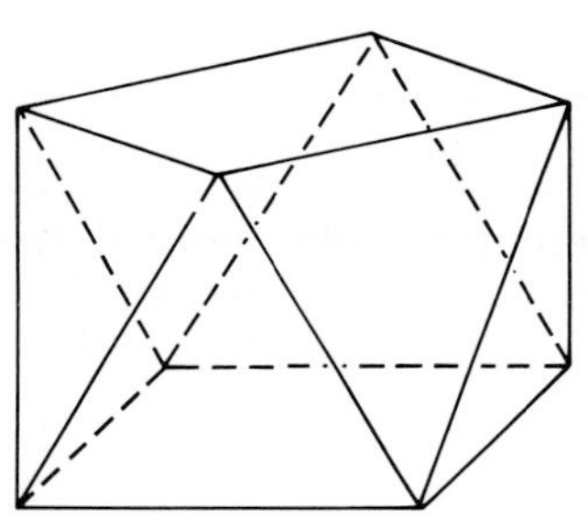

FIGURE 5.65

EXERCISE SET 5.4

1. a. Suppose one corner of a cube is truncated (Fig. 5.63). Describe the axes of rotational symmetry of this figure. Describe the planes of reflectional symmetry.
   b. Suppose 2 corners of the cube are truncated. Answer the questions in part a for this figure.
2. Does Euler's Formula (p. 123 and Exercise 6, Exercise Set 5.1) hold for semiregular polyhedra?
3. In a cube octahedron each vertex is surrounded by a triangle, square, triangle, square, in that order. Consequently, this polyhedron is often represented with the notation (3,4,3,4). Write the parenthetic notation for each of the semiregular polyhedra.
4. When a cube is truncated, each of the vertices of the cube becomes 3 vertices of the truncated cube (Fig. 5.64). Consequently, there are $8 \cdot 3 = 24$ vertices of a truncated cube. Likewise, 3 new edges are added to the 12 edges of the cube when a truncated cube is formed. Use this counting strategy to find the number of vertices, edges, and faces for the truncated octahedron and the truncated dodecahedron.
5. a. Sketch a prism with regular hexagon bases and square sides. Do the same for a prism with regular octagons. Are these polyhedra semiregular?
   b. How many prisms are there that are semiregular polyhedra?
6. An *antiprism* is a prism-type solid with triangular rather than square sides (Fig. 5.65).
   a. Construct with rubber bands and poster board an antiprism that is a semiregular polyhedron.
   b. How many different such antiprisms are there?
7. Put 8 of the solids from the puzzle in Exercise 15 in Exercise Set 5.2 together to form a truncated octahedron.

TESSELLATIONS OF SPACE (Optional)

In Investigation 5.1, the experience with wall building suggested that the cuboid could be used to "fill" space. That is, in much the same way as we considered polygonal tessellations of the plane in Chapter 4, we can consider polyhedral tessellations of space.

Obviously, the cube can also be used to tessellate space. The simple packing of boxes into larger cartons, which occurs every day, illustrates the type of packing used for these polyhedra. Should we desire a more artistic packing or should our curiosity be aroused, a resulting natural question is, *What other*

INVESTIGATION 5.6

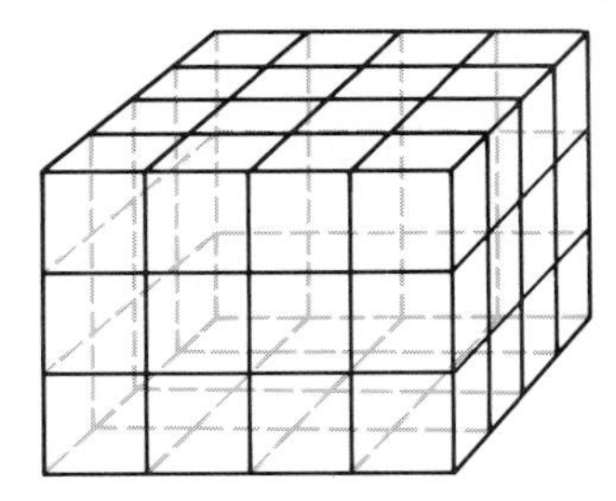

A. Since a square will tessellate the plane, a right square prism (cuboid) will tessellate space. Can you use your knowledge of plane tessellations and experiment with models to decide which other prisms will tessellate space?

B. The cube is a regular polyhedron that tessellates space. Use models of the other regular polyhedra and decide which ones will fill space. Are there any combinations of regular polyhedra that will fill space? Make careful observations and be ready to support your conclusions.

polyhedra can be used to tessellate space? If there are models available to you, completing Investigation 5.6 may help answer this question.

In deciding whether or not regular tetrahedra can be used to fill space, the experimenter often tries to fit five tetrahedra about a common edge, as shown in Figure 5.66. Upon seeing that these five tetrahedra seem to fit, the observer assumes that clusters of five polyhedra like this can be fitted together to tessellate space.

The five tetrahedra will fit around an edge only if the dihedral angle (the angle between faces) of a tetrahedron is one-fifth of 360°, or 72°. We see from the method illustrated in Figure 5.67 that the tetrahedron dihedral angle is actually

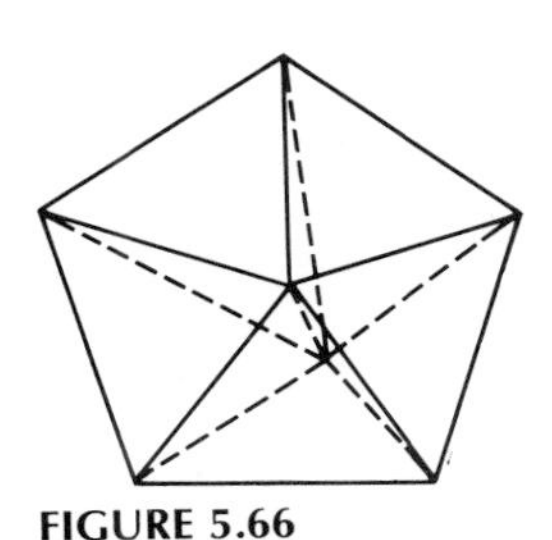

FIGURE 5.66

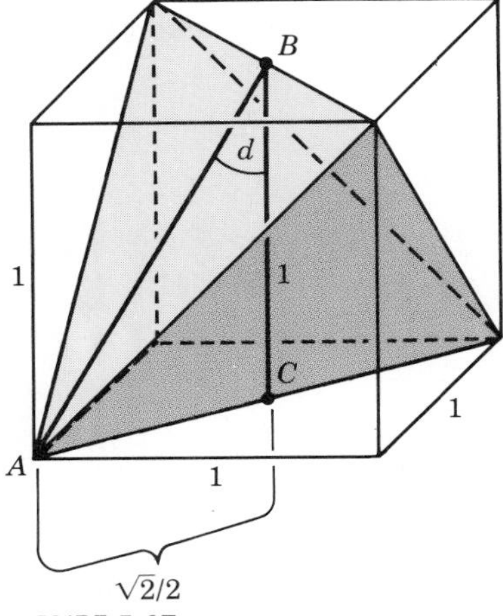

FIGURE 5.67

tan d = $\sqrt{r}/2$
tan d = .707
d = 35° 16′
$2d$ = 70° 32′

TABLE 5.8

Regular Polyhedron	Degree Measure of Dihedral Angle
Cube	90°
Tetrahedron	70° 32′
Octahedron	109° 28′
Dodecahedron	116° 34′
Icosahedron	138° 11′

70° 32′, five tetrahedra do not fit around an edge, and the tetrahedra, therefore, do not tessellate space. The measures of the dihedral angles of the remaining convex polyhedra can be found in a similar manner. From Table 5.8 we see that the cube is the only regular polyhedron that has a dihedral angle with a multiple equal to 360°, and it is the only regular polyhedron that tessellates space.

Are there combinations of regular polyhedra that tessellate space? Since 70° 32′ + 109° 28′ = 180°, we see that two tetrahedra and two octahedra completely surround an edge. We suspect that a combination of tetrahedra and octahedra tessellates space.

To illustrate that this statement is true, we consider space from two points of view. Figure 5.68 shows a way to place two tetrahedra and an octahedron together to make a rhombohedron. A **rhombohedron** is a cube that has been distorted by changing the faces from squares to rhombi and, like the cube, will fill space. Hence, combinations of octahedra and tetrahedra will tessellate space. (Note that *ABCD* is one tetrahedron in Figure 5.68, *PQRS* is the other, and *QBRDCS* is the octahedron.) However, notice that the dihedral angles of the tetrahedra and octahedra are 70° 32′ and 109° 28′ and that 70° 32′ + 109° 28′ = 180°. This suggests that two of each of these polyhedra can be arranged to completely surround an edge and perhaps tessellate space. We shall see this is true in the following discussion.

Consider the space tessellation of cubes. Imagine that each cube has four pyramids truncated like *ABCD* in Figure 5.69. Then, each cube has a regular tetrahedron left at its center that is surrounded by these four pyramids. Each one of these pyramids has seven neighbors just like it. Imagine that these

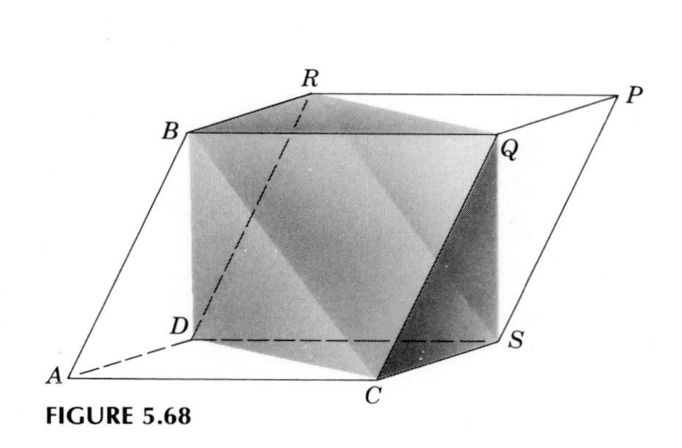

FIGURE 5.68

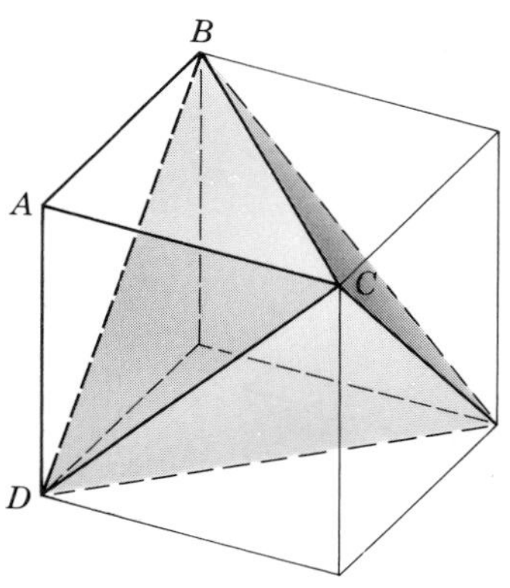

FIGURE 5.69

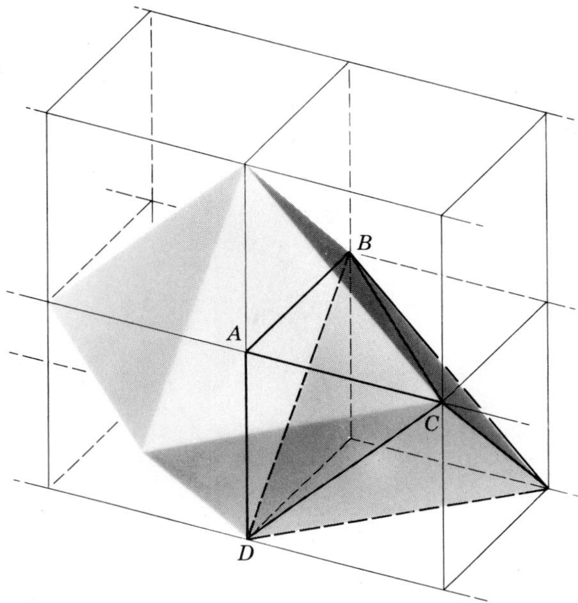

FIGURE 5.70

groups of eight pyramids are glued together. They form a regular octahedron, as shown in Figure 5.70. So we see that the cubes can be cut apart, and the pieces can be realigned to form a space tessellation of regular tetrahedra and regular octahedra.

Still other tessellations are revealed by imagining the space tessellation of cubes cut apart in still different ways. For example, as shown in Figure 5.71, a cube can be cut into a cube

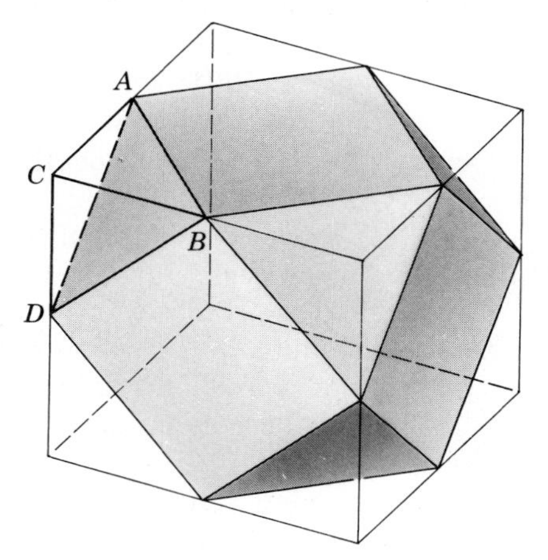

FIGURE 5.71

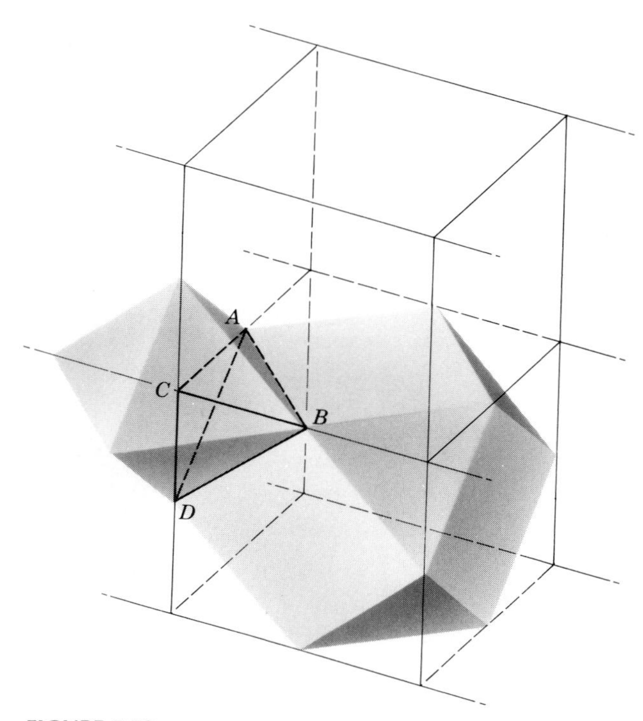

FIGURE 5.72

octahedron and eight right pyramids (like *ABCD* in Fig. 5.71). If we consider a filling of space with cubes and focus on the inscribed cube octahedra, we observe in Figure 5.72 that adjoining cubes contribute eight right pyramids (like *ABCD* in Figure 5.71) to form an octahedron surrounding each vertex of the cube tessellation. Thus, we can view the cubes in this cube tessellation as being cut apart, with the pieces reglued to form a tessellation of cube octahedra and regular octahedra, as shown in Figure 5.72.

Still another method of cutting apart the cubes in the cube tessellation yields a third tessellation by polyhedra that are semiregular. Suppose that we slice a cube with a plane that is the perpendicular bisector of the segment joining a pair of opposite vertices. This plane cuts the cube into two identical pieces, each of which has a regular hexagon face, as shown in Figure 5.73.

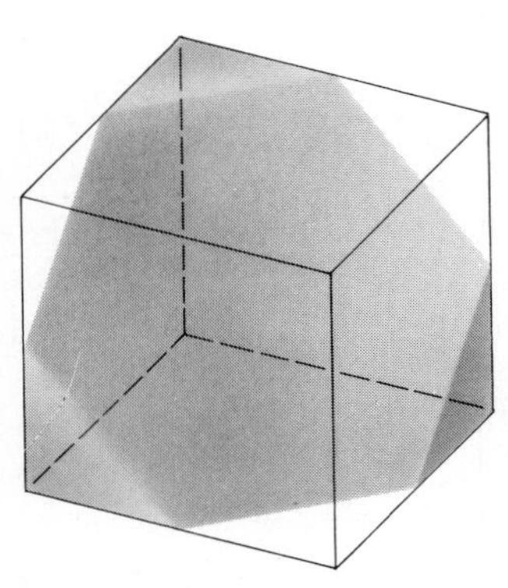

FIGURE 5.73

If we view a filling of space with cubes and focus on these halves, we observe that eight adjacent cubes contribute sufficient halves to form one truncated octahedron, as shown in

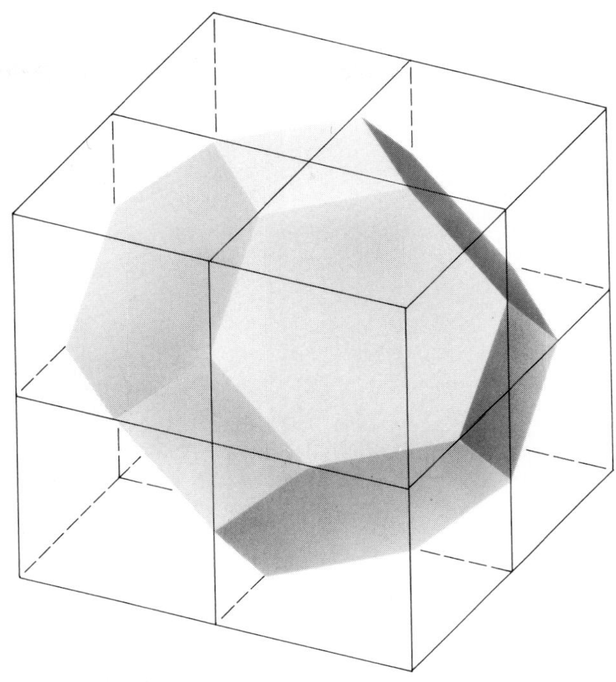

FIGURE 5.74

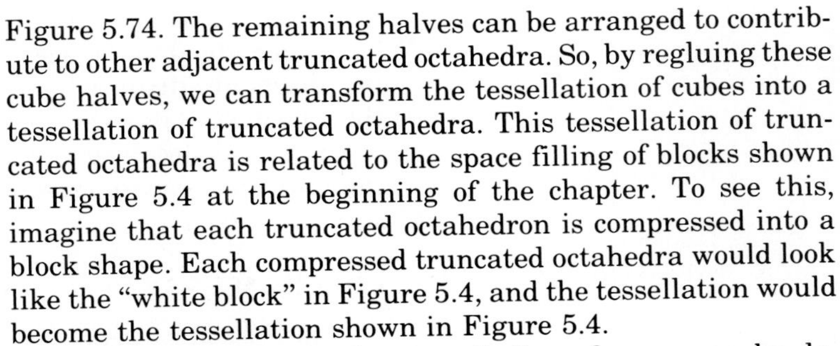

Figure 5.74. The remaining halves can be arranged to contribute to other adjacent truncated octahedra. So, by regluing these cube halves, we can transform the tessellation of cubes into a tessellation of truncated octahedra. This tessellation of truncated octahedra is related to the space filling of blocks shown in Figure 5.4 at the beginning of the chapter. To see this, imagine that each truncated octahedron is compressed into a block shape. Each compressed truncated octahedra would look like the "white block" in Figure 5.4, and the tessellation would become the tessellation shown in Figure 5.4.

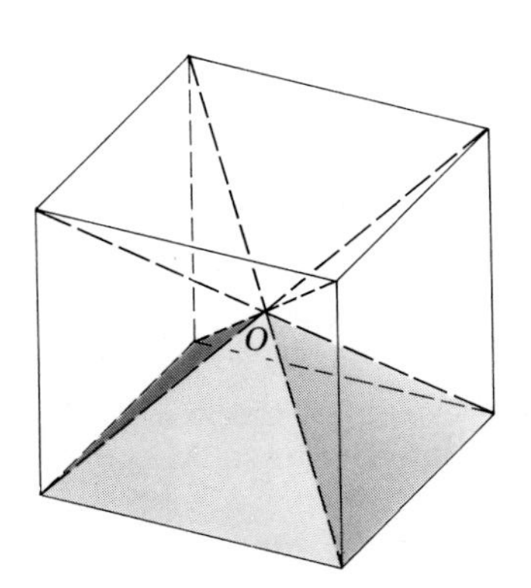

FIGURE 5.75

Still another interesting tessellation of space can be described again considering the basic tessellation of cubes. Note that each cube can be viewed as six square-based pyramids, with bases that are the faces of the cube and with vertices meeting at the center O of the cube, as in Figure 5.75. Imagine that space has been filled with cubes and that they are alternately painted white and black like a three-dimensional checkerboard. Now imagine that all of the white cubes have been cut into six pyramids, as shown in Figure 5.75, and that the

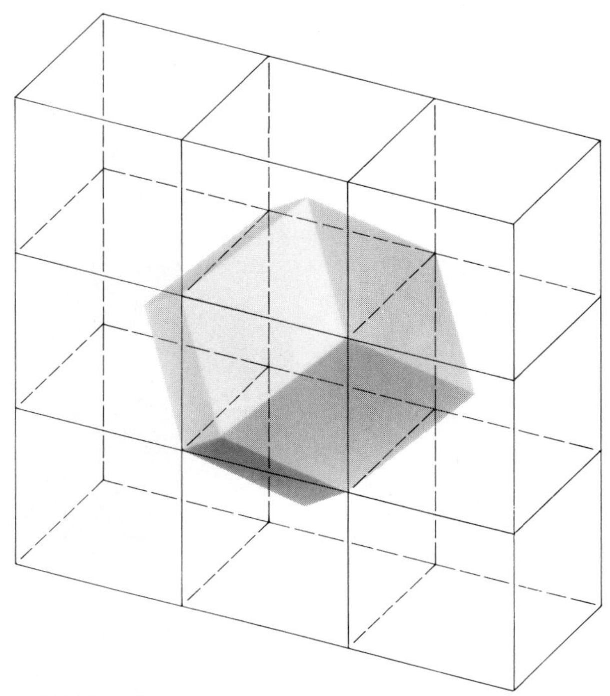

FIGURE 5.76

square face of each pyramid is glued to the neighboring black cube. Now each black cube, together with the six pyramids that have been glued to its faces, forms a polyhedra called a **rhombic dodecahedron**, which is pictured in Figure 5.76. Thus, the cube tessellation has been transformed into a tessellation of rhombic dodecahedra (Fig. 5.77). Clearly, as in the plane, the universe of space tessellation is rich and varied.

EXERCISE SET 5.5

1. Since 4 four-sided polygons meet at each edge, the cube tessellation is denoted by 4^4. Since 4 equilateral triangles meet at each edge of the tessellation of octahedra and tetrahedra, it is denoted by 3^4. Write the symbol for the tessellations described in Figure 5.72.
2. Use the information in Table 5.9 to verify that the sum of the dihedral angles around each edge in the five tessellations described in Figure 5.72 is 360°.
3. Use any standard reference on polyhedra to find the dual of the rhombic dodecahedron.

FIGURE 5.77

TABLE 5.9

Polyhedron	Dihedral Angles
Tetrahedron	70° 32°
Octahedron	109° 28′
Cube	90°
Cube octahedron	125° 16′
Truncated octahedron	109° 28′, 125° 16′
Truncated tetrahedron	70° 32′, 109° 28′

PEDAGOGY FOR TEACHERS

ACTIVITIES

1. Analyze a basic textbook series for the elementary school to find the extent to which the program suggests experiences for the students with space figures. Make a list of the grade level and the page on which this introduction occurs. Suggest specific places where additional work with space figures might be included at a given level of your choice.

2. Suppose you were making out a requisition requesting the purchase of solid models for your classroom. Refer to current catalogs from mathematics laboratory materials suppliers and list the items you would purchase and what they would cost. Then suppose you have less than $50 to spend. Which items would you buy?

3. Procure models of the five regular polyhedra and talk with primary school children about these solids. Are they interested? Can they count the number of faces? The number of edges? The number of vertices? Can they think of objects in their world that remind them of the space figures? Write a summary of the children's reaction to the polyhedra.

4. Look at the book *Aha! Insight* (Gardner, 1978), and find at least two more puzzles that involve solid figures and are different from those in this chapter. Include these puzzles in your card file of geometric recreations.

5. A durable, attractively colored set of regular and Archimedean polyhedra are valuable materials to have in a classroom at any level. Refer to pages 188–192 in *Build Your Own Polyhedra* (Hilton and Pederson, 1988) for a method of making such models and make a set for your classroom. A carefully made set of models, even with extensive student handling, will last several years.

6. In order to encourage children to become actively involved in the exploration of a geometrical idea, an investigation card is often used. An example of such a card is shown in Figure 5.78.

a. Use the following criteria to evaluate the investigation card (Fig. 5.78).
   - The objective(s) for the investigation card is clearly specified and appropriate.
   - The card has the potential to involve the student in an investigation of the ideas specified in the objectives.
   - The questions on the card are simple and clear and motivate the student to become involved.
   - The card utilizes physical materials when appropriate and other materials that effectively aid in giving meaning to the ideas.
   - At least some part of the card has potential to create a situation in which a student is involved in sustained activity centered about the important idea.
   - The card allows possibilities for student choices both in investigation procedures and in modes of recording findings.

On a 5 by 5 geoboard, how many triangles with different shapes can you make that have no nails inside?

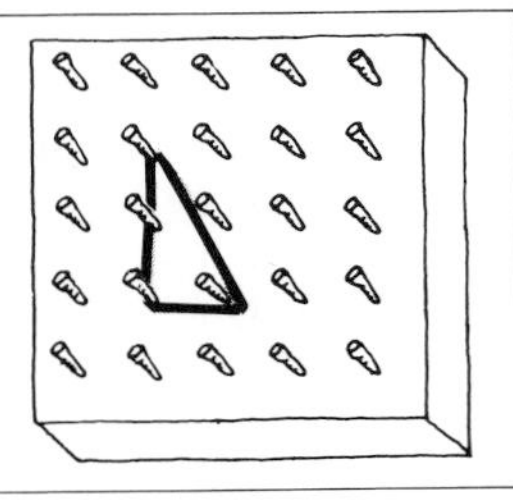

FIGURE 5.78

b. Create an investigation card of your own that would involve students at a given grade level in the exploration of a basic geometric idea about solid figures. Try your card with students and try to improve its effectiveness.

COOPERATIVE GROUP LEARNING DISCUSSION

Some curriculum developers believe that young children should first consider solid figures and that concepts of points, lines, and plane figures should grow out of consideration of the faces, edges, and vertices of the solid figures. Others believe that beginning experiences should be in simple two-dimensional settings involving squares, triangles, rectangles, circles, etc., and that three-dimensional ideas should evolve later. Discuss this in your group and develop a group position on this issue, along with reasons for your decision, to present to the class.

CHAPTER REFERENCES

Gardner, Martin. 1978. *Aha! Insight*. New York/San Francisco: Scientific American, Inc., W. H. Freeman and Co.

Hilton, Peter, and Pederson, Jean. 1988. *Build Your Own Polyhedra*. Menlo Park, CA: Addison-Wesley Publishing Co.

Holden, Alan. 1971. *Shapes, Space, and Symmetry*. New York: Columbia University Press.

Loeb, Arthur L. 1976. *Space Structures, Their Harmony and Counterpoint*. Reading, MA: Addison-Wesley Publishing Co.

Weyl, Hermann. 1952. *Symmetry*. Princeton, NJ: Princeton University Press.

Wenninger, Marcus. 1970. *Polyhedron Models*. New York: Cambridge University Press.

SUGGESTED READINGS

Consult the following sources listed in the Bibliography for additional readings: Carroll (1988), Izard (1990), Lindquist and Shulte (1987), Reys (1989), Thiessen and Matthias (1989), and Winter (1986).

CHAPTER SIX

Measurement—Length, Area, and Volume

INTRODUCTION

From the time of the baked clay tablets of Mesopotamia (dated prior to 3000 B.C.) to the present day, we find records of the importance of measurement. The Babylonians and Egyptians and their contemporaries appear to have viewed all of geometry as related to the practical process of measurement. In fact, as mentioned in Chapter 1, the word *geometry* (meaning "measurement of the earth") probably originated in these cultures. The importance of measurement in cultures continues to be emphasized in present-day writings (NCTM, 1976; Steen, 1990).

The measurement process is a physical process. We go through five steps whenever we measure:

1. Choose an object or event and select a measurable property *p*, (such as length, mass, time, capacity,) to be measured.
2. Select an appropriate unit, depending upon the precision of measurement required.
3. Use as many units as necessary, in an orderly way, to "reach the end," "cover," "fill up," or otherwise assess the property *p*.
4. Ascertain the number of units used.
5. Assert the number of units used to be the measure of *p*.

Steps 3 and 4, by their very nature, require the measurer to make approximations. Consequently, a measurement is always an approximation.

This chapter focuses on the measurable properties of length, area, and volume. But, there are many measurable properties. For example, if in step 1 you would select your own body as the object to be measured, you could choose any of these measurable properties: height, weight, shoe size, waist size, sleeve length, or ring size. In Investigation 6.1, you will gain some experience with the five steps of the general measurement process.

INVESTIGATION 6.1

Work in groups on at least one of the following questions. Be creative!

A. In how many different ways can you measure a ball?
B. Which measurable properties of the human body can you measure with the following:
   1. A large piece of graph paper
   2. A piece of string
   3. Two containers, some water
   4. A stop watch
   5. A balloon

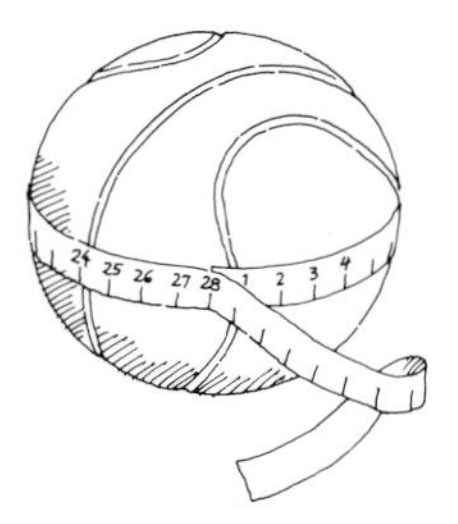

FIGURE 6.1

As we further consider the five steps of the measurement process, we notice the following key elements in the process:

— Choosing measurable properties
— Selecting appropriate units
— Developing and using measuring devices
— Associating a number with the measurable properties

For example, as illustrated in Figure 6.1, in Investigation 6.1, a person may have chosen to measure the distance around a ball (the choice of the measurable property) with a centimeter tape measure (the choice of a unit and a measuring device). As the tape is wrapped around the ball (the use of the measuring device), the tape meets at 28 and we say the circumference of the ball is 28 centimeters (the association of a number with the measurable property). Thus, the measurement process is essentially a physical means of comparing the measurable property with a unit of measure.

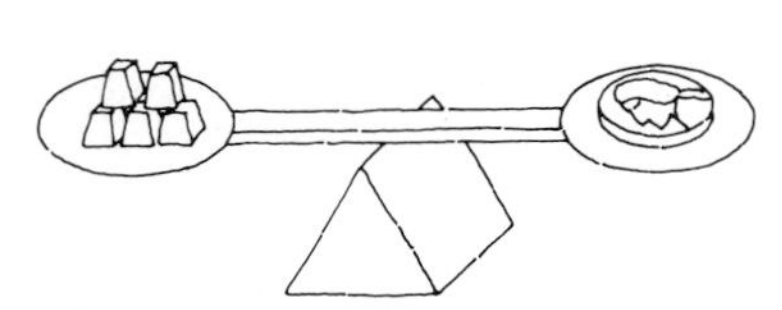

FIGURE 6.2

A device is often used to aid in this comparison, and a number is used to describe this comparison. These basic ideas are further illustrated in Figures 6.2 and 6.3. In Figure 6.2, the nickel balances 5 standard gram units, and the mass of the nickel is 5 grams. In Figure 6.3, the paper clip is as long as 3 centimeter units, and the length of the paper clip is 3 centimeters.

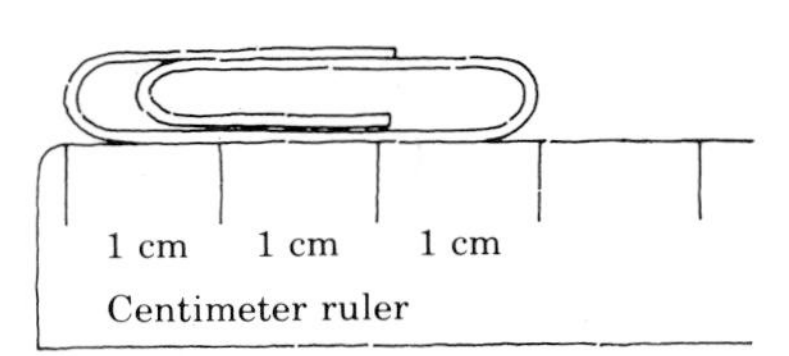

FIGURE 6.3

The reader should focus on each aspect of the measurement process in these illustrations. The following exercises provide an opportunity for further exploration of the key elements of the measurement process.

EXERCISE SET 6.1

1. Name 5 objects in your classroom that each have a different measurable property. For each object, name a property that can be measured.
2. If you wanted to measure the property called "the amount of bounce" for a basketball, what might you use as a measuring device?
3. A commercial baker has a large quantity of flour. What units might he use to measure the "amount of flour" in his warehouse?
4. Name a substance whose quantity might be measured with any of the units quart, liter, or ounces.
5. If a length measurement is declared to be about 3.5 feet or about 40 1/2 inches, which measurement has been made with the greater precision? Explain your answer.
6. Look up in a dictionary the words "precision" and "accuracy." Do they mean the same thing? Based on the information given in Exercise 5, can we know which measurement in Exercise 5 is the most accurate? Why?
7. When you are "reading a ruler" or "reading the dial on a speedometer," which of the 5 steps of the measurement process are you performing?
8. Which of the 5 steps of the measurement process are you performing when you declare "I am 66 inches tall?"
9. When you stand a meter stick on its end up against a table in order to measure the height of the table, which of the 5 steps of the measurement process are you performing?

In Exercises 10 through 13, select all statements that correctly describe the situation. There may be more than one correct choice.

10. Three persons used the same measuring device and measured the same object. Their measurements were 23.56, 23.62, and 23.59.
    a. The measurements differ in accuracy.
    b. The measurements differ in precision.
    c. The precision of the measurements is identical.
    d. One of these measurements might be more accurate than the other two.
11. Measurement is
    a. Exact if precise measuring devices are used
    b. Always exact
    c. Sometimes exact and sometimes approximate
    d. Always approximate
12. Which of these statements are true?
    a. Precision is the same as accuracy.
    b. Accuracy is improved by measuring very carefully.
    c. Precision is improved by using more refined measuring devices.

d. A measurement of 5.1 units could be more accurate than a measurement of 5.008 units.

13. Which of these statements are true?

    a. A centimeter is a frequently used standard unit of length.
    b. A nonstandard unit of length, such as a "thumb width," can never be used in the measurement process.
    c. If two different units are used when measuring an object, the measurement will be different in the two cases.

14. a. List as many different measurement devices as you can without consulting a reference. Describe the measurable property that is measured by each of these devices.
    b. List as many commonly used units of measure as you can without consulting a reference.

15. Describe as many measurable properties as you can think of for which a "tire pump" could be used as a measuring device.

16. **Critical Thinking.** How could you use a stopwatch with the second as the unit to measure each of the following?

    a. Speed of a runner
    b. Distance between 2 geographical points
    c. Temperature of a hot plate
    d. Length of a swinging pendulum
    e. Force of gravity near the earth's surface

Problem Solving: Developing Skills and Strategies

An important problem-solving skill is **formulating a problem**. Consider the following situation and formulate a problem about using the situation to measure some measurable property of the human body. Give a complete solution to the problem you have formulated.

Situation: A jug is partially filled with water, turned upside down in another container of water, and a rubber tube is inserted (Fig. 6.4). If a person blows on the tube, the water level in the jug will recede.

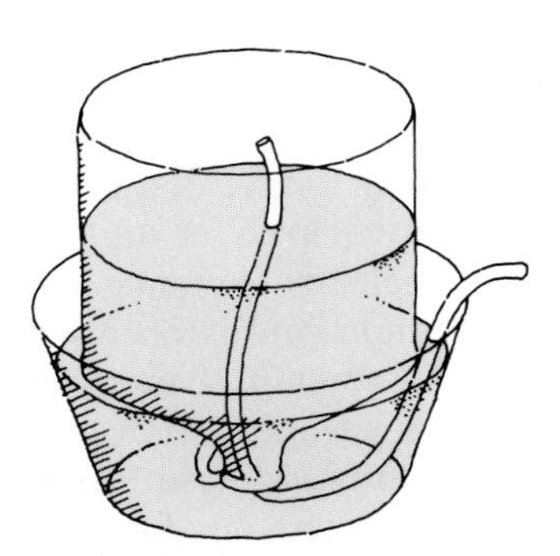

FIGURE 6.4

INVESTIGATION 6.2

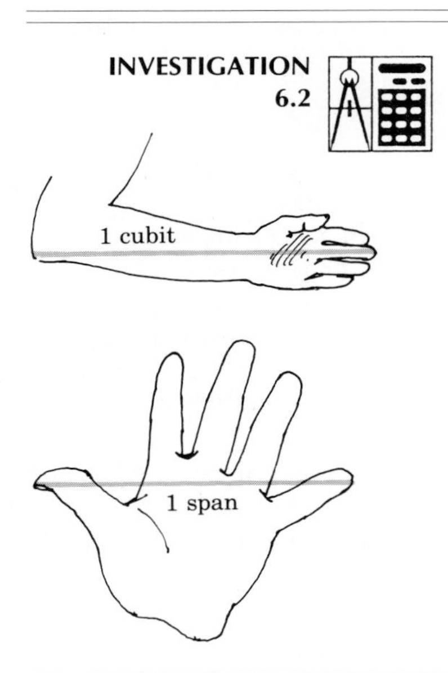

"And there went out a champion out of the camp of the Philistines, named Goliath of Gath, whose height was six cubits and a span." (I *Samuel* 17:4)

A. Using your cubit and span, cut a string that is as long as Goliath was tall. Compare your string length with the strings of others in your group. Discuss any difficulties that might arise from choosing and using units determined by each person's own body.

B. Work in groups and invent an original unit for measuring length, based on part of the body, an object in the environment, and so forth. Invent some smaller and larger related units and make a ruler using your unit. Measure some things.

   1. Communicate with another group about your unit and compare it *as accurately as possible* with their unit. If you measure with your unit and they measure with their unit, would you have any difficulty communicating about your measurements? Try it.
   2. Discuss any difficulties that might arise if each nation had a local unit of measure different from others.

NEED FOR STANDARD UNITS IN MEASURING

A basic step in the measurement process is the choice of a unit of measure. Investigation 6.2 may help the reader capture the spirit of this importance and help establish criteria for choosing units. Investigation 6.2 suggests that we have complete freedom in choosing units of measure, but that indiscriminate use of this freedom can often make accurate communication of the results of our measurements very difficult. For example, the units pictured in Figure 6.5 were chosen freely from common natural phenomena that varied considerably from situation to situation. Until some of them were defined more accurately, they created historical misunderstandings among both people and nations that had far-reaching implications.

As commerce developed and trade between nations became possible over longer distances, sociological and economic issues motivated a search for highly standardized, uniformly accepted units of measure. Scientists who needed to communicate ideas without undo confusion were among the first groups to adopt a common system of measurement units. Manufacturers and other tradespeople followed suit in many countries in order to eliminate misunderstandings and to provide for easily interchangeable parts on the products traded.

TEACHING NOTE

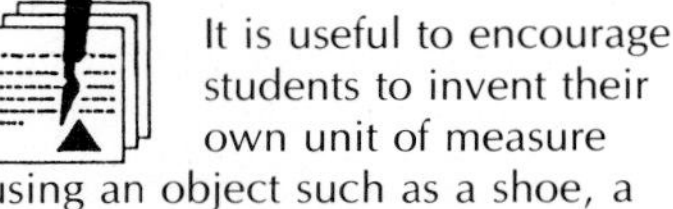

It is useful to encourage students to invent their own unit of measure using an object such as a shoe, a pencil, or a part of their body. If a group of students use the width of the palm of their hand as a unit and find they get different numbers for the measure of the same thing, the need for the standard unit becomes vivid.

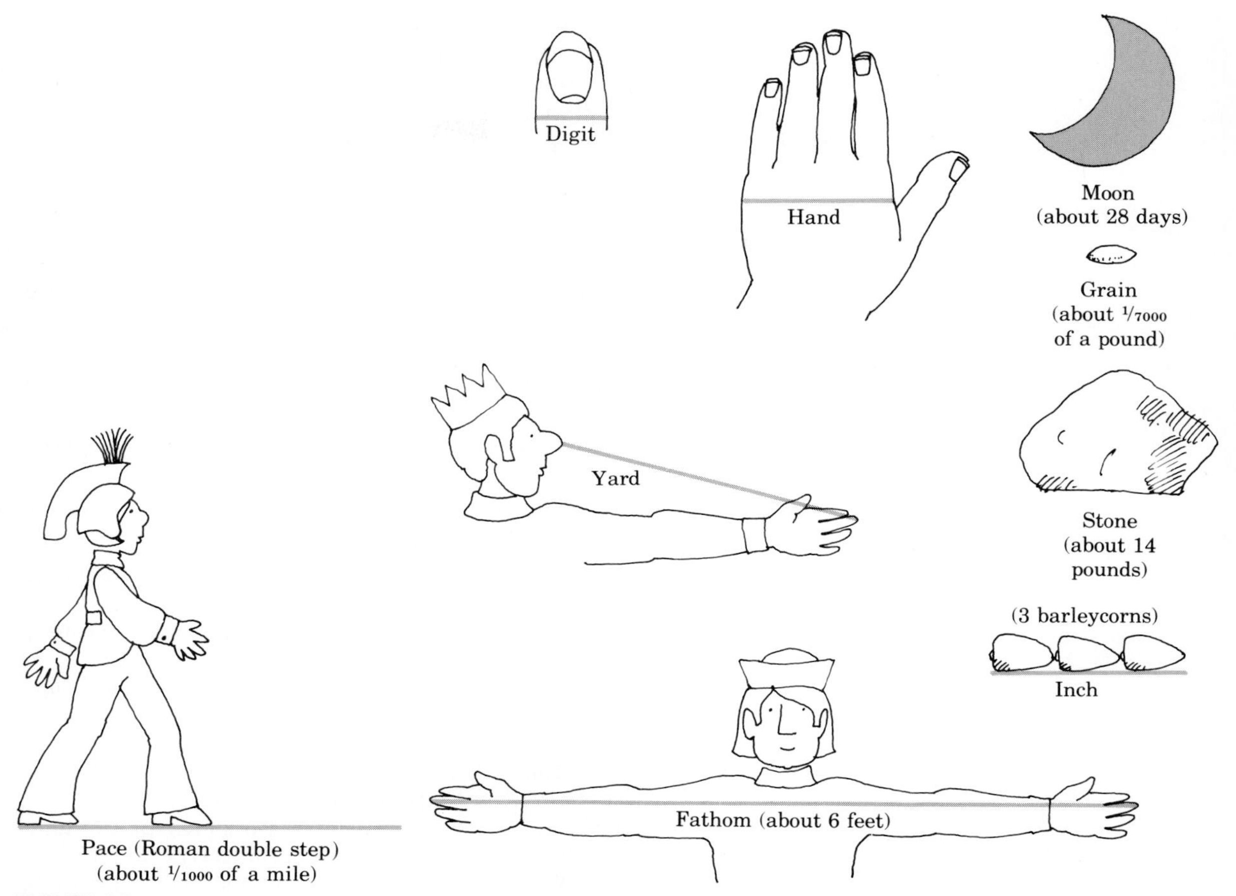

FIGURE 6.5

METRIC SYSTEM OF MEASUREMENT UNITS

In a recent year it was recorded that almost all the nations in the world use a standard system of measurement called the **International Metric System**, or *The Système International d'Unités* (SI System). Some basic units for this system are the **meter** (m) for measuring length, the **liter** (l) for measuring capacity, and the **gram** (g) for measuring mass. These basic units and a few prefixes [such as milli- (1/1000), centi- (1/100), deci- (1/10), and kilo- (1000)] are used to name smaller and larger units. These units are explained more fully in the sections that follow.

A few countries, such as the United States, also use the **Customary**, or **English**, **System** of units. Some basic units in this system are the **inch** (in), **foot** (ft), **yard** (yd), and **mile** (mi) for measuring length; the **cup**, **quart** (qt), and **gallon** (gal) for measuring capacity; and the **ounce** (oz), **pound** (lb), and **ton** for measuring mass.

EXERCISE SET 6.2

1. Explain why there is a need for a worldwide system of standard units.
2. Use a reference book to find how a standard unit, such as the meter, is accurately defined so that it will be the same around the world.
3. Explain why prefixes are useful in simplifying a system of metric units.
4. In each of the following, study the example of a basic unit in the metric system, and estimate the measurements described.
   a. Ten straight pins have a mass of about 1 g (Fig. 6.6). Estimate the mass of a pencil in grams.
   b. A *meter* is about as long as a long pace (Fig. 6.7). Estimate the length of a city block in meters.
   c. A *second* (s) is about the length of time it takes to say "one thousand one." Estimate the time it takes to write your name in seconds.
   d. A *liter* is about 1.1 quart. Estimate the number of liters it would take to fill a 10-gallon gasoline tank.
   e. Degree Celsius (°C) is a commonly used metric unit of temperature that is derived from the basic Kelvin unit of temperature. Water freezes at 0°C and boils at 100°C. Estimate the degrees Celsius of normal body temperature.
5. **Critical Thinking**. Complete the following generalizations. How would you convince someone that they are true?
   a. The smaller a measurement unit, the ______ the measure.
   b. The larger a measurement unit, the ______ the measure.
   c. No measure is exact, but if you want to measure more accurately, use a ______ unit.

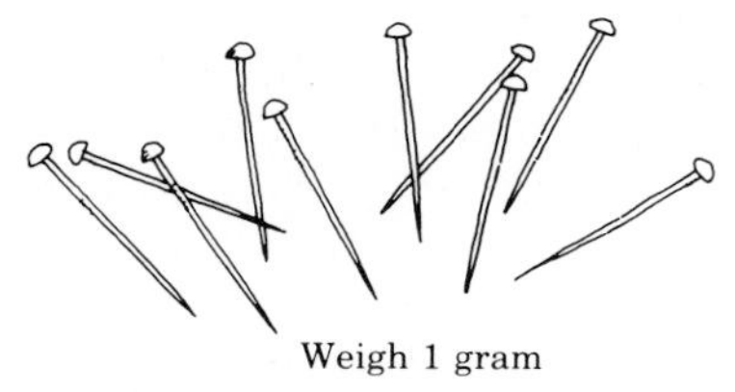

FIGURE 6.6

To extend these ideas, use Activity 1 in Supplementary Exercises: Length, Appendix C, Metric Units.

Problem Solving: Developing Skills and Strategies

Sometimes a problem asks you to become involved in a creative activity. Here is such a problem for you to try.

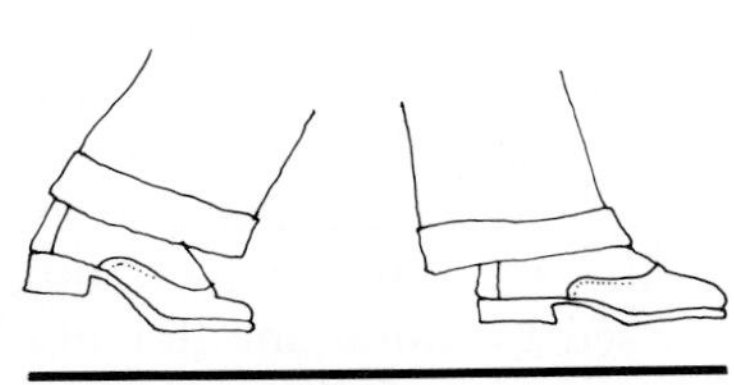

FIGURE 6.7

Problem: How could you use body lengths as units and create a system of measure? Describe your choice of units and how you could make a

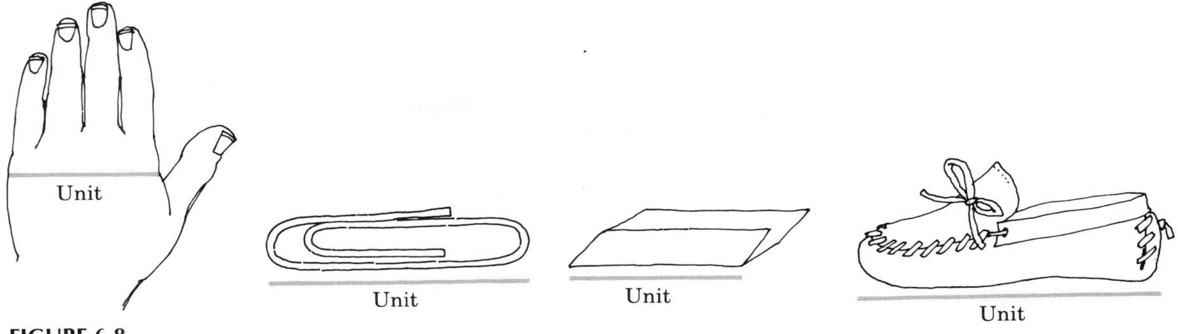

FIGURE 6.8

ruler or measuring device using your units. Discuss the advantages and disadvantages of the system of measure you have created.

TEACHING NOTE

Most children before the age of six or seven are not ready to measure because they are still operating at the topological level and are not aware that a stick or ruler remains the same length regardless of a change in position. For example, if two sticks the same length are moved as shown in Figure 1, the child will assert that stick *B* is the longer. If they are moved as in Figure 2, the child will assert that stick *A* is the longer. This happens because the child looks only at the endpoint, which is now farther away. Children do not consider the rigidity of the stick and the movement of the other endpoint. Piaget has observed that the necessary concepts for length measurement are achieved on the average around eight years old.

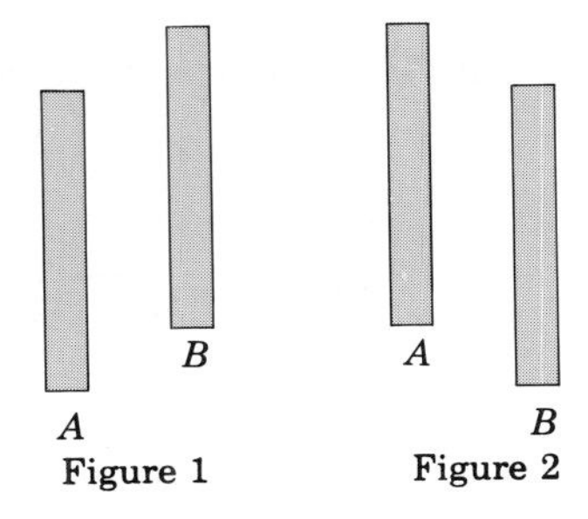

Figure 1 Figure 2

MEASURING LENGTH

The physical process of measurement described at the beginning of this chapter suggested that we first choose a measurable property of an object. When this property can be loosely described as "extension in one direction," we are meaning **length**. To measure length, we choose a unit that has this property and compare it with the object to be measured. This is usually done by placing copies of the unit end to end along the object until we reach or surpass the end and counting the number of units used. The unit of length is arbitrarily chosen.

To help children understand this idea, we encourage them to use unorthodox units, such as those shown in Figure 6.8, in their initial experiences with the measurement process. From these experiences children recognize the need for a standard, accepted unit that is the same in all situations.

The basic standard unit of length used in the metric system is the **meter**. Table 6.1 shows this unit and the subunits gen-

TABLE 6.1 Table of Metric Units for Measuring Length

Unit	Relation to Basic Unit
*kilo*meter (km)	1000 meters
*hecto*meter (hm)	100 meters
*deka*meter (dam)	10 meters
meter (m)	**Basic unit**
*deci*meter (dm)	.1 meter
*centi*meter (cm)	.01 meter
*milli*meter (mm)	.001 meter

This distance on a person might be close to 1 meter.

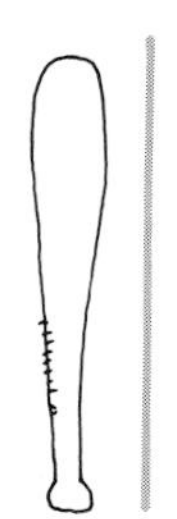

A baseball bat might be just a little shorter than a meter.

FIGURE 6.9

erated by using some of the standard metric prefixes. Figure 6.9 compares the meter with two common distances.

A **decimeter** is 1/10 of a meter. An orange Cuisenaire rod is used in many classrooms to demonstrate the length of 1 decimeter. A **centimeter** is 1/100 of a meter long. Figure 6.10 compares this unit with two common objects. A **millimeter** is 1/1000 of a meter. The edge of the cardboard back from a tablet is about 1 millimeter thick, and so is the wire of a paper clip, as shown in Figure 6.11, or the thickness of a dime. A **kilometer** (km) is a unit 1000 meters long. A football field is approximately 110 m from goal post to goal post, so nine football fields (including end zones) laid end to end would extend about 1 kilometer. A train of about 72 cars would be close to 1 km long. Investigation 6.3 will help you develop a better feeling for the size of each of these units.

Decimals are often used when expressing practical measurements of length in metric units. For example, a person might say, "I am 1.73 meters tall." This statement is easily

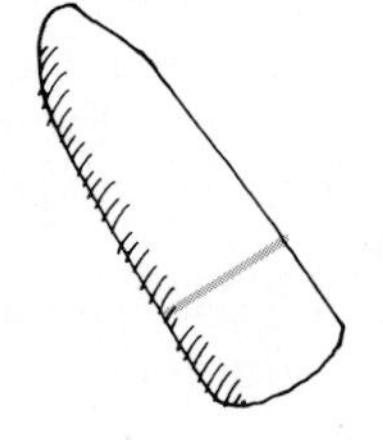

The width of a piece of chalk is about 1 centimeter.

The diameter of the head of a thumbtack is about 1 centimeter.

FIGURE 6.10

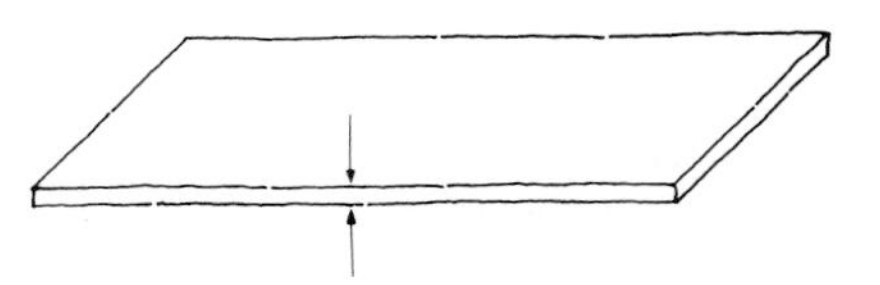

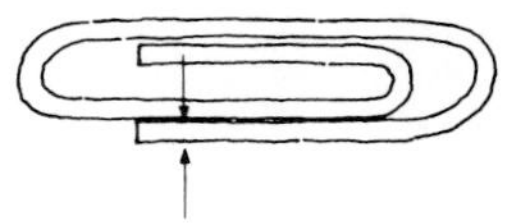

FIGURE 6.11

INVESTIGATION 6.3

Use the descriptions of the metric units of length given in this section and *no calculations or comparisons with other standard units* to individually estimate the following. Then, compare your estimates with others. Finally, check your estimates by calculating or measuring.

A. Estimate, in *centimeters* (cm): (1) your waist measurement, (2) your shoe length, (3) your arm span, and (4) another length of your choice.
B. Estimate, in *millimeters* (mm): (1) the width of your hand, (2) the width of a pencil, (3) the length of a paper clip, and (4) another length of your choice.
C. Estimate in *decimeters* (dm): (1) your height, (2) the height of your classroom, (3) the dimensions of a chalkboard, and (4) another length of your choice.
D. Estimate, in *meters* (m): (1) the height of a flagpole, (2) the length of your classroom, (3) the length of a tennis court, and (4) another length of your choice.
E. Estimate, in *kilometers* (km): (1) the distance to a nearby city, (2) the height of the tallest of the Rocky Mountains, (3) the height of the highest-flying airplane, and (4) another distance of your choice.

To extend these ideas, select from Activities 2 through 9, in Supplementary Exercises: Length, Appendix C.

TEACHING NOTE

It is crucial that students "get a feel" for metric units. One of the best ways to achieve this is to "think metric." If physical objects are selected to illustrate each metric unit and the student is asked to do extensive estimation of other lengths using these units, the student will gradually develop a real concept of the metric unit. Estimation is the key.

According to some learning theorists, the concepts of distance and length are not the same psychologically and must be treated separately. Distance refers to the linear separations of objects; length refers to a property of the objects themselves.

interpreted as, "I am 1 meter and 73 centimeters tall." It might also be interpreted as, "I am 1 meter, 7 decimeters, and 3 centimeters tall," but this interpretation is not usually made in practical situations.

Thus, the use of the decimal system of numeration in the metric system makes interpretation of measurements quite easy. Since each metric unit of length is 10 times as large as the next smallest unit, the decimal system also makes it convenient to convert from one unit of length to another. This is done by simply moving the decimal point to the left or right. Study this example, reading from left to right, and vice versa:

$$\begin{aligned} 0.732 \text{ km} &= 7.32 \text{ hm} = 73.2 \text{ dam} \\ &= 732.0 \text{ m} = 7320.0 \text{ dm} \\ &= 73200.0 \text{ cm} = 732000.0 \text{ mm} \end{aligned}$$

A measurable property of simple, closed curves closely related to length is called **perimeter**, or "distance around." For example, to find the perimeter of polygon *ABCDE*, you add the

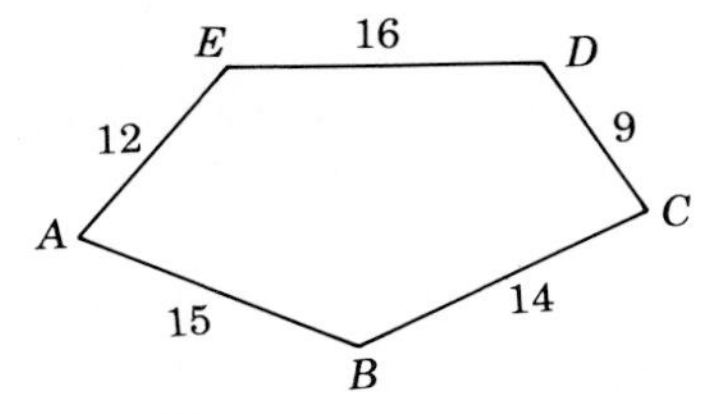

FIGURE 6.12

lengths of each side of the polygon (Fig. 6.12). Notice that in the figure the sides have lengths 12, 15, 14, 9, and 16 units long. In this case, the units of length are not specified. In that case, we record the perimeter as a certain number of "units." For example, the perimeter P of $ABCDE$ is

$$P = 12 + 15 + 14 + 9 + 16 = 66 \text{ units}$$

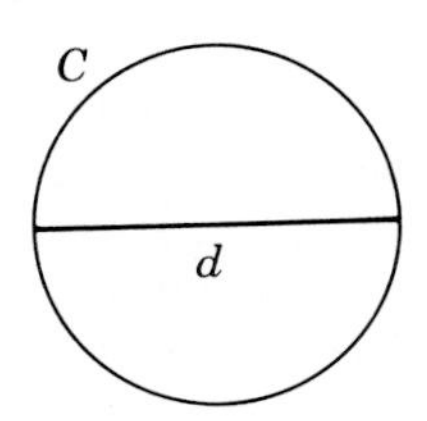

FIGURE 6.13

When the figure is a circle, we use the term **circumference** instead of perimeter. That is, the circumference of a circle is the "distance around" the circle. An important property of a circle is that the ratio of the circumference (C) and the diameter (d) is an infinite, nonrepeating decimal called π (pronounced "pi") that is often approximated by 3.14, or 3.14159, or 22/7. For example, given a circle with circumference C and diameter d (Fig. 6.13), the following equations are true:

$$\frac{C}{d} = \pi \qquad \text{or} \qquad C = \pi d$$

The formula $C = \pi d$ is used to find the circumference of a circle. For example, find the circumference of a circle with diameter 12 cm as follows:

$$\begin{aligned} C &= \pi d \\ &= (\pi)(12 \text{ cm}) \\ &= 12\ \pi\text{cm} \approx (12)(3.14) \text{ cm, or } 37.68 \text{ cm} \end{aligned}$$

Note that when π is replaced by its approximation 3.14, the $=$ sign is replaced by $\approx$, which means "approximately equal to."

EXPLORING WITH THE COMPUTER 6.1

Use geometry exploration software to make and test conjectures about the following questions concerning lengths and distances. Use at least 3 different figures when testing each conjecture. Record your results.

1. A student said that in a right triangle, the product of the lengths of the 2 legs is always equal to the product of the altitude to the hypotenuse and the hypotenuse. Do you think the student was correct?
2. Four cities, W, X, Y, and Z are located as shown in the figure. A factory is to be built at any point P inside the rectangle. The surveyor stated that the relationship $a^2 + b^2 = c^2 + d^2$ between the distances from the factory to the cities *always* holds true, regardless of where P is located inside the rectangle. Do you believe the surveyor was correct?

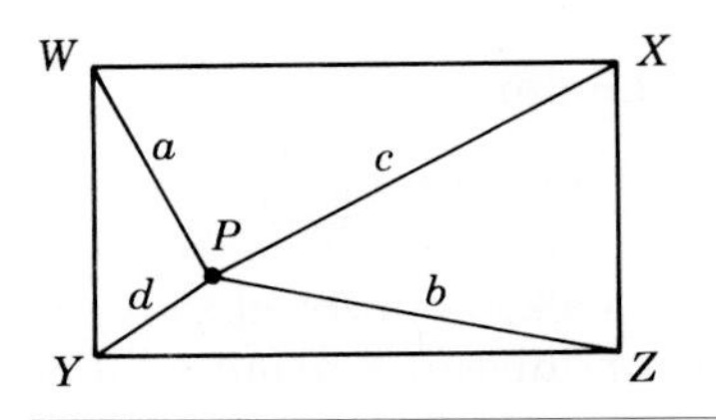

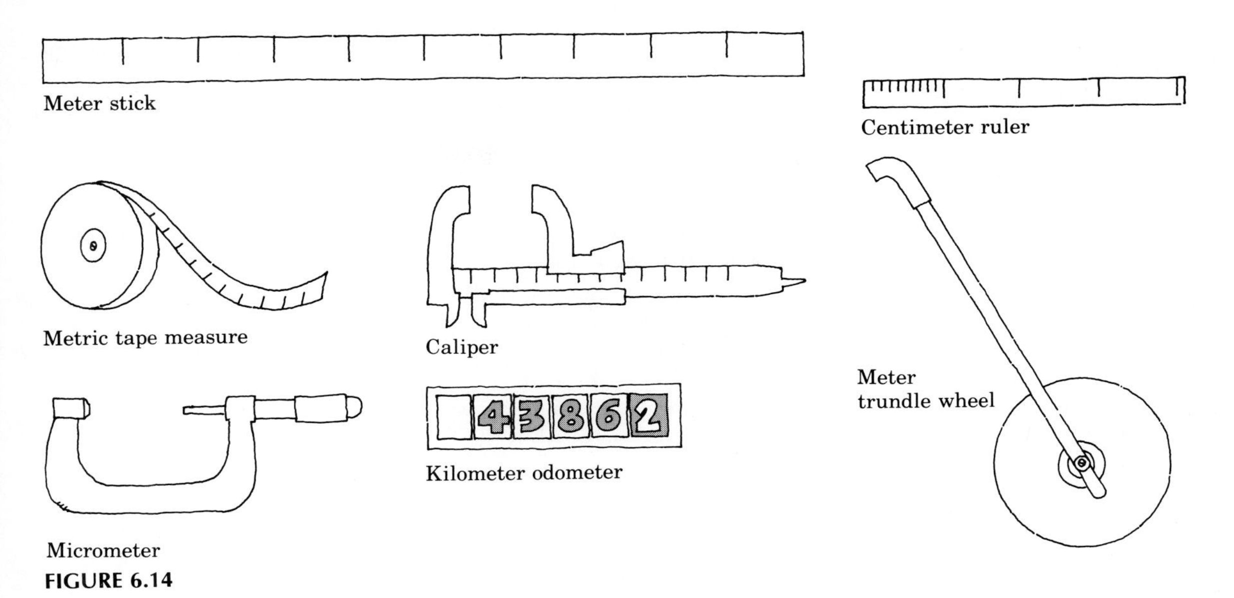

FIGURE 6.14

Several devices are available to aid in the measurement of length using the metric units. Some of them are pictured in Figure 6.14. The reader should develop an understanding of how each of these devices is used.

EXERCISE SET 6.3

1. A millimeter ruler is pictured in Figure 6.15. The length between 2 neighboring marks on this ruler is 1 mm. The length between adjacent marks with numerals indicate 1 cm. Find the length from the end of the ruler to each of the following points.

   a. *A* is ______ mm or ______ cm
   b. *B* is ______ mm or ______ cm
   c. *C* is ______ mm or ______ cm
   d. *D* is ______ mm or ______ cm
   e. *E* is ______ mm or ______ cm

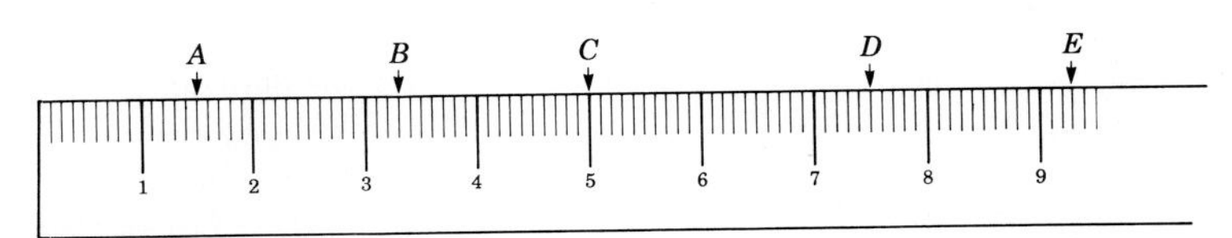

FIGURE 6.15

TEACHING NOTE

For children who can count, the trundle wheel is a very nice way to begin to develop the concept of measurement. Instead of having to cope with the concepts necessary to place a ruler end to end a repeated number of times, the child simply pushes the trundle wheel and counts the number of clicks. Thus, a number is associated with a distance even though the child may not yet have the concepts necessary to effectively use a ruler to find the number.

APPLICATION

Brass Band Problem: A sculpture apprentice made a metal model of the earth with a diameter of 3 feet and with a brass band fitting tightly around it. When the master sculptor saw the work, he suggested that it might be more striking if the band would still encircle the earth at the circumference, but stand out from the earth about 4 inches at every point. "That would be easy," said the apprentice, "I'll just insert a few more inches into the band—10 inches or less would probably do it." "I don't think so," said another apprentice, "I'd guess 10 to 20 inches." "You're both wrong," said the third apprentice, "I'll bet it'll take over 2 feet." Which of the apprentices do you think is correct? Guess first. Then help the apprentice by calculating the length of the piece needed to expand the band.

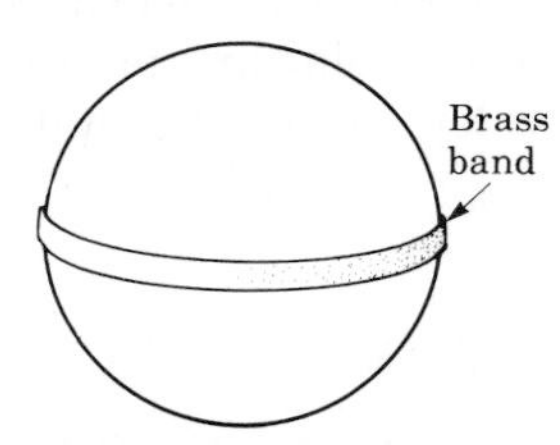

Earth model
diameter 36 in.

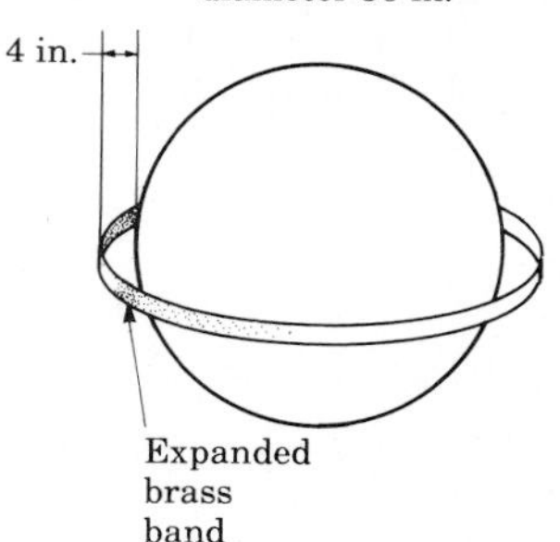

2. Place the decimal point in the number to make the following sentences reasonable.
   a. The length of a school desk might be 683 cm.
   b. The height of a 3-story building is 346 m.
   c. The length from one goal to another goal on a soccer field is 1158 m.
3. Find at least one common object or distance not mentioned in this text with each of the following lengths:
   a. 1 mm b. 2 mm
   c. 5 mm d. 1 cm
   e. 2 cm f. 5 cm
   g. 1 dm h. 1 m
   i. 2 m j. 5 m
   k. 1 km l. 2 km
4. Draw segments that you *estimate* to be each of the following lengths. Then, measure the segments and find out by how many millimeters your estimate was off. The total number of millimeters "off" is your "score." Compare your score with others.
   a. 7 cm b. 3.5 cm c. 2 cm d. 8 mm
   e. 10 cm f. 5.2 cm g. 4.8 cm h. 9.3 cm
5. Draw a "broken line" that you estimate to have a total length of 1 m on a sheet of notebook paper (Fig. 6.16). Measure the line to the nearest centimeter. By how many centimeters did the length of the line differ from 1 m?
6. Choose an appropriate metric unit and estimate the following. Check your estimate by measuring if possible.
   a. The diameter of a penny
   b. The thickness of a nickel
   c. The length of a new lead pencil
   d. The distance around your wrist
   e. The length of a city block
   f. The distance the winner travels in the Indianapolis 500
   g. The length of your "cubit"
   h. The width of a wristwatch
   i. The length of a study desk
7. Choose a pace that you think is 1 m long and "step off" an estimated 10 m. Measure the actual distance you stepped off and calculate the average length of your pace. By how much did your pace differ from 1 m?
8. Estimate the following speed limits in kilometers per hour:
   a. In a school zone
   b. In a residential area
   c. On a super highway
   d. Down main street
   e. On a superhighway for cars with trailers
9. A trip of 5 mi equals a trip of 8 km. Use only this information to

FIGURE 6.16

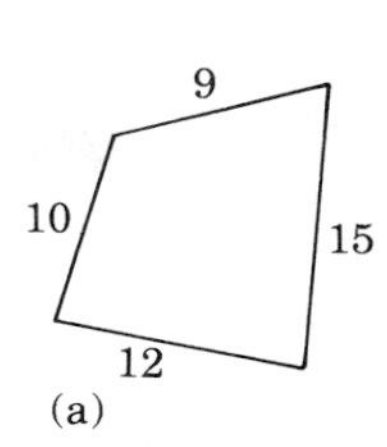

(a)

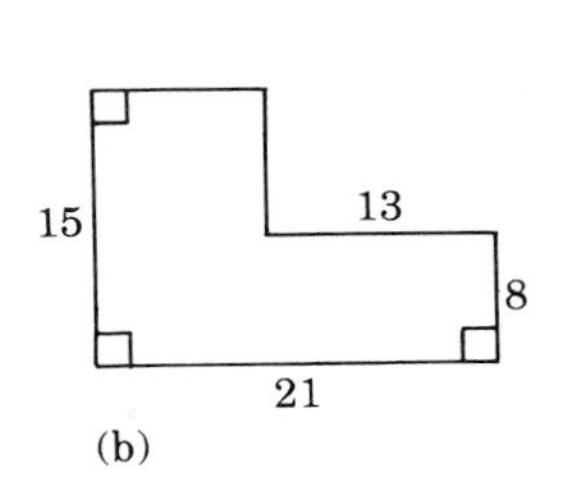

(b)

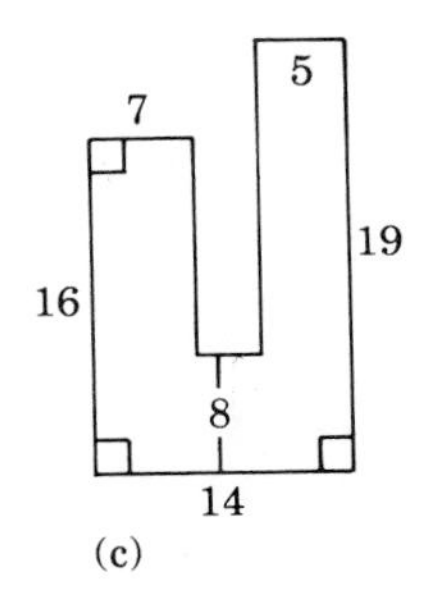

(c)

FIGURE 6.17

give a close estimate of the length of the following trips in kilometers.

a. Chicago to Denver (1038 mi)
b. Chicago to Washington (698 mi)
c. Los Angeles to Houston (1540 mi)
d. Houston to St. Louis (821 mi)
e. Detroit to Boston (735 mi)
f. Washington to New York (233 mi)

10. Complete each statement.

a. 1 m = ______ dm	b. 1 m = ______ cm
c. 1 m = ______ mm	d. 1 dm = ______ m
e. 1 cm = ______ m	f. 1 mm = ______ m
g. 1 cm = ______ mm	h. 1 dm = ______ cm
i. 1 mm = ______ cm	j. 1 cm = ______ dm
k. 1 m = ______ km	l. 1 m = ______ hm

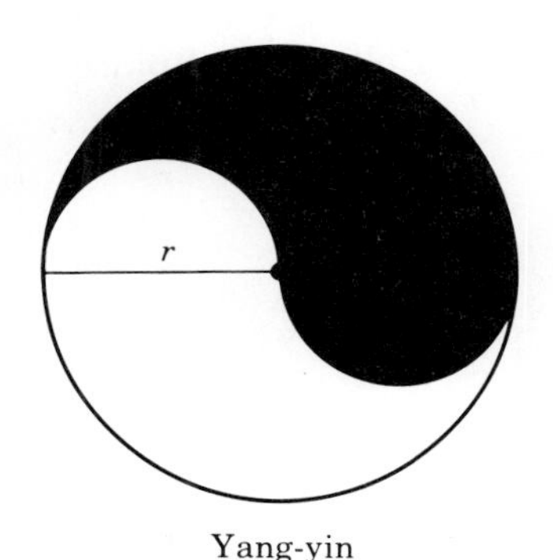

Yang-yin

FIGURE 6.18

11. Complete the following expressions.

a. 867 cm = ______ m	b. 26,428 m = ______ km
c. 632 m = ______ cm	d. 367 mm = ______ m
e. 867 m = ______ hm	f. 647 cm = ______ dm

12. Find the perimeters of the polygons in Figure 6.17.
13. Find the circumference of a circle with a diameter of 6 cm.
14. Find the circumference of a circle with a diameter of 9 cm.
15. The famous Yang-Yin symbol in Figure 6.18 has been constructed from a circle of radius r. Compute the length of the arc that divides the black and the white regions.
16. Estimate the perimeter of the figure shown in Figure 6.19.
17. **Critical Thinking**. Suppose that you need to measure the distance between two points, A and B, and that there is a large tree on the line from A to B (Fig. 6.20). Describe in detail a procedure that could be used to find the distance AB. Include in your description a listing of all the measuring devices and other equipment that you might need.

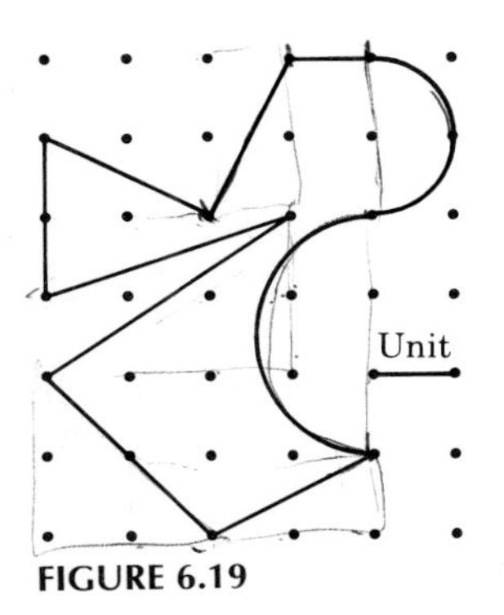

FIGURE 6.19

B •

A • Tree

FIGURE 6.20

To extend these ideas, select from Supplementary Exercises: Length, Appendix C.

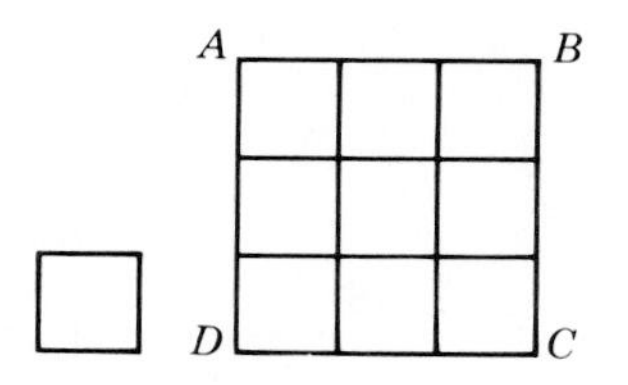

Unit The area of $ABCD$ is 9

FIGURE 6.21

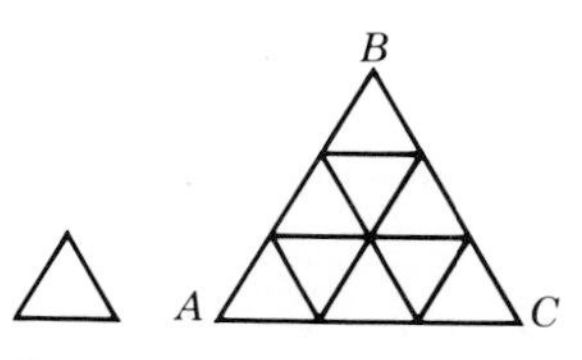

Unit The area of ABC is 9

FIGURE 6.22

MEASURING AREA

When the measurable property can be loosely described as "the extent of surface enclosed," we are measuring the **area** of the surface.

Children often think of area in terms of the "amount covered" and count the number of unit squares needed to cover a surface to estimate its area (Fig. 6.21). As with length, the unit of area is arbitrary, and other figures, such as triangles, are often used as units to illustrate this point (Fig. 6.22). Investigation 6.4 provides some experiences that give further meaning to the idea of measuring area.

Thus, at the beginning stages, the process of measuring area is accomplished by counting the number of units, half-units, and so on, that are needed to "cover up" a region. Children sometimes draw a grid on a sheet of plastic and place a

INVESTIGATION 6.4

If the unit of area is as shown, the areas of the other regions are as indicated. Be sure you see why. (The shading might help.) Use these ideas to complete the following.

A. Seven squares, each with different whole number areas, can be shown on a 5 × 5 geoboard or a 5 × 5 dot paper grid. How many of these can you find and show on dot paper? (The vertices of the squares must be points on the grid.)

B. Can you find triangles with areas ½, 1, 1½, 2, 2½, 3, 3½, 4, 4½, 5, 5½, 6, 6½, 7, 7½, and 8 on 5 × 5 dot paper? Show each one you find.

C. Make a figure on 5 × 5 dot paper for which it is difficult to find its area. Challenge another member of your group to determine the area of your figure.

TEACHING NOTE

A Piagetian test a teacher might use to see whether a young child has developed the conservation ideas necessary for beginning to measure area is as follows: Two rectangular sheets of cardboard colored green, both the same shape and size, are presented. A model of a cow is placed on each and the child is told that the farmers who own the pastures have decided to put a house on each. A model of a house is placed near the middle of one pasture and in the corner of the other, and the child is asked whether the two cows still have the same amount of grass to eat. Houses are added in pairs, one to each pasture, but with the houses on one pasture closer together than the houses on the other. The question is repeated. If the child knows that when the pastures contain the same number of houses, regardless of their arrangement, the grass areas must be the same, the child is becoming aware of the concepts necessary for finding area.

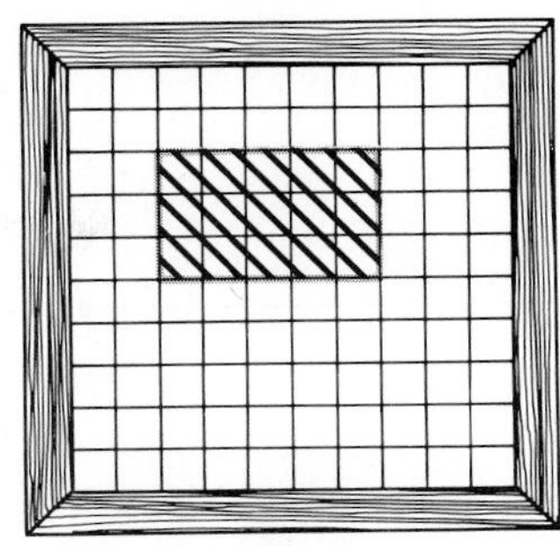

FIGURE 6.23

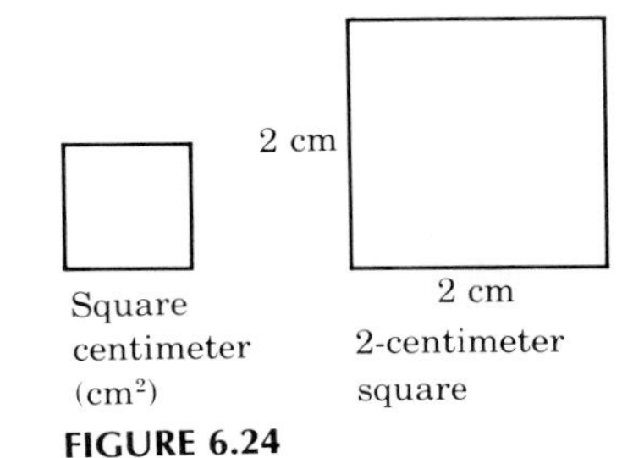

FIGURE 6.24

frame around it to make a device to help them find the area of a figure (Fig. 6.23). This device is placed over a region, such as the hatched rectangle, and the units needed to cover the region are easily counted. To extend these ideas, use Supplementary Activities: Area, Activities 1 and 2, in Appendix C.

In the metric system, the basic unit of area is the **square meter** (m^2). For classroom work, the square centimeter (cm^2) or the 2-centimeter square are useful units of area (Fig. 6.24). Table 6.2 shows other units that are used to measure area.

In Investigation 6.4 you found the area of regions by counting the number of square units in a rectangle and considering what fraction of the shaded rectangle contained the particular triangle in question. The pattern of counting unit squares in a rectangle leads to a formula for finding the area of a rectangle.

Formula 1: The *area of a rectangle* with length l and width w (Fig. 6.25) is found using the formula

$$A = l \times w$$

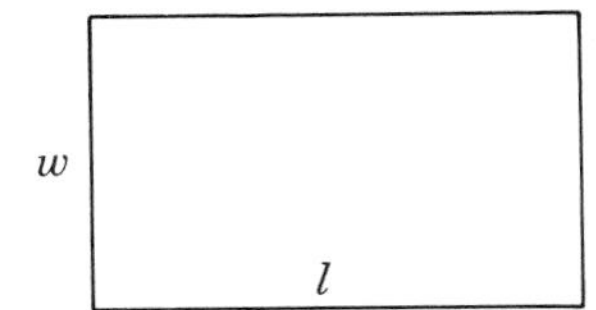

FIGURE 6.25

TABLE 6.2 Metric Units for Measuring Area

Unit	Symbol	Relation to Square Meter
Square kilometer	km^2	1,000,000 m^2
Square hectometer	hm^2	10,000 m^2
Square dekameter	dam^2	100 m^2
Square meter	m^2	**1 m^2 (basic unit)**
Square decimeter	dm^2	0.1 m^2
Square centimeter	cm^2	0.0001 m^2
Square millimeter	mm^2	0.000001 m^2

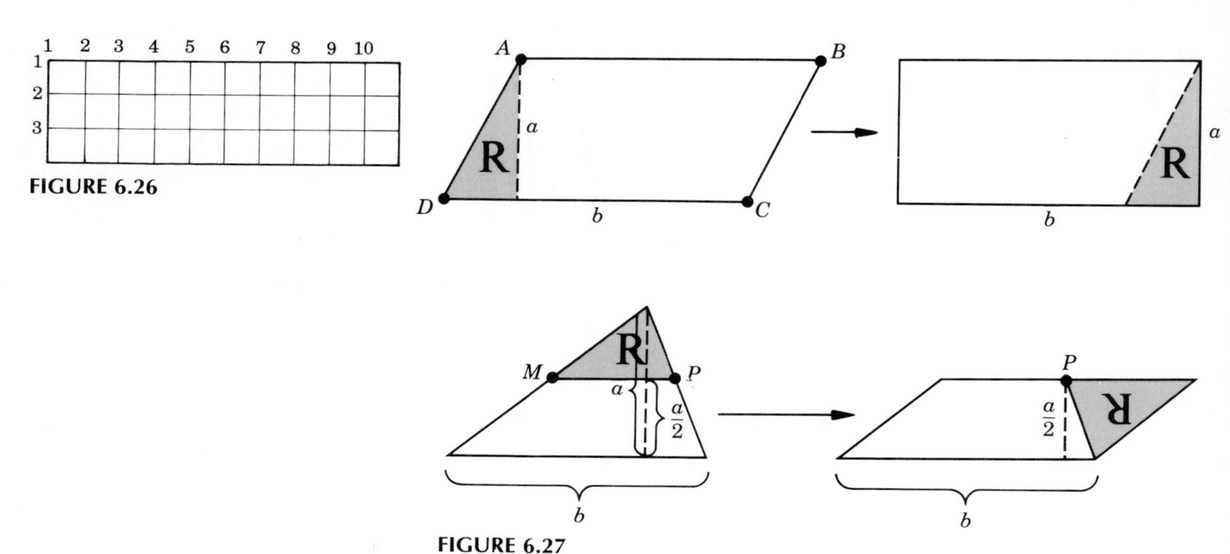

FIGURE 6.26

FIGURE 6.27

In Figure 6.26, for example, a rectangle that is 10 units long and 3 units wide has an area of 3 × 10 or 30 square units.

Formulas for finding the area of other figures are derived from this formula for the area of a rectangle. For example, consider parallelogram *ABCD* in Figure 6.27. A triangle **R** can be cut from one end of the parallelogram and placed on the other end to form a rectangle with the same area as the parallelogram. Recall that the height of a parallelogram (or triangle) is the length of the segment from a vertex perpendicular to the opposite side. Thus, the base and height of the parallelogram are the same as for the rectangle, so the area of both figures is $a \times b$. In a similar manner, a triangle **R** can be cut from the triangle in Figure 6.27 and placed to form a parallelogram. The area of the triangle can be found by finding the area of the parallelogram, $a/2 \times b$. These ideas lead to Formula 2 and Formula 3.

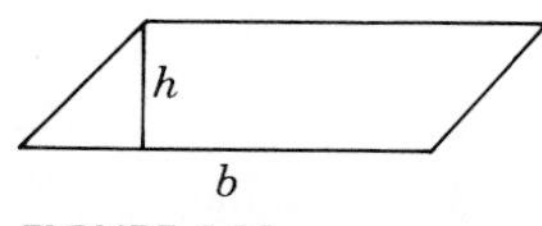

FIGURE 6.28

Formula 2: *The area of a parallelogram* with height h and base length b (Fig. 6.28) is found using the formula

$$A = h \times b$$

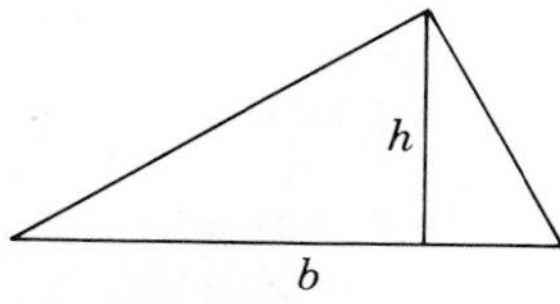

FIGURE 6.29

Formula 3: The *area of a triangle* with height h and base length b (Fig. 6.29) is found using the formula

$$A = \frac{1}{2}h \times b$$

Since the formula for finding the area of a triangle relies on the formula for the area of a parallelogram, and the formula for the area of a parallelogram relies on the length-times-width formula for the area of a rectangle, then any formula that relies on the formula for the triangle ultimately relies on the formula for the area of a rectangle. For example, exercise 12 in Exercise Set 6.4 derives a formula for the area of a trapezoid. Investigation 6.5 explores a method of finding the area of polygons that uses graph paper.

INVESTIGATION 6.5

Try to discover a formula, called **Pick's Formula**, for finding the area of a polygon drawn on dot paper. You can find the areas of the polygons shown in the figure by counting squares and part squares.

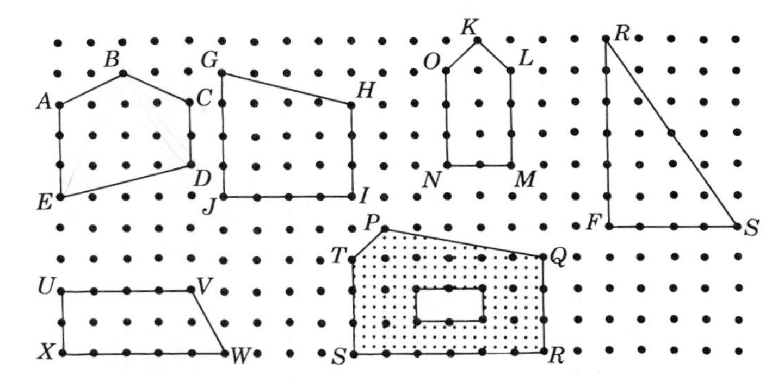

A. Complete the following table, where b = number of dots on the boundary, i = number of dots inside the polygon, and area = area of the polygon. The first row has been filled in for polygon *ABCDE*.

Polygon	b	i	$b/2 + i$	Area
ABCDE	8	9	13	12
GHIJ				
KLMNO				
RSF				

B. State Pick's Formula, and use it to find the area of polygon *UVWX*. Check by finding the area of *UVWX* by adding the areas of the smaller regions.

C. Try to find a Revised Pick's Formula for finding the area polygons with holes, such as *PQRST*. Write a description of what you did, and then explain your revised formula or tell about the difficulty that kept you from finding the formula.

EXPLORING WITH THE COMPUTER 6.2

Use geometry exploration software to make and test conjectures. Use at least 3 different figures when testing each conjecture. Record your results.

1. Does the formula $(a + c)(b + d)/4$ give the area of a quadrilateral with side lengths a, b, c, and d? Test it with rectangles, parallelograms, trapezoids, and general quadrilaterals.
2. Does the formula $\sqrt{s(s - a)(s - b)(s - c)}$ give the area of any triangle with a, b, and c being the lengths of the sides and with s being equal to 1/2 the perimeter? Test with all types of triangles.
3. A student thought that the formula $A = 1/2\ d_1d_2$, where d_1 and d_2 are the diagonals, could be used to find the area of a certain type of quadrilateral, but she couldn't remember which type. Test some quadrilaterals to determine when the formula works.
4. In the figure shown here, the sides of the triangle are divided into 3 parts, and the points connected.

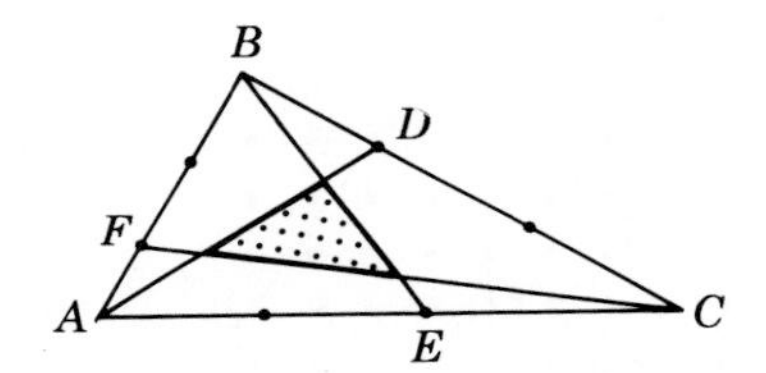

   a. What is your conjecture about the ratio of the area of the shaded center triangle to the original triangle? Test your conjecture and revise it if necessary.
   b. What if the sides were divided into 4 or 5 or 6 parts? Make and test conjectures. Make a generalization if possible.
   c. How do you think the areas of the 3 small "corner triangles" in the figure compare? Make and test a conjecture.

The area of a circle uses the value of π just as the formula for the circumference of a circle did. Your experiences with Exercises 9 and 13 in Exercise Set 6.4 will help you understand the formula for the area of a circle.

Formula 4: The *area of a circle* with radius r (Fig. 6.30) is found using the formula

$$A = \pi r^2$$

FIGURE 6.30

EXERCISE SET 6.4

1. Find the area of each polygon in Figure 6.31 if the smallest square is one square unit.

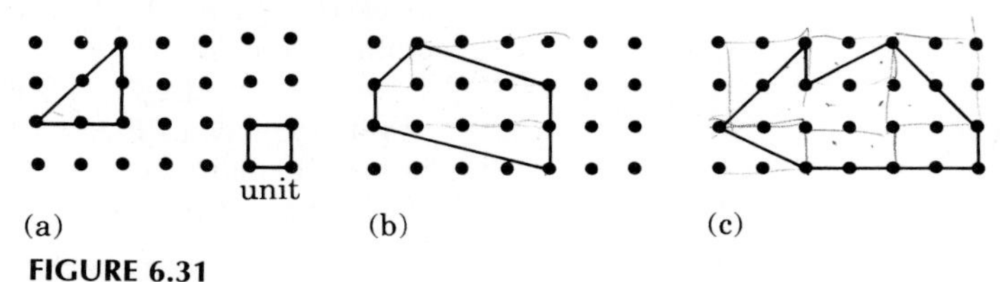

FIGURE 6.31

APPLICATION

Earth Problem: A science student in Oklahoma City had a friend in St. Louis. One day, at high noon in Oklahoma City, the sun's rays made an angle of 0° with a vertical pole. The student called his friend 500 miles away in St. Louis and asked him to find what angle the sun's rays made with a vertical pole. He checked and found the angle to be 7 1/5°. The science student said he could use this information to set up a proportion and calculate the circumference of the earth. Do you believe this is possible? If so, explain how it might be done.

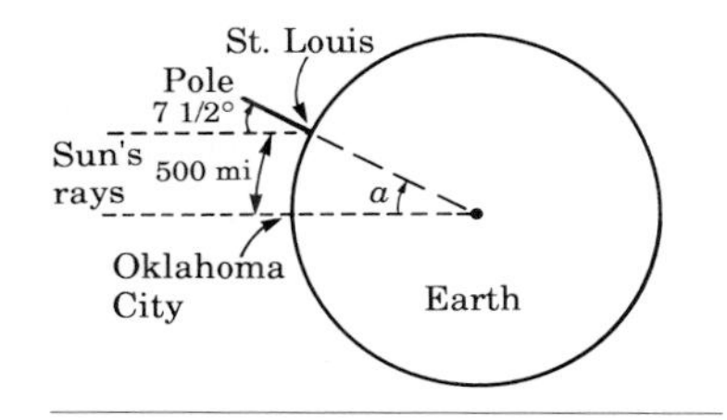

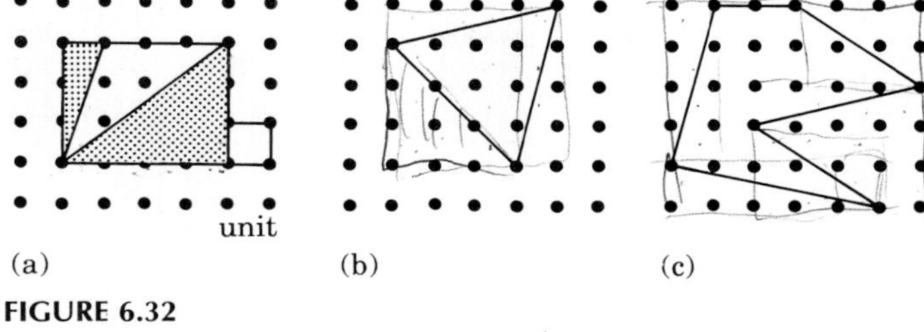

FIGURE 6.32

2. Figure 6.32 illustrates that to find the area of the unshaded triangle, subtract the area of the 2 shaded triangles from the area of the rectangle. Use this method to find the area of each of the polygons in Figure 6.32.
3. Choose the most appropriate metric unit area (cm^2, m^2, or km^2) for finding each of the following areas:
   a. Area of the cover of your textook
   b. Area of a basketball floor
   c. Area of a pin head
   d. Area of a wheat field
4. Estimate the area of each figure of Exercise 3 in square centimeters. Then, make a grid on plastic or tracing paper like the one shown in Figure 6.33, using the square centimeter as the unit. Use the grid to check your estimate.
5. Complete the conversions in Table 6.3.

FIGURE 6.33

TABLE 6.3

mm^2	cm^2	m^2
458.0	4.58	—
—	35.9	—
—	?	12.5
—	0.25	—
8463	?	—

APPLICATION

Pen Problem: A gardener wants to use 36 feet of fence to make a rectangular pen with whole number dimensions for some rabbits. What should the dimensions of the pen be to take up the least amount of his yard? What dimensions give the largest possible grass area? Make a table to help with your decision. If the gardener were willing to use non-whole numbers or shapes other than a rectangle, could he build a pen with larger area?

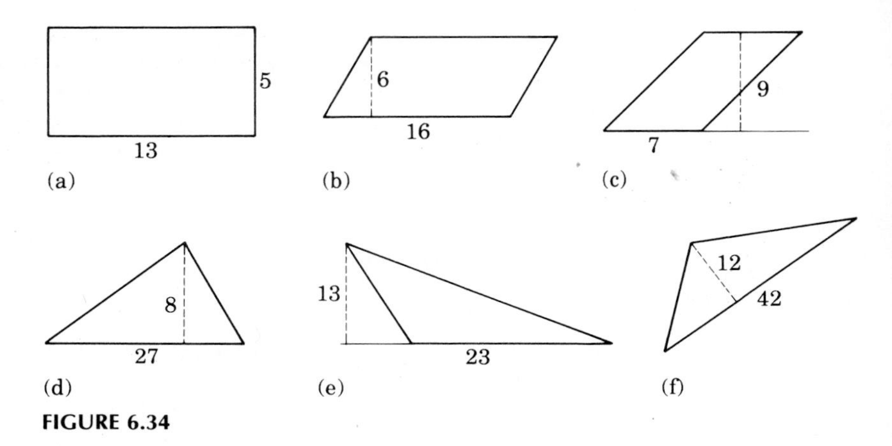

FIGURE 6.34

6. Find the area of each polygon shown in Figure 6.34.
7. The 7 pieces of the tangram puzzle are shown in Figure 6.35. If the area of the small square is 1, what is the area of each of the other pieces of the puzzle?
8. Find the area of each figure in Figure 6.36. Curved portions of the figures are 1/2 or 1/4 of a circle.
9. a. Draw circles on graph paper and count squares (estimate parts of squares, etc.), as shown in Figure 6.37, to complete Table 6.4.
   b. What did you discover in the last column of Table 6.4?
10. Find the area of the shaded ring in Figure 6.38.

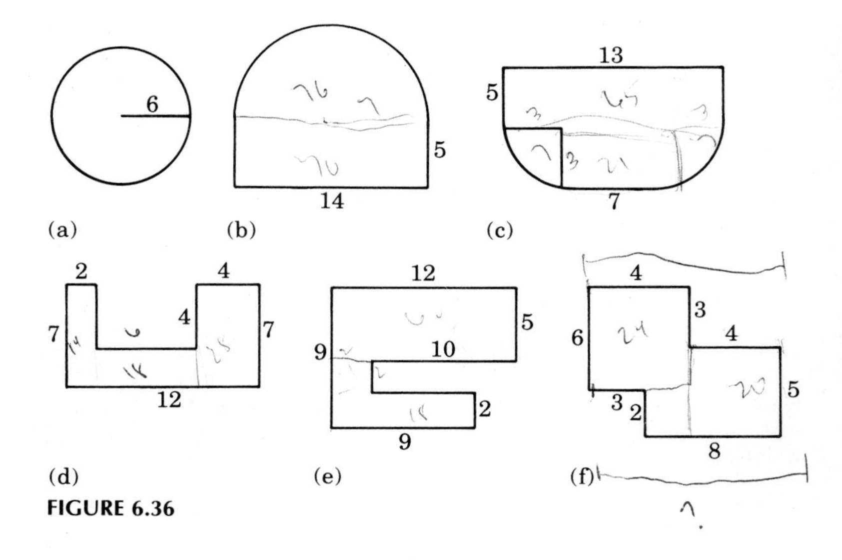

FIGURE 6.36

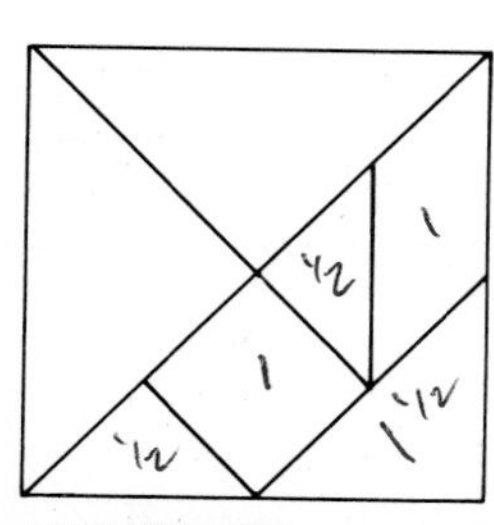
FIGURE 6.35

APPLICATION

Hungry Cow Problem: A farmer tied her cow at the corner of a small barn with plenty of water and left for a five-day trip. When she returned, the cow had eaten all the grass it could reach and had become sick. Before prescribing medicine for the cow, the veterinarian wants to know how much grass it has eaten. The grass in the pasture was 3 inches high. Estimate the number of pounds of grass eaten per square foot to answer the veterinarian's question.

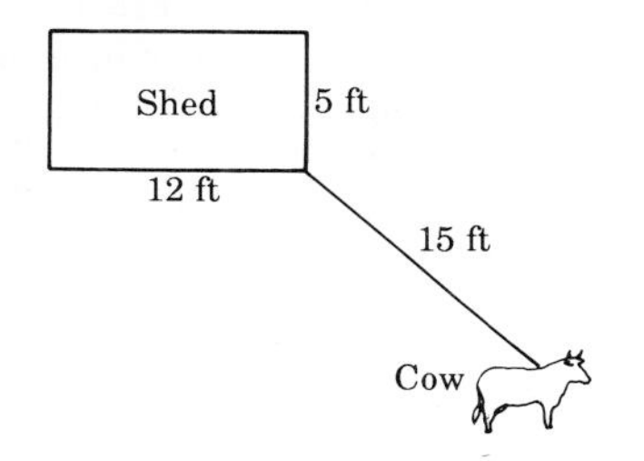

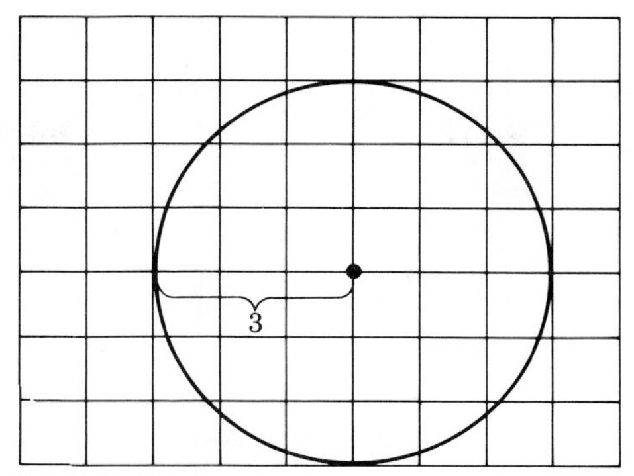

FIGURE 6.37

TABLE 6.4

Radius (r)	n^2	Estimated Area (A)	A/r^2
3			
4			
5			
6			
7			
8			
9			
10			

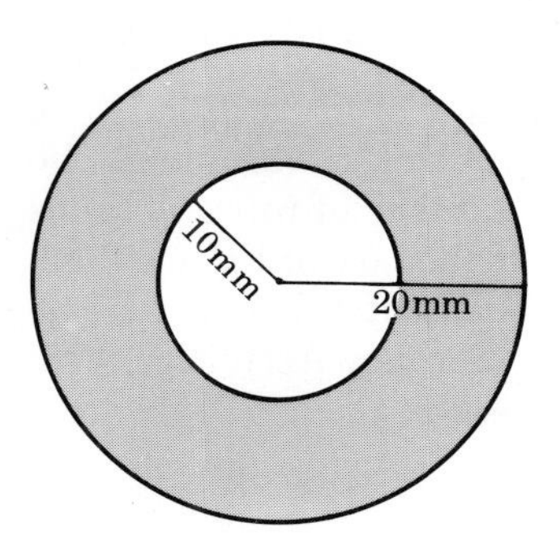

FIGURE 6.38

11. Show by dissecting and reassembling region R in Figure 6.39 that the area of an isosceles trapezoid is $b \times a$. You may want to use scissors and construction paper.
12. Every trapezoid can be divided into 2 triangles, as indicated by the shaded regions I and II in Figure 6.40. The altitude a of each triangle is the same. Use this information to prove that

$$\text{Area of trapezoid } ABCD = \text{Area } \triangle\text{I} + \text{Area } \triangle\text{II}$$
$$= \frac{1}{2}a(b_1 + b_2)$$

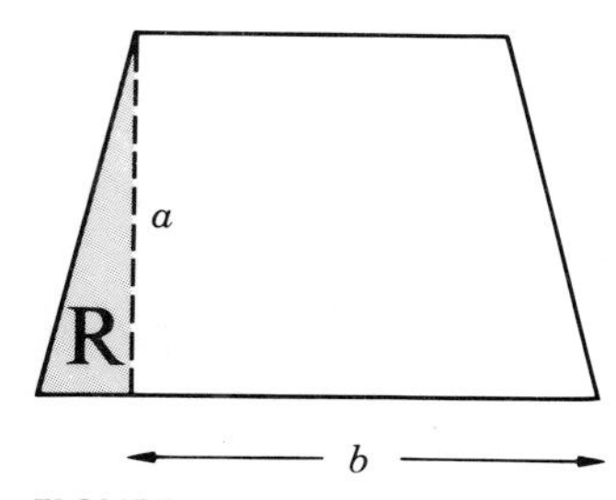

FIGURE 6.39

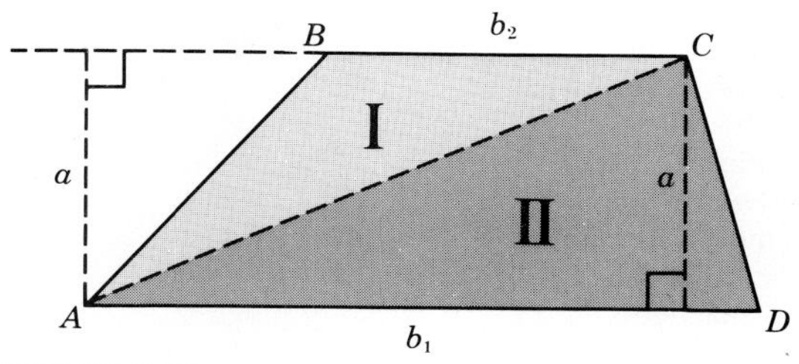

FIGURE 6.40

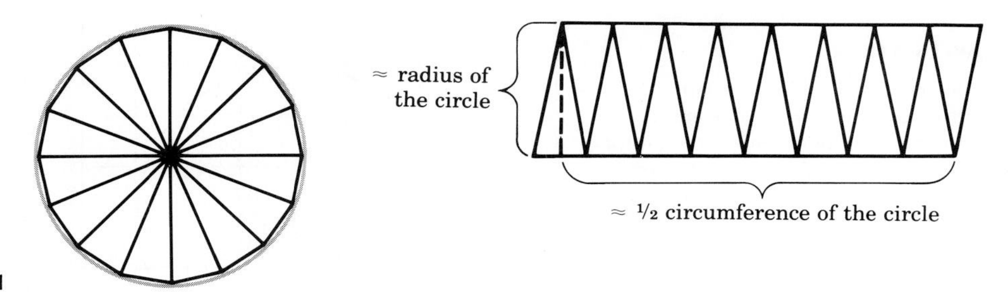

FIGURE 6.41

13. **Critical Thinking**. The area of a circle can be approximated quite accurately by dividing it into triangles, as shown in Figure 6.41. When these triangles are reassembled as shown, they form an "approximate" rectangle. Assuming that the area of this "rectangle" is equal to the area of a circle, use this information to produce the formula for the area of a circle. Write a paragraph that gives a convincing argument that the formula is correct.

    For a related Critical Thinking activity, see Activity 15, Appendix D. To extend these exercises, select from Supplementary Exercises: Area in Appendix C.

MEASURING VOLUME AND CAPACITY

When the "measurable property," as referred to in the description of the measurement process given at the beginning of this chapter, can be described as the "amount of space filled," we are measuring **volume**. Children often think of volume as the "amount the object will hold" or the number of units it takes to "fill the object." Thus, the volume of the box in Figure 6.42, to a child, would be the number of blocks it takes to fill the box. To extend these ideas, use Supplementary Activities: Volume and Capacity, Activity 1, in Appendix C.

TEACHING NOTE

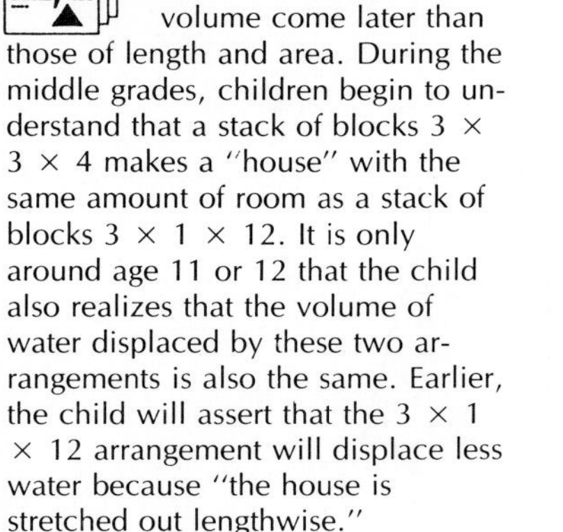

The concepts needed to have an understanding of volume come later than those of length and area. During the middle grades, children begin to understand that a stack of blocks 3 × 3 × 4 makes a "house" with the same amount of room as a stack of blocks 3 × 1 × 12. It is only around age 11 or 12 that the child also realizes that the volume of water displaced by these two arrangements is also the same. Earlier, the child will assert that the 3 × 1 × 12 arrangement will displace less water because "the house is stretched out lengthwise."

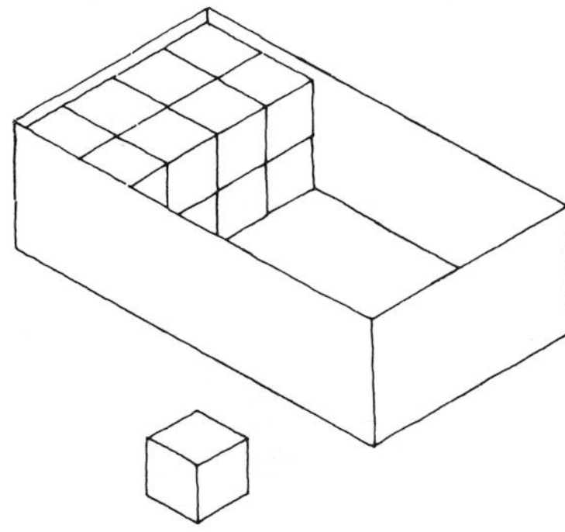

FIGURE 6.42

TABLE 6.5 Metric Units for Measuring Volume

Unit	Symbol	Relation to Cubic Meter
Cubic kilometer	km^3	1,000,000,000 m^3
Cubic hectometer	hm^3	1,000,000 m^3
Cubic dekameter	dam^3	1,000 m^3
Cubic meter	m^3	**1 m^3 (basic unit)**
Cubic decimeter	dm^3	0.001 m^3
Cubic centimeter	cm^3	0.000001 m^3
Cubic millimeter	mm^3	0.000000001 m^3

As with length and area, the unit of volume is arbitrary. In the metric system the basic unit of volume is the **cubic meter** (m^3). Table 6.5 shows this unit and the units related to it using some of the metric prefixes. The units in Table 6.5 are not often used in practical measurement situations, but they are important to know about because of their use in science.

The practical counterpart to the more technical measurement of volume is the everyday measurement of "how much some container will hold," called the **capacity** of the container. As shown in Figure 6.43, the cubic meter is somewhat large for ordinary volume measurement and the cubic centimeter is a bit too small. Thus, a unit cube 1 decimeter on a side—the cubic decimeter (Fig. 6.44)—is often used. Investigation 6.6 will help you focus on a unit that is commonly used to measure capacity.

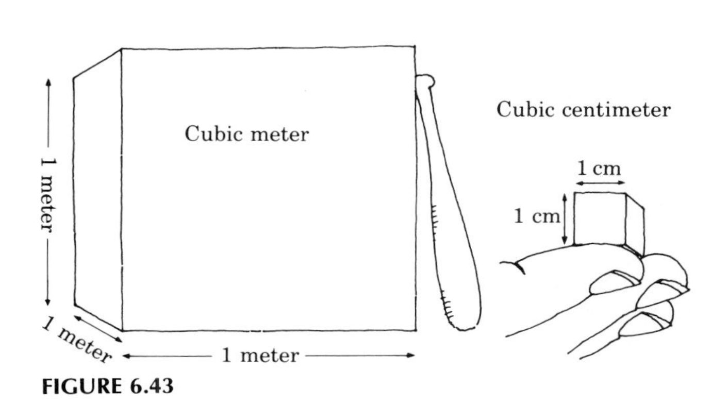

FIGURE 6.43

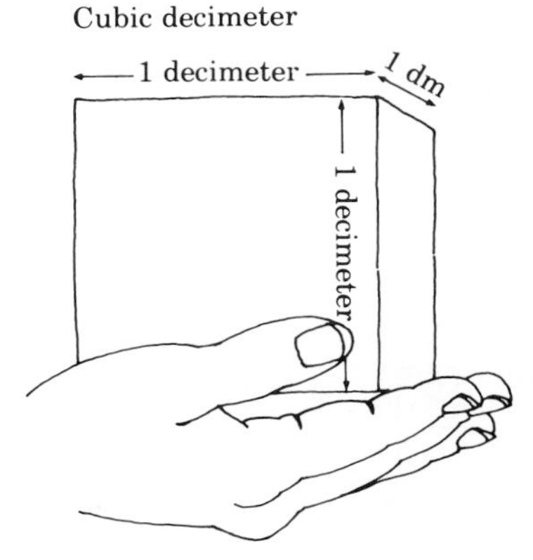

FIGURE 6.44

INVESTIGATION 6.6

Devise a way to construct a cubic decimeter container with both inside and outside covered with centimeter graph paper. Can you answer these questions about your container?

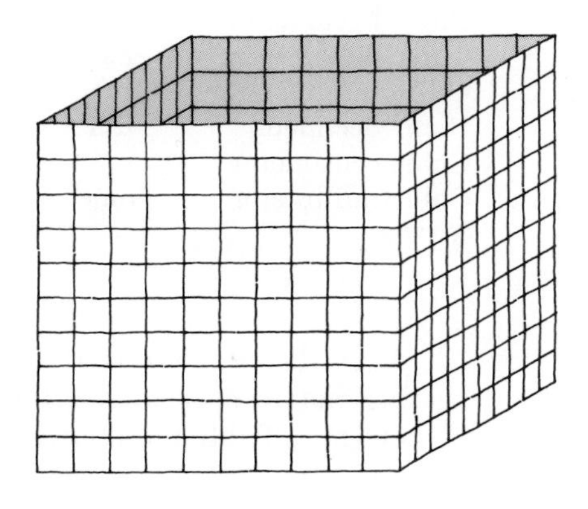

A. How many cubic centimeter containers full of liquid would it contain?
B. How many of your containers full of liquid would it take to fill a cubic meter tank?
C. How does the capacity of your container compare with the capacity of a quart jar?

Since common measures of capacity are often needed, the amount of water an "open top" cubic decimeter will hold is called 1 **liter** (l) of water (Fig. 6.45). Since it takes 1000 cm^3 containers full of water to fill a liter, the capacity of such a container is said to be 1 **milliliter** (ml). These two units of capacity are the ones commonly used in everyday measurements. A 2 l milk carton or a 360 ml soda can are useful examples. To extend these ideas, use Supplementary Activities: Volume and Capacity, Activity 2, in Appendix C. Table 6.6 shows the related metric units for measuring capacity.

To convert from one metric unit of volume to another, it should be noted that while units of length differ by multiples of 10 and units of area differ by multiples of 100, the units of volume differ by multiples of 1000. Thus, to convert from cubic meters to cubic dekameters, one would multiply by 0.001, or move the decimal point three places to the left. To convert from cubic meters to cubic decimeters, one must multiply by 1000, or move the decimal point three places to the right.

TEACHING NOTE

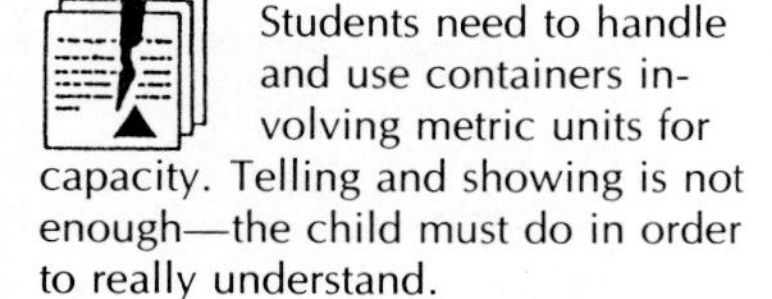

Students need to handle and use containers involving metric units for capacity. Telling and showing is not enough—the child must do in order to really understand.

1 cm 1 cm

1 cm

1 milliliter

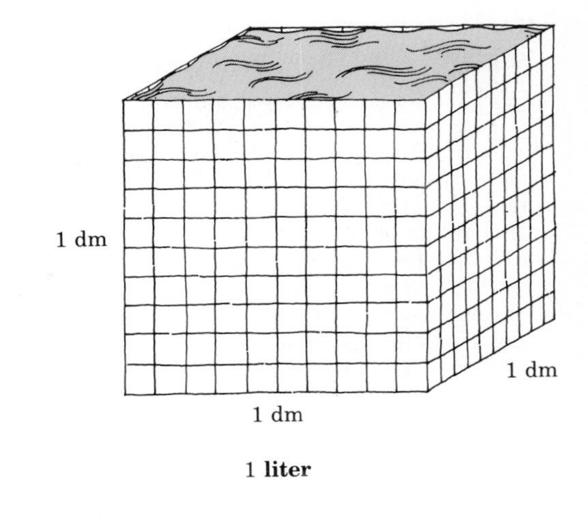

1 **liter**

FIGURE 6.45

TABLE 6.6 Metric Units for Measuring Capacity

Unit	Symbol	Relation to Liter
Kiloliter	kl	1000 l
Hectoliter	hl	100 l
Dekaliter	dal	10 l
Liter	l	**1 l (basic unit = 1 cubic decimeter)**
Deciliter	dl	0.1 l
Centiliter	cl	0.01 l
Milliliter	ml	0.001 l

As with the measurement of area, the basic means of measuring volume in the preliminary stages is that of counting units. However, there are no standard measuring devices for counting the number of unit cubes required to "fill an object" to find its volume like the ruler or square grid paper are used for finding length and area. Consequently, to find the volume of a solid, certain lengths are measured and mathematical formulas are used to calculate the volume of solids. Some useful ideas for developing formulas are found in an insightful book that focuses on key mathematical ideas (Steen, 1990).

A basic formula for finding the volume of many solids uses the concept of cross section. A **cross section** of a solid is a region that is common to a plane and the solid.

TEACHING NOTE

An interesting experiment for children involves the use of sand, rice, crystal sugar, or liquid, and open-top containers as shown in the following figure. Children are asked how many pyramids-full of the material it will take to fill the prism. They are surprised to find that it takes 3 and that the volume of a pyramid is ⅓ the volume of a prism with the same base and height. This idea is also true for cones and cylinders. An experimental approach actually using the materials—the material and the containers—can help the students have a meaningful understanding of this formula.

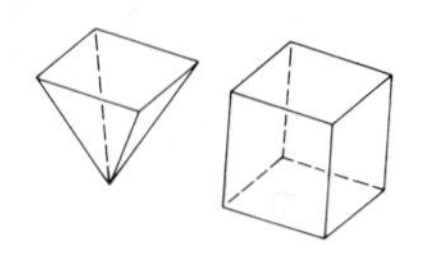

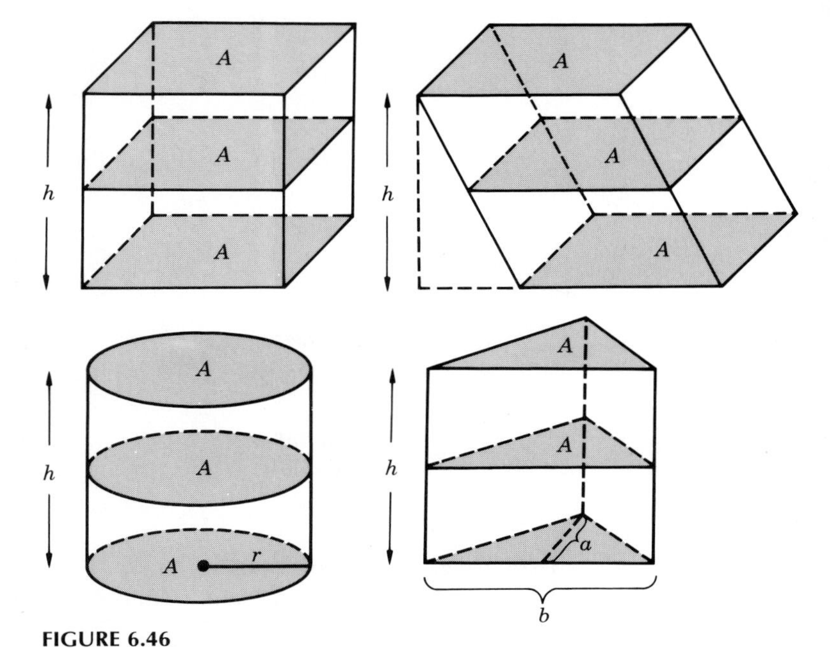

FIGURE 6.46

Formula 1: The *volume of a prism or cylinder* (Fig. 6.46) is found by the following formula, if the area of the base of such a space figure is A and if its height is h, and if all cross sections of the figure parallel to the base also have area A.

$$V = A \times h$$

This formula applies to all space figures, or solids. The method of finding the area of the base depends upon the specific space figure.

Example 1. Find the volume of the prism in Figure 6.47. The base is a rectangle, so its area is length times width.

$$\begin{aligned} V &= A \times h \\ &= (8 \times 15) \times h \\ &= 120 \times 5 \\ &= 600 \text{ cubic units} \end{aligned}$$

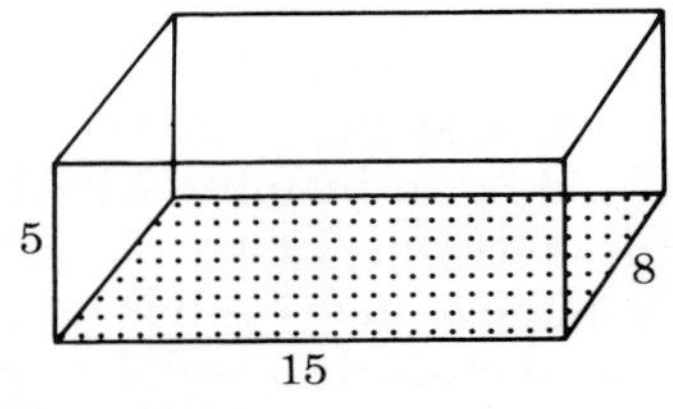

FIGURE 6.47

In this example we see that if the solid is a cuboid with di-

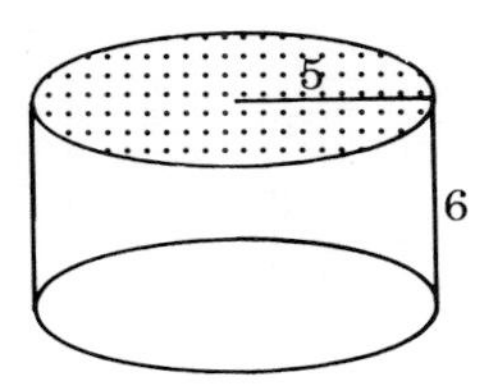

FIGURE 6.48

mensions l, w, and h, Formula 1 becomes $V = l \times w \times h$. If the prism is triangular (Fig. 6.46), Formula 1 becomes $V = 1/2a \times b \times h$.

Example 2. Find the volume of the cylinder in Figure 6.48. Note that the base of a cylinder is a circle. Its area is πr^2.

$$\begin{aligned} V &= A \times h \\ &= \pi r^2 \times \text{h} \\ &= \pi 25 \times 6 \\ &= 150\pi \text{ cubic units} \end{aligned}$$

Here, Formula 1 becomes $V = \pi \times r^2 \times h$.

Suppose a prism and a pyramid have identical bases and the same height. How do their volumes compare? It is possible to demonstrate experimentally that the volume of the prism is three times the volume of the pyramid (Fig. 6.49). The same relationship exists between the volume of a cylinder and a cone that have the same height and the same base radius (Fig. 6.49).

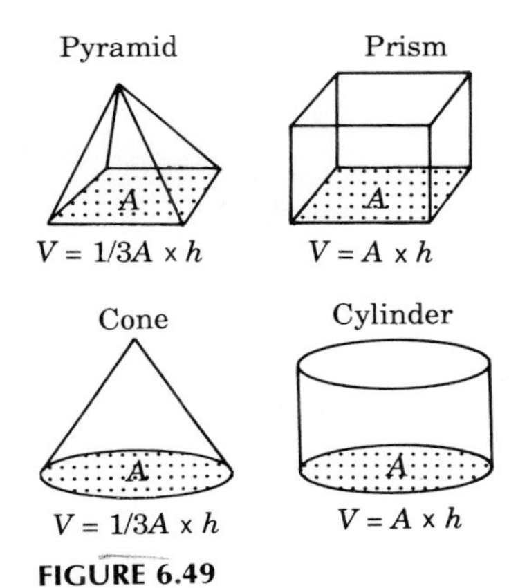

FIGURE 6.49

Formula 2. The *volume of a cone or pyramid* with base area A and height h is found by using the formula

$$V = \frac{1}{3} A \times h$$

APPLICATION

Box Problem: A box manufacture wants to produce an open box for which the volume is 2 cubic meters and the length is twice the width, giving the relationships among the dimensions indicated in the figure. What actual dimensions for the box will require the least amount of material? [*Hint:* Use a calculator and make a table for $x = 0.25$ m to 2 m in 0.25 increments. Narrow the range and make another table, if needed, to close in on the minimum surface area.]

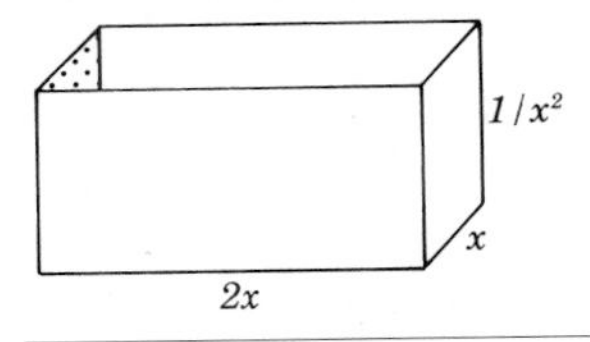

EXERCISE SET 6.5

1. Name objects with volume of
   a. 1 cubic meter
   b. 1 cubic decimeter
   c. 1 cubic centimeter
2. Name objects with capacity of
   a. 1 liter
   b. 1 milliliter
3. Estimate the capacity of the following in liters or milliliters.
   a. A cup
   b. A can of pop
   c. A gallon can
   d. A quart jar
   e. A thimble
   f. A pint carton
   g. An economy car gas tank
   h. A teaspoon
4. Estimate the volume of the following in cubic meters, cubic decimeters, or cubic centimeters.
   a. A classroom
   b. A shoebox
   c. A desk drawer
   d. A 3-drawer filing cabinet
   e. A child's lunch pail
   f. A refrigerator

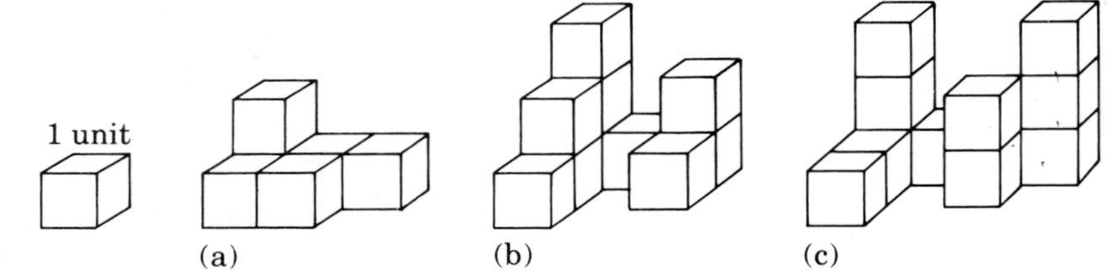

FIGURE 6.50

5. Complete the following:

   a. ______ $cm^3 = 1\ m^3$ b. ______ $dm = 1\ m^3$
   c. ______ $mm^3 = 1\ m^3$ d. ______ $m^3 = 1\ km^3$
   e. ______ $dam^3 = 1\ km^3$ f. ______ $hm^3 = 1\ km^3$
   g. ______ $dm^3 = 1\ dam^3$ h. ______ $mm^3 = 1\ km^3$

6. Complete the following:

   a. ______ ml = 1 l b. ______ cl = 1 l
   c. ______ dl = 1 l d. ______ l = 1 kl
   e. ______ dal = 1 kl f. ______ hl = 1 kl

7. Complete the following:

   a. 947 ml = ______ cl b. 947 ml = ______ dl
   c. 947 ml = ______ l d. 6.496 kl = ______ hl
   e. 6.496 kl = ______ dal f. 6.496 kl = ______ l

8. Find the volume of the solids in Figure 6.50 made by gluing faces of unit cubes together.
9. Find the volume of each of the solids in Figure 6.51.
10. a. Suppose each edge (length, width, and height) of a cuboid is doubled. How does this affect the volume—that is, is the volume double also?
    b. How is the volume changed if each dimension is multiplied by 10?
    c. Suppose the values of l, w, and h are multiplied by a factor k. By what factor is the volume multiplied?
11. a. Complete a table showing the ratio of the volume of a cube to its surface area for cube edge lengths from 1 to 10 cm.
    b. What conclusions can you draw about the ratio of volume to surface area of a cube?
12. If the volume of a cube is 1, estimate the volume of a sphere inscribed in the cube.
13. It can be shown that the formula for the volume of a sphere is $V = 4/3\ \pi r^3$. Use this formula to check your estimate in Exercise 12.
14. One sphere has twice the radius of another. How do the volumes of the spheres compare?

APPLICATION

Duct Choice Problem: A furnace contractor needs to install a duct or ducts that will carry as much air as possible. The largest possible duct that he can use is 7″ × 14″. The smallest is 5″ × 10″. He also wants to do the job as economically as possible. He knows that the larger the cross-sectional area is, the greater the amount of air moved. He also knows that the larger the cross-sectional perimeter is, the more material needed to make the duct. Predict whether the contractor should use one large duct or two smaller ducts. Then calculate to check your answer. Explain your reasoning.

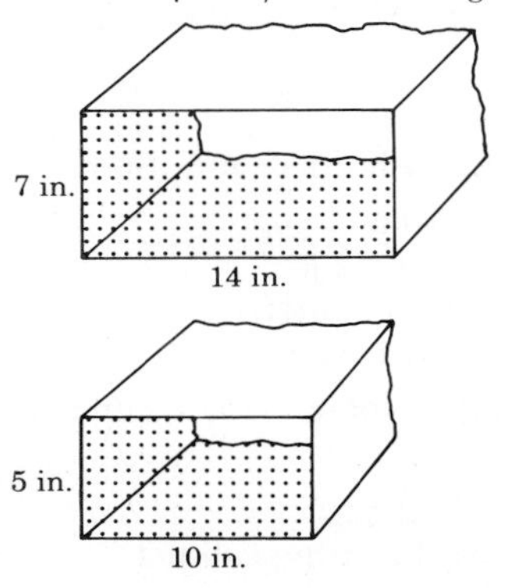

APPLICATION

Pencil Problem: A golf specialty company contracts a pencil manufacturing company to make small score-keeping pencils like the one shown. The pencil manufacturer wants to determine how much graphite they will need to make 500,000 of these pencils. Also, they want to see how much graphite they would save if they reduced the diameter of the graphite part in each pencil to 0.115 inches, and whether it would be worth the $500 it would take to change the machines to accomplish this. Tell as much about a solution to this as you can.

0.125" diameter

3"

0.25"

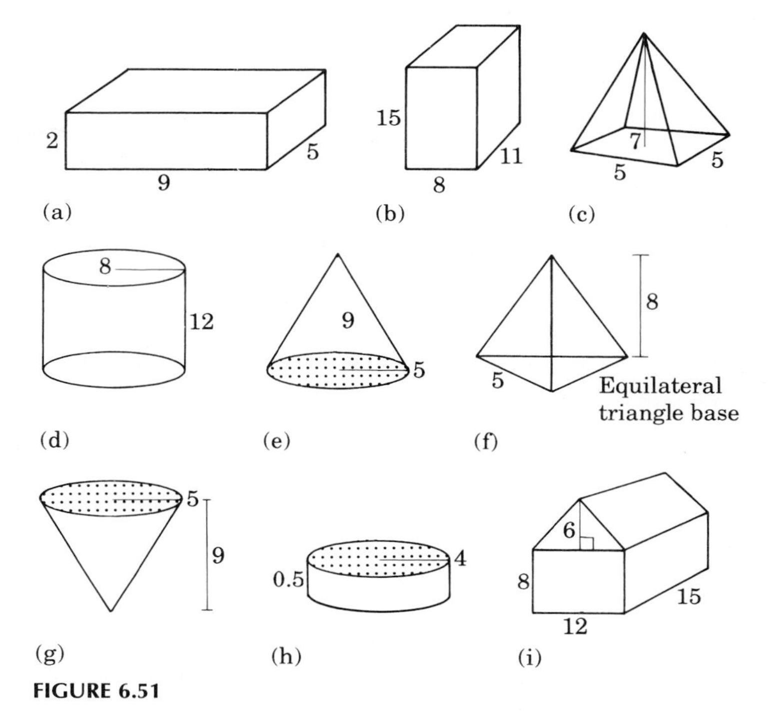

FIGURE 6.51

15. **Critical Thinking**. Make a solid out of clay, wood, or other material that completely plugs up each of the holes in Figure 6.52 and that can be pushed through each with no change in shape.

To extend these exercises, select from Supplementary Exercises: Volume and Capacity in Appendix E.

FIGURE 6.52

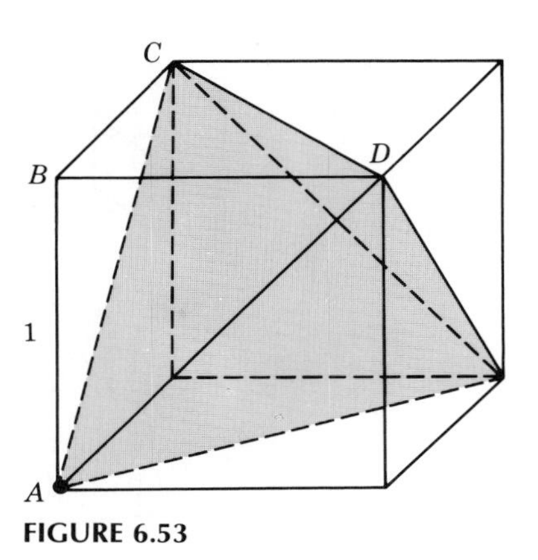

FIGURE 6.53

Problem Solving: Developing Skills and Strategies

To solve some problems, you must **use data from a diagram**. Try the problem below.

Problem: Suppose the length of the edge of a cube (Fig. 6.53) is 1. What is the volume of the inscribed tetrahedron? [*Hint:* Consider the volume of the pyramid *ABCD*.]

MEASURING MASS AND WEIGHT*

When the "measurable property," as referred to in the description of the measurement process given at the beginning of this chapter, can be roughly described by "how much stuff (matter) there is in an object," or, more technically, as "the property of an object that resists acceleration," we are measuring **mass**. When the measurable property can be described as "the amount of attraction between two objects," we are measuring **weight**.

To measure weight, we might use a spring scale and see how much pull is exerted on the object (Fig. 6.56). To measure mass, we might use a balance and compare the object to be measured with a standard unit (Fig. 6.55). Thus, the *mass* of a rock remains the same, whether it is on the earth or on the moon, since it will balance with the standard unit in both places. The *weight* of the rock, however, changes depending where the rock is located. A rock on the earth might pull the

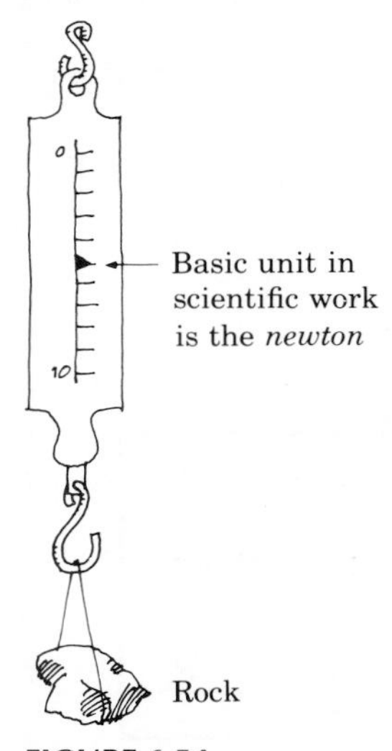

FIGURE 6.54

*These concepts are scientific as opposed to geometrical in nature; consequently, they are not prerequisite to other topics in the text.

TEACHING NOTE

Some interesting experiments for students relating to math as well as to the metric units developed previously in this chapter can be found in *The Activities for Base Ten Blocks* available from Dale Seymour Publications.

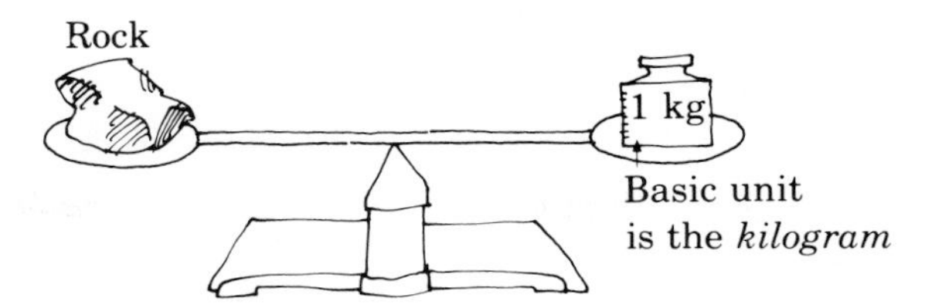

FIGURE 6.55

spring scale down to 36, while the same rock on the moon would pull the same spring scale down to 6.

This difference between mass and weight is important to understand for scientific purposes, but it doesn't significantly affect practical, everyday measurement. This being the case, we will refer to the "weight of an object" and use the standard units of mass for these measurements. Investigation 6.7 provides an opportunity for exploration of the basic metric system unit of mass.

INVESTIGATION 6.7

For this investigation you will need a balance, the cubic decimeter container you made in Investigation 6.6, and a strong plastic food bag. Place the plastic bag in the container and fill it with cold water. Place the filled cubic decimeter on the balance.

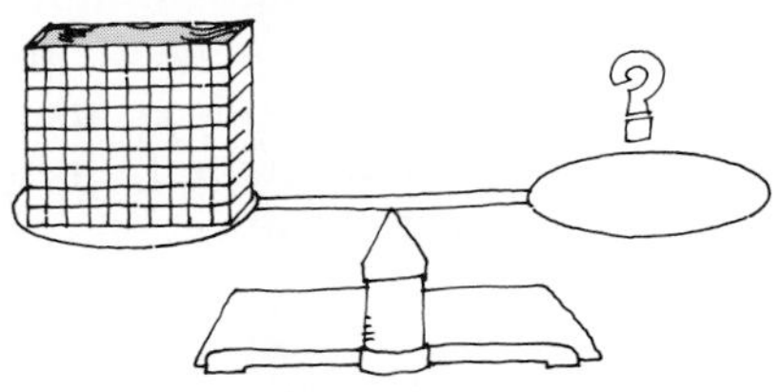

A. Find at least 3 objects or collections of objects that will balance the container of water.
B. About how many objects weighing 1 pound would it take to balance the container of water?
C. Find some metric weights, if possible. Which of these weights or collections of weights will this container of water balance?

For a related activity, see Activity 13 from Appendix D, Critical Thinking.

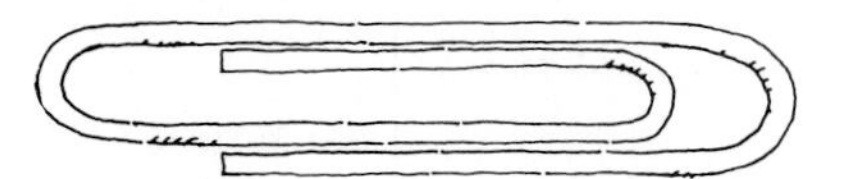

A large paper clip weighs approximately 1 gram.

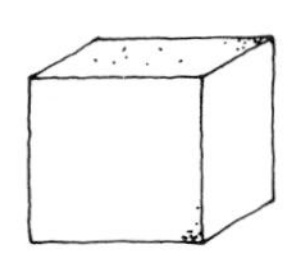

A sugar cube weighs about 2 grams.

A nickel weighs about 5 grams.

FIGURE 6.56

TABLE 6.7 Metric Units for Measuring Mass

Unit	Symbol	Relation to Gram
Kilogram	kg	1000 g
Hectogram	hg	100 g
Dekagram	dag	10 g
Gram	g	**1 g (basic unit)**
Decigram	dg	0.1 g
Centigram	cg	0.01 g
Milligram	mg	0.001 g

The **kilogram** (kg), a metric unit of mass equal to the mass of 1 cubic decimeter (dm^3) of water at a temperature of 4°C, is often used in practical measurement. The basic metric unit of mass is the **gram** (g). It is 1/1000 of a kilogram and is relatively small, as indicated by the sketches shown in Figure 6.56. Table 6.7 describes the gram and related units. For example, a small box of raisins might weigh 34 g, a person might weigh about 70 kg, while a compact car might weigh as much as 900 kg. To extend these ideas, use Supplementary Activities: Mass, Activities 1 through 3, in Appendix C.

In practical situations, the word "kilo" is often used instead of kilogram. If a person said, "I bought a ham that weighed 4.35 kilos," it could be interpreted as, "I bought a ham that weighed 4 kilograms and 350 grams." Since 1 kilogram weighs 1000 grams, this interpretation gives a reasonably clear idea as to just how much ham was purchased.

Since each metric unit of mass is 10 times the next smaller unit (check Table 6.7 to see that this is true), one can convert from one unit of mass to the next larger or next smaller unit by moving the decimal point one place to the left or one place to the right. For example, 453 g equals 0.453 kg and 453,000 mg.

In addition to length, area, volume, capacity, and weight,

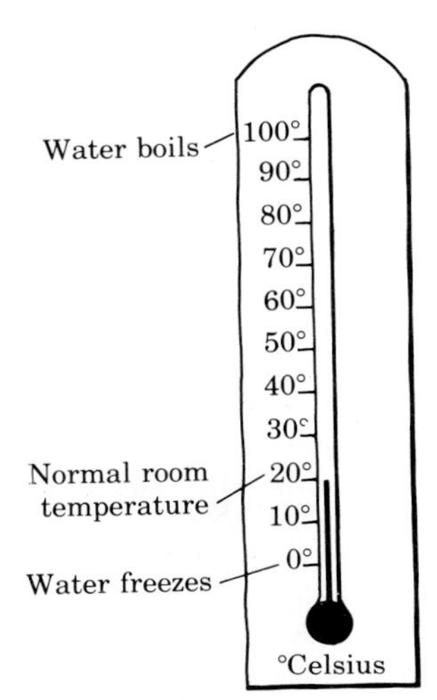

FIGURE 6.57

the measure of time and temperature plays an important role in science and everyday affairs. The unit of time, the **second** (s), is customarily used and is quite familiar. The unit of temperature in the metric system is °**Celsius** (°C) and is defined in terms of the freezing and boiling point of water. At sea level water boils at 100°C and freezes at 0°C (Fig. 6.57).

Other interesting relationships between the metric units are suggested by the diagram in Figure 6.58. The diagram can be used to complete an exercise like the following:

1 cubic centimeter of water at 4°C weighs 1 ____.
1 cubic decimeter of water at 4°C weighs 1 ____.
1 liter of capacity has a volume of 1 ____.
1 cubic centimeter of water has a capacity of 1 ____.
1 milliliter of water at 4°C weighs 1 ____.

We find, then, that the metric system of units is a system in which the basic units are interrelated and in which the interpretation of measurement is simplified through the utilization of the decimal system of numeration.

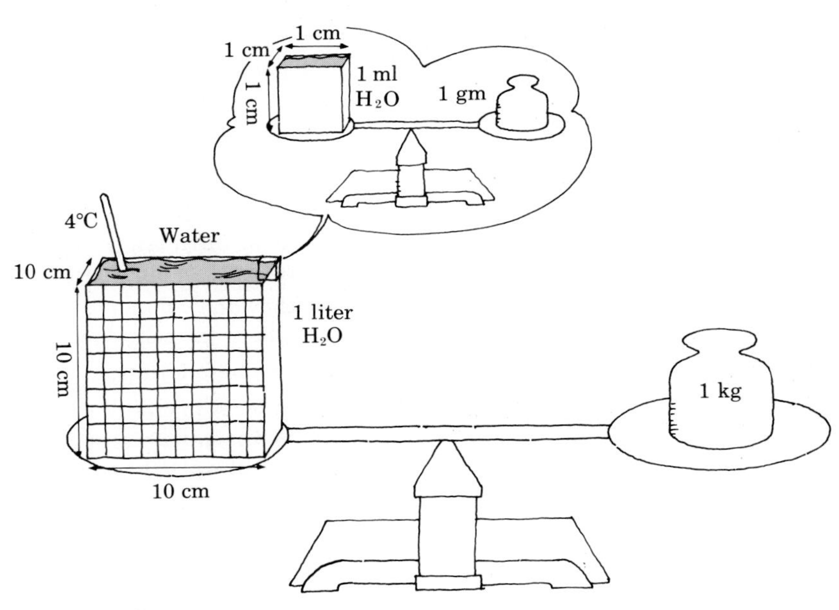

FIGURE 6.58

EXERCISE SET 6.6

1. Select objects or collections of objects, weighing
   a. 1 kg   b. 10 g
2. What does a cubic centimeter of water weigh?
3. Estimate the weight in grams or kilograms of the following. Check your estimates if possible.
   a. A dime
   b. A new piece of chalk
   c. A teaspoon
   d. A paper hole punch
   e. A can of pop
   f. A small chalkboard eraser
   g. Your weight
   h. A small ceramic coffee cup
   i. Your textbook
   j. A penny
   k. A plastic centimeter ruler
   l. An unsharpened lead pencil
   m. A ball-point pen
   n. A newborn baby
   o. A poodle
   p. A "pound of cheese"
4. An object that weighs 11 lb weighs 5 kgs. Use only this information to find or estimate the weight in kilograms of the following:
   a. A 220 lb man
   b. A 4400 lb car
   c. A 55 lb child
   d. A 16 lb turkey
   e. A 100 lb sack of feed
   f. A 450 lb decorative rock
5. Which is more, 1.87 kg or 1869 g?
6. Estimate these temperatures in °Celsius and find their locations on the Celsius thermometer.
   a. Normal body temperature
   b. Cold pop, without ice
   c. Ice cream
   d. Hot soup
   e. Warm bath water
   f. Water from steaming teakettle
   g. Hottest earth temperature on record
   h. A cool spring day ("sweater weather")
7. If you know the volume of a water tank in liters, how do you find the weight of the contained water in kilograms?
8. Complete the following:
   a. 1 cubic decimeter of water weighs ____ gram(s).
   b. 1 liter of water weighs ____ gram(s).
   c. 1 liter of water has volume ____ cubic centimeter(s).
   d. 1 liter has volume ____ cubic meter(s).
   e. 1 liter of water weighs ____ kilogram(s).
   f. 1 milliliter of water weighs ____ gram(s).
   g. 1 milliliter of water weighs ____ kilogram(s).
   h. 1 cubic centimeter of water weighs ____ gram(s).

9. The English system of units has been used in the United States for many years and is of historical interest.
   a. Make a table of length units in the English system and show the comparisons between them. (Use inch, foot, yard, rod, and mile.)
   b. Make a table of area units in the English system and show the comparisons between them. (Use square inch, square foot, square yard, square rod, acre, and square mile.)
   c. Make a table of weight units in the English system and show the comparisons between them. (Use grain, ounce, pound, stone, and ton.)
   d. Make a table of capacity units in the English system and show the comparisons between them. (Use ounce, pint, quart, and gallon.)
10. Compare the relationships in Exercise 9 with the relationships between units in the metric system. What conclusions do you draw?
11. Make a table of conversions from basic metric units to basic English units, and vice versa.
12. Work each of these problems. Show your work.
    a. Given a cubical tank 7 ft 8½ in on a side, find (i) the volume in cubic feet, (ii) the volume in gallons, and (iii) the weight of the water in pounds.
    b. Given a cubical tank 3.08 m on a side, find (i) the volume in cubic meters, (ii) the volume in liters, and (iii) the weight of the water in kilograms.
13. Discuss the relative difficulty of computation in the two solutions to the problem in Exercise 12.
14. Work each of these problems. Show your work.
    a. Given a piece of land 500 yd by 500 yd, find the number of acres of land.
    b. Given a piece of land 500 m by 500 m, find the number of hectares of land.
15. Discuss the relative difficulty of computation in the two solutions to the problem in Exercise 14.
16. a. If a metric ton (1000 kg) costs $150,000, then 1 kg costs ______ and 1 g costs ______.
    b. In the English system, if 1 long ton costs $150,000, then 1 lb costs ______ and 1 oz costs ______.
    c. Discuss the relative difficulty of computation in the two solutions to the problems in parts a and b.
17. If you know the miles per gallon that can be driven for a certain car, how can you easily convert this to liters per kilometer?
18. Write appropriate numbers so that this story makes sense: M(r)(s) Swift had taught (**a**)th grade at Chaos School for (**b**) years. (S)he

was an excellent teacher who enjoyed children and life. (S)he was about (**c**) meters tall and weighed (**d**) kilograms. (S)he loved to play tennis and golf in the summer, but had difficulty deciding which to do. On a hot (**e**)°C day last summer (s)he decided to do both. After 2 sets of tennis and (**f**) milliliters of lemonade, (s)he headed for the golf course. The secret, of course, was to transfer thought from the larger (**g**) centimeter diameter, (**h**) gram tennis ball to the smaller (**i**) millimeter diameter, (**j**) gram golf ball. Naturally, (s)he was successful and lofted a respectable (**k**) meter drive off of the first tee. M(r)(s) Swift lived in a large house with (**l**) square meters of floor space on a (**m**) acre lot. (S)he drove a small, economy car, so (s)he didn't need too large a garage. The car *was* economical, getting (**n**) kilometers/liter on long (**o**) kilometer trips. M(r)(s) Swift has been dreaming of a large, (**p**) hectare wooded area in the country nearby, with a small lake, a cottage, and a cow that gives (**q**) liters of milk a day. Back to reality, M(r)(s) Swift is looking ahead to teaching school next fall and (s)he is making plans to introduce the metric system to the students. (S)he is a devoted teacher and a "real" person.

19. **Critical Thinking**. Suppose you as a teacher received a letter from a parent who felt that the United States should not change to the metric system because it would cause a lot of trouble, it would cost his business a lot of money, it would be difficult for chefs and real estate agents, and it would be impractical for almost everyone. The parent wonders why you as a teacher are spending so much time on the metric system, why you think it is better, and why parents weren't asked to help make the decision about changing to the new system. Write a letter giving your response to this parent.

To extend these exercises, select from Supplementary Exercises: Mass and Supplementary Exercises: Temperature in Appendix C.

PEDAGOGY FOR TEACHERS

ACTIVITIES

1. Write a sequence of discovery exercises that would lead a student to discover the formula for the area of (a) a triangle, (b) a parallelogram, and (c) a trapezoid.
2. Choose one of the formulas relating to area or volume in this chapter and devise a teaching aid that would help students understand why this formula gives the correct area or volume. Field-test your teaching aid and revise it if necessary.
3. Study the strands of measurement in a selected elementary school mathematics textbook series. Outline the topics presented and the grade level at which they are presented. What aspects of the metric system would you present to students at a given grade level that are not presented in the text?

4. Develop an overhead projectural with overlays that could be used to help students understand the formula for the area of a trapezoid. Use color to help make the ideas stand out.
5. Select laboratory aids from companies such as Cuisenaire, Creative Publications, or Dale Seymour Publications that deal with volumes of solid figures or that deal with area and perimeter relationships in the circle. State your criteria and evaluate the appropriateness of these materials for classroom use at a given level.
6. Develop a learning activity for one aspect of (a) length, (b) area, (c) capacity, and (d) mass using metric system units.
7. Refer to catalogues from producers of commerical materials and prepare a requisition ordering the materials that you believe are needed to do an adequate job teaching the metric system at a selected grade level. After you have prepared the desirable list of necessary materials, assume that your school system can provide only $50 for your classroom. Reduce your list to accommodate this budget restriction.
8. Devise an interview that could be used with students at a given grade level, or with adults, to determine the extent of their current knowledge about the metric system. Make the interview brief, containing only about 4 or 5 questions. Interview various persons and display your collected information graphically.

COOPERATIVE LEARNING GROUP DISCUSSION

Suppose that you are on a committee of teachers who is to make a presentation at the next Parent–Teacher Association meeting on the subject "Why the United States should adopt the metric system." Decide whether you agree or disagree with the following reasons and prepare a detailed outline of the presentation that you will make.

1. The confusion caused by certain groups in our country using the metric system (certain sports, film manufacturers, pharmaceutical suppliers, the medical profession, the optical manufacturers, etc.) and other groups using the English system would be eliminated.
2. Time would be saved in the elementary school mathematics and science curriculum (some persons estimate as much as two years) because
   a. The metric system is easier to learn; fewer, more related units.
   b. Calculations take less time; mostly whole numbers and decimals instead of fractions and mixed numerals.
   c. Teaching a dual system of measurement (English for practical use and metric for science use) would be unnecessary.

CHAPTER REFERENCES

NCTM. 1976. *Measurement in School Mathematics*. Reston, VA: National Council of Teachers of Mathematics.

Steen, Lynn Arthur, Ed. 1990. *On the Shoulders of Giants—New Approaches to Numeracy*. Washington, DC: National Academy Press.

SUGGESTED READINGS

Consult the following sources listed in the Bibliography for related readings: Neufeld, 1989; Shaw and Cliatt, 1989; Shroyer and Fitzgerald, 1986; and Smith, 1990.

CHAPTER SEVEN Motions in Geometry

TEACHING NOTE

The concept of function (as a unique type of correspondence) is very important in mathematics. The intuitive work with slides, flips, and turns in the elementary school provides a valuable background for an understanding of the idea of transformation, which is a geometric type of function. Thus, early physical experiences provide the foundation for an important mathematical idea.

INTRODUCTION

For centuries, people have been fascinated by motion, and many have sought to develop perpetual motion machines. Figure 7.1 shows one such endeavor, which is a variation of Leonardo Da Vinci's perpetual motion machine based on freely moving spheres.

Motion plays a significant role in the experience of persons of all ages. We live in a world of all kinds of motion, and our perceptions of the properties of these motions are continually being refined and broadened. Because mathematics is useful in describing and analyzing motion, the mathematics of motion is an appropriate topic of study for students at various levels.

Three simple types of motion will be analyzed in this chapter. These motions—slides, flips, and turns—leave the size and shape of objects unchanged. For example, a glass elevator ride would ordinarily not be thought of as a "slide," but it is an example of this type of motion (Fig. 7.2). The slide projector is a useful example of the flip motion. When a slide-projected picture appears reversed on the screen, we remove the slide, flip it over, and place it back in the machine (Fig. 7.3). A ride on a ferris wheel is a vivid example of the turn motion (Fig. 7.4). The mathematical terms for slide, turn, and flip are **translation, rotation**, and **reflection**, respectively.

FIGURE 7.1

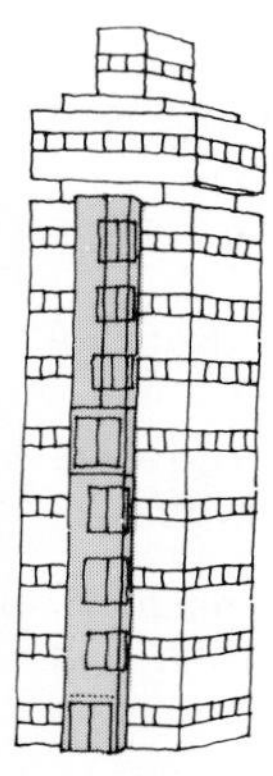

FIGURE 7.2

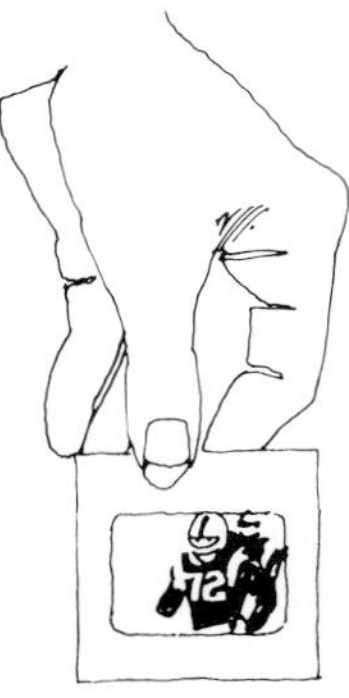

FIGURE 7.3

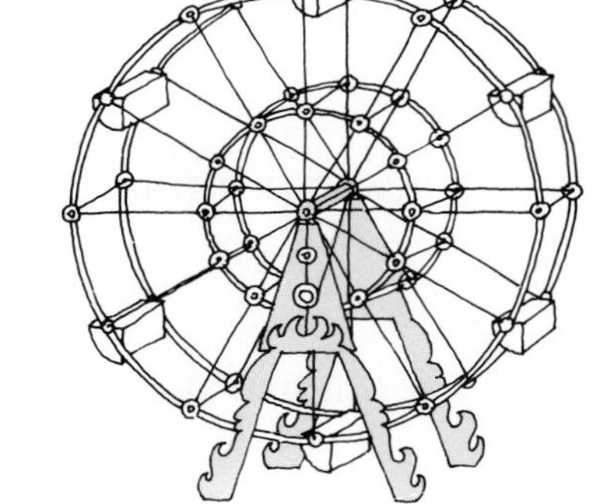

FIGURE 7.4

TEACHING NOTE

Small children have a primitive concept of motion and may perceive their mother moved to a different position as a different mother. Later, they describe change of position of objects in terms of the end position only, disregarding the beginning position. Piaget considers the "change of position" developments in children in a chapter of *The Childs Conception of Geometry* (Piaget, Inhelder, Szminska, 1964).

FIGURE 7.5

SLIDE OR TRANSLATION

A sliding motion is familiar to all of us. We experience it in our homes with sliding doors and sliding drawers; we have experienced it at play with the sled and the sliding board. We may have seen instances of how a job can be made a bit easier by sliding objects on an inclined plane (Fig. 7.5).

In order to study the slide motion and its relationship to geometry more thoroughly, we shall consider slides in the plane. Our geometric ideas will be suggested and reinforced by our experiences with physical objects. Consider, for example, placing a rectangular cutout exactly on the rectangle A in Figure 7.6. This cutout can be made to coincide with the rectangle B by simply *sliding* it along the paper without any turn or twist. We call rectangle B the **slide image** of rectangle A. The cutout was slid in a particular *direction* through a particular *distance*. The direction and distance of the translation can be specified by using an arrow. For example, in Figure 7.7, the points P', Q', and R', are the slide, or translation, images of points P, Q, and R for the slide or translation associated with arrow AB. Investigation 7.1 utilizes tessellations of the plane to provide a situation to help you develop your intuition about slides.

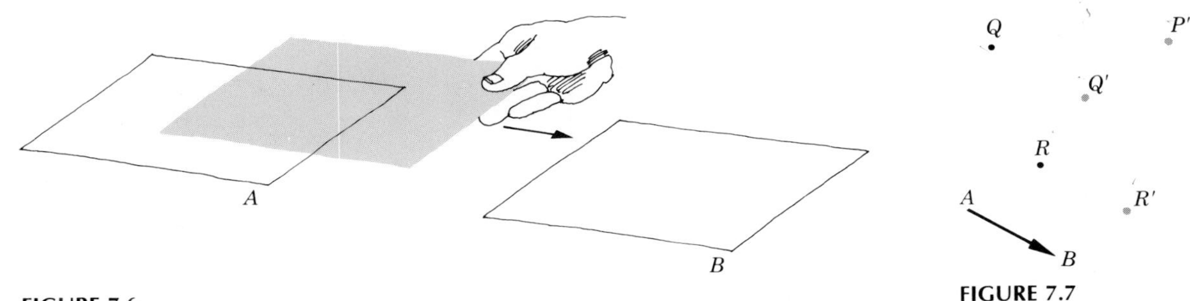

FIGURE 7.6

FIGURE 7.7

INVESTIGATION 7.1

Make a copy of the following tessellation on drawing paper. Then place tracing paper on this tessellation and trace it. If the tracing is slid (no turns or flips) from A to B, the tracing will "fit back on" the tessellation (assuming the original tessellation covers the complete plane, rather than just the portion shown).

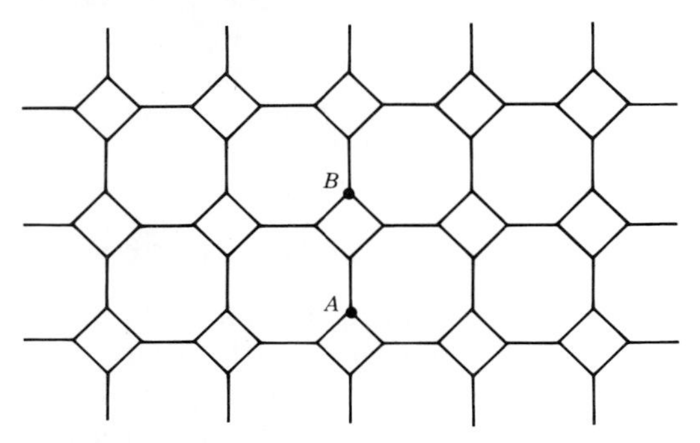

A. Starting with point A, find at least 8 different *directions* along which the tracing can be slid to fit back on the tessellation. Describe a slide in each of these directions by marking a starting point and an ending point for the slide on the tracing paper. Describe at least one more slide in a direction different from those already described that will enable the tracing to fit back on the tessellation.

B. For each tessellation (a), (b), and (c), find at least 6 different directions along which a tracing can be slid to fit back on the original tessellations. Give the starting and ending points to show one slide in each of these directions.

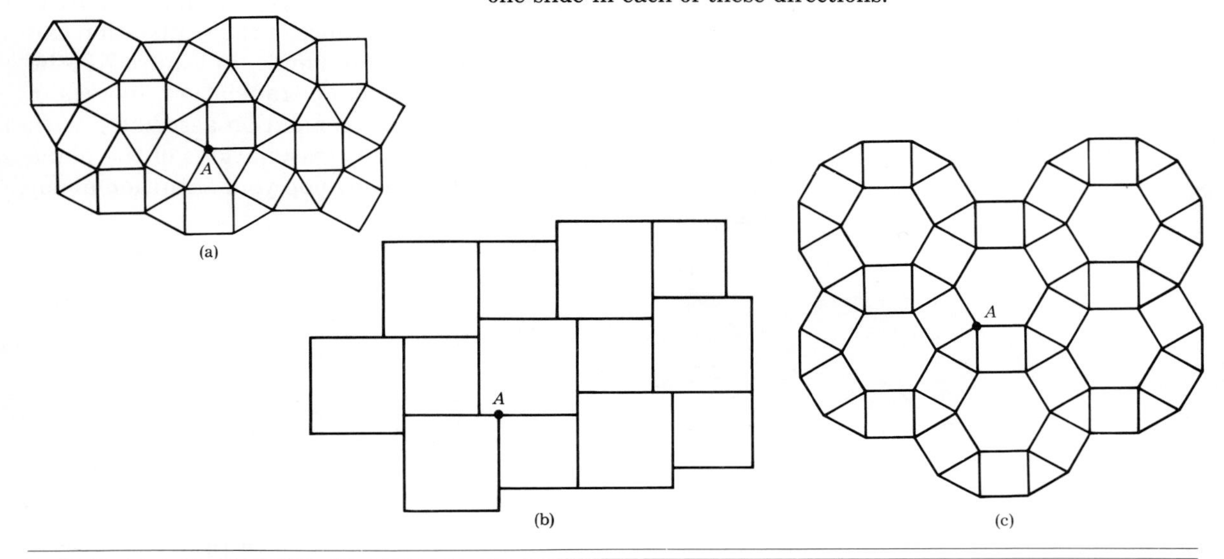

(a) (b) (c)

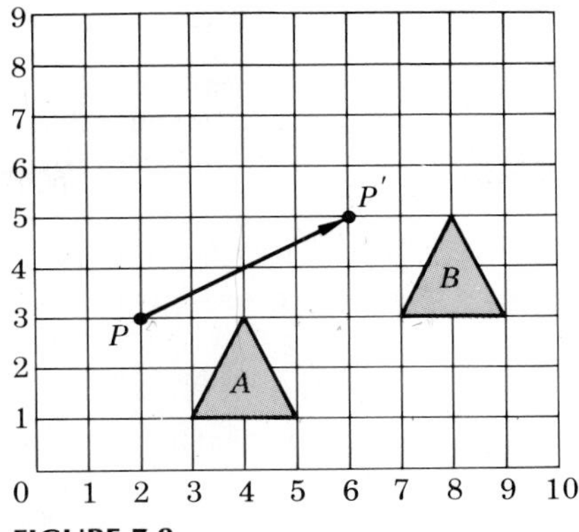

FIGURE 7.8

Since an arrow indicates both a specific *direction* and a specific *distance*, it completely determines a particular slide. In Figure 7.8, the point P' with coordinates (6,5) is the *slide image* of point P with coordinates (2,3). We also say that triangle B is the slide image of triangle A for the slide PP'.

In Figure 7.9, the slide images for the translation arrow AA' of the figures in black are shown in gray. The reader should use tracing paper to show that this is true. Note that if A is marked on the tracing paper and the paper is slid so that A moves to A', then every other figure that has been traced on the tracing paper also moves through the distance and direction AA'.

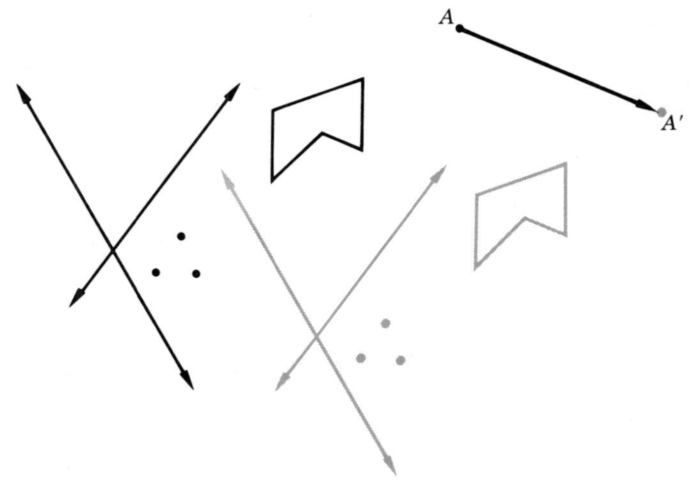

FIGURE 7.9

EXPLORING WITH THE COMPUTER 7.1

Use LOGO software to explore the following: A slide or translation requires a direction and a distance. The procedure SLIDE requires 2 inputs. The first indicates the direction, and the second the distance to move.

1. Try SLIDE with different angle values and distances. Also, try the same direction and distance 2 or 3 times without clearing the screen. Watch the arrow and make note of its orientation. (A REPEAT statement can be used to do the operation several times.)

```
TO SLIDE :A :D          TO ARROW
ARROW                   FD 25 RT 25 BK 10 FD 10 LT 25 BK 25
PU RT :A FD :D PD       END
ARROW
END
```

2. Write a procedure that will draw a figure that can be moved using SLIDE. Be sure that it has some feature that will allow you to watch the orientation of the figure—such as a diagonal or an arrow on one side. Redefine SLIDE to use your figure again. Try several different directions and distances.
3. Use SLIDE to test the following conjectures:
   a. Slide AB followed by slide CD gives the same image of a figure as slide CD followed by slide AB.
   b. Sliding a triangle does not change its shape or size or orientation.
   c. The image of a figure resulting when one slide is followed by a second slide can always be produced by a single slide.
4. Make and test other conjectures based on your observation of SLIDE.

EXERCISE SET 7.1

1. In Figure 7.10, line l is parallel to arrow AA'. Use line l and your compass to find the slide image of point B after the slide AA'.

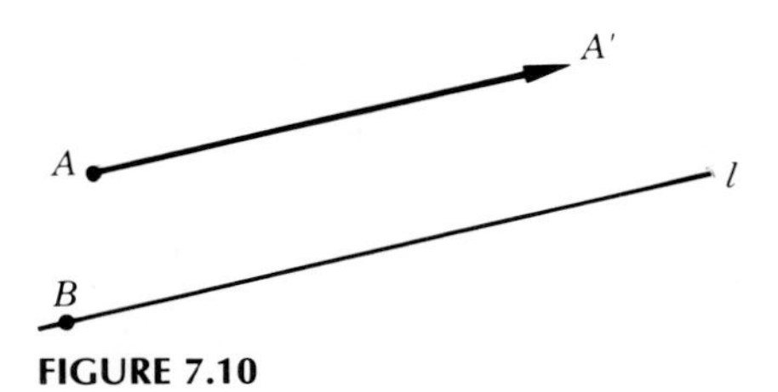

FIGURE 7.10

FIGURE 7.11

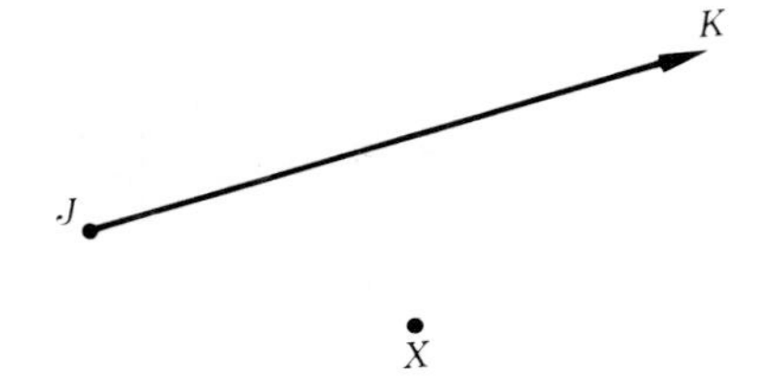

FIGURE 7.12

2. In Figure 7.11, which of the figures on the right is a slide image of the figure on the left?
3. Draw arrow JK and point X on your paper (Fig. 7.12). Construct a line through X that is parallel to a line through arrow JK. Use this line and your compass to find the slide image of point X after slide JK.
4. Draw triangle ABC and arrow RS on your paper (Fig. 7.13). Lay a sheet of tracing paper over this paper and trace triangle ABC and point R. Slide the tracing paper so that the traced point R moves to point S on the original arrow. Push your pencil tip down on points A, B, and C so that impressions of these points show up on the underneath paper. Mark these impression points A', B', and C' and connect them to form the slide image of triangle ABC after the slide RS.
5. Draw quadrilateral $ABCD$ and arrow PQ on your paper (Fig. 7.14). Find the slide image of the quadrilateral after the slide PQ.

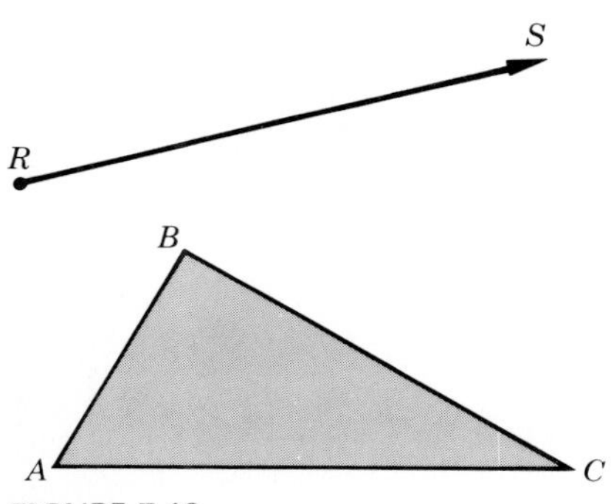

FIGURE 7.13

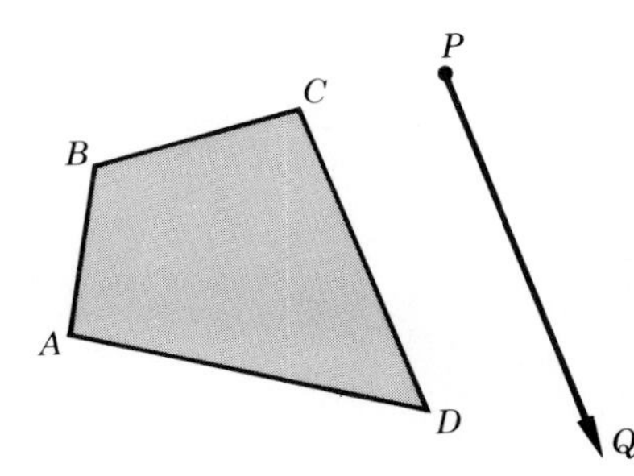

FIGURE 7.14

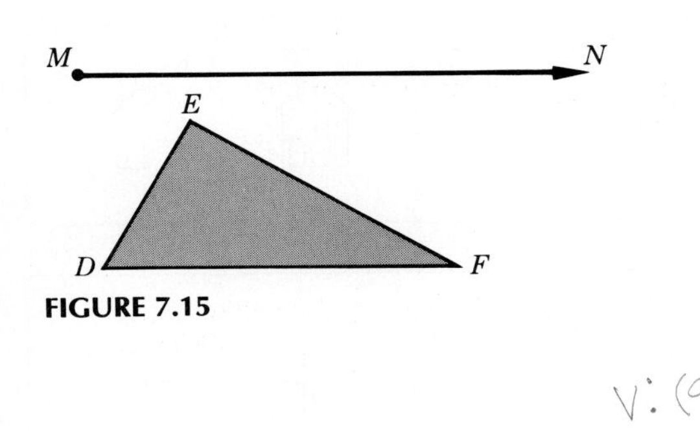

FIGURE 7.15

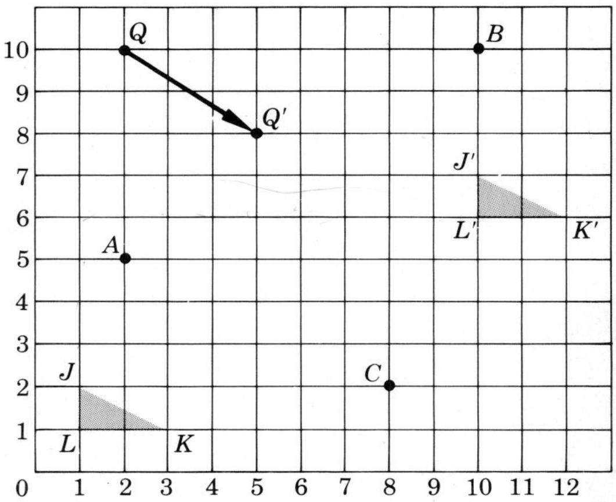

FIGURE 7.16

6. Use the idea in Exercise 4 three times to construct with ruler and compass the slide image of triangle DEF after slide MN (Fig. 7.15).
7. Copy Figure 7.16 on a sheet of graph paper. Draw 5 more arrows that describe the slide QQ'.
8. Give the coordinates of the slide image of points A, B, and C for the slide determined by arrow QQ'.
9. Draw an arrow on graph paper that describes the slide whose slide image of $\triangle JKL$ is $\triangle J'K'L'$.
10. Draw Figure 7.17 on graph or dot paper and do the following:
    a. Draw the slide image of the quadrilateral for slide AA'.
    b. Draw the slide image of the triangle for slide AA'.
    c. Draw the slide image of line p for slide AA'.
11. Draw Figure 7.18 on graph or dot paper and do the following:
    a. Draw the slide image of line p for slide AA'. Does it pass through the point P?
    b. Draw a line q whose slide image for slide AA' does pass through point P.

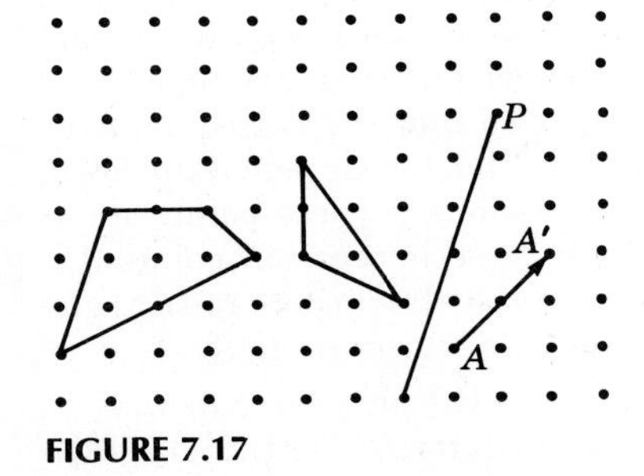

FIGURE 7.17

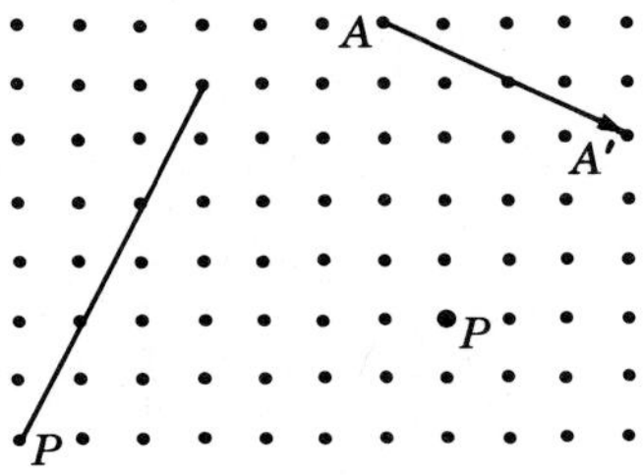

FIGURE 7.18

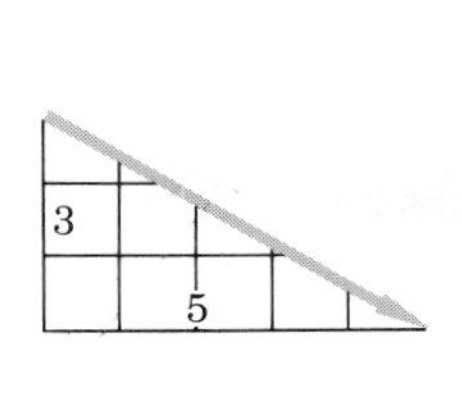

Down 3, right 5
"point slider"

(a)

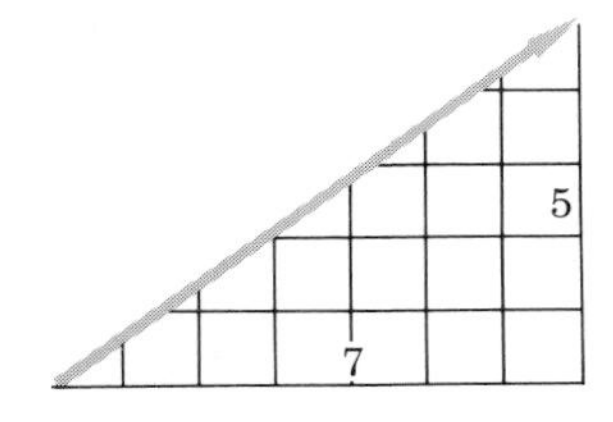

Right 7, up 5
"point slider"

(b)

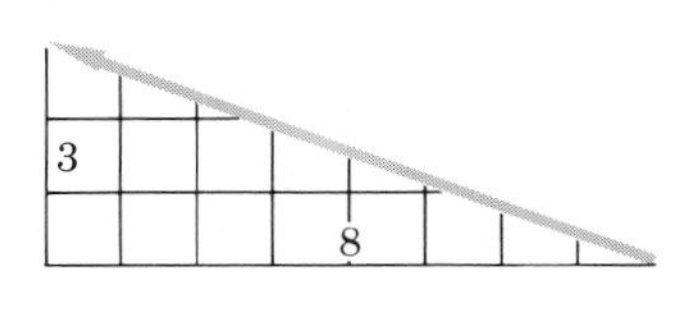

Left 8, up 3
"point slider"

(c)

FIGURE 7.19

12. A student cut "point sliders" from graph paper to describe 3 slides (Fig. 7.19) and used them to find the slide images of certain figures. Make similar slide pointers on graph paper and do the following:
    a. The ends of a line segment AB have these coordinates: $A(3,7)$ and $B(6,4)$. Find the coordinates of the endpoints of the slide image of this segment after the slide pictured in Figure 7.19a.
    b. The vertices of triangle PQR have these coordinates: $P(2,3)$, $Q(4,-2)$, and $R(-1,-3)$. Find the coordinates of the vertices of the slide image of this triangle after the slide pictured in Figure 7.19b.
    c. Two points on line l have coordinates as follows: $R(3,4)$ and $S(2,6)$. Find the coordinates of 2 points on the slide image of line l after the slide pictured in Figure 7.19c.
13. Find the coordinates of the slide image of the given point for the slide that moves point $O(0,0)$ to point $A(3,2)$:
    a. $(-1, 1)$ b. $(4, -2)$ c. $(7, 5)$ d. (x, y)
14. Find the coordinates of the point whose slide image is the given point for the slide that moves point $O(0,0)$ to point $A(3,2)$:
    a. $(1, 3)$ b. $(-2, -6)$ c. $(4, 6)$ d. (x, y)
15. **Critical Thinking**. Consider this generalization: The slide image of a line s is sometimes the line s itself.
    a. Is this generalization true? How would you convince someone?
    b. In Figure 7.20, is slide AD followed by slide DC the same motion as slide AB followed by slide BC? Explain.

FIGURE 7.20

Problem Solving: Developing Skills and Strategies

Solve the following problem: A Koala bear started climbing up a 25 ft tree. It climbed 7 ft each day, but during each night it slid back a certain same distance. If the bear got to the top of the three on the seventh day, how far did it slide back each night? Which strategy—draw a picture; guess, check, and revise; work backward; or look for a pattern—do you think is most useful? Why?

FIGURE 7.21

TURNS OR ROTATIONS

A turning motion, just as a slide motion, is one that we have experienced since childhood and that we experience daily. We experience it in the door knob and the door key; we experience it when we roll up an automobile window, wind a watch, or wind a toy truck. We may have seen instances of how a job can be made easier by using a winding reel or a turning gear (Fig. 7.21).

Our concept of the turning motion is refined and extended as we continue to experience turns not only in everyday life, but in the worlds of science and technology as well. The Shinto symbol shown in Figure 7.22 is symbolic of a turning motion and is said to represent the revolving universe.

TEACHING NOTE

Our experience indicates that perhaps because of the potential for hands-on activity it provides, the topic of motions is exciting for students of all ages. The *Geometric Connectors: Transformations* software from Sunburst Publications should be considered for use with junior high school and secondary level students.

FIGURE 7.22

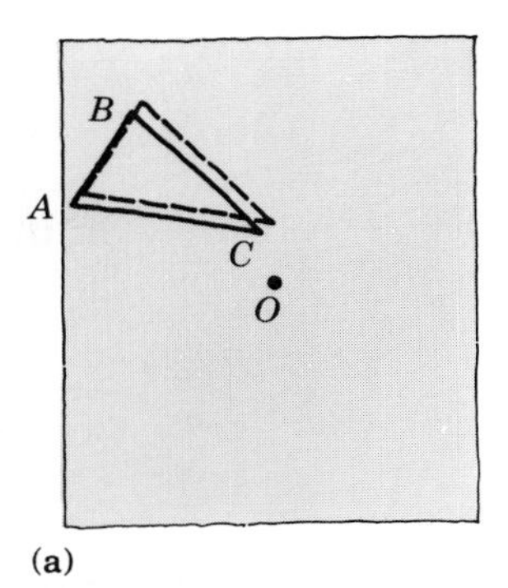

(a)

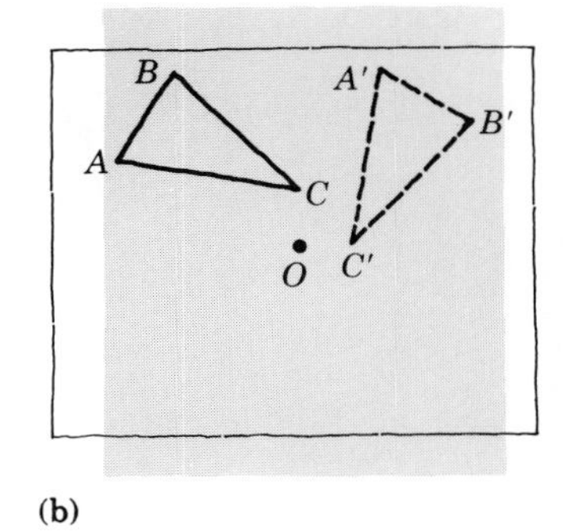

(b)

FIGURE 7.23

In order to illustrate the turning motion using physical materials, consider two sheets of acetate fastened with a paper fastener. The $\triangle ABC$ is drawn on the underneath sheet and traced on the top sheet, as shown in Figure 7.23a. If the underneath sheet is turned as shown in Figure 7.23b, $\triangle A'B'C'$ can be traced on the top sheet.

We call $\triangle A'B'C'$ the **turn image** of $\triangle ABC$. The triangle was turned around point O (called the center of the turn) through a given angle (90°), which is called the **angle of the turn**. Investigation 7.2 utilizes tessellations of the plane to help you develop your intuition about turns (Fig. 7.24).

FIGURE 7.24

"You're fired!"

INVESTIGATION 7.2

Make a copy of the following tessellation on drawing paper. Then, place tracing paper on this tessellation and trace it. If the tracing is held firmly at point F by the tip of a pencil and is rotated 180°, the tracing will "fit back on" the tessellation (assuming the original tessellation covers the complete plane, rather than just the portion shown).

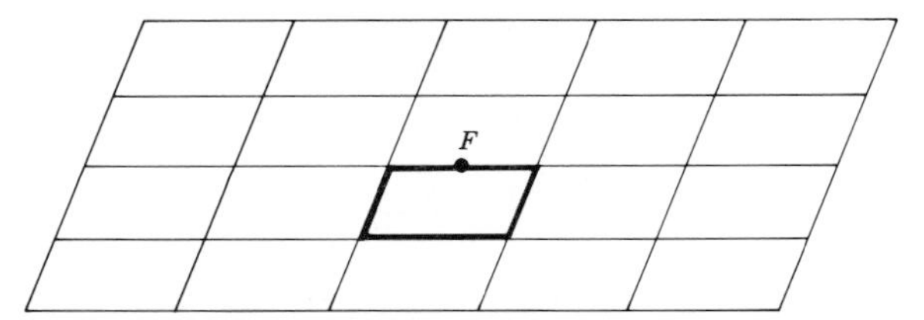

A. How many other points can you find on or inside the parallelogram with darker lines for which this is true? Draw a portion of the tessellation and label these points.
B. If rotations other than 180° are allowed, how many points can you find?
C. Describe all possible centers for turns (such as point F) and the angles of the turns (such as 180°) that will make tracings of tessellations (a), (b), and (c) fit back on themselves.

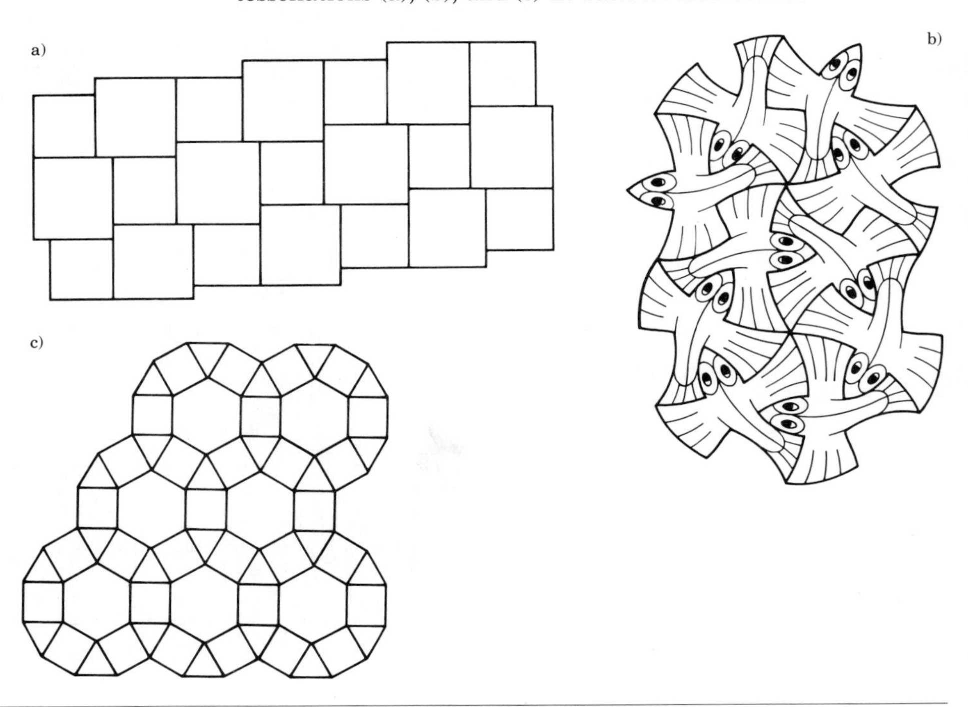

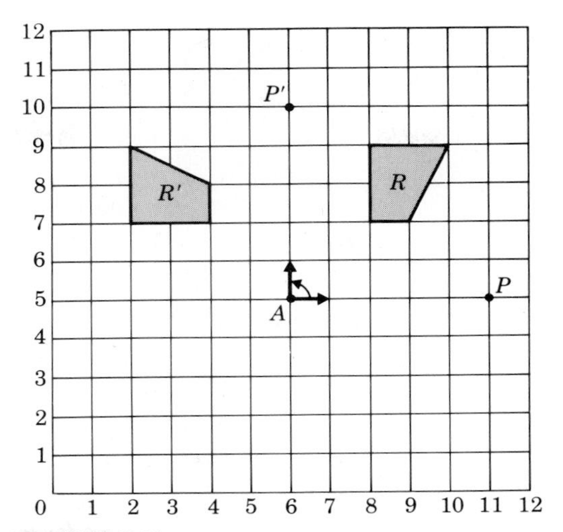

FIGURE 7.25

Since a turn is determined by specifying a center and an angle, the double arrow notation in Figures 7.25 and 7.26 is used to describe turns. The notation in Figure 7.25 indicates a 90° counterclockwise turn, or rotation, centered at point A, while a 180° clockwise turn centered at point B is indicated in Figure 7.26.

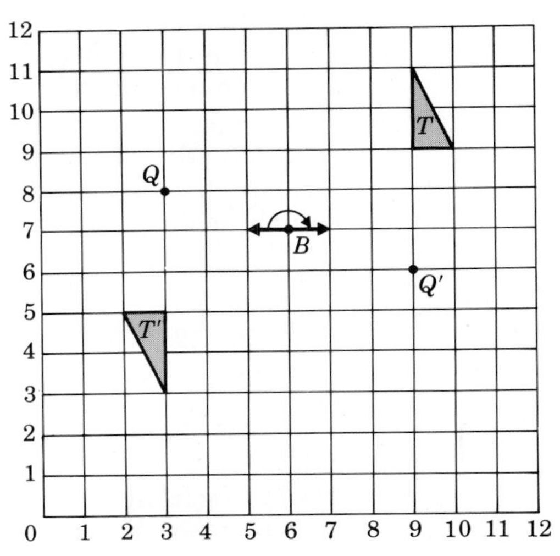

FIGURE 7.26

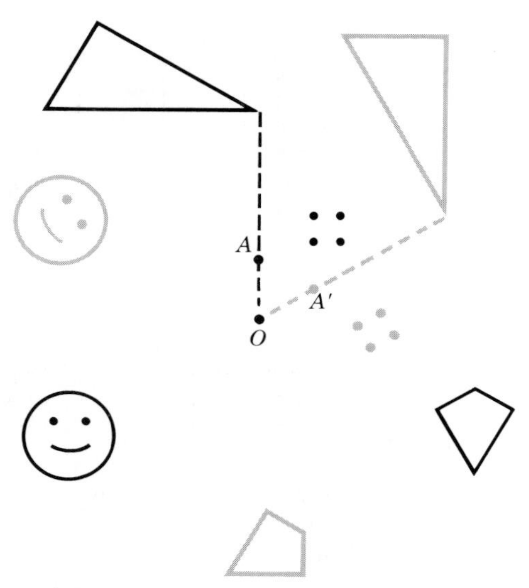

FIGURE 7.27

In Figure 7.25, point P' is the 90° counterclockwise *turn image*, or *rotation image*, of point P around point A. Note that distance AP is equal to distance AP'. We also say that quadrilateral R' is the 90° counterclockwise turn image, or rotation image, of quadrilateral R around point A. In Figure 7.26, point Q' is the 180° clockwise turn image of point Q around point B. We also say that triangle T' is the 180° clockwise turn image of triangle T around point B.

The figures in gray in Figure 7.27 are the 60° clockwise turn images around the point O of the figures in black. The reader should use tracing paper to show that this is true. Note that if tracing paper is held stationary at point O with the tip of a pencil and the tracing of point A is turned to fit upon point A', then every other figure on the tracing paper is turned through an angle of 60° around point O.

EXERCISE SET 7.2

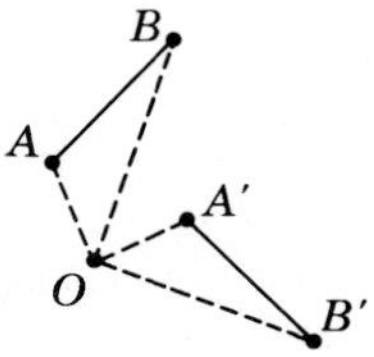

FIGURE 7.28

1. Refer to Figure 7.28 and name the 90° clockwise image about O of the following:

   a. Point A b. Point B c. $\overline{AB}$

2. Refer to Figure 7.28 and name the 90° counterclockwise image about O of the following:

   a. Point A' b. Point B' c. $\overline{A'B'}$

EXPLORING WITH THE COMPUTER 7.2

Use LOGO software to explore the following.

1. A turn or rotation requires an angle of rotation and a distance from the center of rotation. TURN requires 2 inputs. The first tells the turtle how far it is to the center, and the second tells the turtle how far to turn. The turtle will draw a mark at the center of the turn.

```
TO TURN :D :A
ARROW
PU BK :D PD FD 2 BK 2 PU
RT :A FD :D PD
ARROW
END
```

```
TO ARROW
FD 25 RT 25 BK 10 FD 10 LT 25 BK 25
END
```

2. Write a procedure that will draw a figure that can be moved using TURN. Be sure that it has some feature that will allow you to watch the orientation of the figure—such as a diagonal or an arrow on one side. Redefine TURN to use your figure and again try several different directions and distances.
3. Use TURN to test the following conjectures:
   a. Turn A followed by a turn B about the same point gives the same image of a figure as turn B followed by turn A.
   b. Turning a triangle does not change its shape or size or orientation.
   c. The image of a figure resulting when one turn is followed by a second turn can always be produced by a single turn.
4. Make and test other conjectures based on your observation of TURN.
5. Use TURN and SLIDE to create an interesting design.

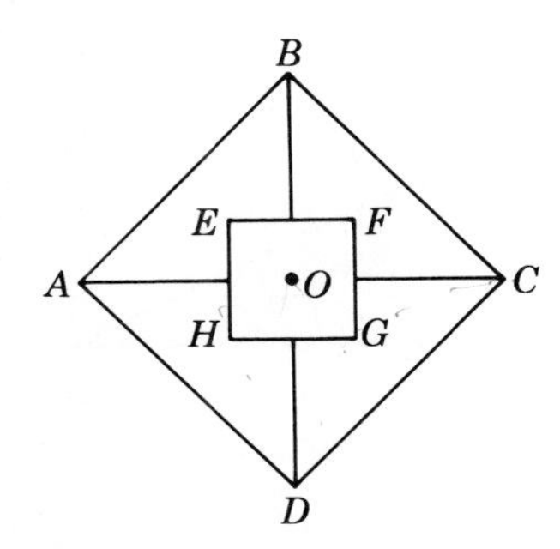

FIGURE 7.29

3. In Figure 7.29, *ABCD* and *EFGH* are squares. The 90° clockwise rotation image about *O* of
   a. *A* is ____.
   b. *D* is ____.
   c. *H* is ____.
   d. *F* is ____.
   e. $\overline{BC}$ is ____.
   f. $\overline{BE}$ is ____.
   g. $\overline{CG}$ is ____.
   h. $\overline{FC}$ is ____.

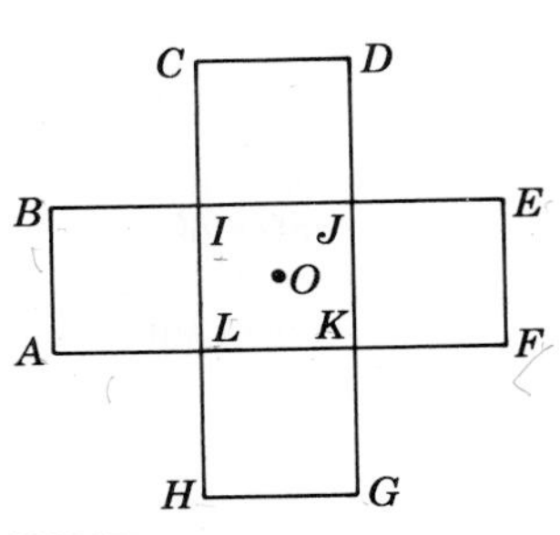

FIGURE 7.30

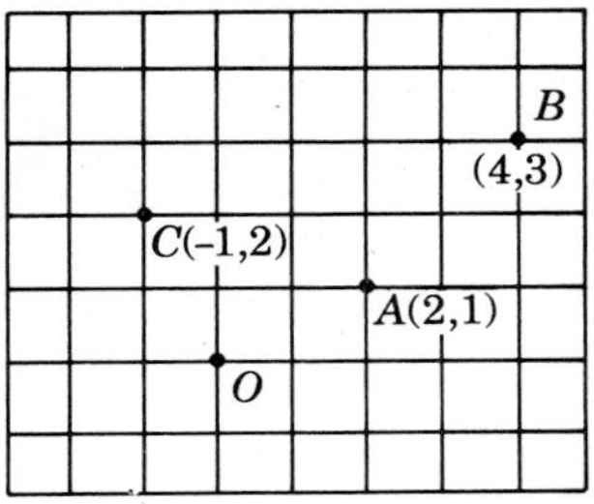

B (4,3)
C(−1,2)
A(2,1)
O

FIGURE 7.31

4. The pattern of 5 squares is centered at O in Figure 7.30. The 90° counterclockwise rotation image about O of
   a. A is ______.
   b. Square $ABIL$ is ______.
   c. Rectangle $ABEF$ is ______.
   d. Rectangle $CDLK$ is ______.
   e. Rectangle $IJKL$ is ______.
   f. $\overline{AE}$ is ______.

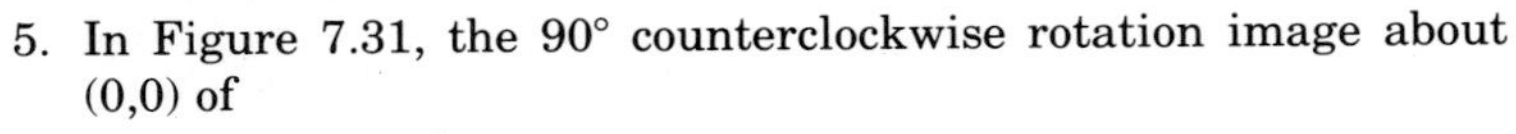

5. In Figure 7.31, the 90° counterclockwise rotation image about (0,0) of
   a. $A(2,1)$ is ______.
   b. $B(4,3)$ is ______.
   c. $C(-1,2)$ is ______.

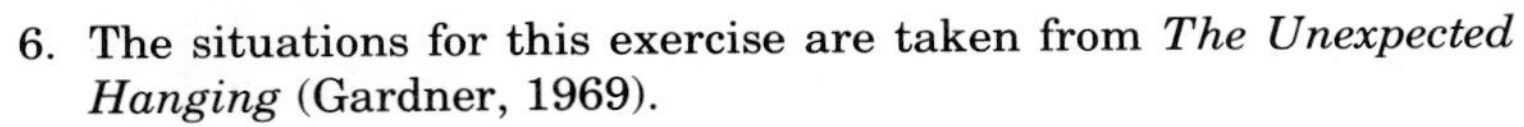

6. The situations for this exercise are taken from *The Unexpected Hanging* (Gardner, 1969).
   a. Turn the card in Figure 7.32 counterclockwise. What do you see?
   b. A sign like the one in Figure 7.33 appeared beside a public swimming pool. Turn this card 180°. What do you see?
   c. A slice is missing from the cake in Figure 7.34. Can you turn the card in such a way as to find the slice?
   d. Oliver Lee asked that the number in Figure 7.35 appear on his license plate. Turn the card 180°. Can you explain why he made this request?
7. Draw a triangle ABC and a point O on your paper (Fig. 7.36). Use a protractor and tracing paper to find the 90° counterclockwise turn image of triangle ABC about point O.

FIGURE 7.32

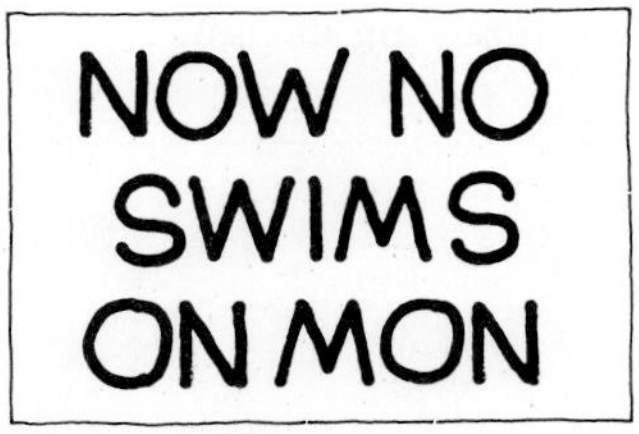

NOW NO
SWIMS
ON MON

FIGURE 7.33

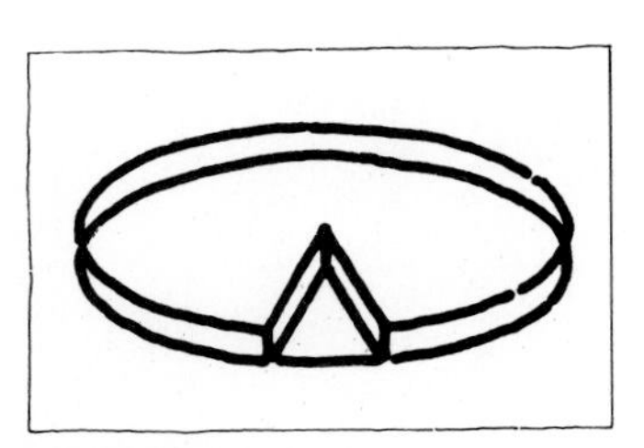

FIGURE 7.34

337-31770

FIGURE 7.35

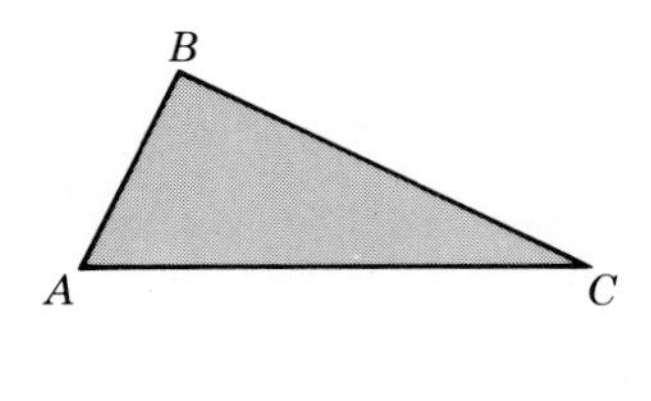

FIGURE 7.36

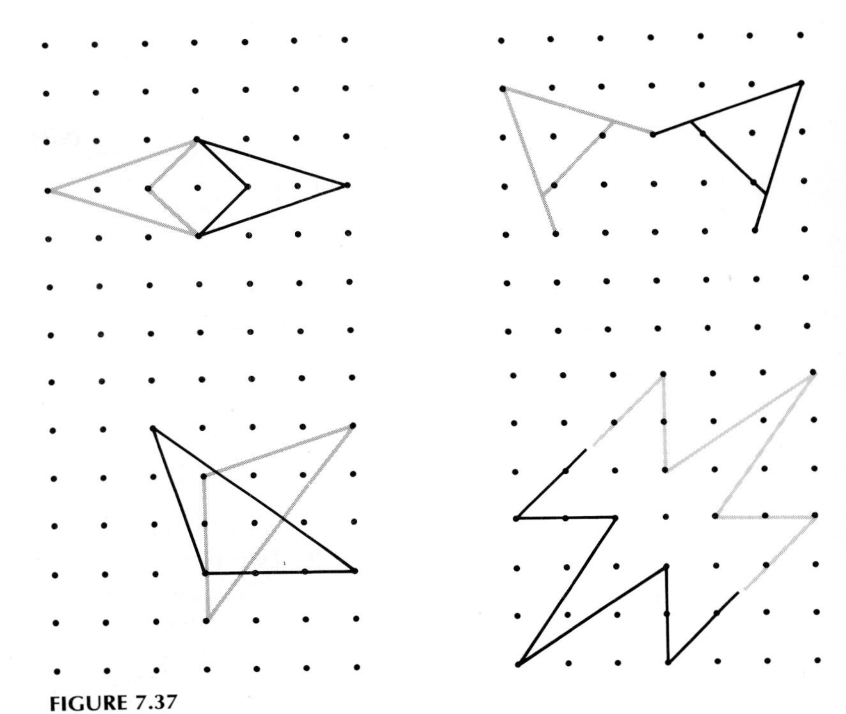

FIGURE 7.37

8. In each of the figures shown in Figure 7.37, the gray figure is the image of the black figure under a turn. Label the center of the turn and indicate the angle measure for each turn. (Tracing paper may be helpful.)
9. Give the coordinates of the 90° counterclockwise turn image centered at O of point A (Fig. 7.38).
10. Give the coordinates of the 90° clockwise rotation image centered at O of points B, C, and D (Fig. 7.38). Draw this rotation image of $\triangle BCD$.

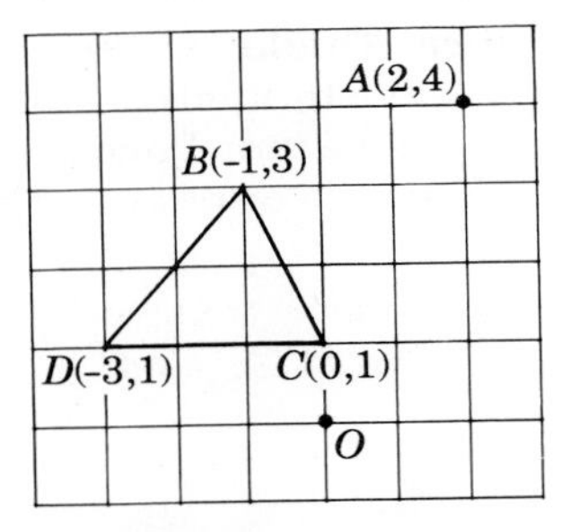

FIGURE 7.38

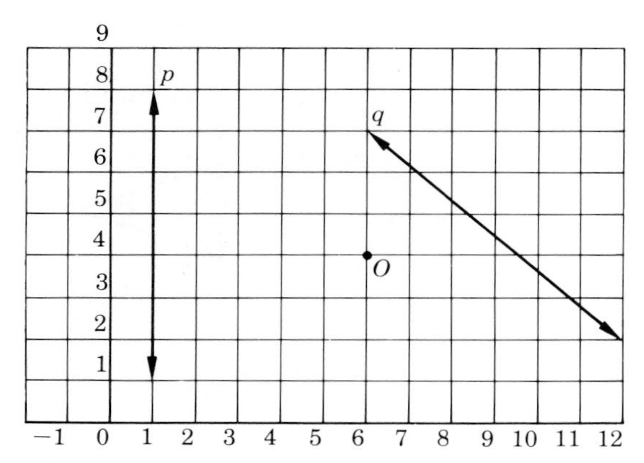

FIGURE 7.39

11. Draw in red the images of lines p and q under a 90° clockwise turn about point O (Fig. 7.39).

*12. If you place your finger on the center nail of the geoboard in Figure 7.40 and turn the board 90° (clockwise or counterclockwise), the gray figure appears exactly as it is now. The figure is said to have 90° turn symmetry about the center of the geoboard. How many different figures like this can you find on a 5 × 5 geoboard? Each figure you find should have the following characteristics.

   a. It divides the geoboard into 4 sections that are the same size and shape.
   b. The figure is formed by drawing 4 identical paths, starting at the center of the geoboard.
   c. The paths are formed by drawing segments from one dot to another. Paths cannot cross or touch. When a path reaches a point on the boundary of the geoboard array, it stops.
   d. If a figure can be flipped or turned to look like another, the two figures should be considered the same.

There are 24 such figures. Can you find them all?

13. **Critical Thinking**. Joan says that one of the letters in Figure 7.41 is the turn image of the other one, with point O being the center of the turn. Provide a logical argument to explain why you think Joan is correct or why you think she is incorrect.

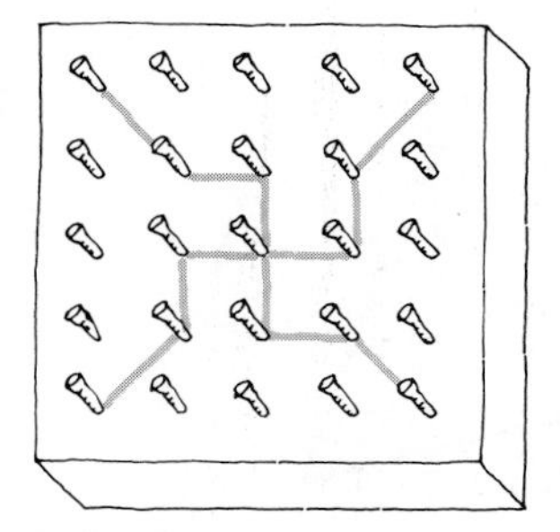
FIGURE 7.40

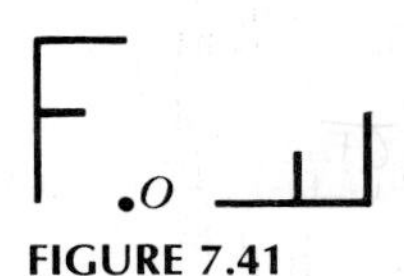

FIGURE 7.41

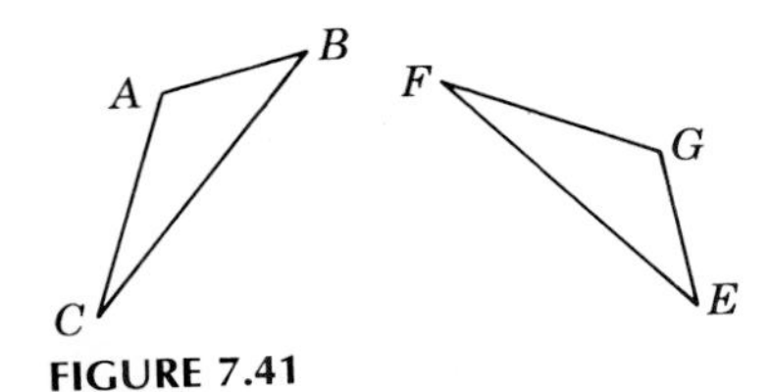

FIGURE 7.41

Problem Solving: Developing Skills and Strategies

Completing a ruler-compass or MIRA construction can be an effective problem-solving strategy for some geometry problems. Use this strategy for the following problem: One of the triangles in Figure 7.42 is the turn image of the other one. A student said that the center of the turn was the intersection of the perpendicular bisector of BE with the perpendicular bisector of CF. Do you think this is true? If not, how can you find it? If so, why does this procedure work?

FLIPS OR REFLECTIONS

A flipping motion, just as the slide and the turn motion, is experienced in the real world. We flip the pages of a book or of a desk calendar. The admonition "flip it over" is appropriate for activity with a wide variety of objects, such as pancakes, cards, or records.

The main motivation for studying the motion of flipping, however, comes from its relationship to mirrors. This is so because the mirror seems to "flip" the object in front of it over and shows the result as the image in the mirror.

Scientists have used mirrors to build telescopes that probe deeply into outer space, and architects and decorators have learned to use mirrors to achieve many interesting effects (Fig. 7.43). We take mirrors for granted, and many of us do not

TEACHING NOTE

Earlier, we suggested that the 3-mirror kaleidoscope provides exciting creative experiences for students. Other mirror problems such as those in the materials on MIRA math available from Dale Seymour Publications, can also broaden the students' horizons. Also, students universally enjoy mirror writing, so don't forget the mirror as an interesting geometrical object to explore.

FIGURE 7.43

FIGURE 7.44

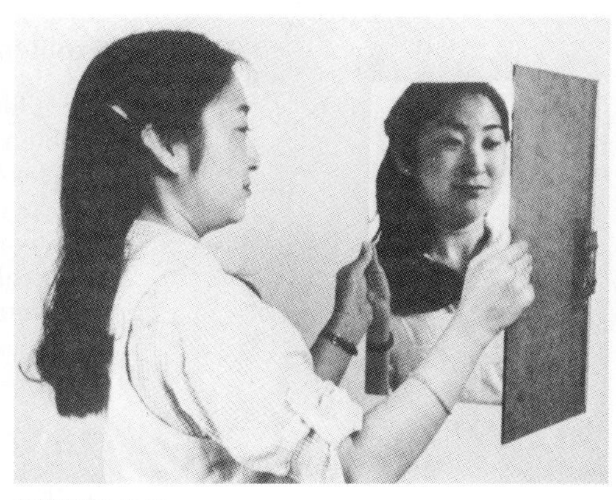

FIGURE 7.45

understand their basic properties. Many people find it difficult to visualize what happens to an image reflected by a mirror. This can be illustrated by trying the following experiments.

Hinge two mirrors at a 90° angle and look into them as shown in Figure 7.44. How does your image differ from your usual image in a flat mirror? Second, hinge two mirrors at a 90° angle and look into them as shown in Figure 7.45. Are you surprised by what you see? Can you explain your observation?

A handy instrument for experimenting with mirrors is a MIRA (Fig. 7.46), described in Chapter 2. For example, place a MIRA along a line in a plane that is perpendicular to a sheet of paper (Fig. 7.47). Then, any geometric figure on one side of

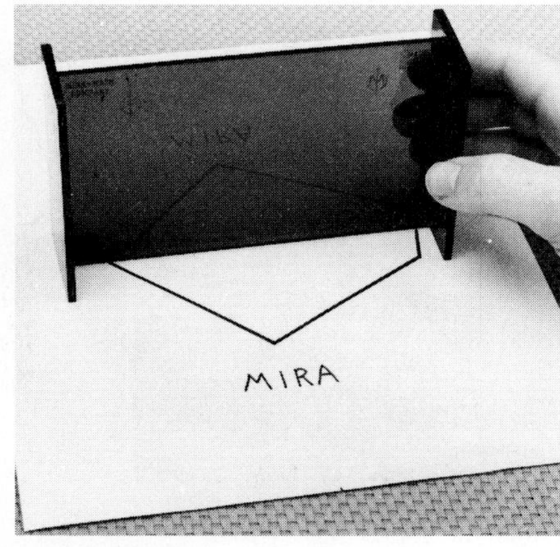

FIGURE 7.46

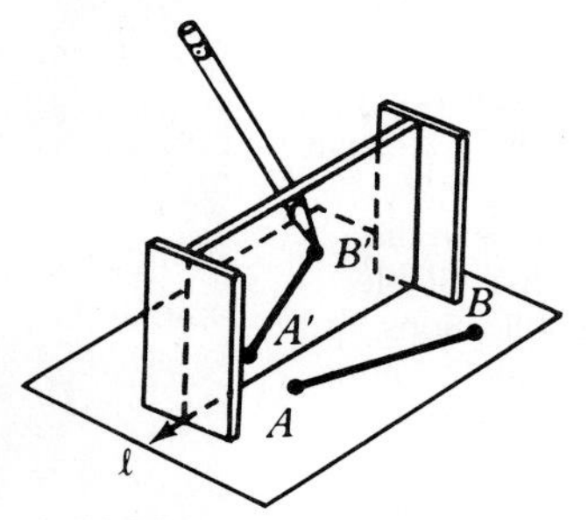

FIGURE 7.47

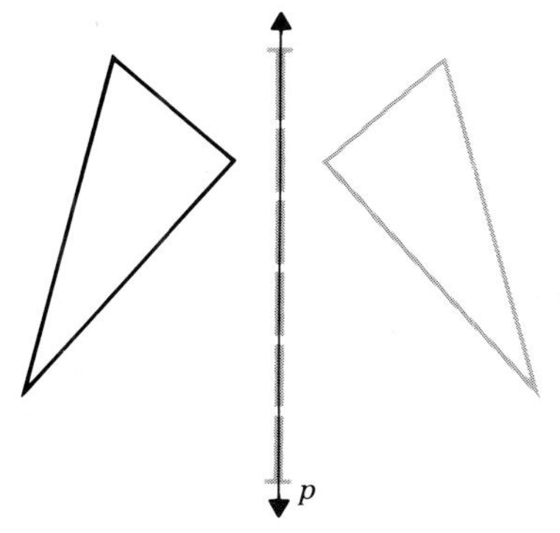

FIGURE 7.48

the MIRA has a "flip" image on the opposite side. By reaching around behind the MIRA, we can trace this image. (See Appendix B for activities that introduce the MIRA.)

This idea is illustrated in Figure 7.48. Suppose that the MIRA is placed along line p. Then, the gray triangle would be the image of the black triangle. If a cardboard cutout is placed to coincide with the black triangle, it needs to be "flipped" in order to make it coincide with the gray triangle. Hence, it is natural to refer to the gray triangle as the "mirror image," the "reflection image," or the "flip image" of the black triangle. The line p is called both the "mirror line" and the "flip line."

The defining characteristic of a reflection is described as follows: A point P' is the **reflection image**, or the **flip image**, in a line p of point P if p is the perpendicular bisector of segment PP'. The stimulating drawing by M.C. Escher reproduced in Figure 7.49 utilizes a tessellation of the plane and a mirror to make the notion of a mirror image "come alive."

Magic Mirror, M. C. Escher

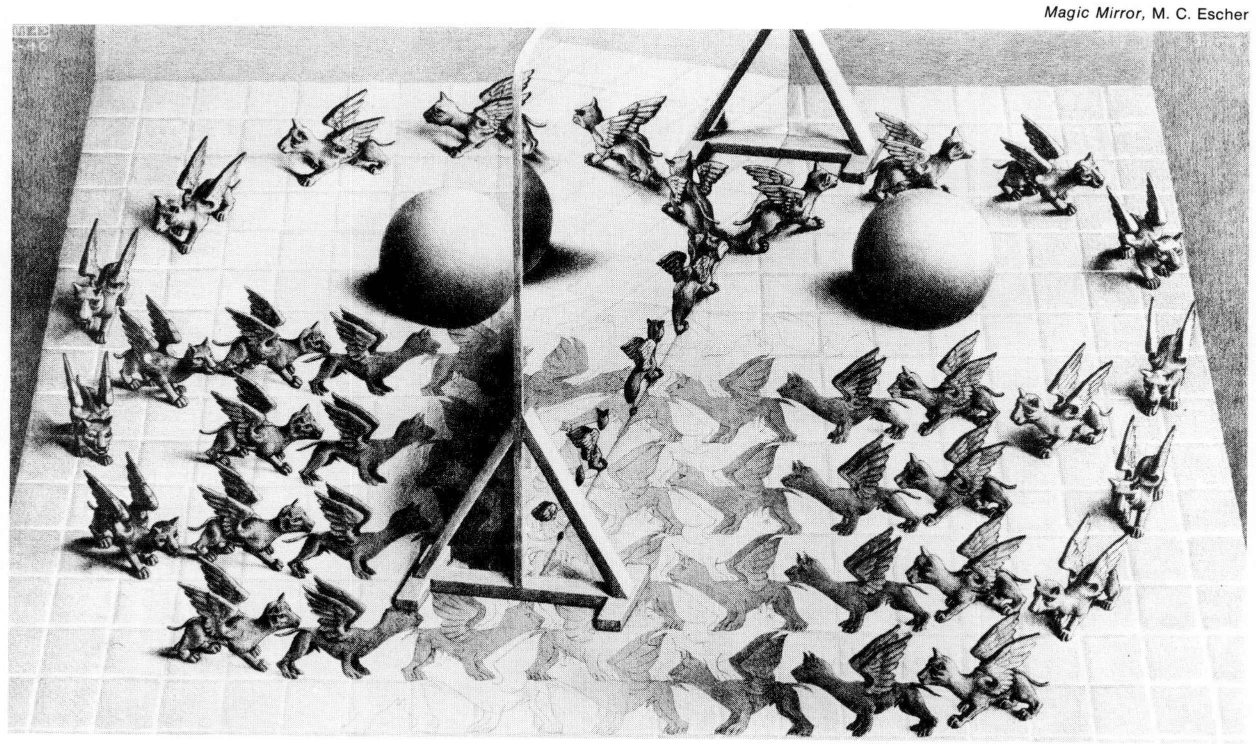

FIGURE 7.49

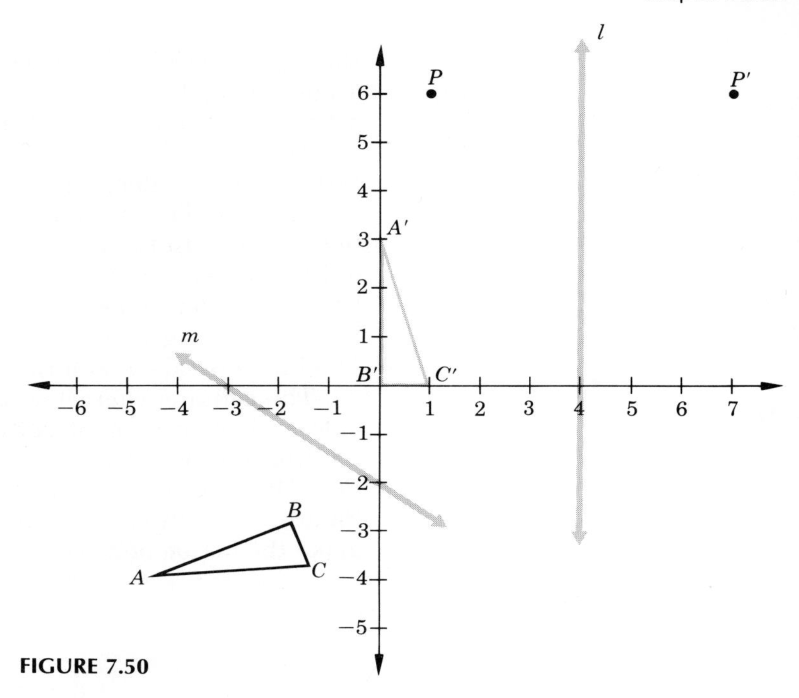

FIGURE 7.50

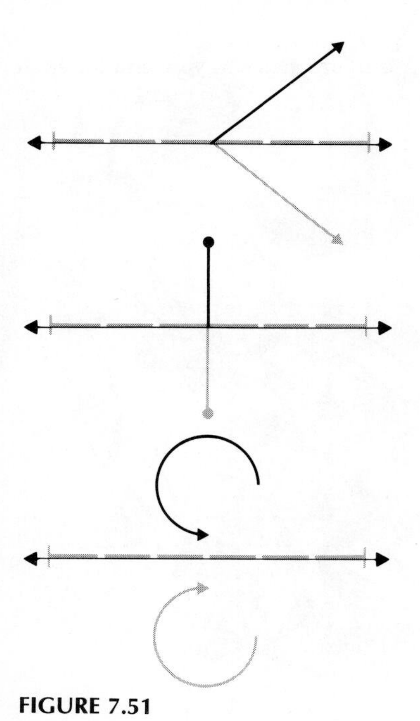

In a slightly less dramatic setting than that depicted by Escher, Investigation 7.3 will encourage you to use tessellations and mirrors to help develop your intuition about flips. Since each reflection involves a reflection line, we usually describe a particular flip by indicating the reflection line.

For example, in Figure 7.50, point P' is the *reflection image* of point P about reflection line l, and P is the reflection image of point P' about reflection line l. Also, $\triangle A'B'C'$ is the reflection image of $\triangle ABC$ about reflection line m, and $\triangle ABC$ is the reflection image of $\triangle A'B'C'$ about reflection line m.

In Figure 7.51, the gray figures are the flip images of the black figures (and vice versa) about the flip line. The reader should check this, using both a mirror and tracing paper.

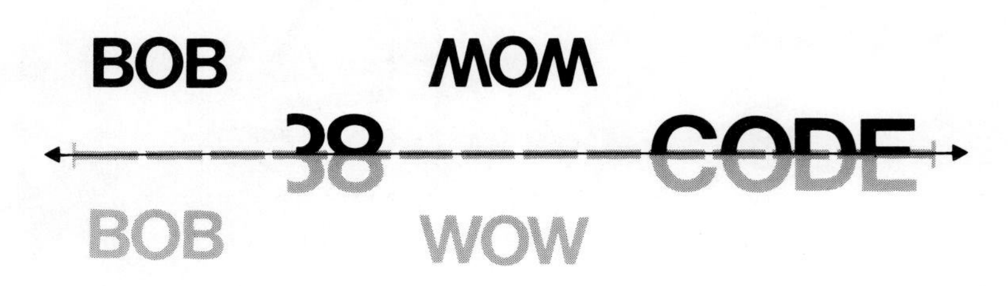

FIGURE 7.51

A. What will you see if

1. A piece of tinted plexiglass is placed along the flip line and you look at one side of it?
2. A mirror is placed along the flip line and you look in the mirror?
3. The tessellation is traced, folded together along the flip line, and you look through the folded paper at a bright light?

B. Copy the following tessellation. Can you show other mirror lines on the tessellation that are essentially different from the one shown? (You may need to supply your own definition of the phrase "essentially different.")

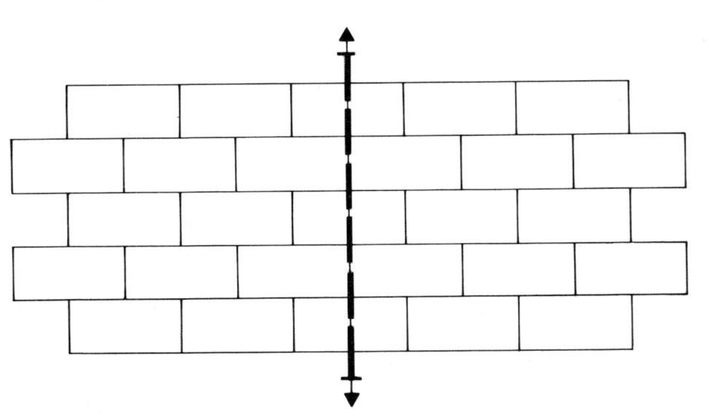

C. How many essentially different mirror lines can you find for each of tessellations (a), (b), and (c)? (Show each line you find on a tracing of the tessellation.)

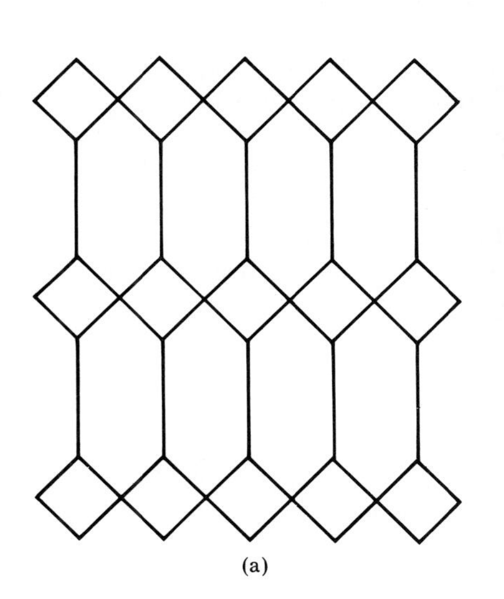
(a)

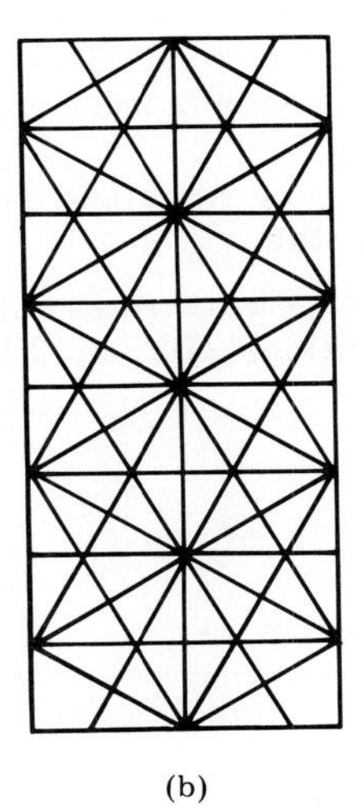
(b)

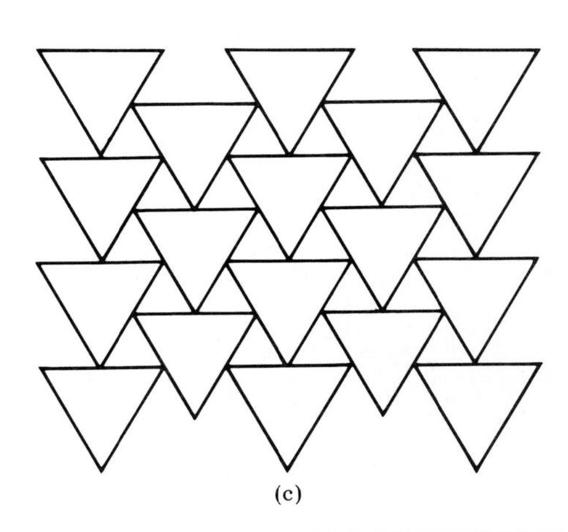
(c)

EXPLORING WITH THE COMPUTER 7.3

Use LOGO software to explore the following.

1. A flip or reflection requires a direction and a distance in this procedure. Therefore, the procedure FLIP requires 2 inputs. The first indicates the direction angle, and the second the distance to move. The procedure will also draw a part of the reflection line. Note that the procedure ARRO is slightly different than ARROW that you used in previous sections. Try the FLIP procedure with different angle values and distances. Also, try the same direction and distance two or three times without clearing the screen. Watch the arrow and make note of its orientation. (A REPEAT statement can be used to do the operation several times.)

```
TO FLIP :A :L
ARROW
PU LT :A BK :L RT 90 PD
FD 60 BK 120 FD 60 LT 90 PU BK :L RT 180 - :A PD
ARRO HT
END
```

```
TO ARRO
FD 25 LT 25 BK 10 FD 10 RT 25
END
```

```
TO ARROW
FD 25 RT 25 BK 10 FD 10 LT 25 BK 25
END
```

2. Write a procedure that will draw a figure that can be moved using FLIP. Be sure that it has some feature that will allow you to watch the orientation of the figure—such as a diagonal or an arrow on one side. Redefine FLIP, remembering to write a procedure to draw the reflection image of your figure in order to use it again. Try several different directions and distances.
3. Use FLIP to test the following conjectures:
   a. The image of a figure resulting when one flip is followed by a second flip can always be produced by a single flip.
   b. A flip over line r followed by a flip over line s gives the same image of a figure as a flip over line s followed by a flip over line r.
   c. Flipping a triangle does not change its shape or size or orientation.
4. Make other conjectures and test them with FLIP.
5. Use TURN, SLIDE, and FLIP to create an interesting design.

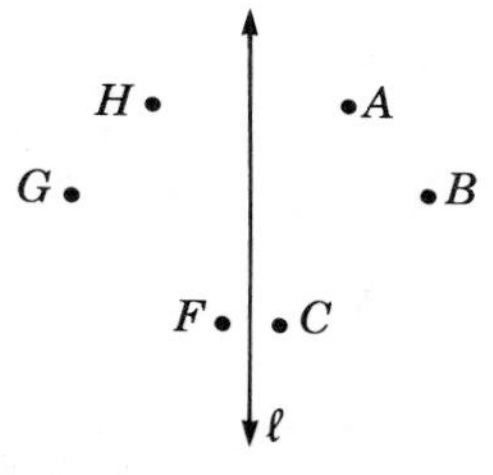

FIGURE 7.52

EXERCISE SET 7.3

1. Refer to Figure 7.52 and answer the following:
   a. The reflection image in ℓ of A is ____.
   b. The reflection image in ℓ of G is ____.
   c. The reflection image in ℓ of $\overline{FG}$ is ____.
   d. The reflection image in ℓ of $\triangle ABC$ is ____.

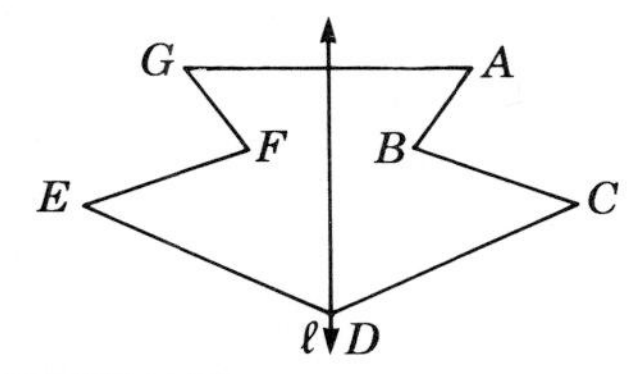

FIGURE 7.53

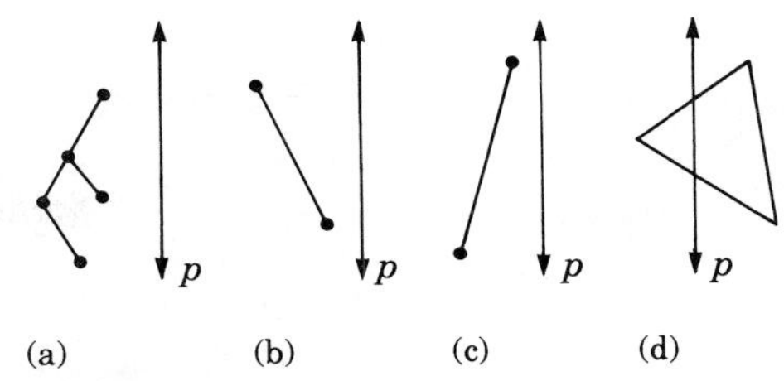

FIGURE 7.54

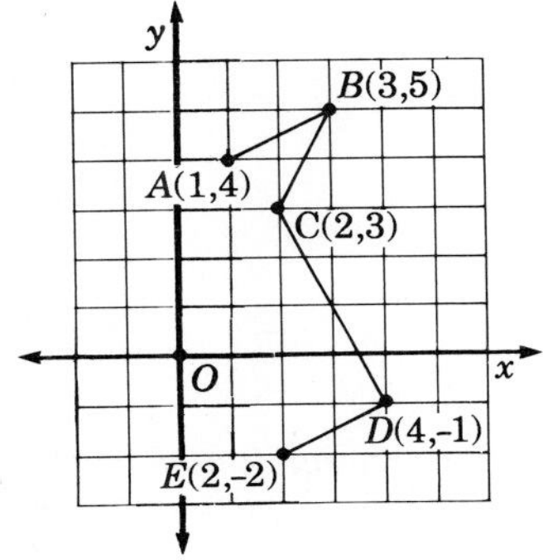

FIGURE 7.55

2. Refer to Figure 7.53 and answer the following:
   a. The reflection image in ℓ of AB is ____.
   b. The reflection image in ℓ of D is ____.
   c. The reflection image in ℓ of GF is ____.
   d. The reflection image in ℓ of $ABCDEFG$ is ____.
3. Copy the figures in Figure 7.54. Use a MIRA, tracing paper, or a compass and straight edge to draw as accurately as possible the reflection image in line p of each figure.
4. Refer to Figure 7.55 and find the coordinates of the reflection image in the y axis of the following:
   a. Point A b. Point C
   c. Point E d. $\overline{CD}$
   e. $\overline{BC}$
5. Copy the points and figures of Figure 7.56 on *square dot paper*. Using only the dot paper (and a straight edge), draw the mirror image in line p of each of the figures shown.

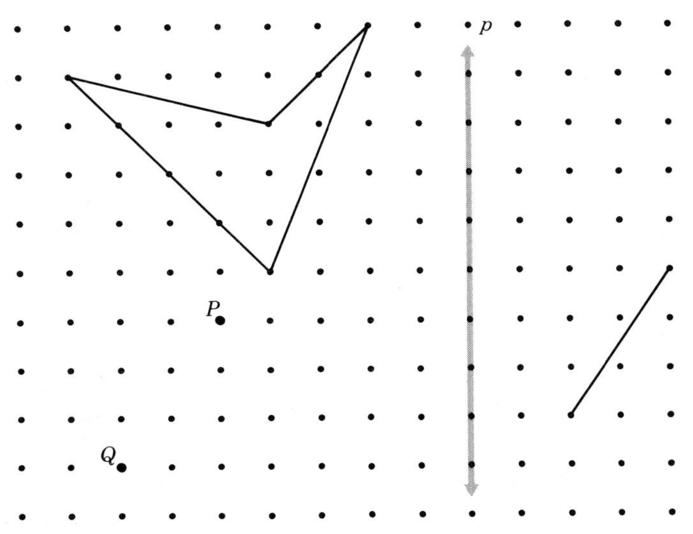

FIGURE 7.56

APPLICATION

Dock Problem: The owner of two canneries wants to make a loading dock on a nearby river so that his products can be shipped on the river to other locations. Where should the loading dock be placed so that the owners trucks would use minimum gasoline traveling to and from the dock? A surveyor working for the owner claimed that he could use the reflection image of one of the canneries in the line of the river bank to help find the solution. Do you agree? Draw a diagram to show your solution.

Cannery A

Cannery B

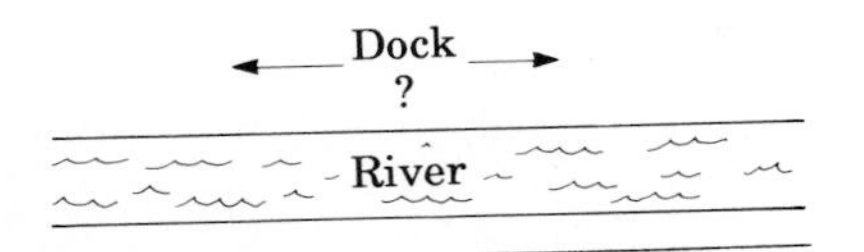

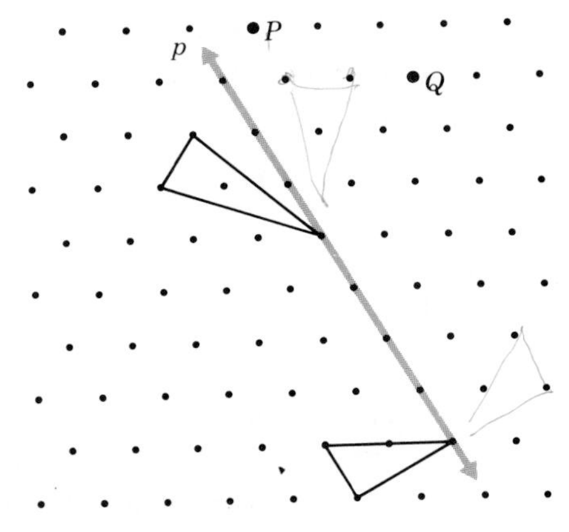

FIGURE 7.57

6. Copy the points and figures of Figure 7.57 on *triangular dot paper*. Using only the dot paper (and a straight edge), draw the mirror image in line p of each of the figures shown.
7. Reproduce Figure 7.58 on dot paper and do the following:
   a. Draw a line q whose reflection image in line p is parallel to p. How many such lines are there?
   b. Draw a line r whose reflection image in line p is the line r again.
8. The situations for this exercise and Exercises 9 through 12 are taken from *The Ambidextrous Universe* (Gardner, 1964). Place a MIRA on the dashed line in Figure 7.59. How can you explain what you see?

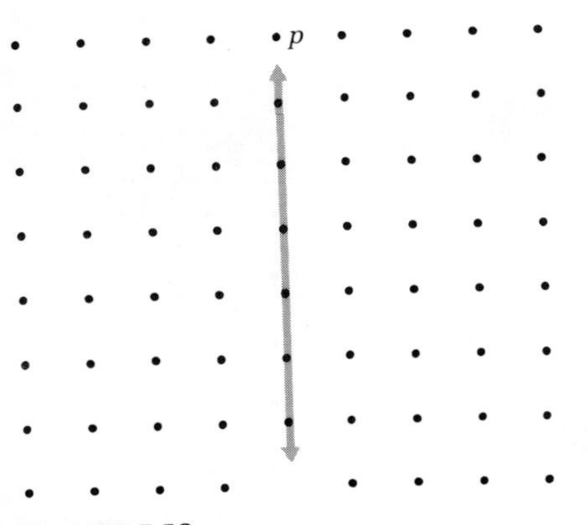

FIGURE 7.58

CHOICE QUALITY

FIGURE 7.59

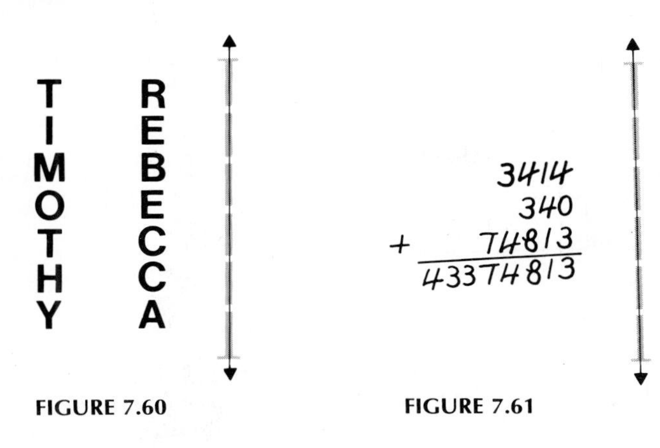

FIGURE 7.60

FIGURE 7.61

The reason this looks so strange is that it was written using a mirror.

FIGURE 7.62

FIGURE 7.63

9. "A MIRA is unbelievable, since it reverses girls' names, but doesn't reverse boy's names." Place your MIRA on the dashed line in Figure 7.60. Now do you believe that this is true? Explain.
10. Is the sum in Figure 7.61 correct? Place your MIRA along the dashed line to check it.
11. Can you decode the message shown in Figure 7.62?
12. If you place a mirror on line l in Figure 7.63, you form a word from the "half-word" shown.
    a. How many such half-words can you find?
    b. How many different half-numerals can you find?
13. Choose a tessellation, draw it, and analyze it with regard to reflectional symmetry.

*14 The patterns (a) through (g) shown in Figure 7.65 are called frieze patterns. Complete Table 7.1 by filling in the words "yes" or "no" for the symmetry types of each of these patterns. How many of these 7 patterns are uniquely characterized by the entries in this table?

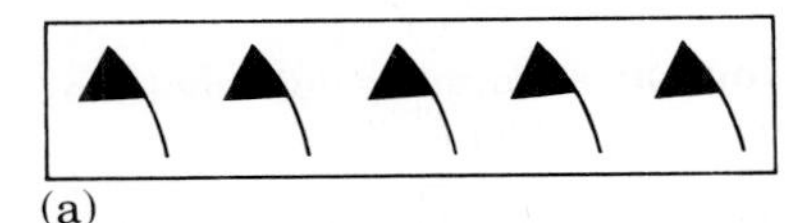

(a)

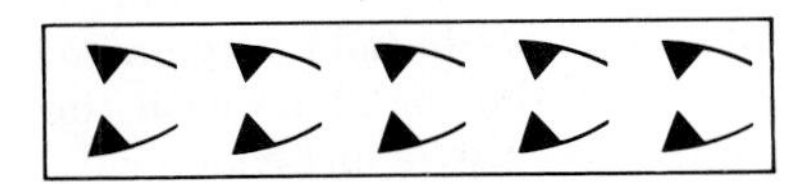

(b)

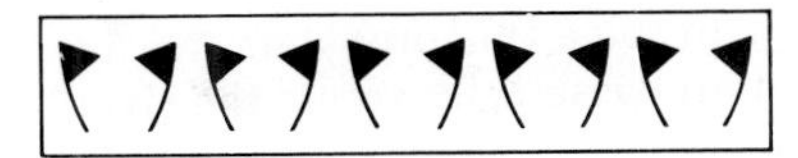

(c)

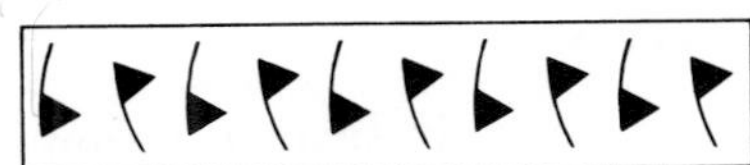

(d)

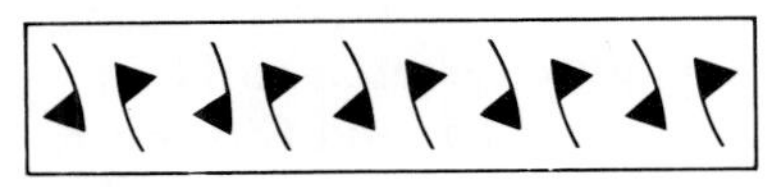

(e)

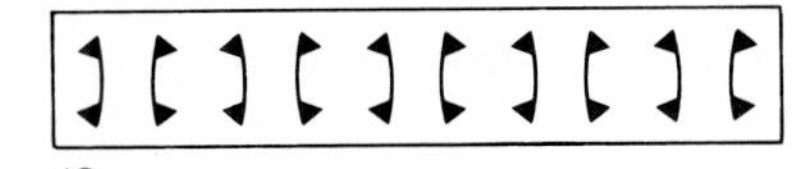

(f)

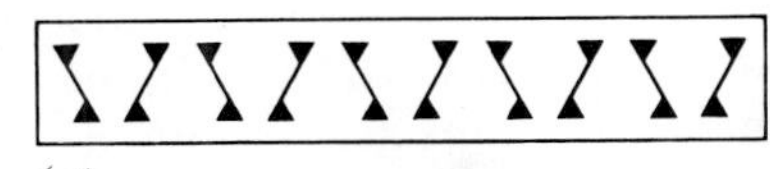

(g)

FIGURE 7.64

TABLE 7.1

	180° Rotational Symmetry	Vertical Lines of Symmetry	Horizontal Lines of Symmetry
(a)	no	no	no
(b)	no	no	yes
(c)	no	yes	no
(d)	no		no
(e)	yes	no	
(f)	yes	no	yes
(g)	no	yes yes	no

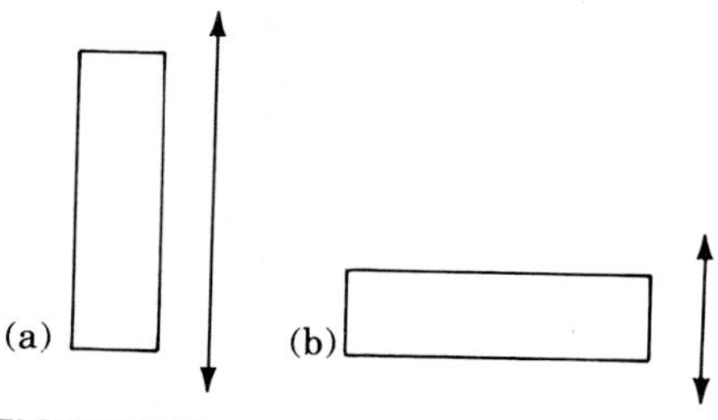

FIGURE 7.65

15. **Critical Thinking**. A reasonable thought process will be more successful than random trial and error in this problem. Explain your thought process. Copy Figure 7.64 and write several words that when printed in rectangle (a) will have a mirror image in line p that is also a word. Do the same for rectangle (b).

Problem Solving: Developing Skills and Strategies

The problem-solving strategy **use logical reasoning** is sometimes used to eliminate possibilities and solve a problem.

Problem: A pet store owner must get a cat, a pet mouse, and a brick of cheese across a river in a boat that has places to hold only one of these three "passengers." Now, if she leaves the cat alone with the mouse, the cat will eat the mouse. If she leaves the mouse alone with the cheese, the mouse will eat the cheese.

1. How would you use this strategy to solve this problem?
2. How does the pet store owner get all passengers across the river?

COMBINING SLIDES, TURNS, AND FLIPS

It is only on rare occasions that we observe a physical situation involving motion in which a slide, a turn, or a flip occurs completely by itself. Instead, we often observe a myriad of combinations of these motions as, for instance, when a child on a sled slides down the hill, makes a sharp turn, and flips over. As in this example, we often miss the intricacy of the combination of motions because of a primary concern about the final result. In laboratory situations, as when a scientist investigates the motions of certain particles in a magnetic field, we would expect a much more careful analysis of the component motions.

Investigation 7.4 provides an opportunity to consider the operation of combining motions and to analyze the results of these combinations. Note that when two motions are combined, the image under the first motion becomes the starting figure for the second motion.

TEACHING NOTE

The early study of combinations of slides, turns, and flips in a physical situation provides an interesting conceptual background for the later study of composition of functions. Many concepts develop slowly over a long period of time, and it is desirable to begin the intuitive development of important concepts early.

INVESTIGATION 7.4

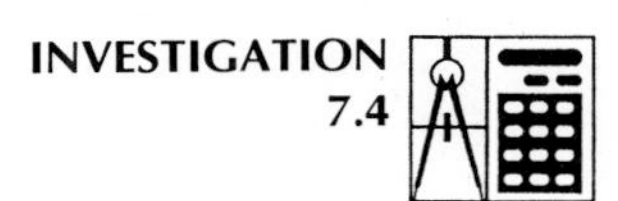

For each situation A through C, draw a triangle on your paper and use tracing paper, compass, and straight edge, or a MIRA to perform first one motion on the triangle and then another.

In each case, can you *carefully* describe a single motion that has the same effect (produces the same final image) as the two motions in succession? Tell as much about the single motion as you can and try to form a generalization about the situation. You may need to construct more examples of your own to formalize these generalizations.

A. Refer to Figure (a). First, find the flip image in line r. Then, find the flip image in line s.

B. Refer to Figure (b). First, find the flip image in line r. Then, find the flip image in line s.

C. Refer to Figure (c). First, find the flip image in line r. Then, find the flip image in line s.

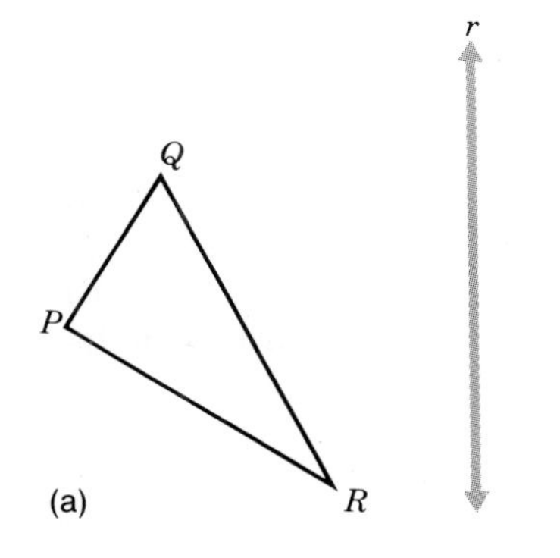

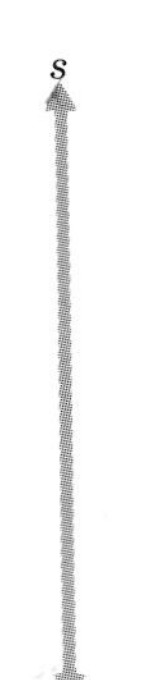

(a)

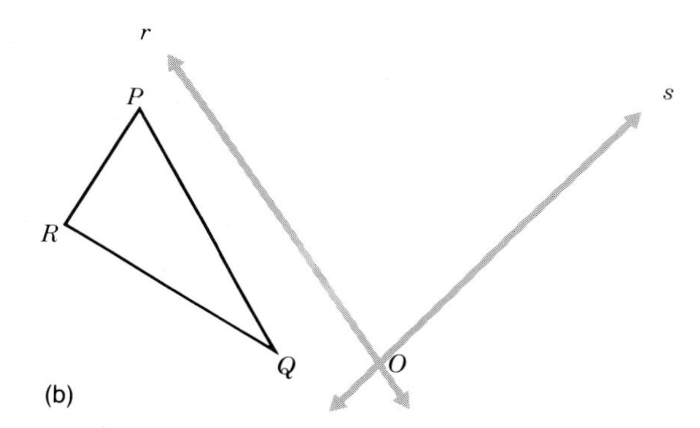

(b)

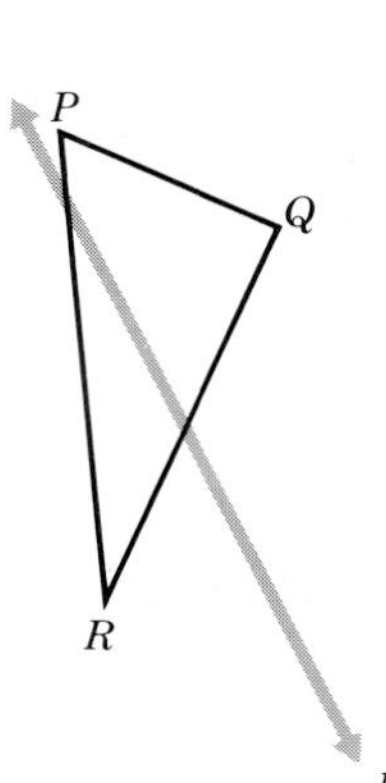

(c)

When two motions are combined, as in Investigation 7.4, we often call the single motion that has the same effect the **product of the two motions**. Just as the product of two whole numbers is another whole number, the product of two motions is another motion. *Which of these conjectures do you think are true about the product of two motions?* Can you change those that are false in such a way as to make them true?

Conjecture 1. The product of two reflections in parallel lines produces the same effect as a single reflection.

Conjecture 2. The product of two reflections in parallel lines produces the same effect as a slide in a direction perpendicular to the lines and through a distance equal to the distance between the lines.

Conjecture 3. The product of two reflections in intersecting lines produces the same effect as a turn about the point of intersection through an angle equal to the angle between the two lines.

After a careful analysis of both your investigation and the above three conjectures, you should decide that all three conjectures are false. Figure 7.66 shows that if lines r and s are parallel, the two flips have the same effect as a slide with direction perpendicular to the lines r and s and through a distance $2d$, with d the distance between the lines.

Figure 7.67 convinces us that if lines r and s intersect at point O, the two flips have the same effect as a turn about point O and through an angle 2α, where α is the measure between

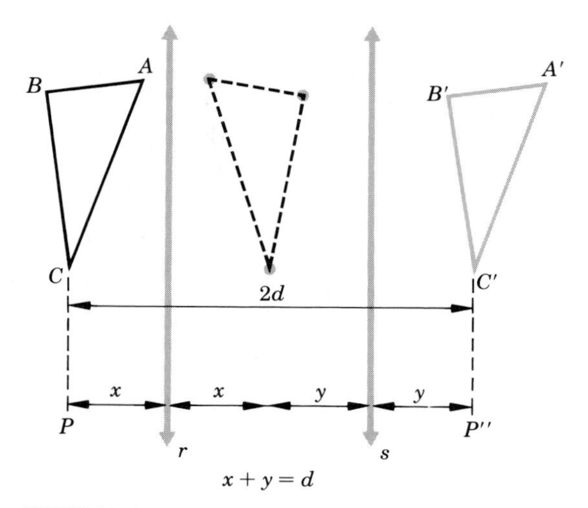

FIGURE 7.66

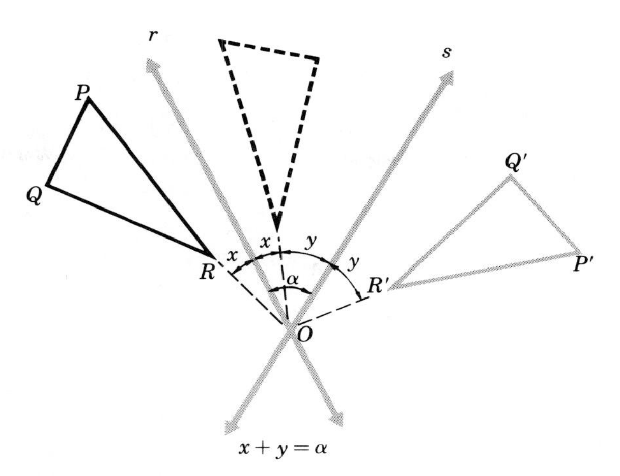

FIGURE 7.67

the two lines. To summarize, we see that the result of two flips is either a slide or a turn, depending upon whether the two flip lines are parallel or intersecting.

So far in this section, we have considered only the possible single motions resulting from the combination of two flips. A detailed analysis of all possible combinations—such as the result of combining two turns with different centers, the result of combining a slide and a turn, and so on—is beyond the scope of this chapter. A few of these possibilities will be considered in Exercise Set 7.4.

It should be mentioned, however, that the motion resulting from the combination of a slide and a flip is usually not itself a slide, a turn, or a flip. This motion, described and exemplified in Figure 7.68, will be called a **slide-flip**, or *glide-reflection*, for the purposes of this chapter. On the left side of the figure, a slide RR′ followed by a flip in line l ($l \parallel RR'$) results in a side-flip, which is different from a slide, a flip, or a turn. On the right side of the figure, the pattern made by a person's footprints in the snow is an example of a sequence of side-flips.

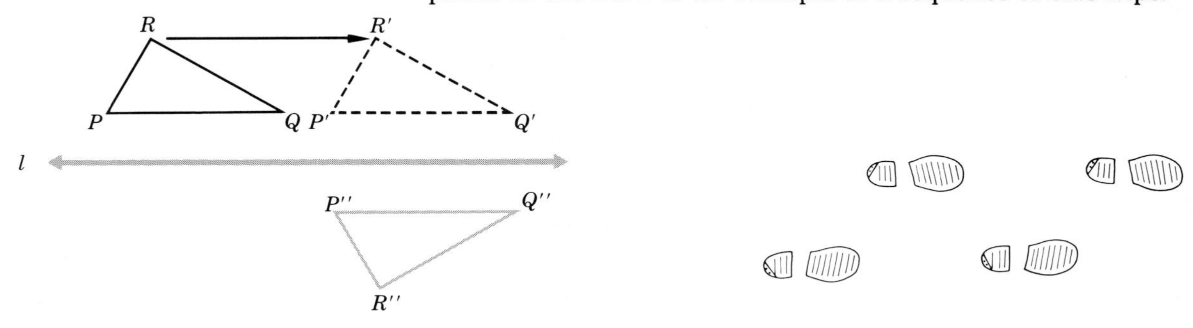

FIGURE 7.68

EXPLORING WITH THE COMPUTER 7.4

Use LOGO software to explore the following.

1. Experiment with the procedure GLIDE. The procedures FLIP and ARRO used within the procedure GLIDE are defined in Exploring with a Computer 7.3. What conjectures can you make about GLIDE?

```
TO GLIDE :A :L :M            TO BSL :R :M
FLIP :A :L                   ARRO PU
BSL 90 + :A :M               RT :R BK :D LT :R PD
END                          ARRO
                             END
```

2. Apply any one motion and then another motion to a triangle. Choose the pairs of motions from SLIDE, TURN, and FLIP. Note that the motions chosen can be the same or different. What single motion could have produced the same image? Discuss and record any generalizations you discover.

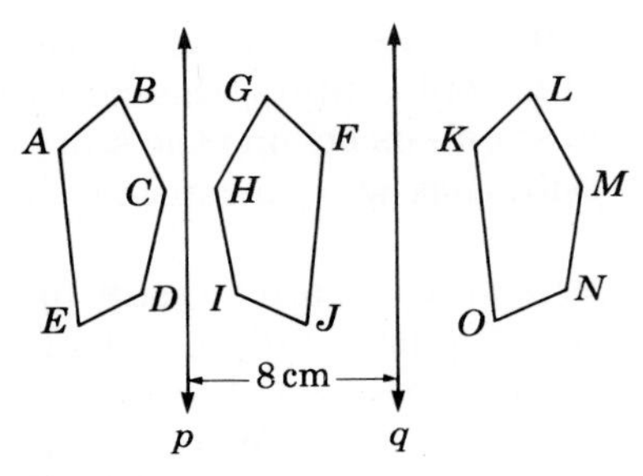

FIGURE 7.69

EXERCISE SET 7.4

1. In Figure 7.69, the product of the reflection in p followed by the reflection in q maps

   a. A to ____. b. E to ____.
   c. ____ to N. d. ____ to L.

2. In Figure 7.70, the reflection in line q followed by the reflection in line s maps

   a. A to ____. b. L to ____. c. O to ____. d. N to ____.

3. In Figure 7.70, the reflection in line p followed by the reflection in line q maps

   a. A to ____. b. B to ____. c. D to ____.
   d. L to ____. e. C to ____. f. M to ____.

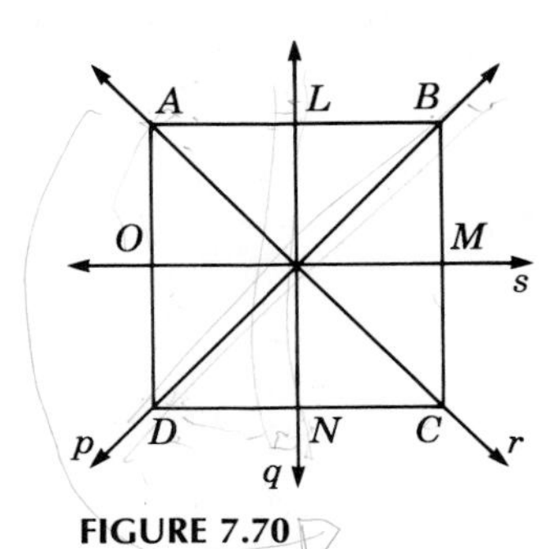

FIGURE 7.70

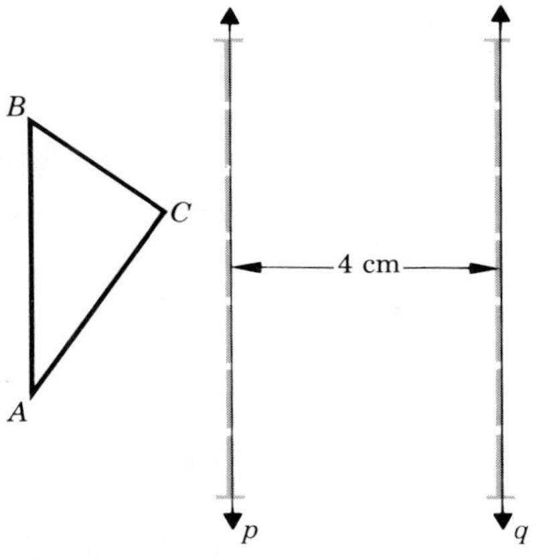

FIGURE 7.71

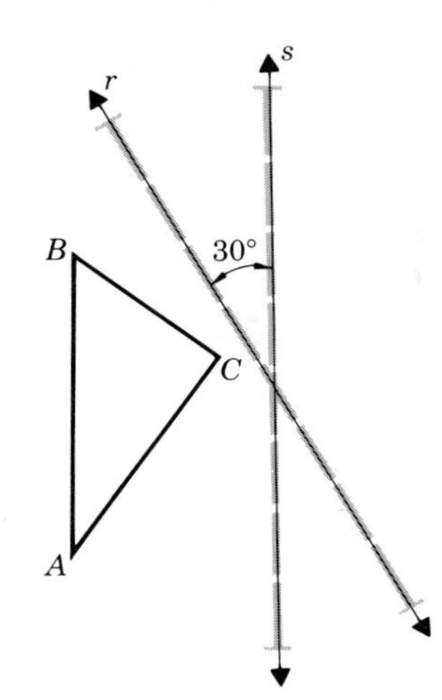

FIGURE 7.72

4. Draw parallel lines p and q on your paper 4 cm apart (Fig. 7.71).
   a. Use the information in this section to describe the slide that is the result of a flip in line p followed by a flip in line q. Give a careful description of both the distance and the direction of the slide.
   b. Use tracing paper or a MIRA to carry out the flips and check to see that the description you gave in part a is correct.
5. Draw intersecting lines r and s on your paper with angle of intersection 30° (Fig. 7.72).
   a. Use the information in this section to describe a turn that is the result of a flip in line r followed by a flip in line s. Give a careful description of the center, angle, and direction of the turn.
   b. Use tracing paper or a MIRA to carry out the flips and check to see that the description you gave in part a is correct.
6. Trace the figures in Figure 7.73 on your paper. Find the half-turn image of $\triangle PQR$ about point C followed by the slide AB. What single motion has the same effect as this sequence of motions? Describe this single motion as carefully as possible.

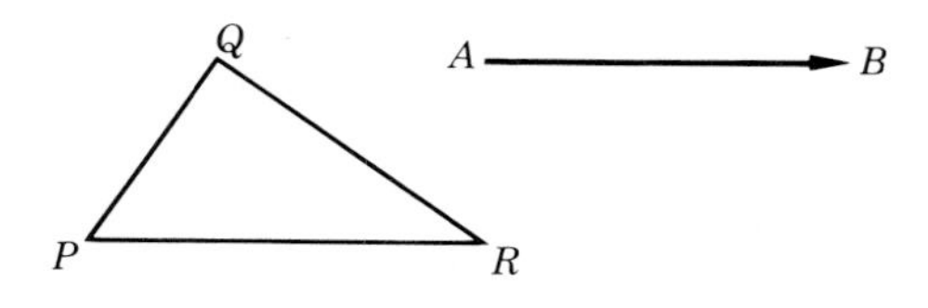

FIGURE 7.73

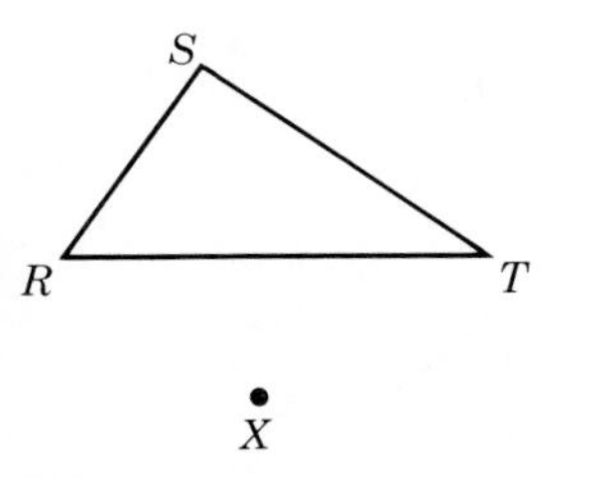

FIGURE 7.74

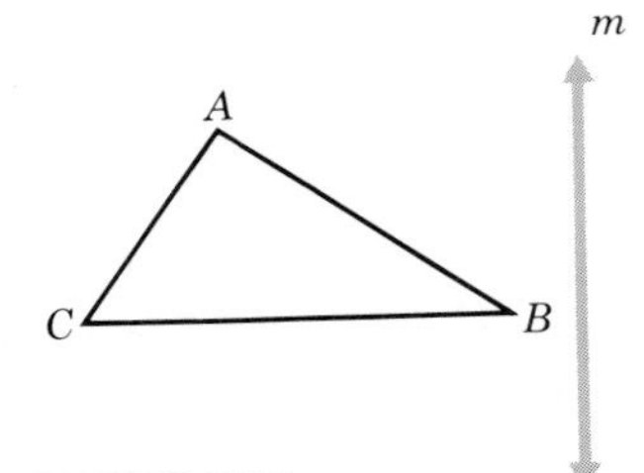

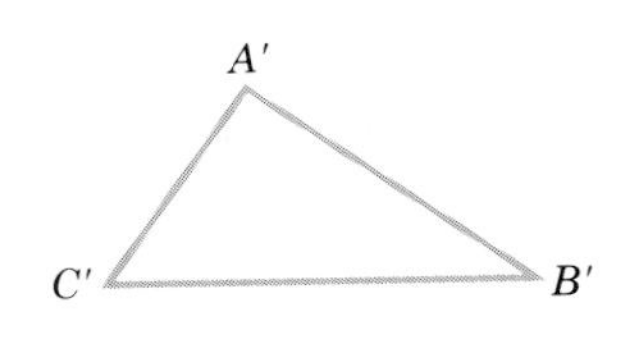

FIGURE 7.75

7. A 180° turn about a point is called a *half-turn*. Mark points X and Y on your paper and draw $\triangle RST$ (Fig. 7.74).
   a. First, complete a half-turn around X. Using the image of this turn, complete a half-turn around Y.
   b. What single motion has the same effect as a half-turn about X followed by a half-turn about Y? Be as specific as possible in describing this motion.
8. $\triangle A'B'C'$ is the image of $\triangle ABC$ under a slide (Fig. 7.75). Trace these figures. Can you construct a second line n so that a flip about the line m followed by a flip about the line n will have the same effect as the slide?
9. $\triangle A'B'C'$ is the image of $\triangle ABC$ under a turn about center Q (Fig. 7.76). (Test this with tracing paper.) Trace these figures. Can you construct line q so that a flip about the line p followed by a flip about the line q will have the same effect as the turn?
10. Draw lines t and u that are perpendicular to each other (Fig. 7.77). Describe the result of a flip in line t followed by a flip in line u. Carry out the flips to see if your description is correct.

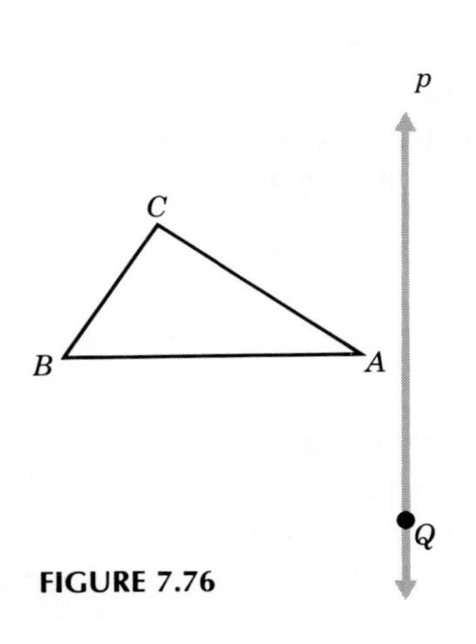

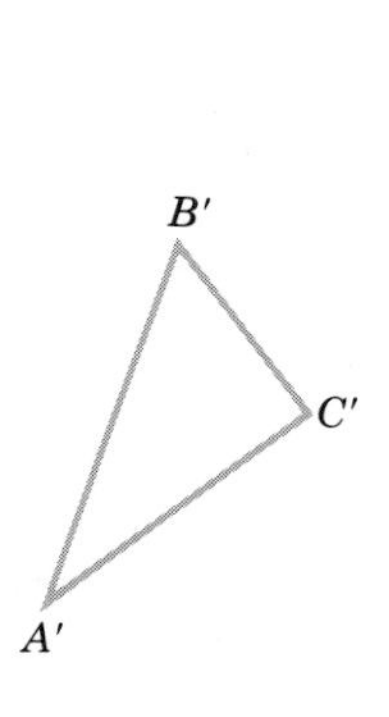

FIGURE 7.76

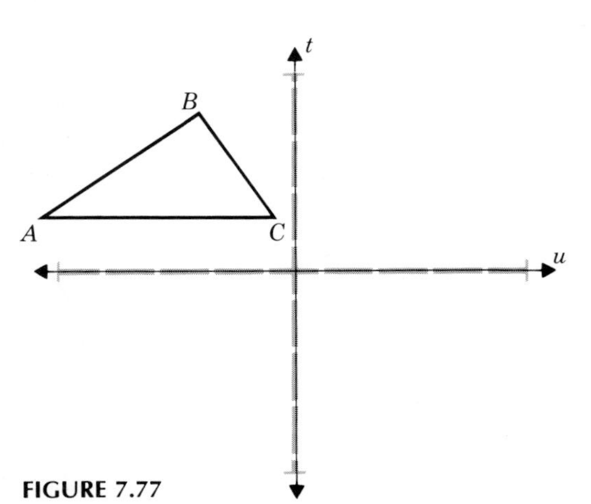

FIGURE 7.77

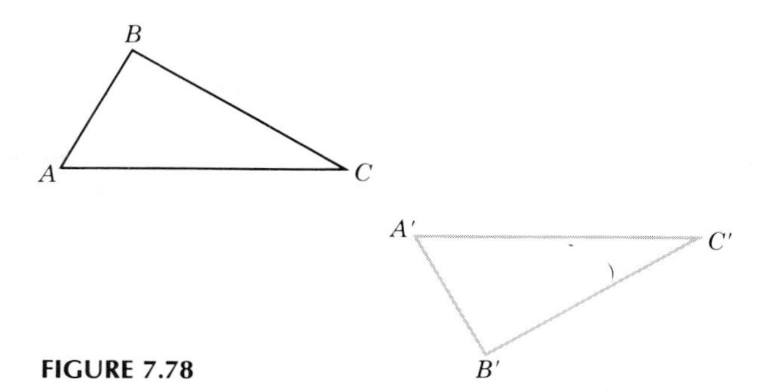

FIGURE 7.78

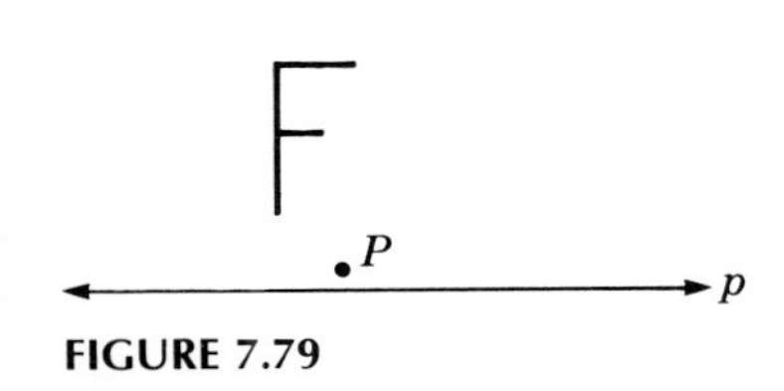

FIGURE 7.79

11. Triangle $A'B'C'$ is the image resulting from first performing a slide and then performing a flip with $\triangle ABC$ (Fig. 7.78). We say that $\triangle A'B'C'$ is the *slide-flip* image of $\triangle ABC$. The flip line can be found by connecting the midpoints of AA' and BB'. Trace the 2 triangles shown in the figure on your paper and find the flip line. Use tracing paper to see if the line you found is correct.
12. **Critical Thinking**. A student claims that a 90° clockwise rotation centered at P followed by a reflection in line p is the same as a reflection in line p followed by a 90° clockwise rotation about point P (Fig. 7.79). Do you agree? Complete a construction drawing to confirm your answer.

Problem Solving: Developing Skills and Strategies

One of the important characteristics of a good problem solver is to "be flexible." Try this problem.

Problem: A coin collector wanted to sell 8 old gold coins. He said they were all the same weight. The buyer suspected that one was lighter than the other 7. He found the lighter coin using a balance scale and only 2 weighings.

1. How did he do it?
2. Explain how the idea of flexibility played a part in arriving at your solution.

MOTIONS AND CONGRUENCE

In Investigation 7.5, you will begin with a pair of congruent figures and search for a reflection line such that the reflection image of one figure is the other. If that is not possible, you will search for two or three lines such that the product of the reflections in these lines maps one figure onto the other figure.

INVESTIGATION 7.5

For each situation shown, construct a line p such that the flip image of the "black figure" in p is the "gray figure." If you cannot find a single line p, search for 2 lines p and q such that the gray figure is the product of the flip in p and q. If 2 lines are not sufficient, use 3 lines.

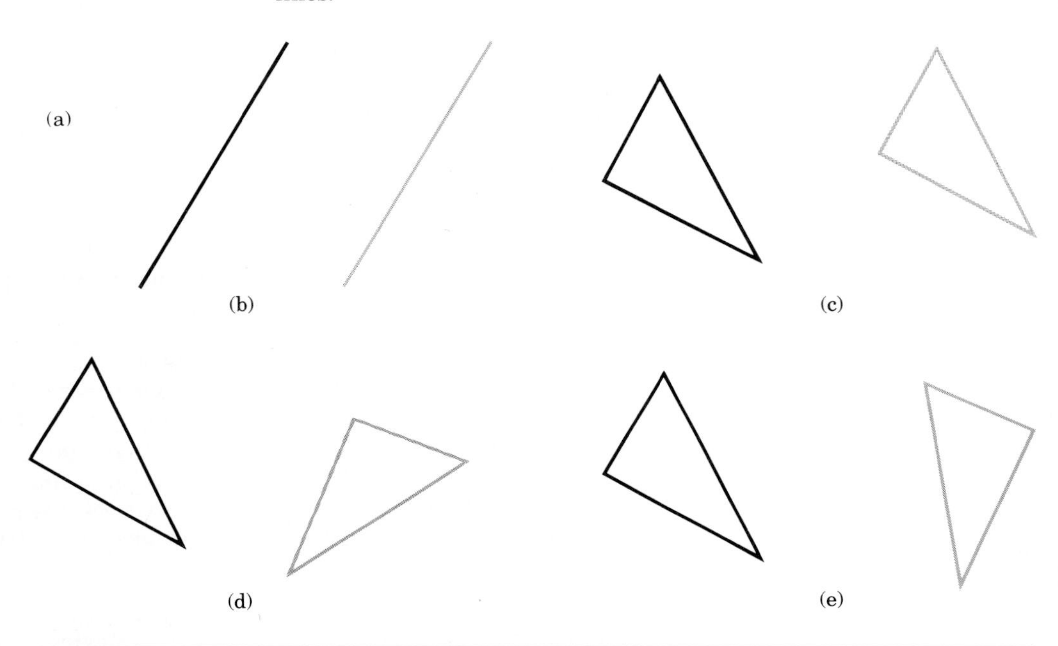

The experience with this investigation illustrates the fact that if two figures are **congruent** (same size and shape), then one of the figures is the reflection image of the other or is the reflection image of a product of two or three reflections.

After you have completed Investigation 7.5, you will undoubtedly notice that certain parts were easier to complete than others. Part (b) simply required the construction of the perpendicular bisector of the segment determined by the gray and black points. The theme of "perpendicular" can be used to complete parts (c), (d), and (e). Figure 7.80 illustrates the following solution to part (e):

Step 1. Construct perpendicular bisector p of $\overline{AA'''}$.
Step 2. Construct perpendicular bisector q of $\overline{C'C'''}$.
Step 3. Construct perpendicular bisector r of $\overline{B''B'''}$.

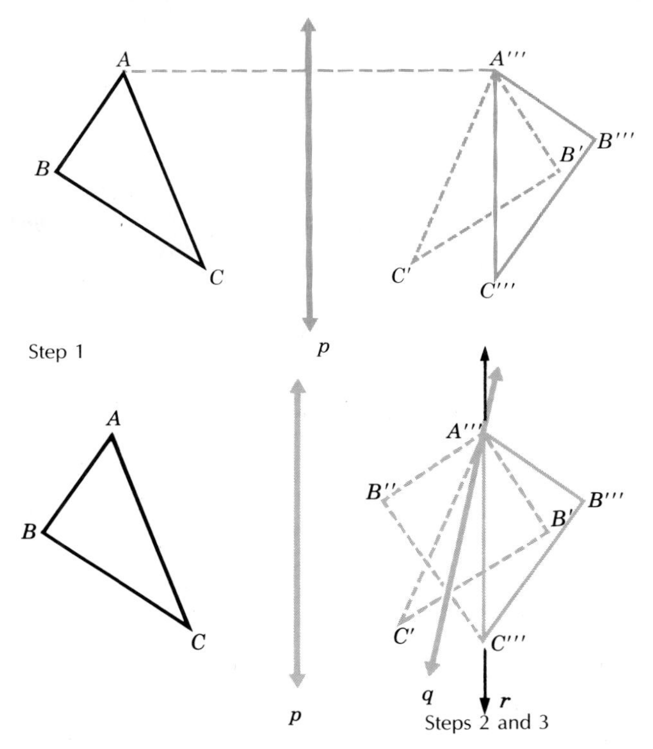

FIGURE 7.80

Investigation 7.5 and Figure 7.80 suggest the following generalizations:

1. Given a geometric figure (e.g., a triangle) and its image under a *slide*: A pair of parallel lines p and q can be found so that the image under flip in lines p and q takes the original figure onto the slide image of the figures.
2. Given a geometric figure and its image under a *turn*: A pair of lines p, q intersecting in the center of the turn can be found so that the image under flip in lines p and q takes the original figure onto the turn image.
3. Given a geometric figure and its image under a *slide-flip*: A set of three lines p, q, r exist such that the image under flip p, q, and r takes the original figure to the final position under the given motion.

These generalizations will be accepted as true without proof and may be used in the following exercises.

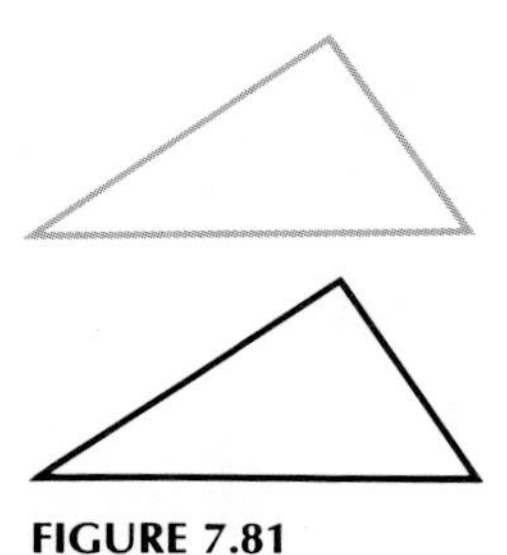

FIGURE 7.81

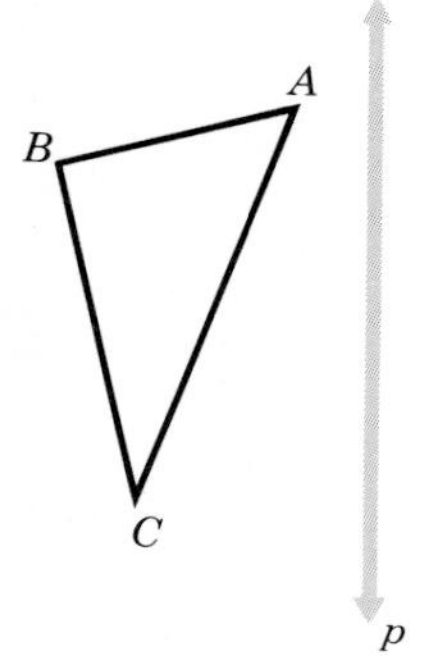

FIGURE 7.82

EXERCISE SET 7.5

1. Consider a triangle and its image under a slide (Fig. 7.81).
   a. Draw a pair of lines so that the product of flips in the 2 lines takes the black triangle onto the gray triangle.
   b. How many different pairs of lines such as those in part a can you find?
2. Draw a triangle and use tracing paper to find its image after a turn.
   a. Draw a pair of lines so that the product of flips in the 2 lines takes the original triangle onto its image.
   b. How many different pairs of lines like those in part a can you find?
3. If p and q are parallel, as shown in Figure 7.82, how does the final position of $\triangle ABC$, after flipping through p, then q, compare to the final position after flipping through q, then p? Is the result ever the same?
4. If p and q intersect, as shown in Figure 7.83, how does the final position of $\triangle ABC$, after flipping through p, then q, compare to the final position after flipping through q, then p? Is the result ever the same?
5. Which of the following conjectures do you think are correct (i.e., have been proven to be true)?
   a. If $\triangle ABC$ and $\triangle A'B'C'$ are the same size and shape, then one of these triangles can be made to coincide with the other using the product of 2 flips. *Note*: The two triangles may not have the same orientation.
   b. Any slide or turn can be described as the product of 2 flips.
   c. If $\triangle ABC$ and $\triangle A'B'C'$ are oppositely oriented, then the two triangles can never be made to coincide after 2 flips.

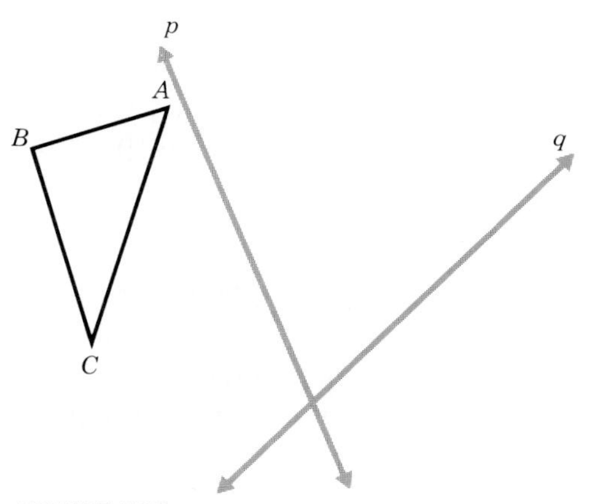

FIGURE 7.83

6. **Critical Thinking**. Translations, rotations, and reflections are called *transformations* (T) and can be described algebraically as follows: We write $T(x,y) = (x + 2, x + 5)$ to show that the image of a point, say (4,1) is $(4 + 2, 1 + 5)$, or (6,6). Graph a triangle and figure out a way to show that the transformation T is a translation. Then decide whether the following transformations are translations, rotations, reflections, or none of these three. Describe centers of rotation and reflection lines whenever they exist.

   a. $T(x,y) = (x + 3, y - 2)$
   b. $T(x,y) = (-y,x)$
   c. $T(x,y) = (y,x)$
   d. $T(x,y) = (-x + 2, -y + 4)$
   e. $T(x,y) = (-2x, \frac{1}{2}y)$
   f. $T(x,y) = (x + 3, -y)$

PEDAGOGY FOR TEACHERS

ACTIVITIES

1. Analyze the *Geometric Connectors: Transformations* software and worksheets available from Sunburst Publications. Write a one-page evaluation of these materials and their potential effectiveness in helping students explore the properties of transformations.

2. Study the material available for exploring transformations with the MIRA available from Dale Seymour Publications and decide at what grade level this material might be appropriate. Give specific reasons for your decision and some ideas about how you might use the material in your classroom.

3. At least one basic text series for the elementary school presents ideas of slides, flips, and turns in the middle grades. Find such a series and develop an outline that specifies what type of material it presents and at what grade levels. Give page numbers of parts of the textbooks where these ideas are developed. Make any additional suggestions as a result of your study of this chapter that you think might improve this development for elementary school children.

4. Use any materials you wish and do the following:
   a. Devise a teaching aid that would enable middle-grade children to easily find the rotation image of a triangle through any number of degrees.
   b. Devise teaching aids that would enable students to perform slides and flips mechanically.

5. Plan a lesson, including all appropriate teaching aids, that could be used to help junior high school or high school students understand how to combine motions and to form generalizations about these combinations. Try your ideas with an individual student or a small group of students and report your results. Suggest specific ways you would improve the activity for use in your classroom.

6. Plan a video tape that would graphically demonstrate the meaning of slides, turns, and flips.

7. Make a list of questions about mirrors that you could ask students at a given level. For ideas, refer to *The Mirror Puzzle Book* (Walter, 1985). Then ask a student or group of students your questions. Provide mirrors so that they might experiment and answer the questions. Report on their knowledge of mirrors and their reactions to your questions.

COOPERATIVE LEARNING GROUP DISCUSSION

Work together to explore and analyze the following situation. Make a cardboard rectangle with a gray arrow on one side and a black arrow on the opposite side. Both arrows should point in the same direction (Fig. 7.84). Suppose you have the cardboard rectangle lying inside a rectangular outline on a piece of paper (Fig. 7.85).

Here are descriptions of possible motions that you can make with the rectangle so it will fit back inside the outline again.

FIGURE 7.84

FIGURE 7.85

1. Flip the rectangle about a vertical line of symmetry. This is a V flip (Fig. 7.86).

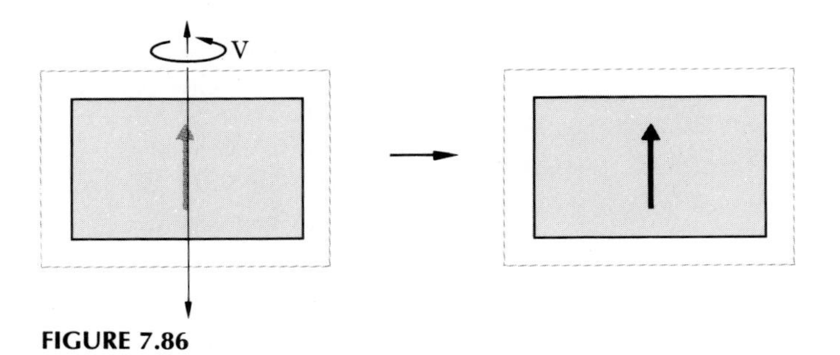

FIGURE 7.86

2. Flip the rectangle about a horizontal line of symmetry. This is an H flip (Fig. 7.87).

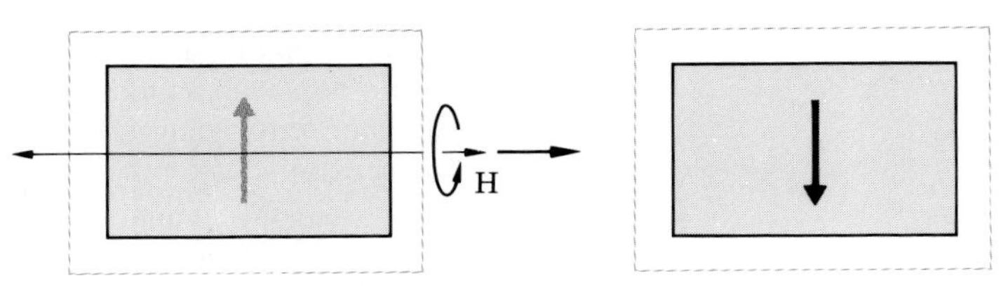

FIGURE 7.87

3. Turn the rectangle 180° about the center (clockwise or counterclockwise). This is the U (upside down) turn (Fig. 7.88).

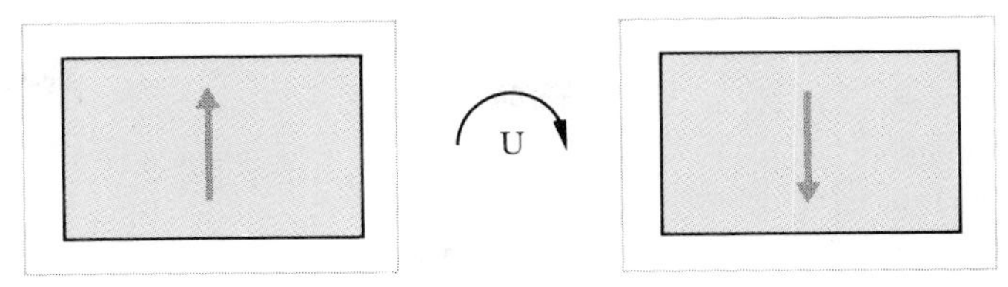

FIGURE 7.88

4. Turn the rectangle 360° (or leave it unchanged). This is the I (identity) turn (Fig. 7.89).

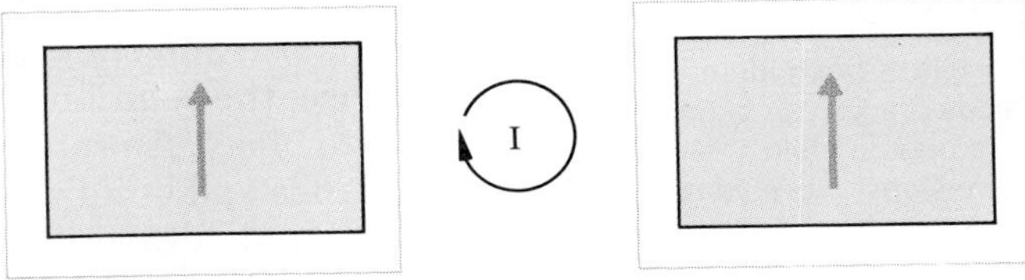

FIGURE 7.89

5. Complete the following product table, to show all possible products of pairs of these motions. What patterns do you see in the table?

Followed by	I	U	H	V
I				
U				
H				U
V				

6. Discuss the value of this activity for students. At what grade level do you think it would be appropriate, if any?

CHAPTER REFERENCES

Gardner, Martin. 1964. *The Ambidextrous Universe*. New York: Basic Books.

Gardner, Martin. 1969. *The Unexpected Hanging*. New York: Simon and Schuster.

Piaget, J., Inhelder, H., and Szminska, Alina. 1964. *The Child's Conception of Geometry*. New York: Harper and Row.

Walter, Marion. 1985. *The Mirror Puzzle Book*. New York: Parkwest Publications.

SUGGESTED READINGS

Consult the following sources listed in the Bibliography for related readings: Bannon, 1991; Del Grande, 1962; Lindquist and Shulte, 1987; and Sanok, 1987.

CHAPTER EIGHT Magnification and Similarity

INTRODUCTION

Congruence, a relation studied in Chapter 7, is at the very heart of modern technology. Mass production along assembly lines is possible only because we can produce many parts congruent to each other. Figure 8.1 shows how automobile parts, identical in size and shape, are produced and assembled.

Likewise, the relation of "same shape" plays an important role in technology. An integral part of large construction projects, airplanes and buildings, for example, is the production of models and scale drawings in which every aspect of the final object is scaled down by a constant factor. These scaled-down versions of the larger structure (see Fig. 8.2) possess the same shape as the larger object and are essential to providing a full understanding of the intricacies of the actual structure itself.

TEACHING NOTE

Teachers of children in early grades can capitalize upon a child's delight in miniature versions of everyday objects. The demand for toy autos and people that are precisely like the real thing show the children's recognition of the correspondence that should exist between part of an object and its model.

FIGURE 8.1

FIGURE 8.2

Photographic enlargement is another method often used to reveal intricate details. It is a useful procedure because the process preserves the shape of objects. For example, in Figure 8.3 we see a photograph taken at 10,000 feet and an enlargement of a small portion of it.

FIGURE 8.3

FIGURE 8.4

This chapter answers two basic questions: (1) *What precisely is meant by the phrase "same shape?"* and (2) *What characteristics of two figures must be alike before the two figures possess the same shape?* For example, which one of the three figures in Figure 8.4 is not the same shape (above the waist) as the other two and why? Which pairs of figures in Figure 8.5 have the same shape, and why? Often our intuition leads us to guess that objects are the same shape, but to explain precisely why two figures are not the same shape may be much more difficult.

MAGNIFICATION

Chapter 7 stated that two figures are the *same size and shape* or are *congruent* if one figure is the reflection image of the other or if one is the reflection image of a product of two or

Which pairs are the same shape?

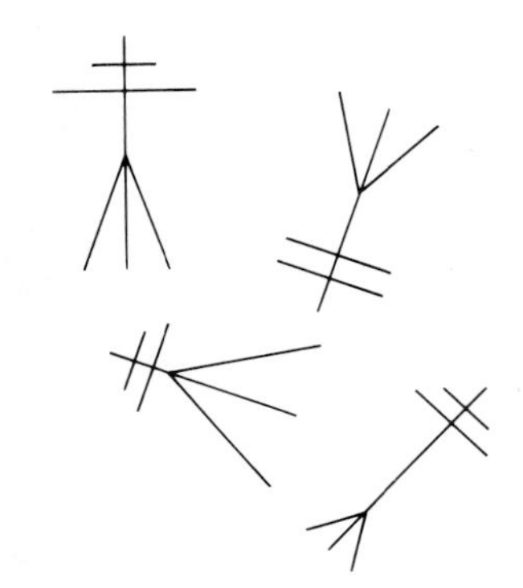

Why not the same shape?

FIGURE 8.5

INVESTIGATION 8.1

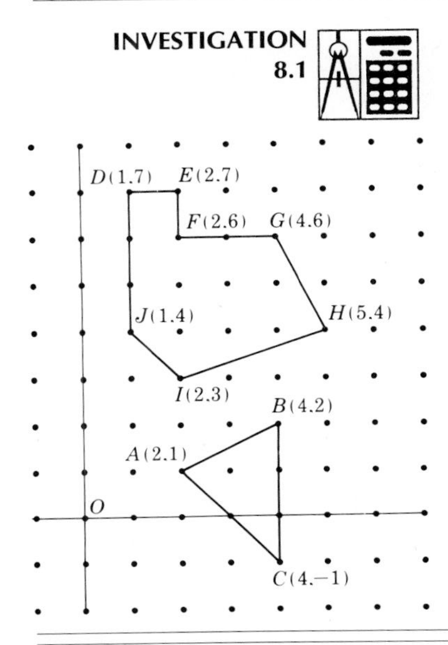

Use dot paper or graph paper for this investigation.

A. 1. Multiply the coordinates of points A, B, and C by 2 to obtain points A', B', and C', respectively, and multiply the coordinates D, E, F, G, H, I, and J by 1/2 to obtain points D', E', ... Show these points on your grid and connect them to form polygons.
2. Do the points A', B', . . . , J' lie on rays OA, OB, . . . , OJ?
3. How do the distances OA' and OA compare? How do the distances OB' and OB compare?
4. Does the $\triangle A'B'C'$ have the same shape as $\triangle ABC$? Do the two heptagons have the same shape?

B. Conjecture what the results in part 2 would be if all coordinates were multiplied by 3; by 4; by 1/2. Check your conjecture through drawings on dot paper.

C. Draw on dot paper a picture of your own choosing. Draw the figure resulting from multiplying all coordinates (relative to some coordinate system) by 2.

three reflection images. We shall see that the concept of **magnification** is essential to a broad understanding of "same shape."

We begin our study of magnification with Investigation 8.1. We see in this investigation that when all coordinates are multiplied by a fixed constant k, the resulting figure has the same shape as the original figure. The perimeter of the resulting figure is k times the perimeter of the original figure, and the area of the resulting figure is k^2 times the area of the original figure. These observations are accurate and are the result of the following basic fact: If the coordinates of the points A, B, and C are multiplied by the positive number k,

$$OA' = k \cdot OA$$

$$OB' = k \cdot OB$$

$$OC' = k \cdot OC$$

This observation motivates the concept we call a magnification with center O and scale factor k.

A **magnification with center O and scale factor $k > 0$**, maps each point $P \neq O$ to the unique point P' on ray $\overline{OP}$, where

TEACHING NOTE

Students enjoy making magnifications of pictures using coordinates on graph paper. A picture such as the cartoon character shown is placed on a small-grid graph paper, and key points are labeled. These points are plotted on larger-grid paper, and the figure is reproduced. This magnification can also be done by doubling or tripling the numbers in the ordered pair for each point.

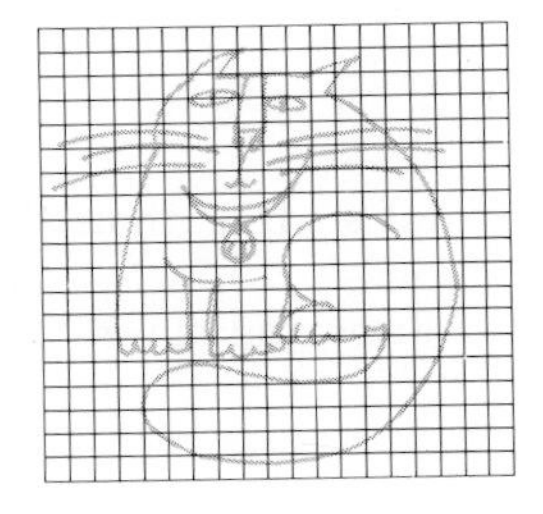

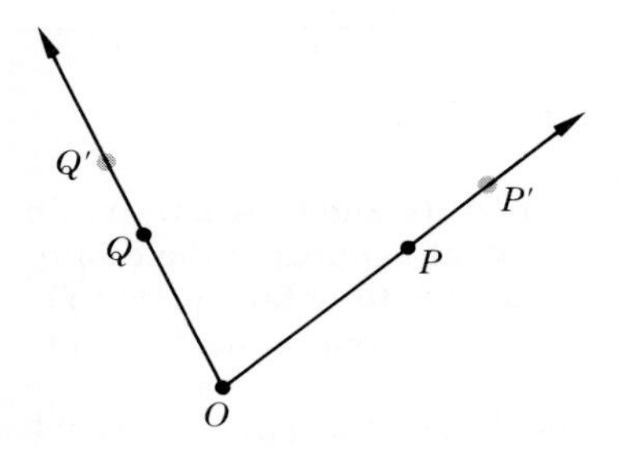

$$\frac{OP'}{OP} = \frac{OQ'}{OQ} = k,\ k > 1$$

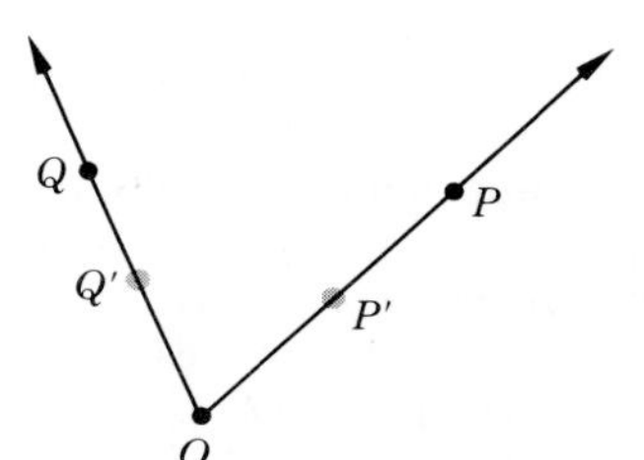

$$\frac{OP'}{OP} = \frac{OQ'}{OQ} = k,\ k < 1$$

FIGURE 8.6

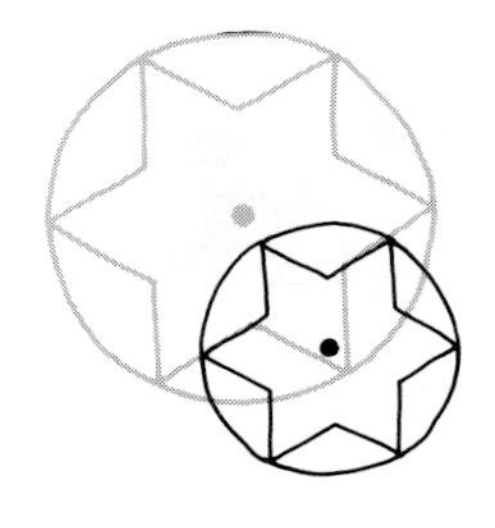

FIGURE 8.7

$OP' = k \cdot OP$, and maps point O to itself (Fig. 8.6). The point P' is called the **image of *P* under the magnification**. (Note that if $k < 1$, the magnification with center O and scale factor $k > 0$ actually has a shrinking effect rather than a magnifying one.)

Note that the definition of magnification does not refer to a coordinate system, even though the definition was motivated by using one. The point O can be any point in the plane and is not restricted to the origin of a coordinate system.

By finding the image of each point P in a set, we obtain what we call the **image set**. In Figure 8.7, we see two figures (in black) and their images (in gray) under the magnification with center O and scale factor 3/2. In Exercise Set 8.1, we continue to explore magnifications.

TEACHING NOTE

A pen light may be used as a point, with a shape held in front of the pen light perpendicular to the light ray. Magnification of the shape will appear in the shadow of the shape on the wall. Children can experiment with the effect of the proximity of the point to the figure on the size of the magnification. A slide projector can also be used for this demonstration of magnification.

EXPLORING WITH THE COMPUTER 8.1

Use LOGO software to explore the procedure for drawing a triangle. Try it, using the command TO TRIANGLE 1.

```
TO TRIANGLE :K
FD 3*:K RT 90
FD 4*:K RT 37
FD 5*:K
END
```

1. What happens when you use values larger and smaller than 1 for *K*?
2. Write and try a procedure that can be used to draw and magnify a parallelogram.

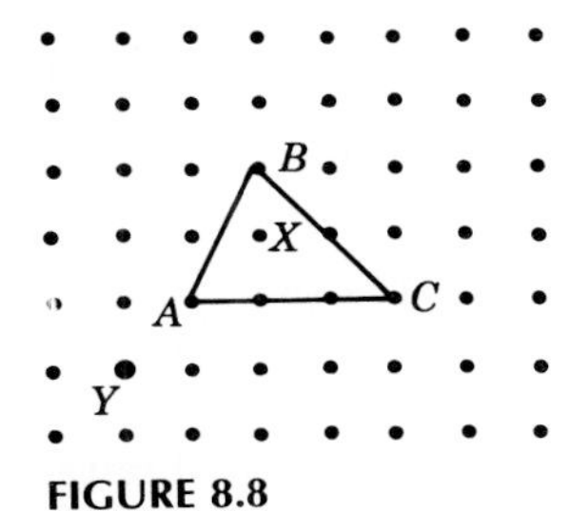

FIGURE 8.8

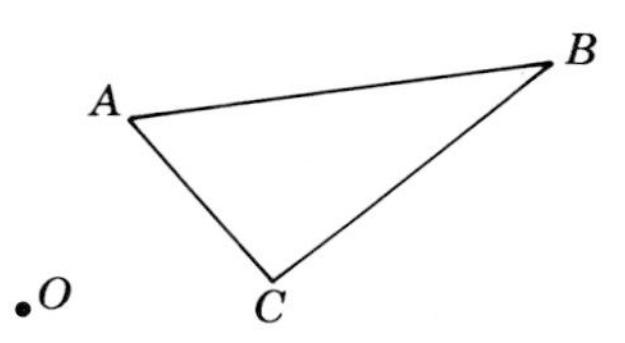

FIGURE 8.9

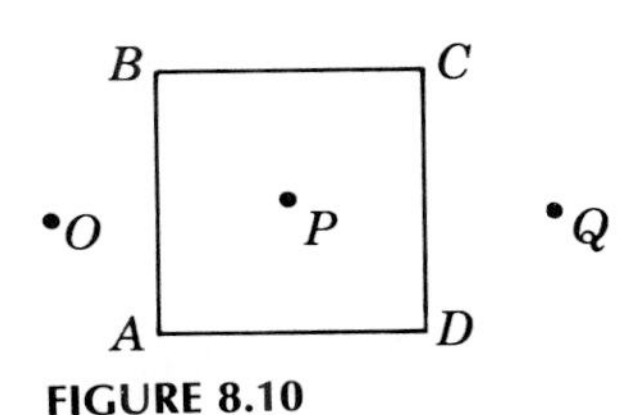

FIGURE 8.10

FIGURE 8.11

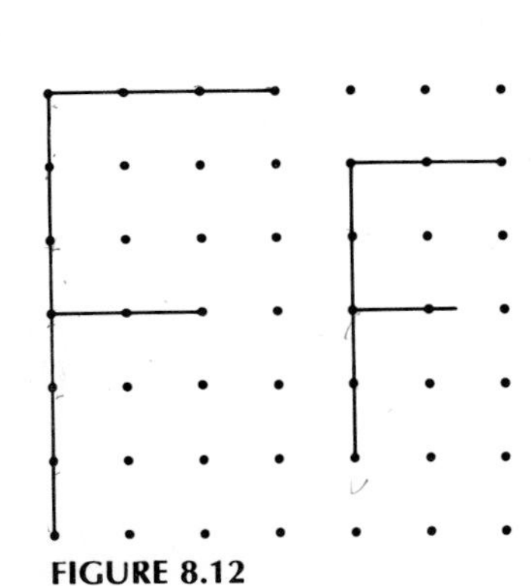

FIGURE 8.12

EXERCISE SET 8.1

Use dot paper or graph paper for these exercises when it is convenient.

1. Draw the image of $\triangle ABC$ in Figure 8.8 for the magnification with center X and scale factor 2.
2. Draw the image of $\triangle ABC$ in Figure 8.8 for the magnification with center Y and scale factor 1/2.
3. Copy the figure in Figure 8.9. Draw the image of $\triangle ABC$ for the following:
   a. The magnification with center O and scale factor 2.
   b. The magnification with center A and scale factor 3.
4. Copy the figure in Figure 8.10. Draw the image of square $ABCD$ for the following:
   a. The magnification with center O and scale factor 2.
   b. The magnification with center P and scale factor 3.
   c. The magnification with center Q and scale factor 1/2.
5. The center of the magnification is point O in Figure 8.11. Complete the following:
   a. For a scale factor of 2, the image of K is ____.
   b. For a scale factor of 2, the image of $IJKL$ is ____.
   c. For a scale factor of 3, the image of $IJKL$ is ____.
   d. For a scale factor of 1/2, the image of H is ____.
   e. For a scale factor of 1/2, the image of $EFGH$ is ____.
   f. For a scale factor of 1/3, the image of $ABCD$ is ____.
6. For the figures in Figure 8.12, find a center O and a scale factor k such that the magnification with center O and scale factor k maps the small F to the large F.

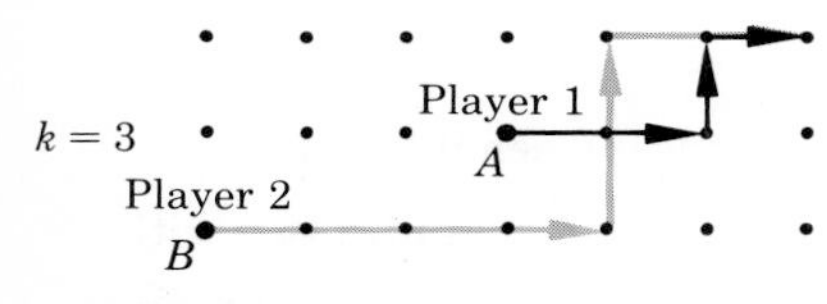

FIGURE 8.13

7. Convince yourself through drawings on dot paper or graph paper that for a magnification with scale factor k,
   a. The image of a line l is a line parallel to l.
   b. The image of an angle is another angle.
   c. The measure of an angle and its image are equal.
   d. The measure of a segment AB and its image segment $A'B'$ satisfy the equation $A'B'/AB = k$.
8. The following game can be played with a partner or as solitaire. Select any pair of points A and B on a sheet of dot paper (Fig. 8.13).

   Player 1 begins at point A and draws a horizontal or vertical line segment to some other dot. Player 2 is obligated to respond by drawing a segment in a parallel direction through twice the distance. Player 1 continues by drawing another horizontal or vertical segment, and Player 2 responds accordingly. The game continues in this fashion.

   Object: Select a positive integer $k > 2$. Player 1 tries to find a sequence of moves so that Player 1 and Player 2 end at the same dot after each player has taken exactly k moves. (You should assume that your sheet of dot paper is of unlimited size.)

   Question: Can the object of the game be met no matter what value of k you select and no matter what pair of points A and B you begin with? Explain.

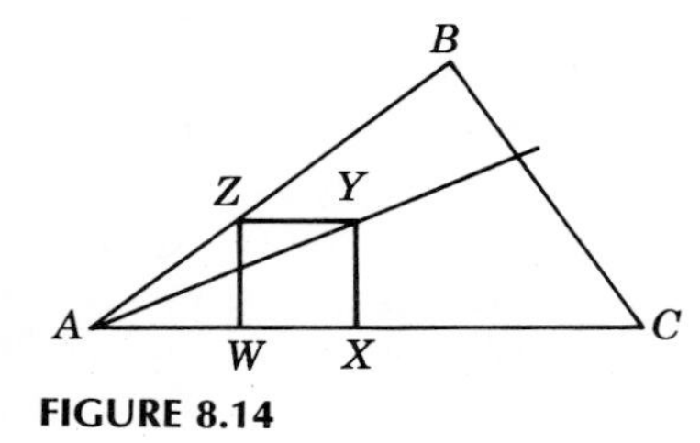

FIGURE 8.14

Problem Solving: Developing Skills and Strategies

The problem-solving strategy of **using logical reasoning** can be used to solve the following problem (Fig. 8.14): Given any $\triangle ABC$, describe how to construct a square $EFGH$ so that E and F are on AC, G is on AB, and H is on BC. [*Hint:* First construct a square that meets the first 3 conditions. Then, use the fact that for a magnification, the image of a square is a square.]

TEACHING NOTE

In his interviews, Piaget has observed that a full-blown concept of ratio does not usually appear until well into the formal operations stage, around ages 11 and 12. Teachers of younger children should create experiences that prepare for this concept acquisition. Teachers of older students should utilize the ideas of ratio to study similar triangles and to solve other problems in geometry.

PROPERTIES OF MAGNIFICATIONS

Through the experience of Investigation 8.1 and the exercises in the previous section, the reader may have already discovered many properties of magnifications. This section summarizes some of these properties, concentrating on geometric properties that are preserved under magnifications.

1. *Betweenness is preserved* (Fig. 8.15). If B is between A and C, then B' is between A' and C'.

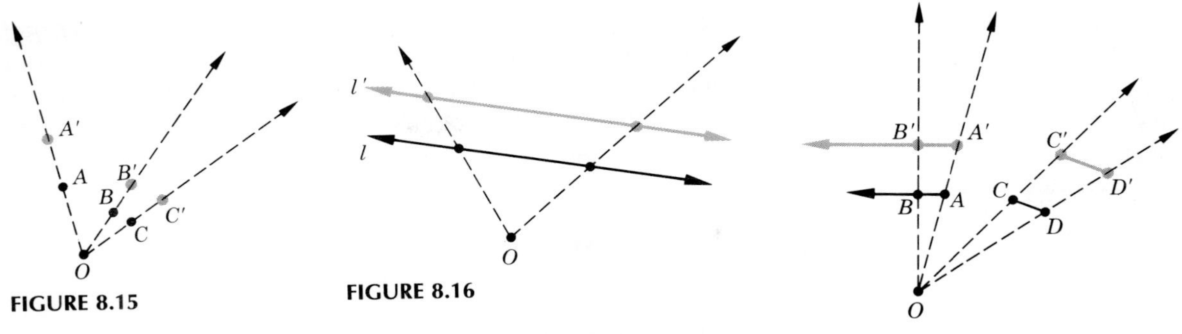

FIGURE 8.15

FIGURE 8.16

FIGURE 8.17

2. *Lines are preserved* (Fig. 8.16). If l is a line, then the set l' of all images of points of l is also a line. In fact, l' is always parallel to l.
3. *Rays and segments are preserved* (Fig. 8.17). The image of a ray $\overline{AB}$ is a ray $\overline{A'B'}$, and the image of a segment $\overline{CD}$ is a segment $\overline{C'D'}$.
4. *Angles and their measure are preserved* (Fig. 8.18). Since the image of a ray is a ray, it follows that the image of an angle is an angle. The measure of an angle and its image are equal. For example, $m(\angle ABC)$ is equal to $m(\angle A'B'C')$.
5. *Polygons are preserved* (Fig. 8.19). Since the image of a segment is a segment, it follows that the image of a polygon is a polygon.

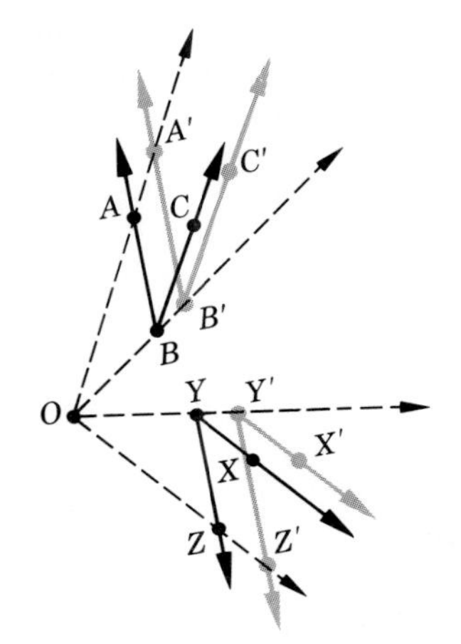

FIGURE 8.18

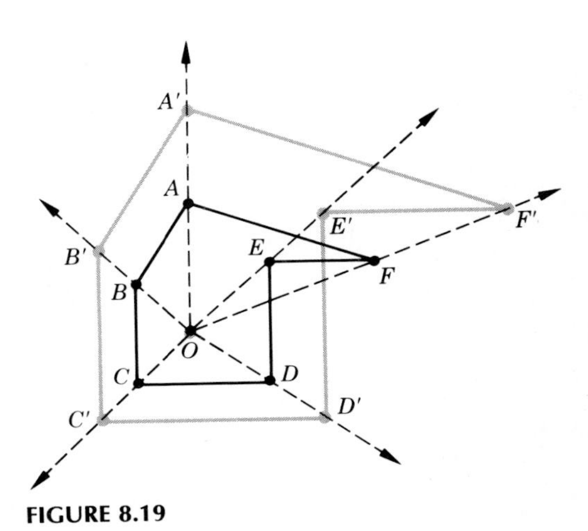

FIGURE 8.19

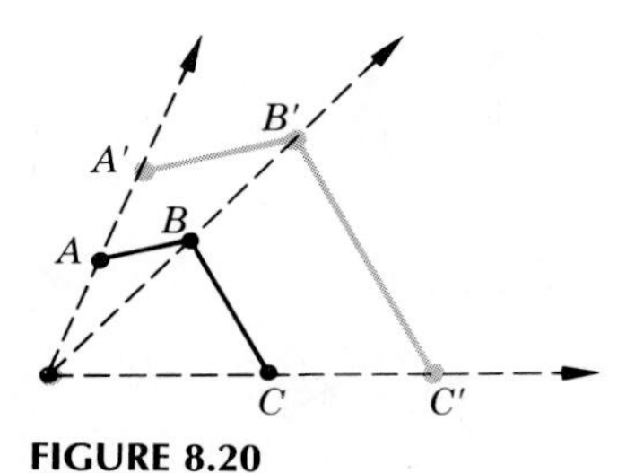

FIGURE 8.20

6. *Ratio of distances is preserved* (Fig. 8.20). The ratio AB/BC is equal to the ratio $A'B'/B'C'$. In general, the ratio of the lengths of two segments is equal to the ratio of the lengths of the image segments. The above equality of ratios can be written

$$\frac{AB}{BC} = \frac{A'B'}{B'C'}$$

The sixth property says that the ratio of two segments of a figure is equal to the ratio of the corresponding pair of segments in the image figure. That is, given points A, B, C, and D,

$$\frac{AB}{CD} = \frac{A'B'}{C'D'}$$

From this equation we can derive the equality

$$\frac{A'B'}{AB} = \frac{C'D'}{CD}$$

which says that the ratio of a segment and its image is the same beginning with AB as when beginning with CD. It can be shown that this latter ratio is equal to the scale factor k of the magnification.

In other words, given a figure and its image under a magnification with center O and scale factor k the distance $A'B'$ is k times the distance AB for each pair of points A and B. This fact, illustrated in Figure 8.21, ensures that *any figure and its image under a magnification are the same shape.*

Scale factor 2
$A'B' = 2 \cdot AB$

FIGURE 8.21

EXPLORING WITH THE COMPUTER 8.2

Use LOGO software to explore the procedure for drawing a square and some magnified squares.

```
TO MAGNIFIED.SQUARE :K                 TO SQUARE :K
IF :K > 100 THEN STOP                  REPEAT 4[FD :K RT 90]
SQUARE :K >-----------------> END
MAGNIFIED.SQUARE (:K * 2)
END
```

1. What happens when you run MAGNIFIED SQUARE?
2. How can you change the procedure to
   a. Produce similar square 1.5 times the size of the first square?
   b. Produce a number of triangles that are magnifications of an original triangle?

EXERCISE SET 8.2

1. Refer to Figure 8.22. $AB = 1$, $AC = \sqrt{5}$, $BC = 2\sqrt{2}$, $\triangle DEF$ is the image of $\triangle ABC$ for the magnification with center O and scale factor 2, and $\triangle GHI$ is the image of $\triangle ABC$ for the magnification with center O' and scale factor 3. Answer the following:
   a. $DE =$ ____ b. $EF =$ ____
   c. $DF =$ ____ d. $GH =$ ____
   e. $GI =$ ____ f. $HI =$ ____
   g. If $\triangle ABC$ is the magnification image of $\triangle DEF$, then the scale factor $k =$ ____.
   h. If $\triangle ABC$ is the magnification image of $\triangle GHI$, then the scale factor $k =$ ____.
   i. If $\triangle GHI$ is the magnification image of $\triangle DEF$, then the scale factor $k =$ ____.
   j. If $\triangle ABC$ is the magnification image of $\triangle DEF$, then $\triangle ABC$ is similar to ____.

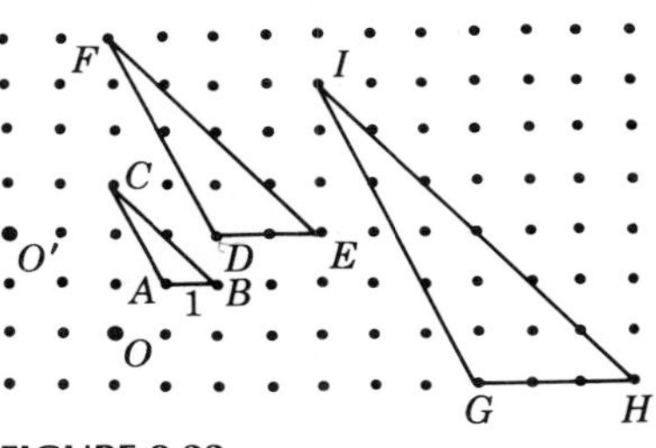

FIGURE 8.22

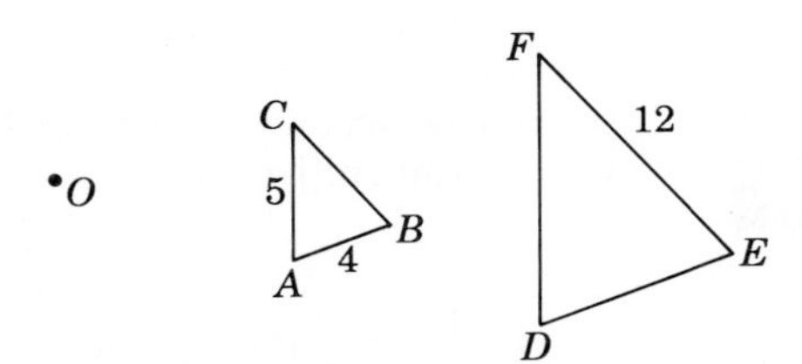

FIGURE 8.23

2. Refer to Figure 8.23. $\triangle DEF$ is the image of $\triangle ABC$ for a magnification with scale factor 2. Complete the following:
   a. DE = ____
   b. DF = ____
   c. BC = ____
3. Suppose that $EFGH$ is the image of $ABCD$ for a magnification with scale factor 5/3 (Fig. 8.24). Complete the following:
   a. EF = ____
   b. FG = ____
   c. GH = ____
   d. EH = ____
4. Suppose that $\triangle XYZ$ is the image of $\triangle ABC$ for a dilation with scale factor 4. Complete the following:
   a. If $AB = 25$, then XY = ____.
   b. If $AC = 30$, then XZ = ____.
   c. If $YZ = 32$, then BC = ____.
   d. If $XY = 14$, then AB = ____.
5. Find the center of the magnification that maps A to A' and B to B' (Fig. 8.25).
6. Given that the magnification that maps B to B' has its center on line l, describe how to construct the image point A' under the magnification (Fig. 8.26). [*Hint:* $\overleftrightarrow{AB} \parallel \overleftrightarrow{A'B'}$.]

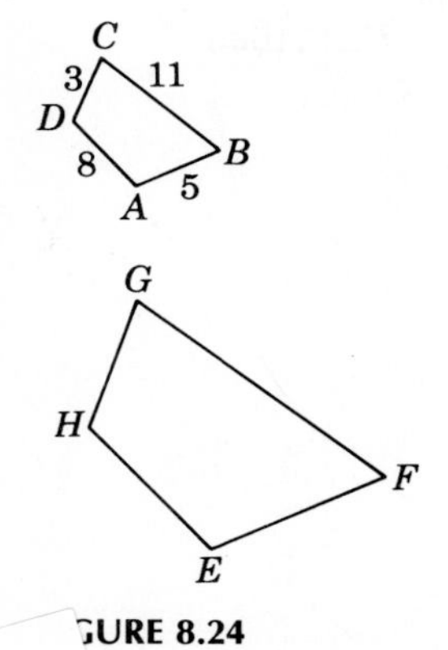

URE 8.24

A'
A
B'
B

FIGURE 8.25

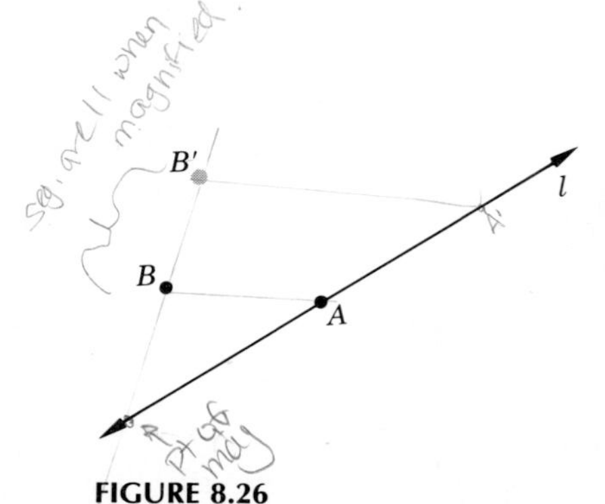

FIGURE 8.26

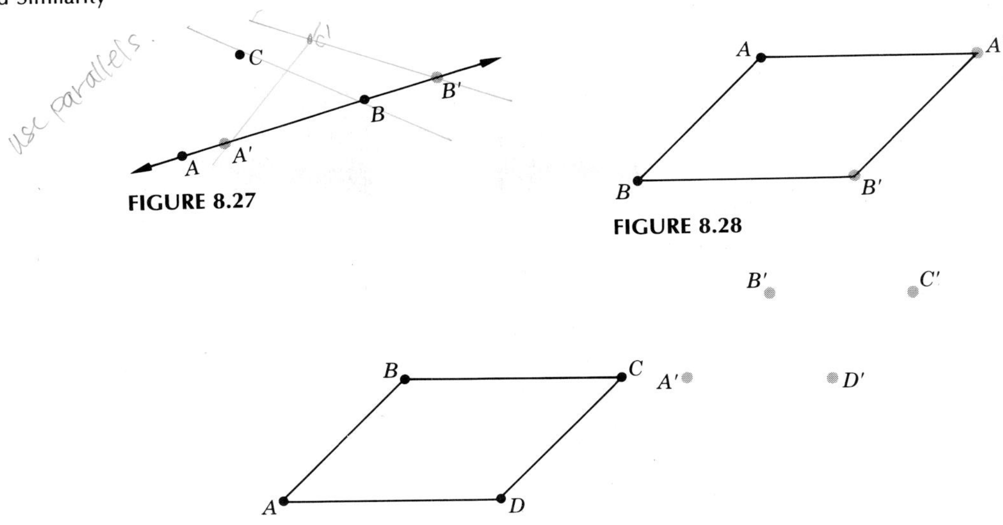

FIGURE 8.27

FIGURE 8.28

FIGURE 8.29

7. Given a magnification that maps A to A' and B to B', construct the image of point C under this magnification (Fig. 8.27).
8. Given a parallelogram $AA'B'B$, explain why there can be no magnification that maps A to A' and B to B' (Fig. 8.28).
9. Suppose $ABCD$ is a parallelogram (Fig. 8.29). If A', B', C', and D' are the images of A, B, C, and D under a magnification, explain why we know that quadrilateral $A'B'C'D'$ is a parallelogram.
10. Given $\triangle ABC$ with altitude $\overline{AM}$ (Fig. 8.30), if A', B', C', and M' are images of A, B, C, and M under a magnification, explain why we can assert that $A'M'$ is an altitude of $\triangle A'B'C'$.
11. Suppose a quadrilateral $ABCD$ is a kite (Fig. 8.31) and that A', B', C', and D' are the images of A, B, C, and D under a magnification. Explain why quadrilateral $A'B'C'D'$ is also a kite.

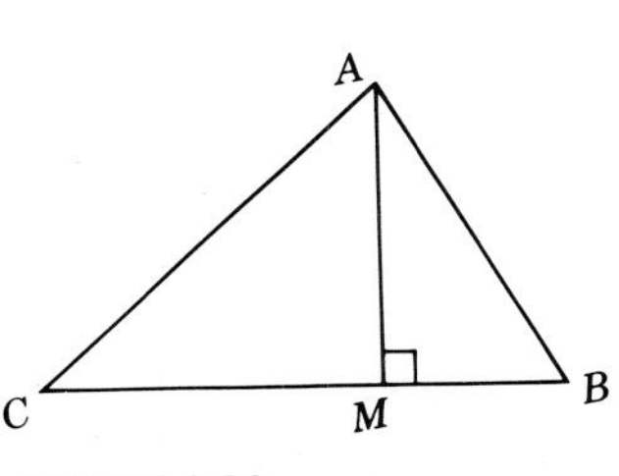

FIGURE 8.30

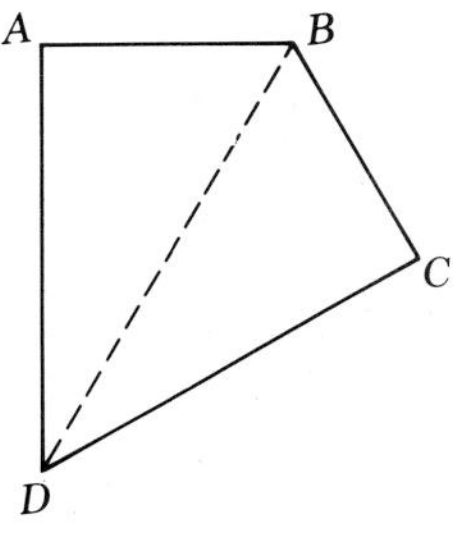

FIGURE 8.31

SIMILARITY

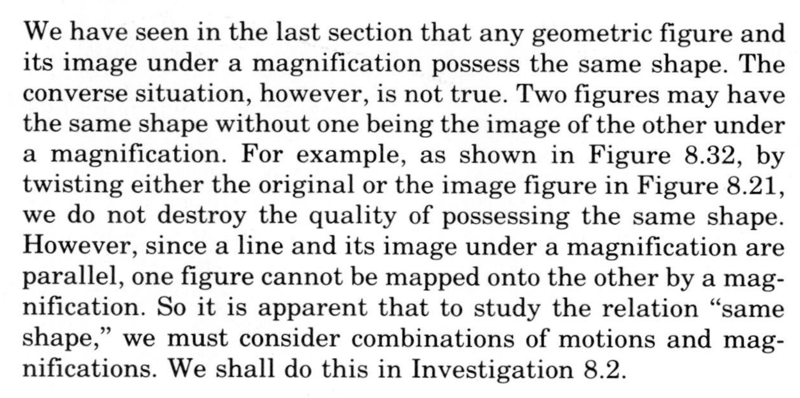

FIGURE 8.32

We have seen in the last section that any geometric figure and its image under a magnification possess the same shape. The converse situation, however, is not true. Two figures may have the same shape without one being the image of the other under a magnification. For example, as shown in Figure 8.32, by twisting either the original or the image figure in Figure 8.21, we do not destroy the quality of possessing the same shape. However, since a line and its image under a magnification are parallel, one figure cannot be mapped onto the other by a magnification. So it is apparent that to study the relation "same shape," we must consider combinations of motions and magnifications. We shall do this in Investigation 8.2.

INVESTIGATION 8.2

Protractor, straight edge, and tracing paper should be an aid for this investigation. For each pair of similarly shaped figures:

A. Find a center of a turn or a line of reflection (whichever is required) and a center O for a magnification such that the motion followed by the magnification maps Figure 1 onto Figure 2. In how many ways can this be done?
B. Repeat this process showing that Figure 1 can be mapped onto Figure 2 by a magnification followed by a motion. Convince yourself by completing accurate constructions that this can be done in more than one way.

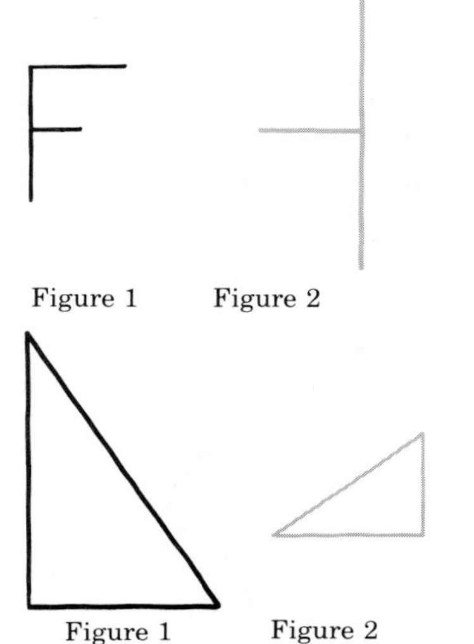

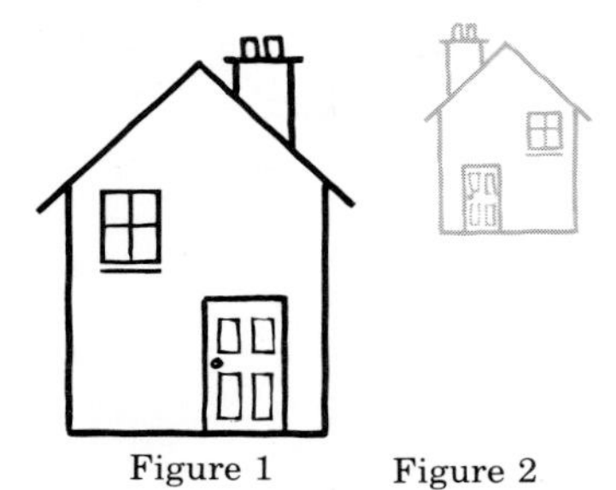

TEACHING NOTE

A pantograph is a useful mechanical device for producing a figure similar to a given figure. Upper-level students might enjoy constructing a pantograph and proving that the figure produced is, indeed, similar to the original figure. Information on a pantograph can be found on p. 107 of *A Survey of Geometry* (Eves, 1972).

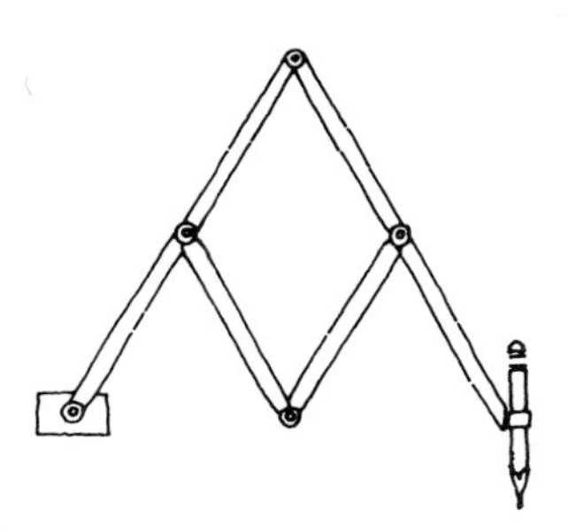

The experience of Investigation 8.2 indicates that given two figures, one the enlargement of the other, after one is repositioned appropriately by a turn or flip, the first can be magnified to coincide with the second. This observation is the basis for the following definition of similarity.

> **Definition:** Two geometric figures $\mathcal{F}$ and $\mathcal{F}'$ are similar if there exists a motion followed by a magnification that maps one figure onto the other figure.

Note that this definition of similarity is not restricted only to triangles, nor even to polygons. It applies to all types of geometric figures. We see from this definition that exploring products of motions and magnifications is, in essence, exploring the relation of similarity.

Studying similarity or "same shape" in terms of motions and magnifications is intuitively appealing and provides a broad understanding of this concept. The process of photographic enlargement is a helpful and accurate physical model for thinking about similarity. On the other hand, there are some very practical problems for which it is more convenient to describe similarity in terms of angle measures and ratios of segments. For example, suppose a drafter is asked to complete an enlarged drawing with scale factor of 9/4 of the floor plan in Figure 8.33. What types of measurements must she make and how many? Or, consider the job of surveying a field and producing a map like the one appearing in Figure 8.34. What step-by-step procedure does the surveyor use in constructing this map?

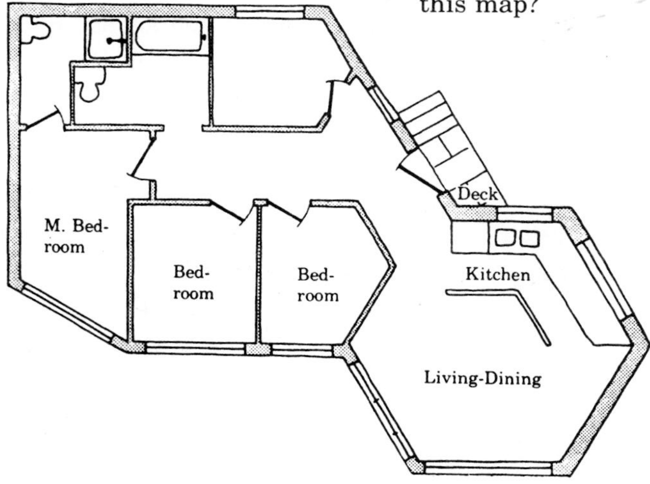

FIGURE 8.33

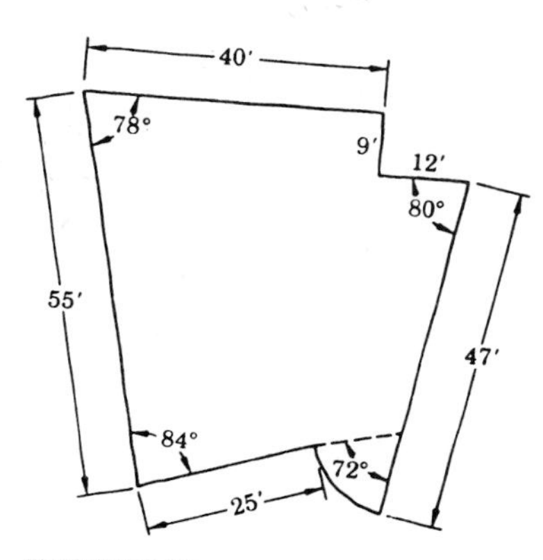

FIGURE 8.34

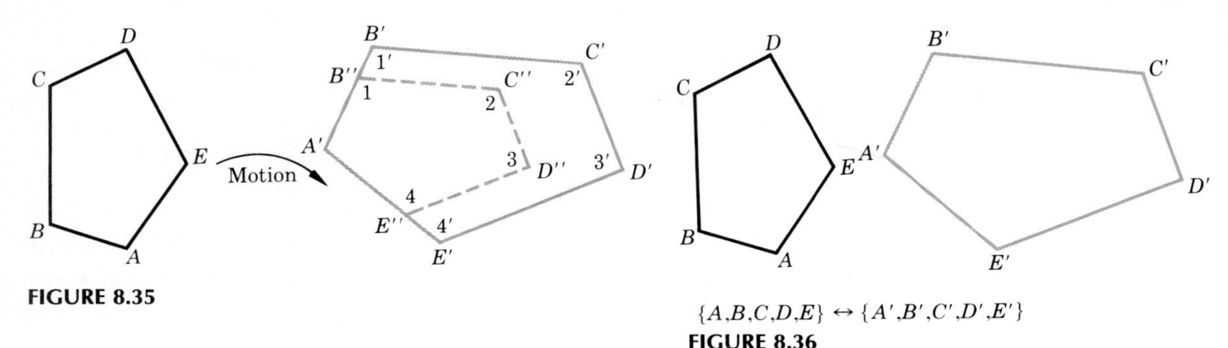

FIGURE 8.35

$\{A,B,C,D,E\} \leftrightarrow \{A',B',C',D',E'\}$

FIGURE 8.36

We have seen that two polygons are similar if, as shown in Figure 8.35, there exists a motion followed by a magnification that maps one of the polygons onto the other. This process establishes a one-to-one correspondence $\{A, B, C, \ldots\} \leftrightarrow \{A', B', C', \ldots\}$ between the vertices of the two polygons, as illustrated in Figure 8.36.

Since a magnification preserves angle measure and the ratio of distances, we have, for similar polygons, the following theorem:

> **Theorem:** Two polygons are similar if and only if there exists a one-to-one correspondence between the vertices of the polygons such that (1) corresponding angles possess equal measures, and (2) ratios of corresponding segments are equal and the same as the scale factor of the similarity transformation.

These ideas are illustrated for a pair of pentagons and a pair of triangles in Figures 8.37 and 8.38. Thus, as the theorem

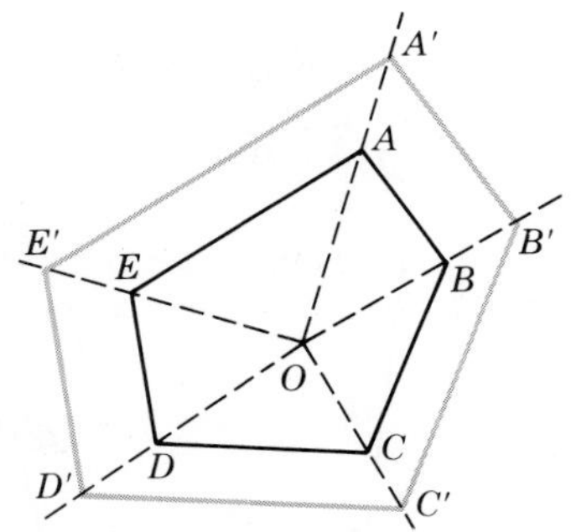

$m\angle A = m\angle A'$, $m\angle B = m\angle B'$, . . . $m\angle E = m\angle E'$

$$\frac{A'B'}{AB} = \frac{B'C'}{BC} = \frac{C'D'}{CD} = \frac{D'E'}{DE} = \frac{A'E'}{AE} = \frac{OA'}{OA}$$

FIGURE 8.37

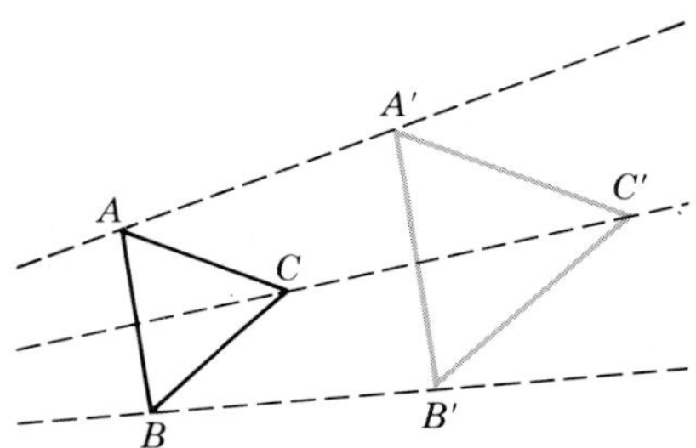

$m(\angle A) = m(\angle A')$; $m(\angle B) = m(\angle B')$; $m(\angle C) = m(\angle C')$

$$\frac{A'B'}{AB} = \frac{B'C'}{BC} = \frac{A'C'}{AC} = \frac{OA'}{OA}$$

FIGURE 8.38

EXPLORING WITH THE COMPUTER 8.3

Use geometry exploration software to make and test conjectures about the following questions. Use at least 3 different pairs of similar figures when testing each conjecture. Record your results.

1. How does the ratio of the perimeters of two similar triangles compare to the ratio of the lengths of two corresponding sides of the two triangles?
2. How does the ratio of the areas of two similar triangles compare to the ratio of the lengths of two corresponding sides of the two triangles?

indicates, if we assume accurate measurements of angles and sides are possible, that given any polygon and any ratio factor, a polygon similar to the given one can be drawn.

FIGURE 8.39

EXERCISE SET 8.3

1. For parts a through d, draw on dot paper or graph paper lines p and q, point O, and the polygon shown in Figure 8.39. Then, draw the image of this polygon for the magnification or combination of magnification and motion that is requested.
   a. The magnification with center O and scale factor 3/2.
   b. The counterclockwise turn of 90° followed by the magnification with center O and scale factor 2.
   c. A flip in line p followed by the magnification with center O and scale factor 2.
   d. A flip in line q followed by the magnification with center O and scale factor 2.
2. For parts a through c, draw on dot paper or graph paper $\triangle ABC$ and point O, as shown in Figure 8.40. Then, draw the desired image of $\triangle ABC$.
   a. Draw the image of $\triangle ABC$ for a 90° clockwise turn about O followed by the magnification with center O and scale factor 2.
   b. Draw the image of $\triangle ABC$ for the magnification with center O and scale factor 2 followed by a 90° clockwise turn about O.
   c. How do you compare the images in parts a and b?

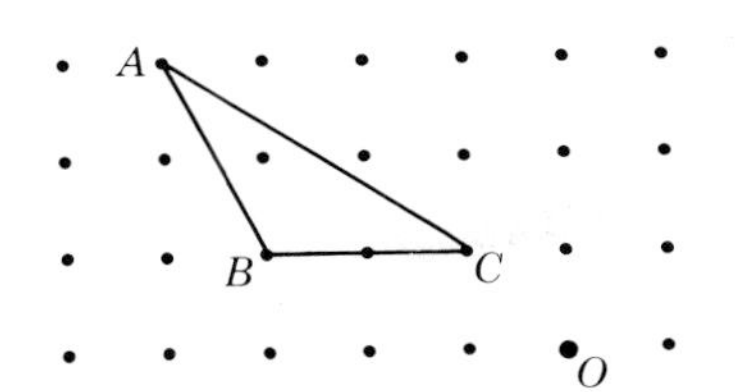

FIGURE 8.40

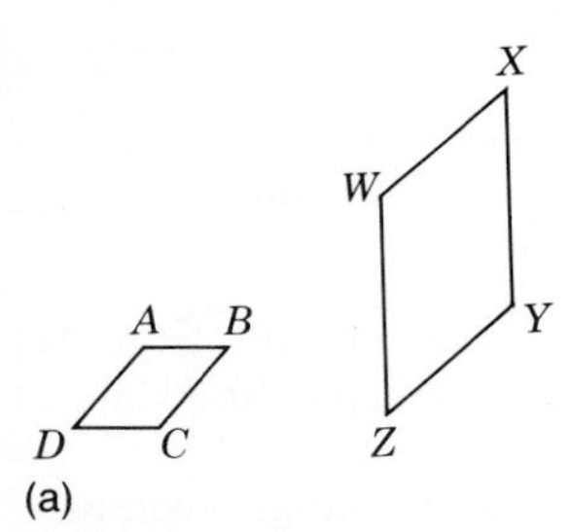

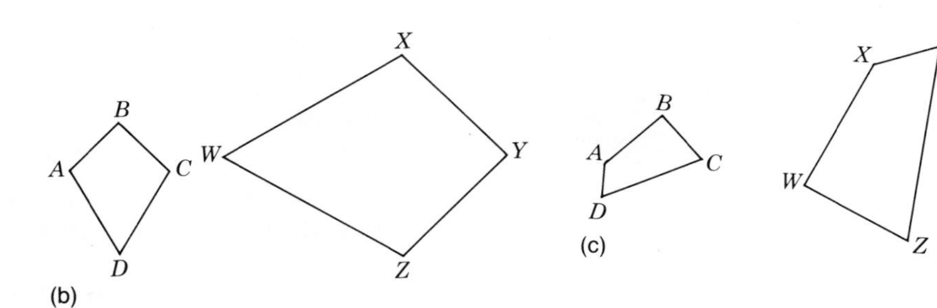

FIGURE 8.41

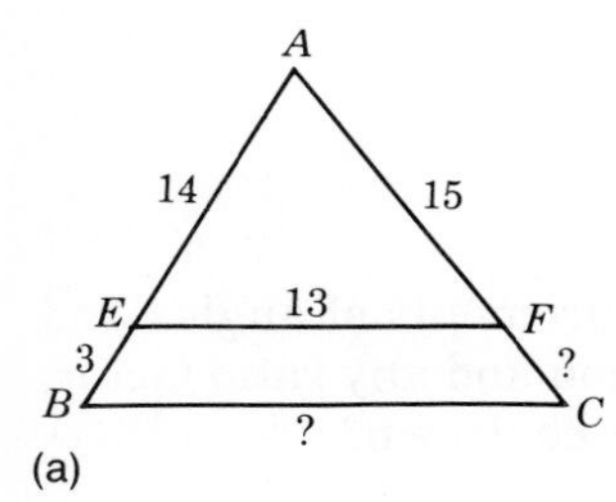

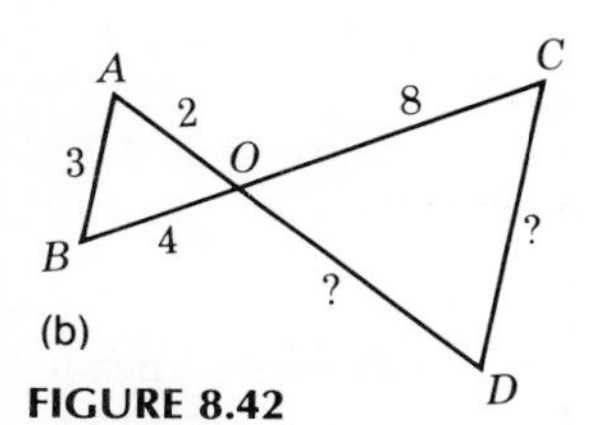

FIGURE 8.42

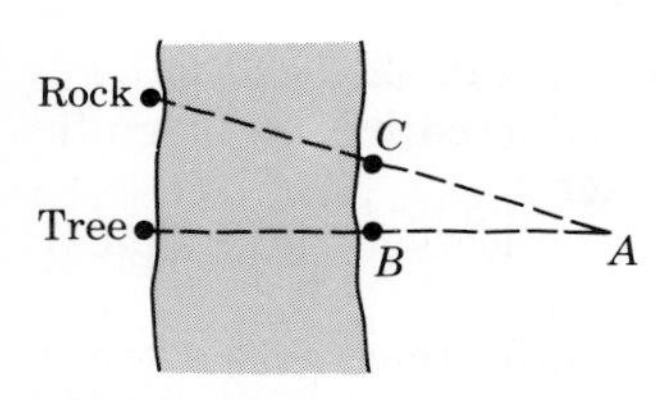

FIGURE 8.43

3. For each pair of similar polygons in Figure 8.41, write down a one-to-one correspondence between the vertices that satisfies the conditions of the theorem in the last section.
4. Each of the pairs of triangles in Figure 8.42 are similar. Find all missing measures of the sides.
5. A camper wishes to estimate the width of a river (Fig. 8.43). He stands at point A directly across from a large tree and drives a stake at point B, on the bank of the river. Next the camper places a stake at point C along his line of sight to a large rock on the opposite side of the river so that $\overline{BC}$ is perpendicular to $\overline{AB}$. The camper measures $\overline{AB}$ to be 3 m, $\overline{BC}$ to be 1.7 m, and the distance from the tree to the rock to be 75 m. How wide is the river?
6. The line of sight across a 2 ft and a 3 ft stick crosses the top of a tree (Fig. 8.44). The 3 ft stick is 100 ft from the base of the tree, and the 2 ft stick is 102 ft from the tree. How tall is the tree?
7. Suppose $\triangle ABC$ and $\triangle A'B'C'$ (Fig. 8.45) satisfy the conditions that $(A'B'/AB) = (B'C'/BC)$ and $m\angle B = m\angle B'$. Show that these triangles are similar by showing that there exists a motion followed by a magnification that maps $\triangle ABC$ onto $\triangle A'B'C'$. (The conditions in this exercise are known as the SAS condition for similarity of triangles since they involve 2 sides (S) and the included angle (A) of a triangle.)

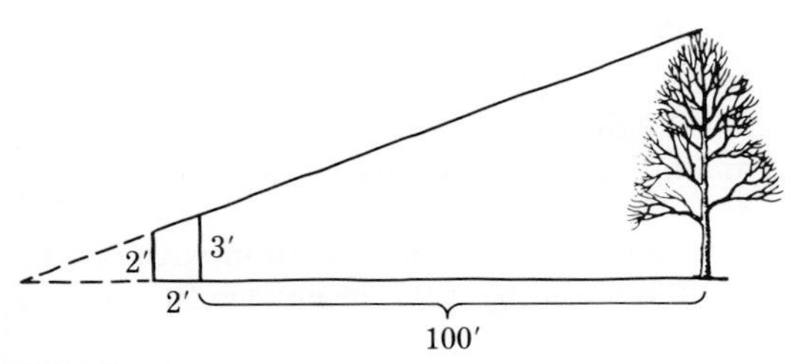

FIGURE 8.44

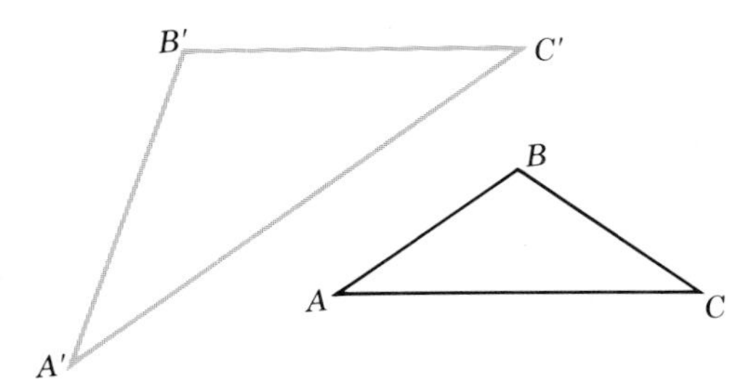

FIGURE 8.45

APPLICATION

Flagpole Problem: To find the height of a flagpole, a student laid a mirror on the ground and stood so that she could look in the mirror and see the reflection of the top of the pole (Fig. 8.49). Describe how this procedure could be used to find the pole's height and explain why it works. Choose some numbers for distances in the figure and write a problem that can be solved using this method.

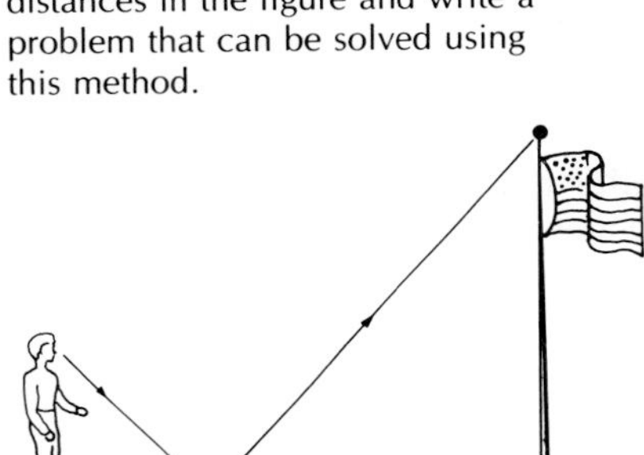

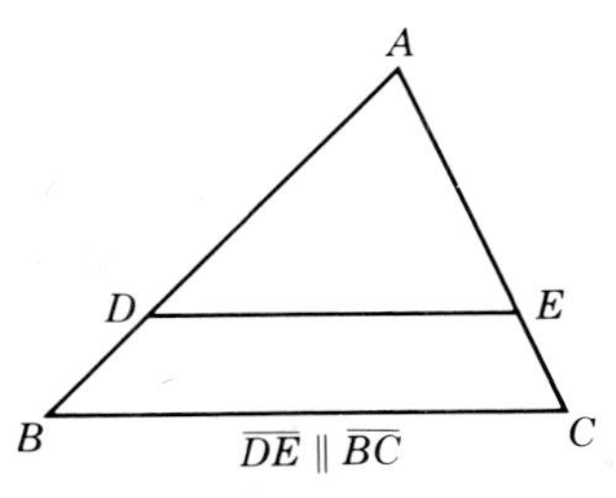

FIGURE 8.46

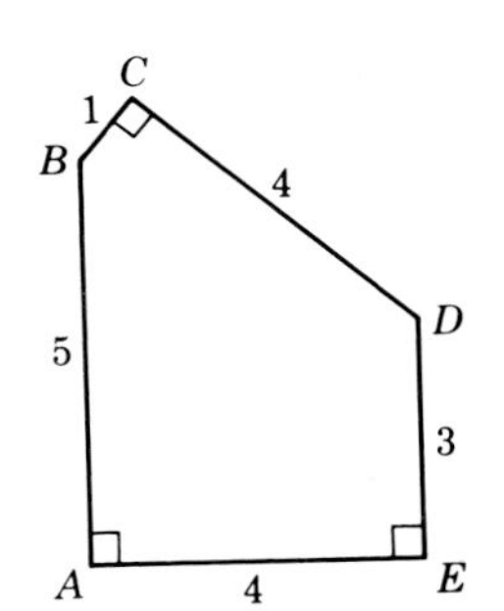

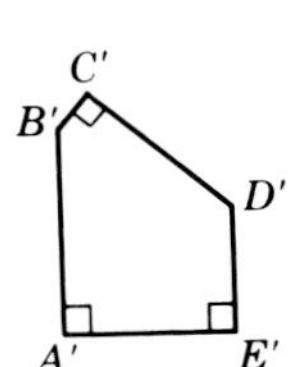

FIGURE 8.47

8. Suppose $\triangle ABC$ and $\triangle ADE$ are related as in Figure 8.46. Use the definition of similarity of geometric figures to show that $\triangle ABC$ and $\triangle ADE$ are similar. [*Hint:* Consider the magnification with center A and scale factor AB/AD.]

9. Given in Figure 8.47 are two pentagons, with angles A, C, E, A', C', and E' all right angles.

   a. If the only additional information is that $m(\angle B) = m(\angle B')$, are these pentagons necessarily similar?
   b. If in addition to the information above, you know that $A'B' = 2\frac{1}{2}$ and $A'E' = 2$, are these two pentagons necessarily similar?

10. Suppose $\triangle ABC$ and $\triangle A'B'C'$ are related by $m\angle A = m\angle A'$, $m\angle B = m\angle B'$, and $m\angle C = m\angle C'$.
    Show that these triangles are similar by showing that there exists a motion followed by a magnification that maps $\triangle ABC$ onto $\triangle A'B'C'$.
    (The condition in this exercise is known as the AAA condition for similarity since it involves the 3 angles of the triangle.)

11. Begin with segment $A'B'$ and use a ruler and protractor to draw a figure $A'B'C'D'E'F'G'$ that is similar to figure $ABCDEFG$ shown in Figure 8.48.

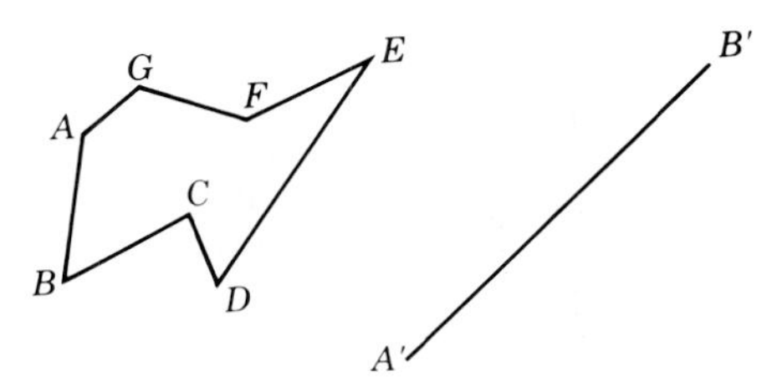

FIGURE 8.48

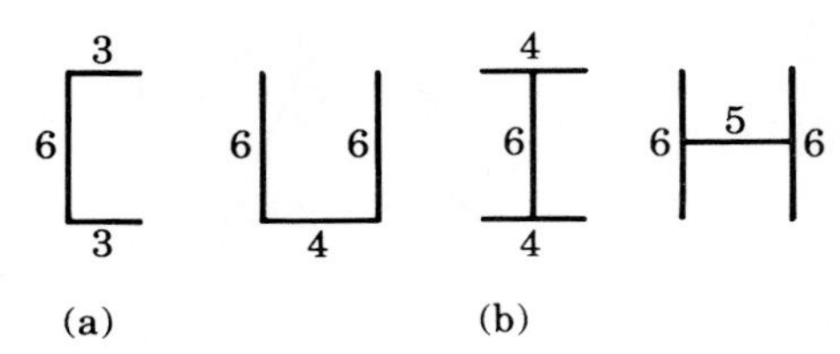

FIGURE 8.49

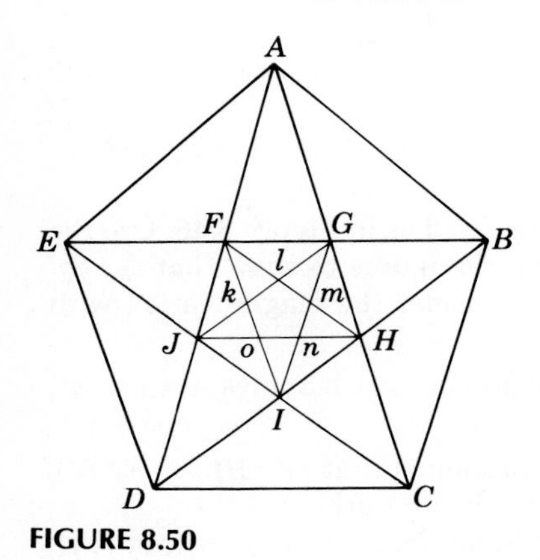

FIGURE 8.50

12. **Critical Thinking**. Explain why the letters C and U (Fig. 8.50a) and letters I and H (Fig. 8.50b) are similar or are not similar.
13. Suppose that figure F_1 is similar to but not congruent to F_2 and F_2 is similar to but not congruent to F_3. Must it be true that F_1 is similar to but not congruent to F_3? Explain.

Problem Solving: Developing Skills and Strategies

Using the problem-solving strategy of **make an organized list** may be helpful in completing this problem: How many different nonconvex polygons can you find in Figure 8.51 so that when you have listed them no pair of polygons in the list are similar?

PEDAGOGY FOR TEACHERS

ACTIVITIES

1. Create a concept card using examples and nonexamples to help a student understand the concept of similar polygons. Try the concept card with a student and revise it to improve its effectiveness.
2. A useful outline for preparing a lesson plan is as follows:
   I. Preparation (reviewing basic ideas and providing motivation for the next phase)
   II. Investigation (an exploratory phase in which the student is actively involved with physical materials, often with a group of other students, and one in which the student takes major responsibility for exploring a central idea of the lesson)
   III. Discussion (a phase in which the result of the investigation is discussed and one in which the teacher shapes the concept of the idea to be developed)
   IV. Utilization (a phase in which the student uses the ideas developed in the two previous phases)
   V. Extension (a phase in which both remedial and enrichment needs are met)

Use this outline and prepare a lesson plan for teaching scale drawing at an appropriate level. Include in your lesson plan:

a. An idea that would arouse the students' curiosity about the lesson during the preparation phase
b. An activity card that could be used by a group of students during the investigation phase to explore basic ideas
c. A list of at least 5 questions you would ask the students during the discussion phase
d. At least 2 sample problems from a ditto sheet that the students could use during the utilization phase
e. An activity that could be used for enrichment during the extension phase

3. Seek a reference on the pantograph and write a set of instructions for making and using this device at a grade level of your choice.

4. Make a sequence of overhead projectuals that could be used to develop some idea of magnification or similarity. You may want to develop these ideas with young children or with junior high or secondary students. Attempt to include an element of the discovery approach in your projectual presentation.

5. Make a list of household items that could be used in a lesson to illustrate the concepts of similarity and nonsimilarity. Plan a brief student activity or teacher demonstration using the materials on your list.

COOPERATIVE LEARNING GROUP DISCUSSION

Are there any real world applications of similarity? Brainstorm in your group to make a list of as many applications as you can. Then discuss the questions, Why, when, and how should we teach the ideas of similarity? Summarize your major points and share them with others in your class.

CHAPTER REFERENCE

Eves, Howard. 1972. *A Survey of Geometry, Rev. Ed.* Boston: Allyn and Bacon.

SUGGESTED READINGS

Consult the following sources listed in the Bibliography for related readings: Lappan, Fitzgerald, Phillips, and Winter, 1986; Lindquist and Shulte, 1987; and Senk and Hirschorn, 1990.

CHAPTER NINE Topology

INTRODUCTION

Paul Bunyan and his mechanic, Ford Fordsen, had started to work a uranium mine in Colorado. The ore was brought out on an endless belt that the manufacturers had made all in one piece, without any splice or lacing, and they had put a half-twist in the return part so that the wear would be the same on both sides. After several months of operation, Paul decided he needed a belt twice as long and half as wide. He told Ford Fordsen to take his chain saw and cut the belt in two lengthwise.

> "That will give us two belts," said Ford Fordsen. "We'll have to cut them in two crosswise and splice them together."
> "No," said Paul. "This belt has a half twist—which makes it what is known in geometry as a Moebius strip" (Johnson and Glenn, 1960).

In Investigation 9.1, we shall explore the Moebius band and discover why Paul said that no splicing would be necessary.

TEACHING NOTE

Topology provides a rich source of interesting problems for students at all levels. For example, in *Mathematical Funfare* (Bolt, 1989) Problems 12, 23, 64, 80, 85, 104, 106, and others utilize topological ideas.

INVESTIGATION 9.1

Take a strip of paper 10″ × 1″, give one end a half-twist, and join the ends together with tape, as shown here.

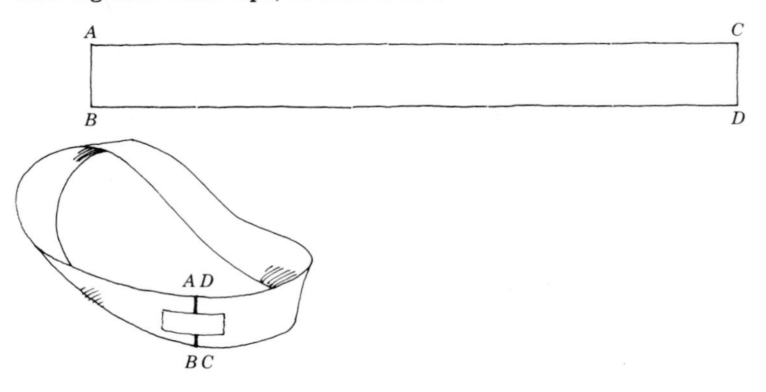

A. Color the edge of the Moebius band until you return to the beginning point. What does this tell you about the number of edges on a Moebius band?
B. Punch a hole in the surface of the band with a sharp object. Label the entrance of the point X and the exit of the point Y. Draw a continuous line (without crossing the edge) from X to Y. What does this tell you about the number of sides to the band?
C. Along the middle of the band, draw a line; predict the outcome, then cut the band in half by cutting along this line.

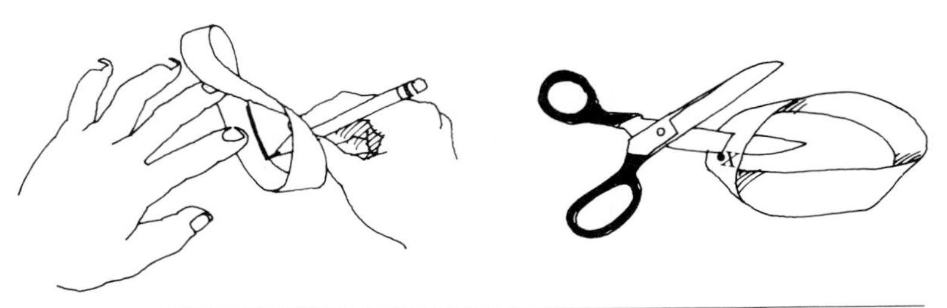

As you discovered in Investigation 9.1, a **Moebius band** is a surface with exactly one edge and one side. Hence, when it is cut in half lengthwise, it remains in one piece. (Notice that in part B of the investigation, the tip of the object that punctured the surface exited the same side which it entered.)

The Moebius band is only one of the interesting and curious surfaces that can easily be made with paper, scissors, and tape. In the following exercises we shall further explore the Moebius band, as well as other surfaces.

EXERCISE SET 9.1

1. In a Paul Bunyan story that appears in *Topology* (Johnson and Glenn, 1960), Loud Mouth Johnson, who was described as Public Blow-Hard Number One, lost $1000 when he bet Ford Fordsen that Paul's belt would end up in two pieces. Loud Mouth subsequently stumbled onto Paul in the process of cutting the belt a second time to obtain a belt four times the length and one-fourth the width. When Loud Mouth was informed that Ford had gone into town to get splicing material, he informed Paul, having learned through the loss of the $1000, that no splicing material would be necessary. Make another paper Moebius band and discover whether Loud Mouth was correct.

TABLE 9.1

Number of Half-Twists	Number of Sides Before Cut	Number of Edges Before Cut	Number of Sides After Cut	Number of Edges After Cut
0				
1				
2				
3				

2. A Moebius band is made by giving a strip of paper one half-twist. Complete Table 9.1 and see what happens when bands constructed with a different number of twists are cut down the middle.
3. Make another Moebius band and draw a line one-third of the width from an edge (Fig. 9.1). Cut along this line, but predict the result before cutting. Were you correct?
4. Refer to Figure 9.2.
   a. Which of the surfaces are *one*-sided?
   b. How many edges does each surface have?
5. Figure out a way of constructing with paper, scissors, and tape each of the surfaces in Exercise 4.

*6. Search for a generalization in parts a through c.

*a. Construct bands with 0,1,2,3, and 4 half-twists (the Moebius band is constructed by one-half twist). Cut each of these bands in half lengthwise to see how many separate bands (which may be interlocking) you obtain. Can you guess at a general pattern?

*b. Construct bands with 0, 1, 2, 3, and 4 half-twists. Cut each of these bands in half lengthwise and check the number of twists in the resulting band or bands. (You may need to cut across the resulting band and count the number of half-twists required to untwist the band.)

*c. Construct bands with 1,3, and 5 half-twists and cut each in half lengthwise. The resulting band will be knotted. Compare the nature of the resulting knots. Are some more knotted than others? Can you find a general pattern?

FIGURE 9.1

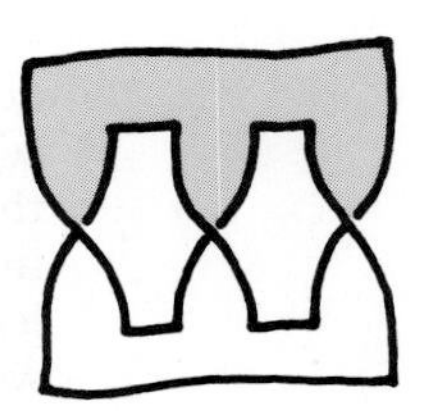
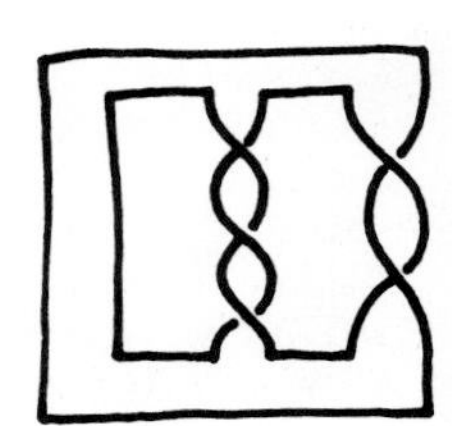

FIGURE 9.2

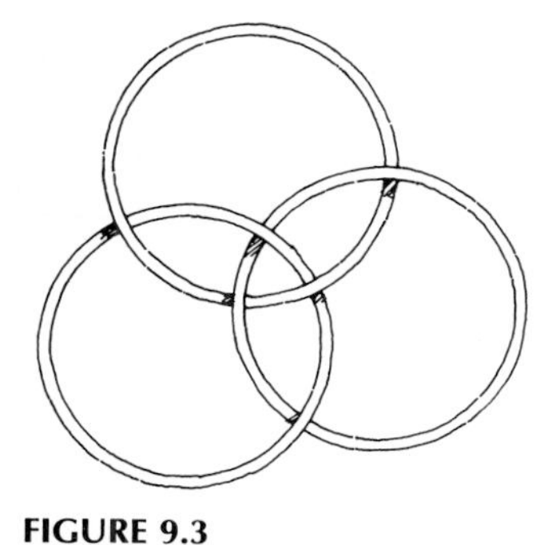

FIGURE 9.3

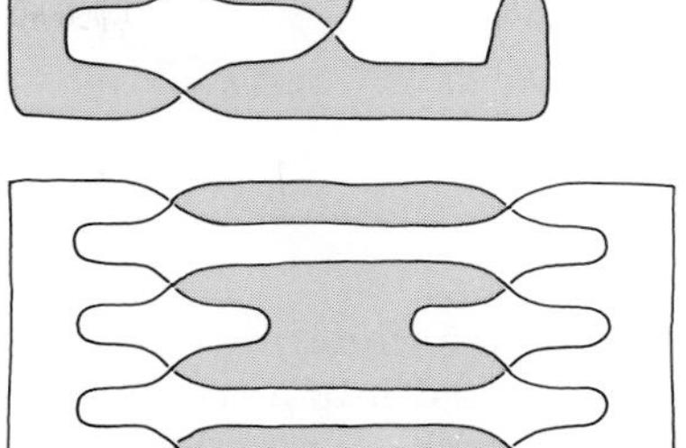

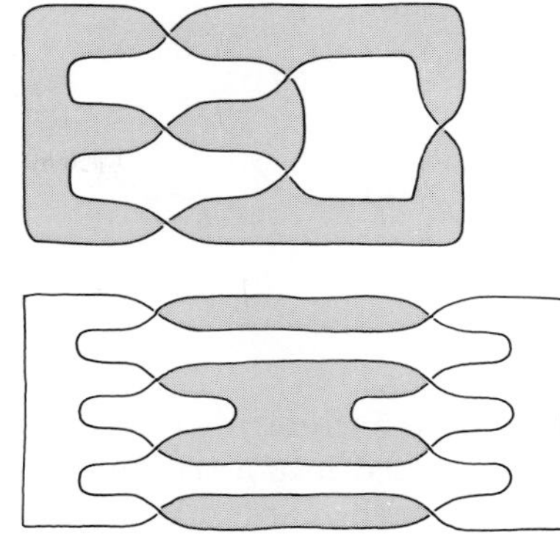

FIGURE 9.4

*7. *The Unexpected Hanging* (Gardner, 1969) includes a chapter on Borromean rings. The 3 rings pictured in Figure 9.3 are Borromean rings. Two of the 3 surfaces in Figure 9.4 have Borromean-ringed edges. Which 2?

8. **Critical Thinking**. How could you draw two dotted lines on a strip of paper so that when a Moebius strip is made from the paper and cut along the lines, two pieces of equal area will result? Can you find more than one way to do it?

TEACHING NOTE

In an article by Piaget on "How Children Form Mathematical Concepts" (*Scientific American*, November 1953), it is asserted that the first geometric notions that a child learns appear to be those of a topological character. For example, at the age of three, a child cannot tell the difference between squares and triangles, but can distinguish between open and closed figures. Topological ideas developed early by children include proximity, separation, order, enclosure, and continuity. These ideas are more carefully described in *How Children Learn Mathematics* (Copeland 1979) in the chapters entitled, "How a Child Begins to Think about Space" and "From Topology to Euclidean Geometry."

TOPOLOGICAL EQUIVALENCE

Topology is a relatively new field in geometry that has come into full bloom since 1920. In contrast to the ideas of congruence and similarity, where we were concerned with measurable quantities such as distance and angle measure, **topology** deals with geometric concepts that can usually be described without using numbers.

A topologist, who is a mathematician with special interest in topology, has been described in the popular press as one who cannot tell the difference between a coffee cup and a donut. How could anyone with the reputed intelligence of mathematicians have such a problem? This chapter will shed some light on this apparant paradox.

Topology is often described as the study of properties of curves and surfaces that are maintained as the curves or surfaces are stretched or distorted. For example, the characteristic "has no ends" is an example of such a property for a circle.

FIGURE 9.5

Figure 9.5 illustrates that this property is not lost as the circle is distorted into a triangle.

It is through this type of distortion that many physical settings can profitably be reduced to a diagram on paper. For example, in Figure 9.6, we see a partial schematic for a radio receiver. In Figure 9.7, we see a diagram of the London Underground system. Each figure represents a distortion of a physical situation that preserves the basic structure of the situation.

Our objectives for the remainder of this chapter are twofold. First, these types of distortions, which shall be called topological transformations, will be more carefully described; and second, we shall discover some of the geometrical properties of surfaces and line figures that remain unchanged by these transformations. The first of these objectives is considered in the remainder of this section.

TEACHING NOTE

When a three-year-old is asked to make a copy of a square or triangle, he may draw both as a simple closed curve, which appears somewhat like a circle. Another child may draw a picture of a person with the eyes outside of the head. As the child approaches age six, these difficulties are corrected, and the basic topological ideas of the child are refined and expanded.

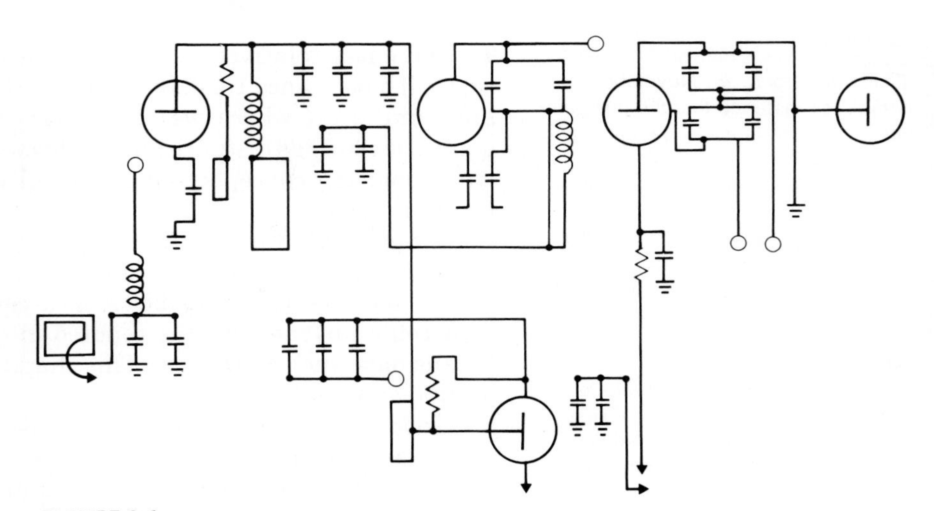

FIGURE 9.6

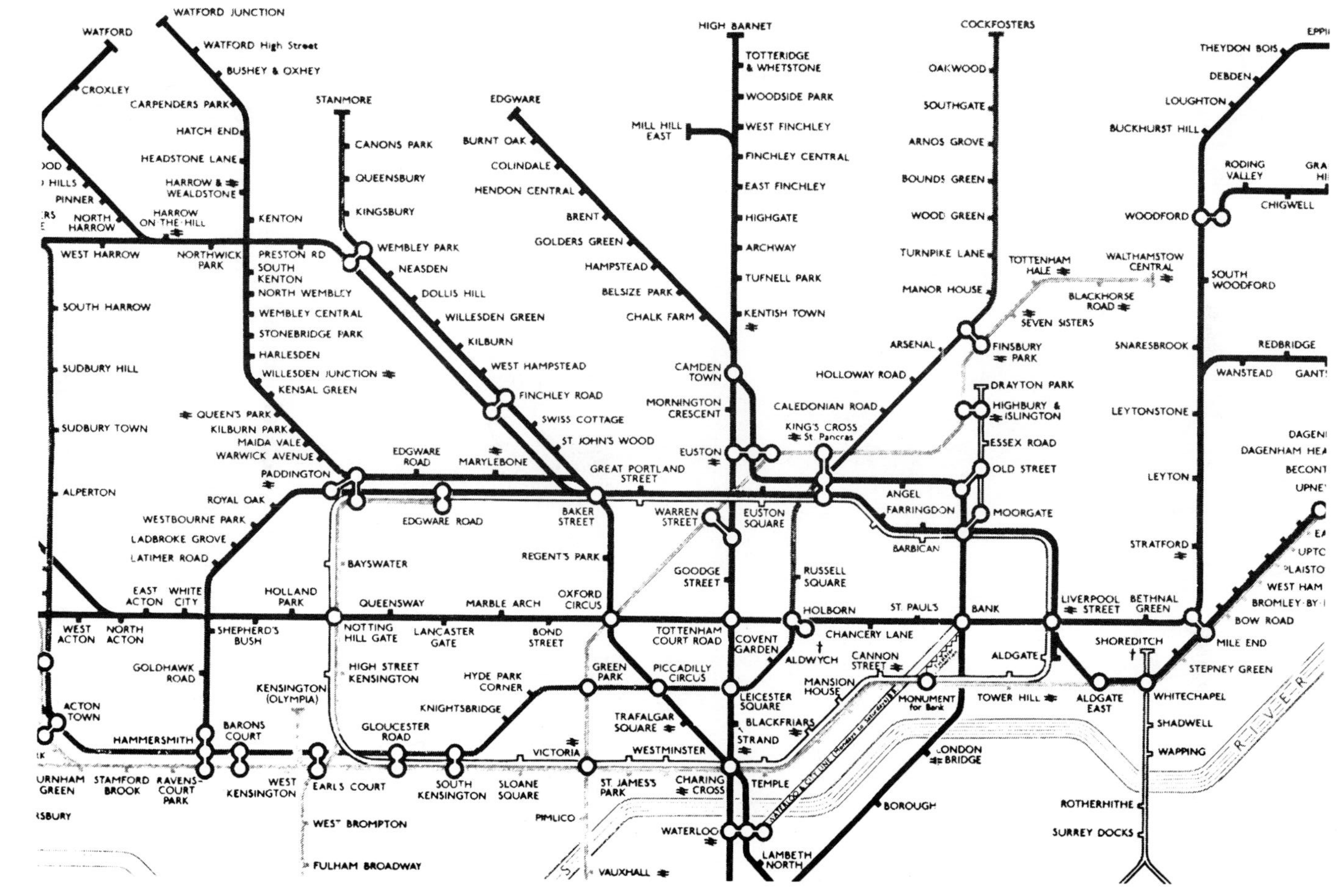
WATFORD
WATFORD JUNCTION
WATFORD High Street
BUSHEY & OXHEY
CROXLEY
CARPENDERS PARK
HATCH END
HEADSTONE LANE
HARROW & WEALDSTONE
PINNER
NORTH HARROW
HARROW ON THE HILL
KENTON
WEST HARROW
NORTHWICK PARK
PRESTON RD
SOUTH KENTON
NORTH WEMBLEY
WEMBLEY CENTRAL
STONEBRIDGE PARK
HARLESDEN
WILLESDEN JUNCTION
KENSAL GREEN
SOUTH HARROW
SUDBURY HILL
SUDBURY TOWN
ALPERTON
QUEEN'S PARK
KILBURN PARK
MAIDA VALE
WARWICK AVENUE
PADDINGTON
ROYAL OAK
WESTBOURNE PARK
LADBROKE GROVE
LATIMER ROAD
EAST ACTON
WHITE CITY
WEST ACTON
NORTH ACTON
SHEPHERD'S BUSH
HOLLAND PARK
GOLDHAWK ROAD
ACTON TOWN
HAMMERSMITH
BARONS COURT
STAMFORD BROOK
RAVENSCOURT PARK
WEST KENSINGTON
KENSINGTON (OLYMPIA)
STANMORE
CANONS PARK
QUEENSBURY
KINGSBURY
WEMBLEY PARK
NEASDEN
DOLLIS HILL
WILLESDEN GREEN
KILBURN
WEST HAMPSTEAD
FINCHLEY ROAD
SWISS COTTAGE
ST JOHN'S WOOD
EDGWARE
BURNT OAK
COLINDALE
HENDON CENTRAL
BRENT
GOLDERS GREEN
HAMPSTEAD
BELSIZE PARK
CHALK FARM
MILL HILL EAST
HIGH BARNET
TOTTERIDGE & WHETSTONE
WOODSIDE PARK
WEST FINCHLEY
FINCHLEY CENTRAL
EAST FINCHLEY
HIGHGATE
ARCHWAY
TUFNELL PARK
KENTISH TOWN
CAMDEN TOWN
MORNINGTON CRESCENT
EUSTON
EDGWARE ROAD
MARYLEBONE
GREAT PORTLAND STREET
BAKER STREET
WARREN STREET
EUSTON SQUARE
BAYSWATER
REGENT'S PARK
GOODGE STREET
RUSSELL SQUARE
QUEENSWAY
MARBLE ARCH
OXFORD CIRCUS
HOLBORN
NOTTING HILL GATE
LANCASTER GATE
BOND STREET
TOTTENHAM COURT ROAD
COVENT GARDEN
ALDWYCH
HIGH STREET KENSINGTON
HYDE PARK CORNER
GREEN PARK
PICCADILLY CIRCUS
LEICESTER SQUARE
KNIGHTSBRIDGE
GLOUCESTER ROAD
VICTORIA
TRAFALGAR SQUARE
WESTMINSTER
BLACKFRIARS
STRAND
EARLS COURT
SOUTH KENSINGTON
SLOANE SQUARE
ST JAMES'S PARK
CHARING CROSS
TEMPLE
WEST BROMPTON
FULHAM BROADWAY
PIMLICO
VAUXHALL
WATERLOO
LAMBETH NORTH
COCKFOSTERS
OAKWOOD
SOUTHGATE
ARNOS GROVE
BOUNDS GREEN
WOOD GREEN
TURNPIKE LANE
MANOR HOUSE
ARSENAL
FINSBURY PARK
TOTTENHAM HALE
SEVEN SISTERS
BLACKHORSE ROAD
WALTHAMSTOW CENTRAL
HOLLOWAY ROAD
CALEDONIAN ROAD
KING'S CROSS St Pancras
DRAYTON PARK
HIGHBURY & ISLINGTON
ESSEX ROAD
OLD STREET
ANGEL
FARRINGDON
MOORGATE
BARBICAN
ST PAUL'S
BANK
CHANCERY LANE
CANNON STREET
MANSION HOUSE
MONUMENT for Bank
LONDON BRIDGE
BOROUGH
TOWER HILL
ALDGATE
ALDGATE EAST
LIVERPOOL STREET
BETHNAL GREEN
SHOREDITCH
WHITECHAPEL
SHADWELL
WAPPING
ROTHERHITHE
SURREY DOCKS
THEYDON BOIS
DEBDEN
LOUGHTON
BUCKHURST HILL
RODING VALLEY
CHIGWELL
WOODFORD
SOUTH WOODFORD
SNARESBROOK
REDBRIDGE
WANSTEAD
LEYTONSTONE
LEYTON
STRATFORD
MILE END
STEPNEY GREEN
BOW ROAD
WEST HAM
R I V E R

FIGURE 9.7

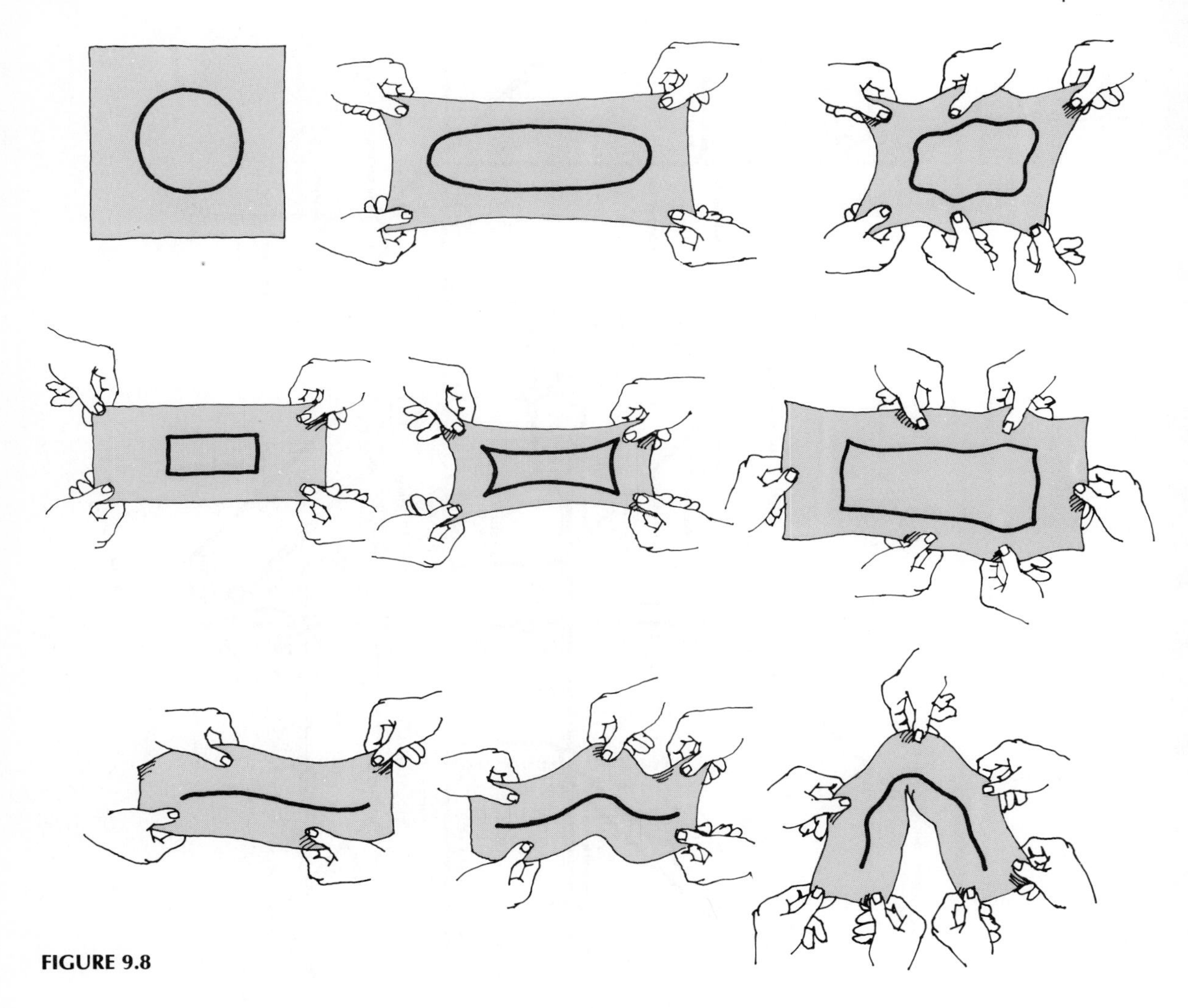

FIGURE 9.8

A **topological transformation** will distort, stretch, shrink, or bend a curve or surface. It will not weld, cement, or join together, and with one type of exception, it does not tear or cut. A stretching rubber sheet is a helpful physical model of a topological transformation. Figure 9.8 illustrates how a circle, a rectangle, and a line segment can be distorted by topological transformations. In fact, as suggested by Figure 9.9, these transformations can distort a line into a knotted line since the quality of being a continuous curve with its ends not joined is not destroyed.

TEACHING NOTE

Some interesting pictures for students regarding topological equivalents appear in the Life Science Library book entitled *Mathematics* (Bergamini, 1963, p. 179). A pictorial description is given showing how to remove one's vest without removing one's coat. This is very easy to do and students enjoy a demonstration.

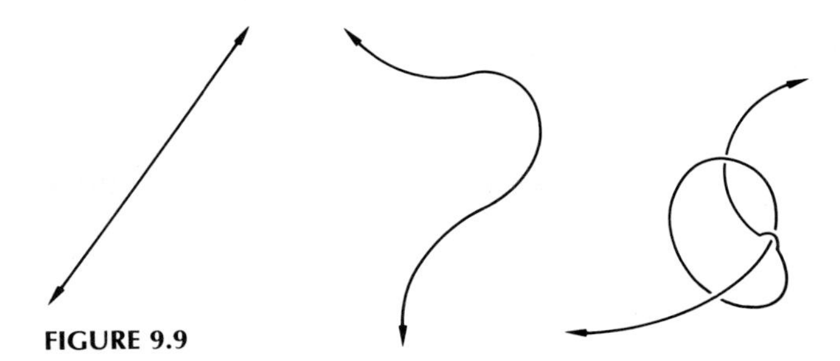

FIGURE 9.9

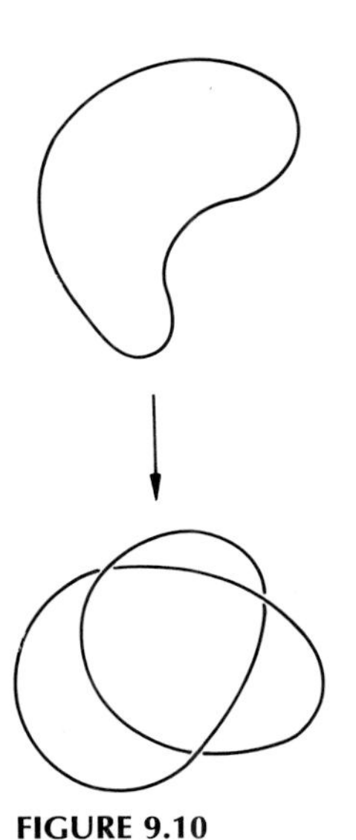

FIGURE 9.10

It is this connectedness, this continuity, of lines and surfaces that is preserved by topological transformations. This means that these transformations do not disconnect what had been connected (i.e., no cutting or puncturing of holes), nor do they connect what had been separated (i.e., no welding, glueing, etc.). The only time tearing or cutting of a surface or curve is permitted is when the surface or curve is rejoined after the cut in such a way that points along the cut "match up" identically to their positions prior to the cutting. For example, a topological transformation may take a simple closed curve and knot it by cutting and rejoining (Fig. 9.10).

When one surface or curve can be transformed by a topological transformation to another surface or figure, we say that the two figures are **topologically equivalent**. That is, within the realm of topology, two topologically equivalent surfaces are indistinguishable—they are topologically alike. Figure 9.5 illustrates that a circle and an equilateral triangle are topologically equivalent. Also, the simple closed curve and the knot of Figure 9.10 are topologically equivalent. Does Figure 9.11 convince you that a length of pipe and a washer are equivalent?

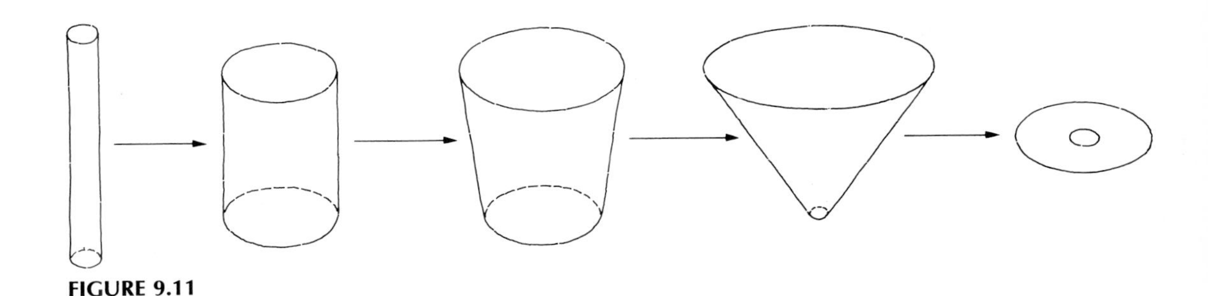

FIGURE 9.11

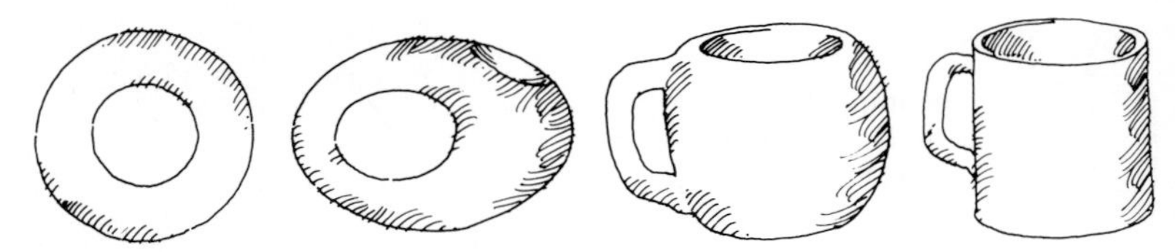

FIGURE 9.12

Finally, in Figure 9.12, we see how topologists got their reputation. A donut and a coffee cup are topologically equivalent.

EXERCISE SET 9.2

1. Which of the figures presented in parts a, b, c, and d of Figure 9.13 are topological distortions of the figure on the extreme left? That is, which of the figures are topologically equivalent to the figure on the left?
2. Which pairs of letters printed in Figure 9.14 are topologically equivalent?

FIGURE 9.13

FIGURE 9.14

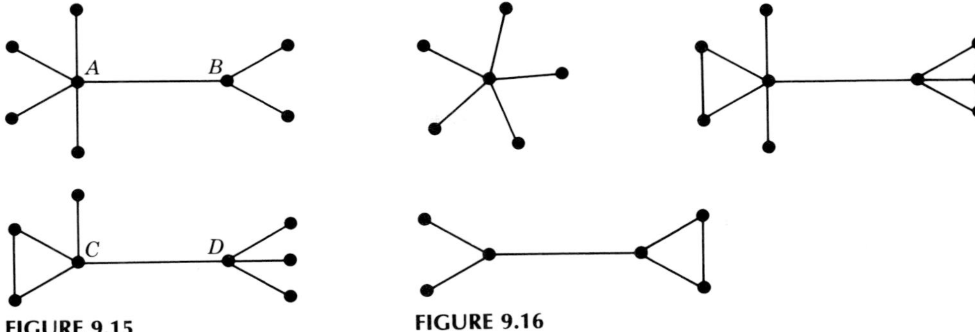

FIGURE 9.15

FIGURE 9.16

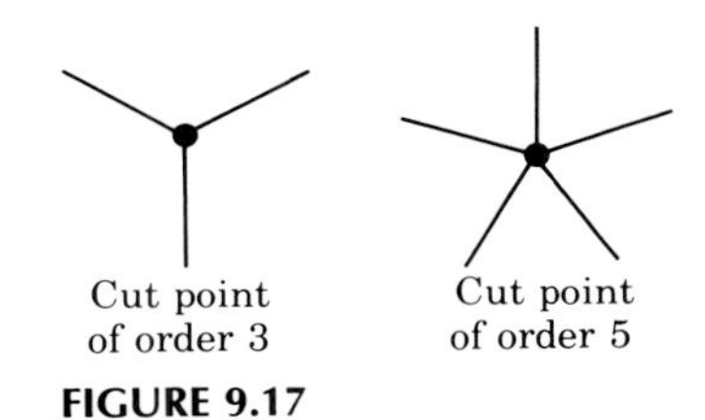

FIGURE 9.17

3. Each of the points A, B, C, and D in Figure 9.15 is an example of what is called a *cut point* because if these points are removed, the figure is separated into several (more than 2) disjoint sets. In each of the figures in Figure 9.16, find all of the cut points.
4. If a cut point P, when removed from a figure, separates the figure into exactly n disjoint pieces, we call it a *cut point of order n* (Fig. 9.17). For each of the cut points found in the figures in Exercise 3, describe the order of cut point.
5. It is intuitive, and possible to prove, that two geometric figures are not topologically equivalent if they do not have the same number of cut points of each order. Using the "cut point of order n" criteria, select the one figure that is not topologically equivalent to the other two in parts a, b, c, and d of Figure 9.18.

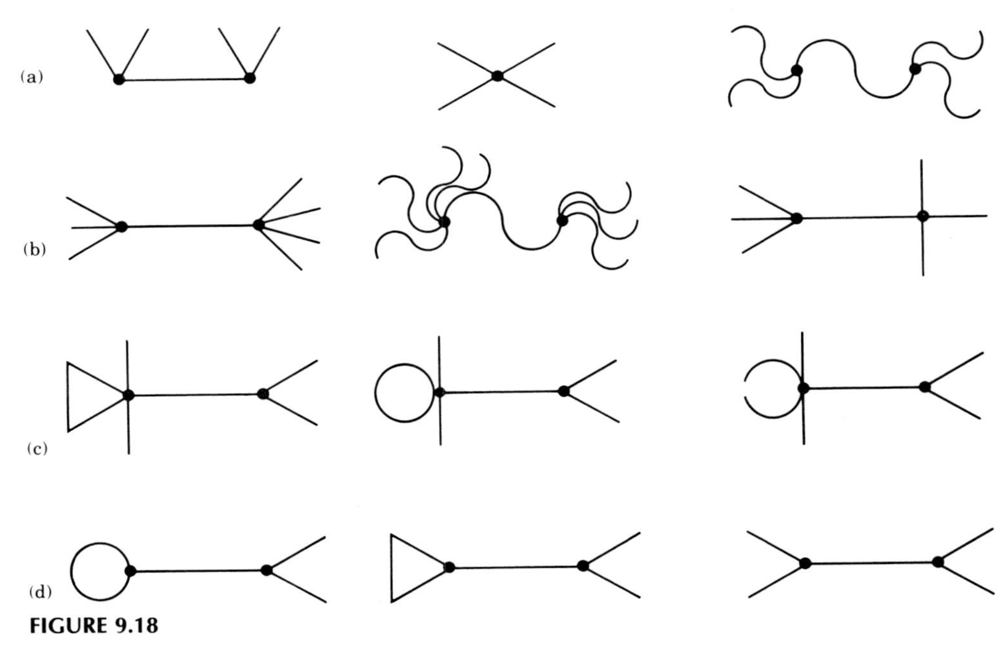

FIGURE 9.18

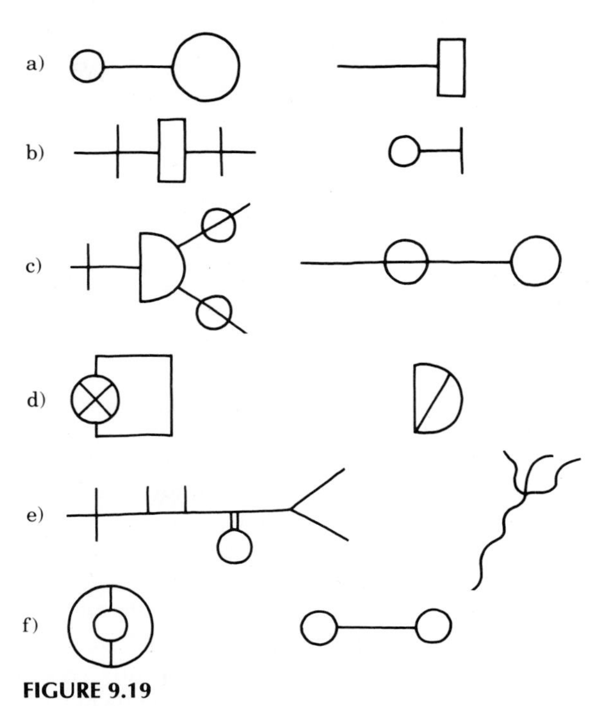

FIGURE 9.19

6. In each part of Figure 9.19, complete the figure on the right so that it is topologically equivalent to the figure on the left.
7. Do you think a soup bowl is topologically equivalent to a donut? Why or why not?

*8. The object in Figure 9.20 is sometimes seen on the shoulder of a uniform. We shall call it a woggle. The 4 stages of Figure 9.21 demonstrate that a woggle is topologically equivalent to a strip with 2 slits made in it. Try to make a woggle.

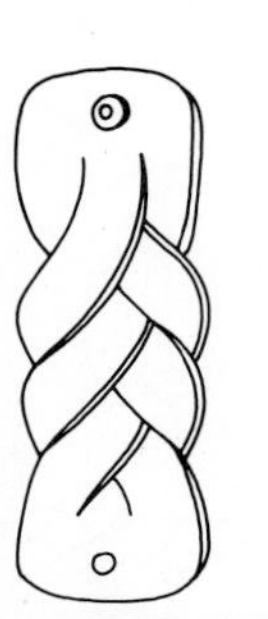

FIGURE 9.20

Stage 1

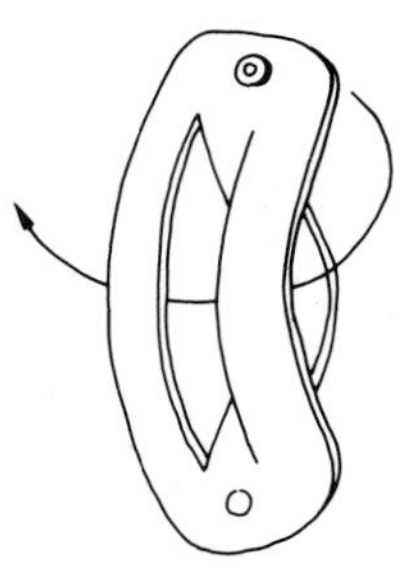

Stage 2

Stage 3

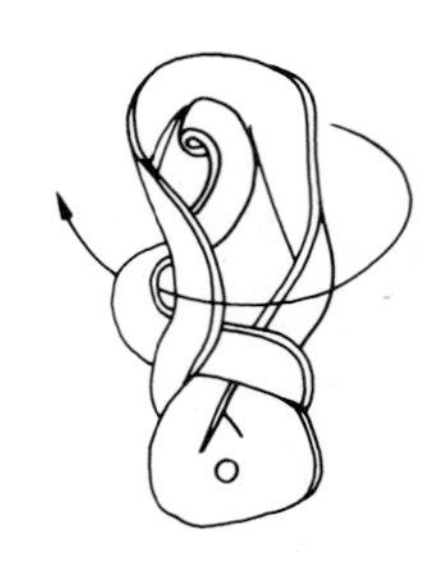

Stage 4

FIGURE 9.21

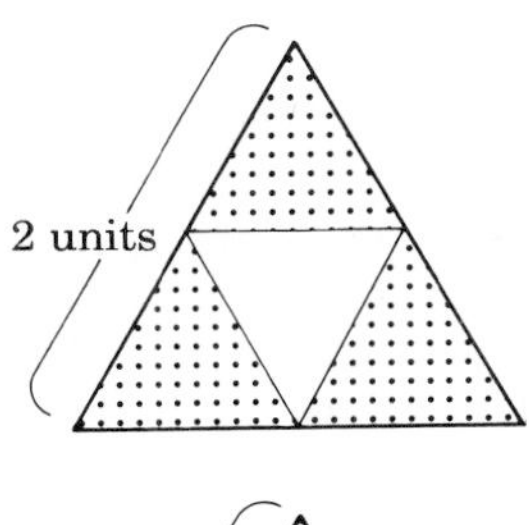

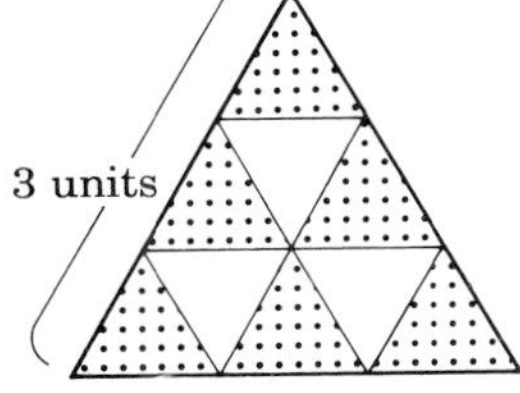

FIGURE 9.22

Problem Solving: Developing Skills and Strategies

The strategies **look for a pattern** and **make a table** are often used together to solve a problem. After a problem is solved, it is useful to analyze the solution and look for a way to generalize. How can these strategies help you solve the following problem? Can you suggest a generalization?

Problem: When an equilateral triangle 2 units on a side is divided into small equilateral triangles and shaded according to the pattern shown in Figure 9.22, 1 of 4, or 25%, is white and 3 of 4, or 75%, is gray. What would be the percentages of white and gray when the triangle has 1000 units on a side?

NETWORKS

In the remaining four sections of this chapter, we shall shift our attention from topological transformations themselves to properties of curves and surfaces that remain unchanged by topological transformations. The properties of "has no ends," "has exactly one edge," and "has exactly one surface" are all examples of properties that are preserved under topological distortion, which we have already encountered in this chapter. Investigation 9.2 introduces a famous problem that originated in the city of Kaliningrad.

What did you find in your investigation of the Konigsberg bridges? Note that the bridge system can be drawn in a much-simplified form by drawing each land mass A,B,C, and D as a point and each bridge as a curve connecting the appropriate points, resulting in Figure 9.23. This diagram is an example of what is commonly called a network. Other examples of networks are shown in Figure 9.24. A **network** is a collection of points, called **vertices**, together with a collection of curves, called **edges**, that join the vertices.

Before answering the Konigsberg bridge question, we shall detour for a brief study of network theory. In the process of this study, we shall find an answer to the Konigsberg problem.

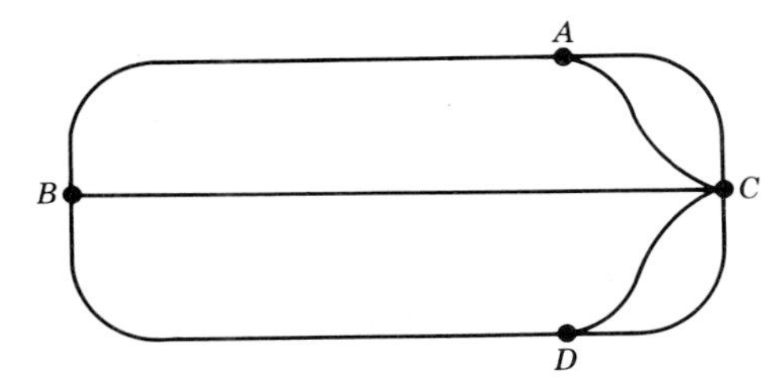

FIGURE 9.23

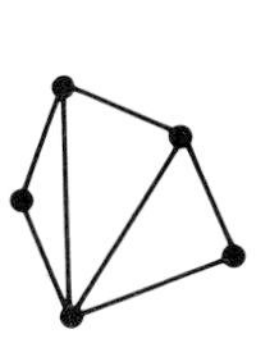

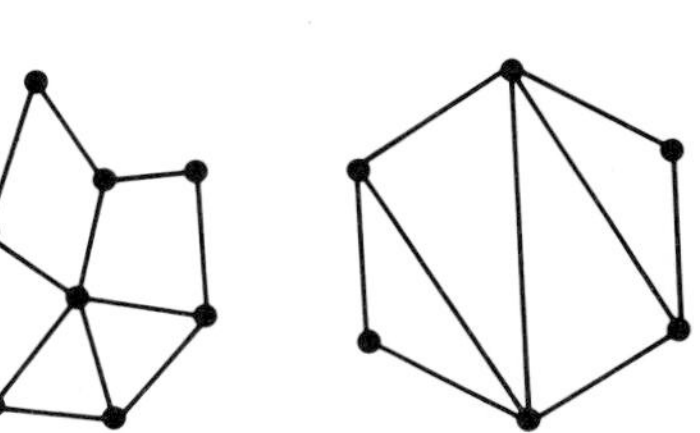

FIGURE 9.24

INVESTIGATION 9.2 

The city of Konigsberg (now Kaliningrad) in East Prussia is located on the banks and 2 islands of the Pregel River. The various regions of the city were connected by 7 bridges, as shown in the figure. A popular question in the early eighteenth century was the following: Is it possible to take a walk around town in such a way that starting from home, one can return there after having crossed each bridge just once? Explore this question.

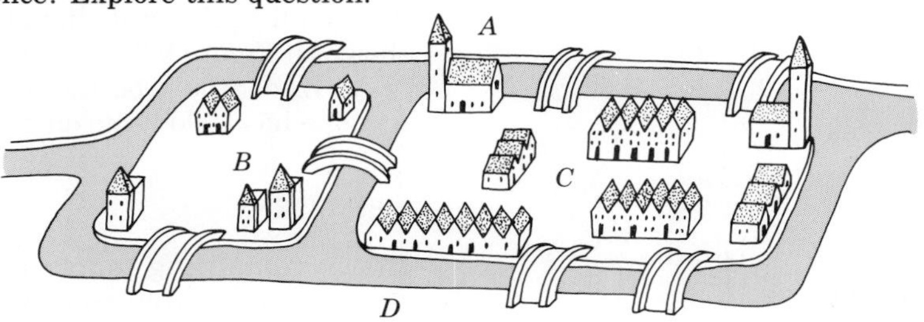

A familiar question asked in network theory is, *Can we trace a path on a given network that traverses each edge once and only once?* For example, for a given network of highways, a highway inspector wants to inspect each of the highways without the wasted effort of traveling any highway more than once. In fact, we see that the Konigsberg bridge question can be restated as: Is it possible to trace a path along the network in Figure 9.23 that traverses each edge once and only once and that ends at its beginning point? If so, the network is called **traversable**. If the beginning point and ending point of such a path coincide, the network is called **traversable type 1**; if the beginning point and ending point do not coincide, the network is called **traversable type 2**. In Figure 9.25, we see an example of a type 1 network, a type 2 network, and a network that is not traversable without an edge being covered twice or more.

In the Konigsberg network in Figure 9.23, there is an odd number of edges joined at vertex A, and so A is called an **odd vertex**. There are five edges joined at vertex C, so C is also an odd vertex. When there is an even number of edges joined at a vertex, it is called an **even vertex**. The number of even and odd vertices to a network influences whether or not it is traversable. See if you can discover in Investigation 9.3 a rule that determines when a network is traversable type 1 and when it is traversable type 2. Apply the rule that you discover to answer the Konigsberg bridge question.

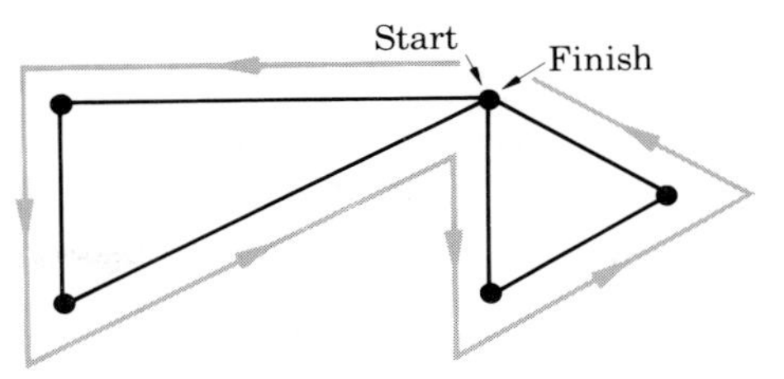

(a) Traversable type 1

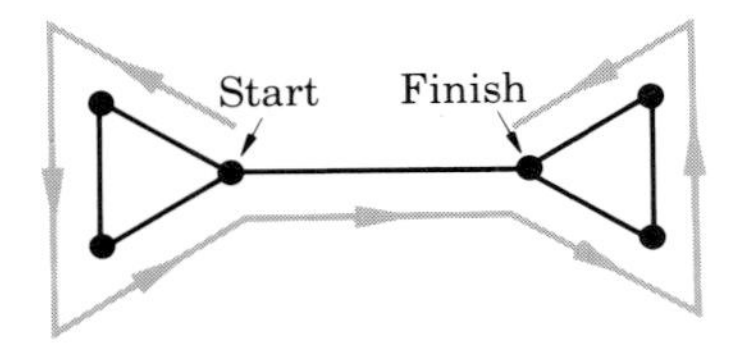

(b) Traversable type 2

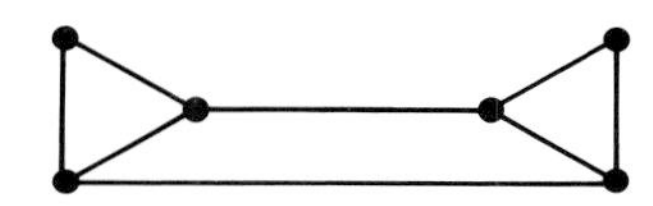

(c) Not traversable

FIGURE 9.25

On the basis of the experience gained in completing Investigation 9.3, which of the following conjectures do you believe are true?

TEACHING NOTE

Some of the ideas on networks presented here can often be couched in more popular language. For example, a student can be asked, "Can you draw this figure without lifting your pencil from the paper?" The conclusions from Investigation 9.3 should give an immediate method for deciding on answers to questions like this.

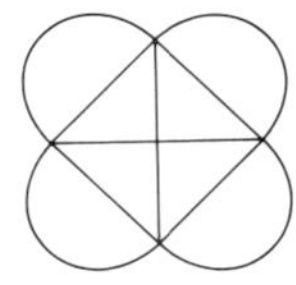

Conjecture 1: If a network has all of its vertices of even order, then the network is traversable type 1. Furthermore, a traversable path may begin at any vertex.

Conjecture 2: If a vertex of a network is an odd vertex, then a traversable path (if it exists) must either begin or end at that vertex.

Conjecture 3: If in a network exactly two vertices are odd vertices, then the network is traversable type 2. Any traversable path must begin at one of these vertices and end at the other vertex.

Conjecture 4: If a network has more than 2 odd vertices, then it is not traversable type 1 or type 2.

INVESTIGATION 9.3

A. Complete the table for networks 1 through 10 in the figure. Note that some may not be either traversable type 1 or traversable type 2.

Network	Number of Even Vertices	Number of Odd Vertices	Traversable Type 1 (Yes or No)	Traversable Type 2 (Yes or No)
1				
2				
3				
4				
5				
6				
7				
8				
9				
10				

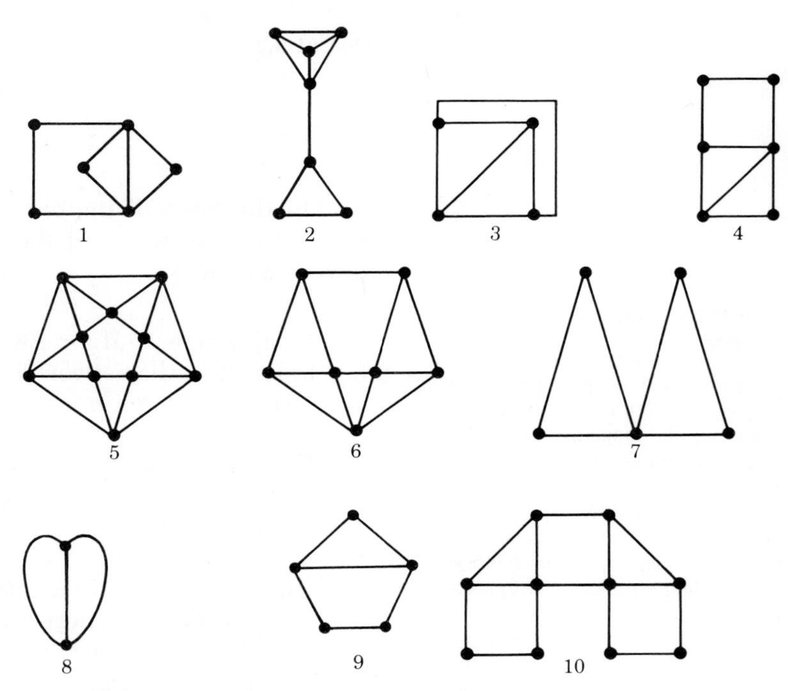

B. Reflect on the table in part A. What generalizations and conjectures are suggested by the relationships expressed in this table?

EXPLORING WITH THE COMPUTER 9.1

Use LOGO software to complete the following.

If possible, write a procedure that instructs the turtle to start at point A on the network shown, traverse each path exactly once, and return to point A. If it isn't possible, change the network so that it is.

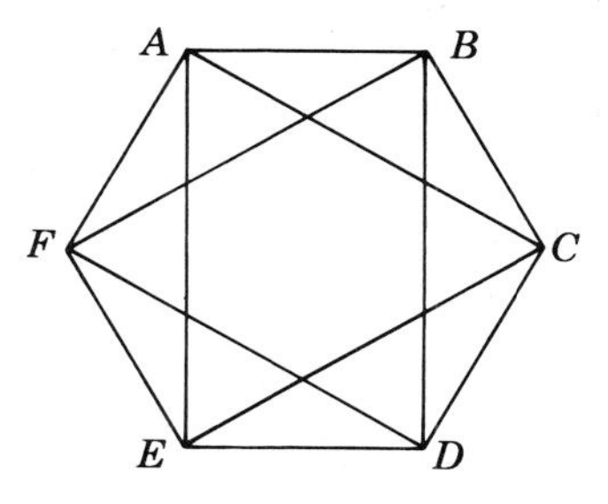

EXERCISE SET 9.3

1. Use the network in Figure 9.23 and explain why a walk cannot be taken in Konigsberg so that each of the bridges is crossed exactly once and the walker returns to the starting place.
2. A city planner commented that any one of the 7 bridges at Konigsberg could be relocated so that a walk could be taken crossing each bridge exactly once. Show that the city planner was correct by drawing networks representing the bridges before and after relocation.
3. The networks in Figure 9.26 are not traversable. What is the fewest number of edges that, when added, makes each network traversable type 2? Traversable type 1?
4. A network and its distorted image under a topological transformation have some properties in common. That is some properties of networks are preserved by topological transformations. List some of these properties.

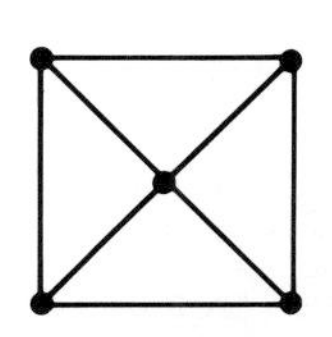
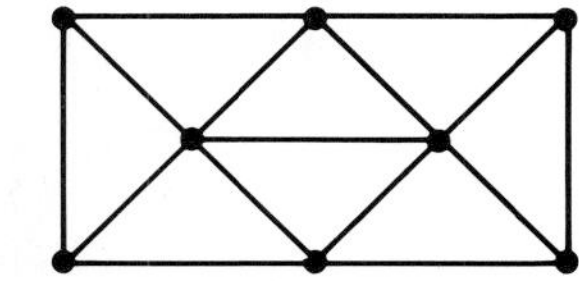
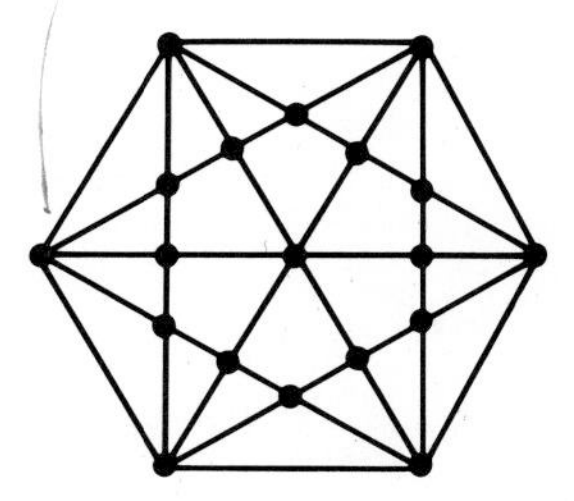

FIGURE 9.26

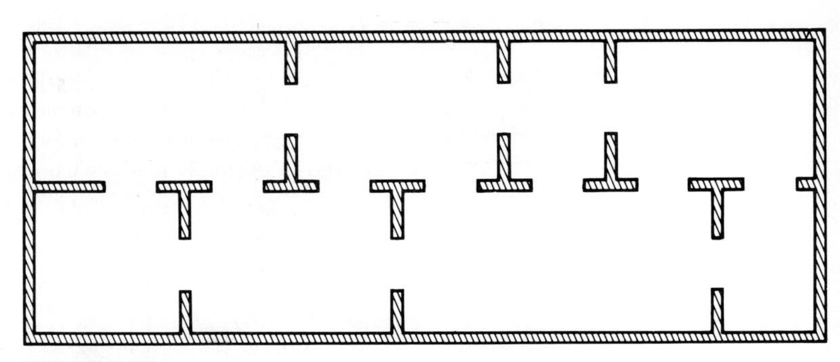

FIGURE 9.27

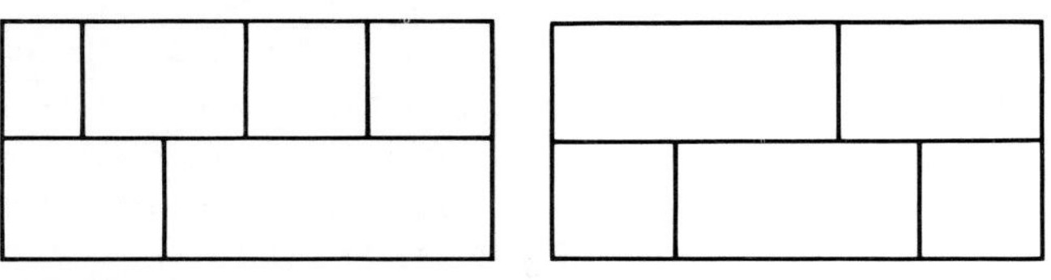

FIGURE 9.28

5. **Critical Thinking**. Can a person find a continuous path in the house of Figure 9.27 that will take her through each door exactly once? Describe a strategy that can be used to solve a problem like this. For a related Critical Thinking activity, see Activity 14 in Appendix D.

Problem Solving: Developing Skills and Strategies

Can you draw a continuous path crossing each edge of the figures in Figure 9.28 just once without going through a corner?

1. Seek an answer by trial and error.
2. Seek an answer by translating this question into a consideration of the traversability of a related network. [*Hint:* Let each region become a vertex and each boundary edge between regions become an edge.]

APPLICATION

Salesman Problem: A salesman is planning his itinerary to visit companies *A* through *F* as shown on the map. One of the companies is in the same city as his home office. He wants to begin there, visit each of the other companies, and return to another city to stay overnight without traveling along any road twice. Decide which city could be his home office and show how he can do this. Then mark roads that would be needed to company *G* so that the salesman could start at *G*, visit every other company, and return to *G* without going over the same road twice. Draw a diagram to show the salesman's routes in each situation.

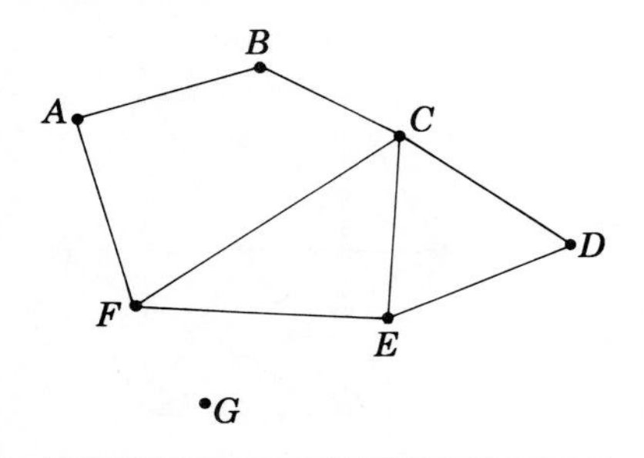

JORDAN CURVE THEOREM

The Jordan Curve Theorem concerns simple closed curves in the plane and is one of the most famous theorems in topology. Although we shall not prove this theorem, we shall experience it on the intuitive level. We begin with Investigation 9.4.

INVESTIGATION 9.4

Given 3 houses A, B, and C and the utility centers of electricity (E), gas (G), and water (W), connect each of the 3 houses to each of the 3 utilities without the lines crossing.

Are you convinced that you have found a solution to the utilities problem? We shall base our analysis of the problem on the concept of a simple closed curve. A **simple closed curve** is that which is the image of a circle under a topological transformation. In Figure 9.29, there are four curves in the plane. Only curves (a), (b), and (d) are simple closed curves.

Suppose that in searching for a solution to the utilities problem, we proceed by joining A with each utility and B with each utility, as indicated in Figure 9.30. This process yields a simple closed curve $AWBGA$ with E on the "inside" and C on the "outside" of this curve. Any path that joins E to C would then cross the simple closed curve $AWBGA$, and, hence, it would cross one of the utility connections. Readers should convince themselves that any attempt at a solution to this problem

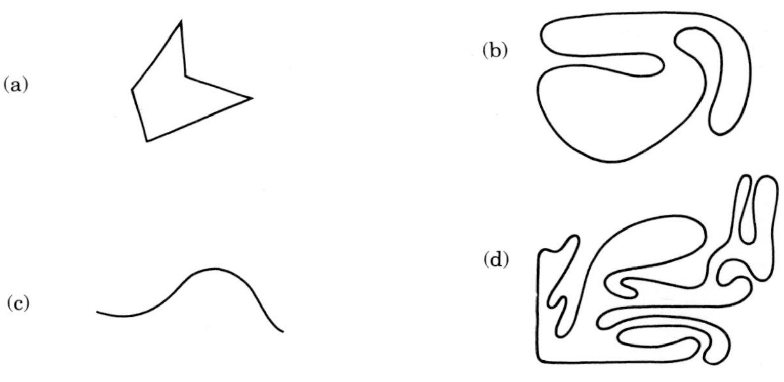

FIGURE 9.29

TEACHING NOTE

Since the notions of inside and outside are available to very young children, ideas of simple closed curves can be presented quite early. Young children like to solve simple mazes involving closed curves or open curves. Older students enjoy inventing mazes. Some challenging mazes occur in *Mazes 2* (Koziakin, 1972).

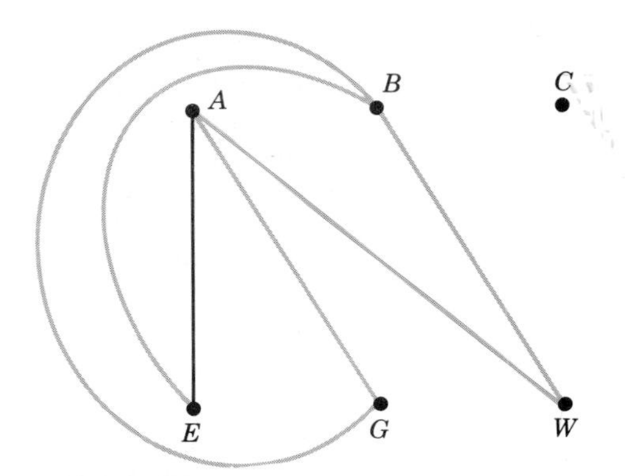

FIGURE 9.30

results in a simple closed curve with a point on its inside and a point on its outside.

We have, in fact, been assuming in this explanation of the utilities problem the famous Jordan Curve Theorem—a theorem easily stated, readily accepted, but surprisingly difficult to prove. The essence of the **Jordan Curve Theorem** is that a simple closed curve separates the plane into two subsets—one we call "inside," the other "outside"—in such a way that each continuous path from any point inside the curve to any point outside the curve crosses the curve.

It is sometimes surprising how difficult it is to determine whether a given point is in what we commonly call the inside or on the outside of a simple closed curve. For example, in Figure 9.31, is point A on the inside or the outside? This type of question is the subject of Investigation 9.5.

FIGURE 9.31

INVESTIGATION 9.5

A. Figure (a) shows a simple closed curve and 2 points, A and B. Which point is outside the curve? Can you reach B from A without crossing the curve?

B. Figures (b) and (c) each show a simple closed curve and 2 points, A and B. Is A inside the curve in each case? Are A and B one inside and one outside the curve? [*Hint:* How many times does the dashed line cross the curve?]

C. Draw some simple closed curves. Find a rule for deciding which points are inside your curves.

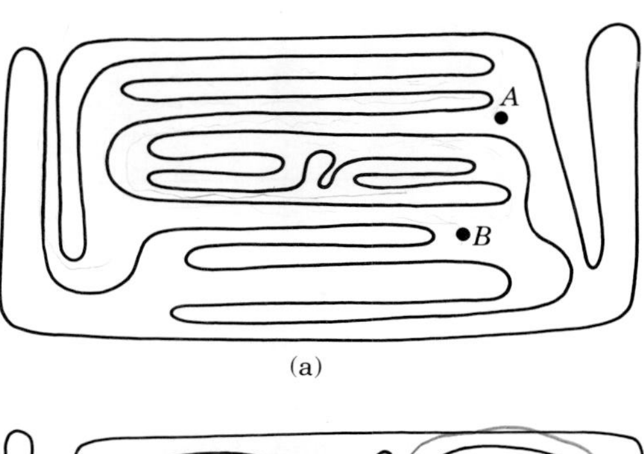

(a)

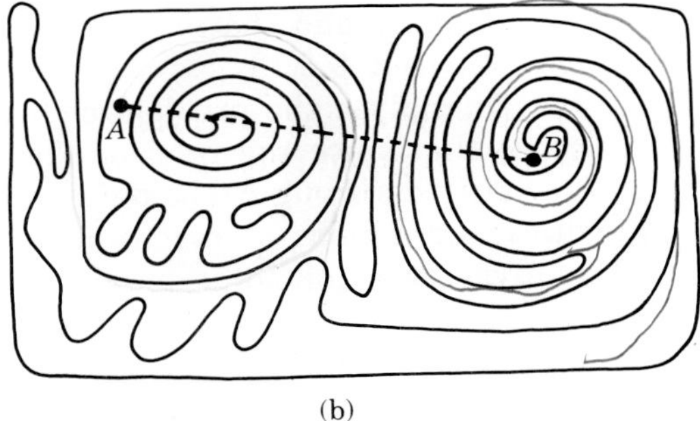

(b)

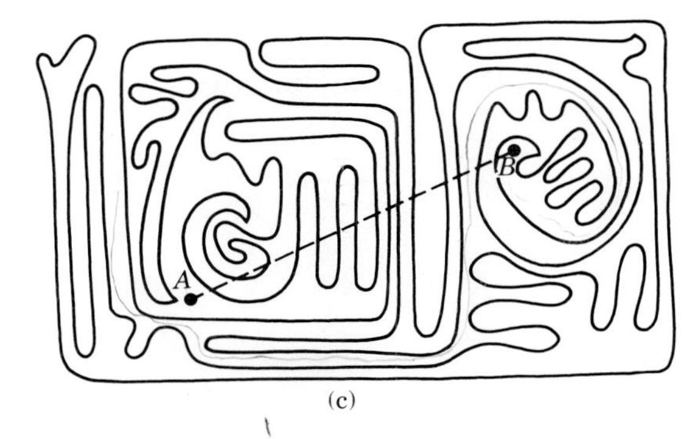

(c)

INVESTIGATION 9.6

Consider the following story. A man with 3 sons stipulated in his will that his land was to be divided among his sons so that every son would be a neighbor to each of his brothers. (Being neighbors means that their land touches along a boundary line, rather than only at a point.) That is, each pair of regions must be contiguous along a boundary line. For example, with 3 sons, the division in the figure satisfies this condition.

Can you find such a division for 4 sons? Five sons?

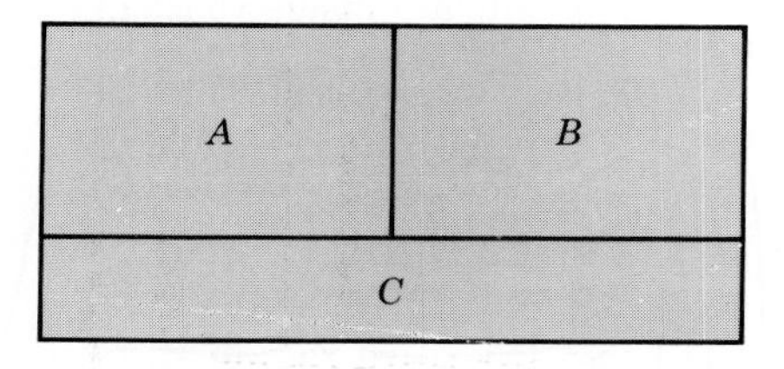

CONTIGUOUS REGIONS

In this section we shall study relationships between a region in the plane and the regions surrounding it. We begin with Investigation 9.6.

To analyze the problem of Investigation 9.6, we first observe that since the relation "neighboring regions" remains unchanged through topological distortion, it is sufficient to draw our regions as rectangular (Fig. 9.32). Continuing our analysis, we find solutions for the case where there are two sons and, hence, only two regions. Then, we solve the problem for the case where there are three regions by considering all the ways one can add a third region to each solution in the case of two regions. We continue this process until we arrive at the case of five regions.

If there are only two sons, there are only two essentially

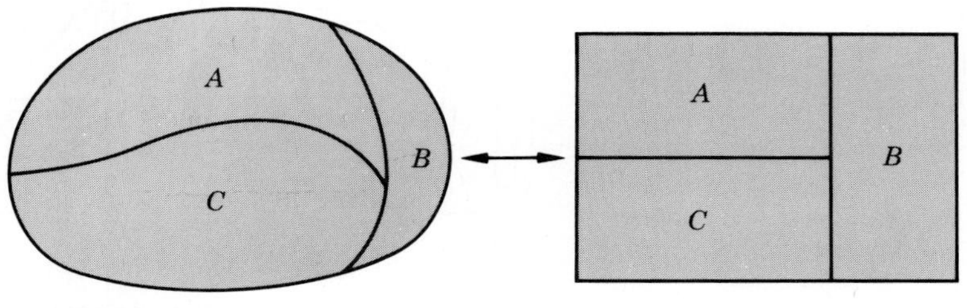

FIGURE 9.32

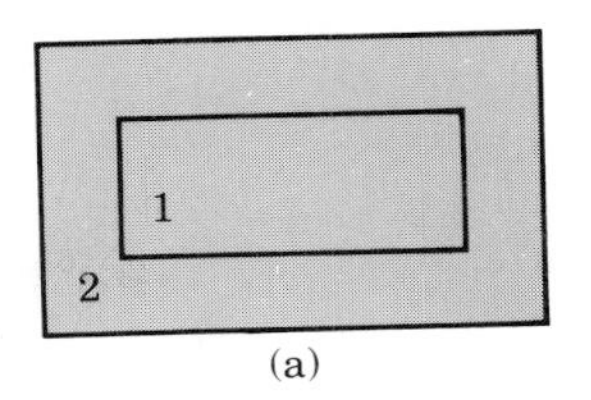

(a)

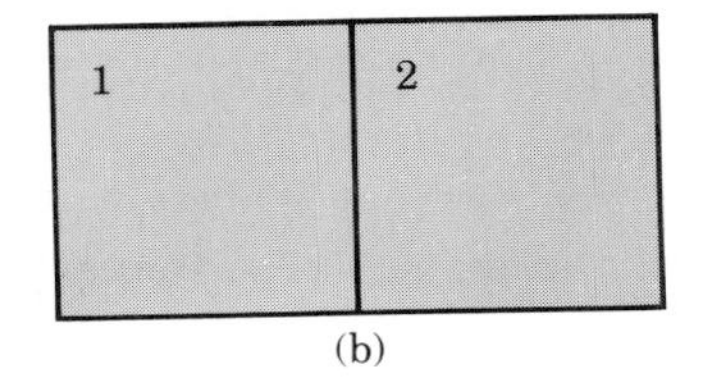

(b)

FIGURE 9.33

different divisions for the land. In one of the divisions, one region completely surrounds the other (Fig. 9.33a).

If there are three sons, there are three possible divisions, one in which two regions are completely surrounded, one with one region completely surrounded, and one with no regions completely surrounded. These three regions correspond to three different ways of adding a third region to Figure 9.33b. Note that any third region added around the outside of Figure 9.33a would not be a neighbor of the inner region.

In Figure 9.34, we see that when any one region becomes completely surrounded, the process of adding regions that are neighboring to all others must stop. Furthermore, in order for the fourth region to be neighboring to each of the first three regions, it must completely surround one region. Consequently, it appears that it is impossible to divide land into five regions with each region neighboring the remaining four.

FOUR-COLOR PROBLEM

A property of maps observed empirically by early English cartographers is that four colors are sufficient to color a map so that neighboring regions (i.e., regions with a common boundary) are colored differently. (Two regions that touch only at a vertex may be colored alike.) In 1850 Francis Guthrie, a student of mathematics at Edinburgh, noted that if this observation were always true, it would be an interesting mathematical theorem. Guthrie's conjecture was that in an arbitrary geographical map where regions represent a political subdivision, four or fewer colors are always sufficient to color the map so that any two regions touching along a boundary line have different colors. In 1878 Arthur Cayley communicated this conjecture, which became known as the **four-color problem**, to the London Mathematical Society. From then on, many mathematicians sought to prove that the conjecture was true. It was not until 1976 that K. Appel and W. Haken of the University of Illinois, with the help of a high-speed computer, verified that the conjecture is true.

TEACHING NOTE

Students at various levels enjoy exploring the coloring problem. Young children might, for example, attempt to color a portion of a map of the United States using as few colors as possible according to the condition that neighboring regions are colored differently. Older students will be interested in the problem and the fact that the four-color conjecture was only recently proven to be true.

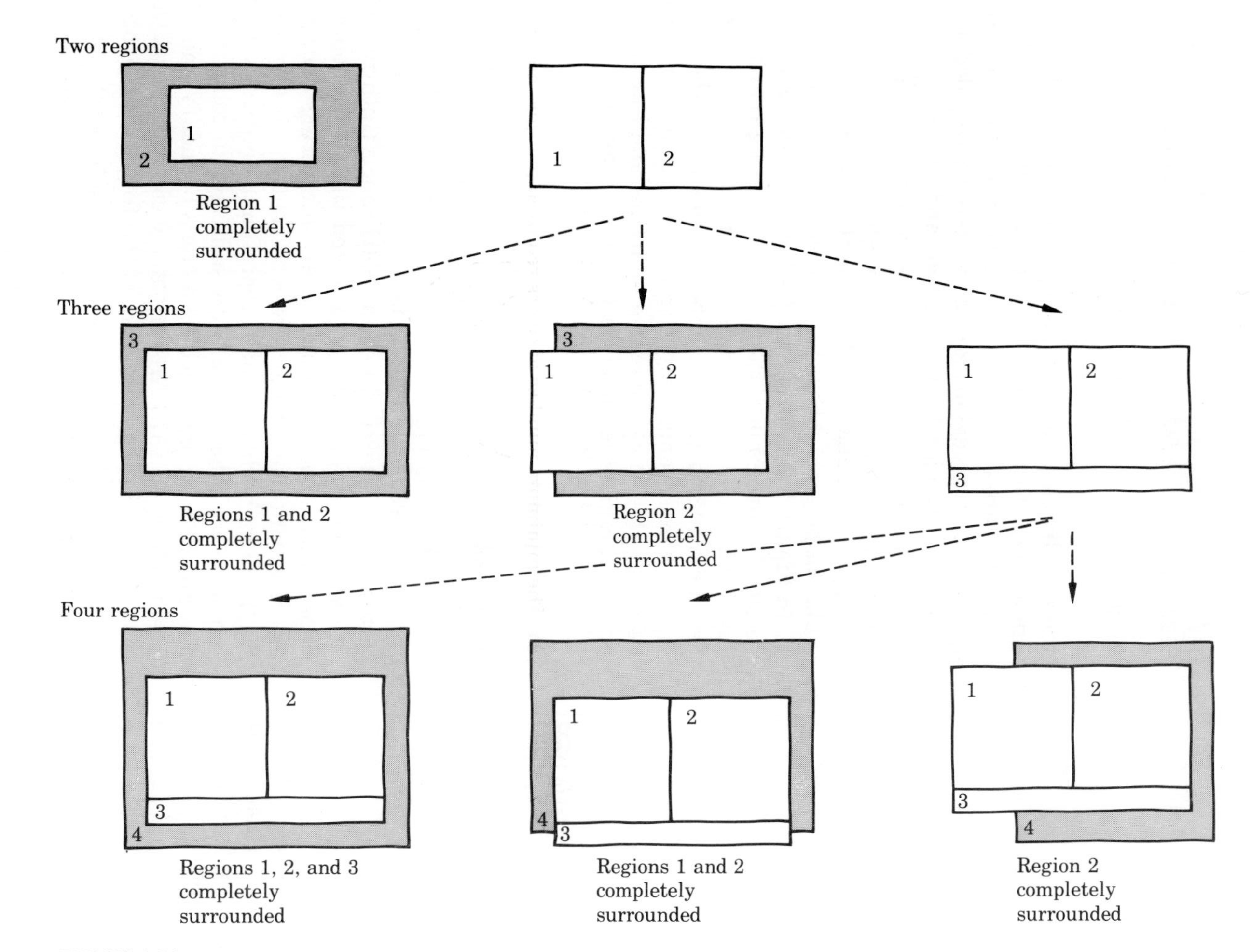
Two regions
1
2
Region 1
completely
surrounded
1
2
Three regions
3
1
2
Regions 1 and 2
completely
surrounded
3
1
2
Region 2
completely
surrounded
1
2
3
Four regions
1
2
3
4
Regions 1, 2, and 3
completely
surrounded
1
2
4
3
Regions 1 and 2
completely
surrounded
1
2
3
4
Region 2
completely
surrounded

FIGURE 9.34

This so-called four-color problem is, of course, related to the contiguous region problem of the last section. For certainly, a map containing four contiguous regions would require four colors—one for each region. In Investigation 9.7 we explore some of the relationships between the contiguous region and four-color problems.

Several observations should be made about Investigation 9.7. Figure (a) is a map with at most two neighboring regions and which requires three colors. Figure (e) is a map with at most three neighboring regions and which requires four colors. So we see that in some cases one more color is required than the number of neighboring regions. Since it is possible to have a map with four neighboring regions, as we have seen in Figure 9.34, doesn't it seem plausible that there might exist a map that requires more than four colors? The four-color conjecture claims that the answer is no. Do you believe it?

INVESTIGATION 9.7

For each of the maps shown find the following.

A. The minimum number of colors required to color the map so that neighboring regions are of a different color
B. The maximum number of regions all neighboring to one another

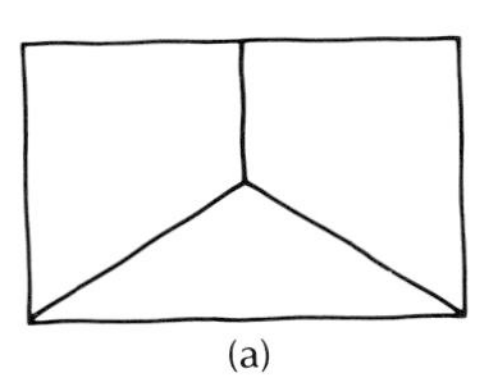
(a)

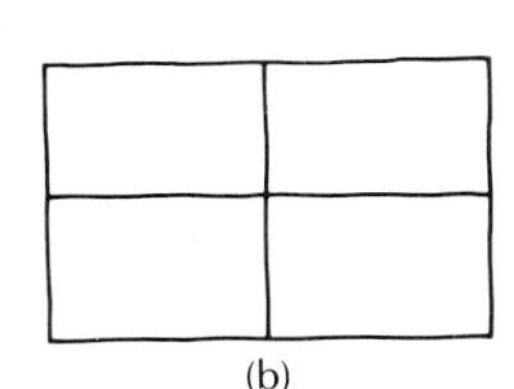
(b)

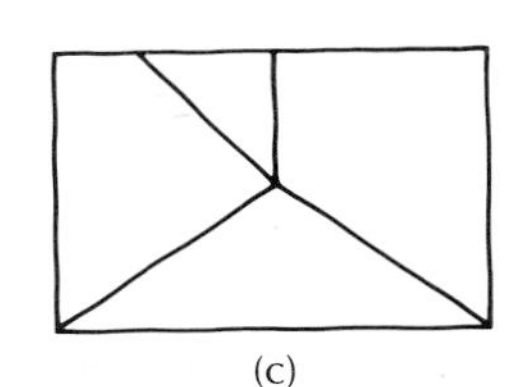
(c)

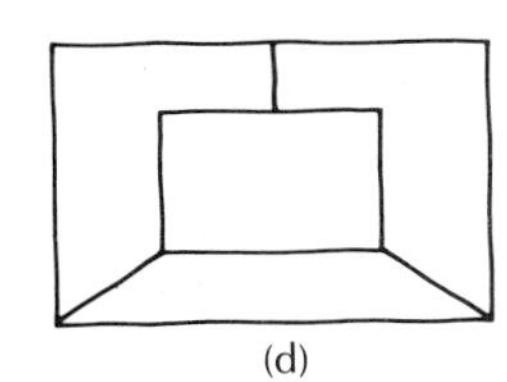
(d)

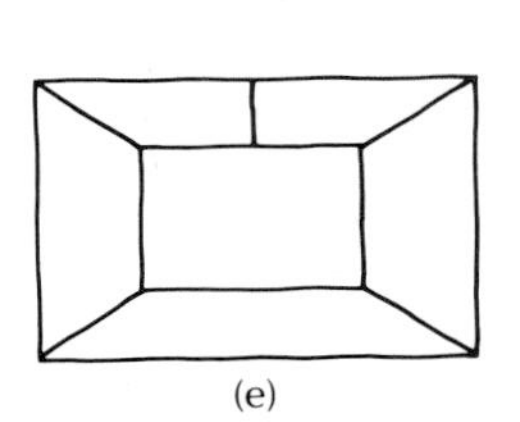
(e)

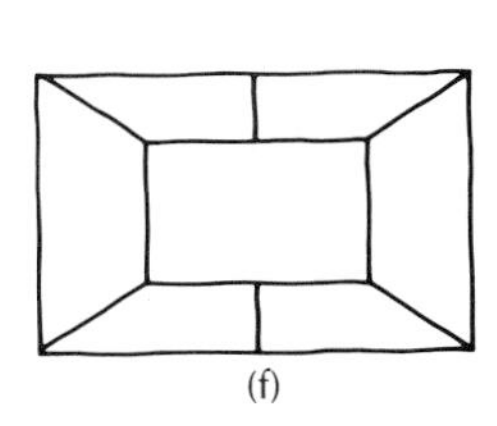
(f)

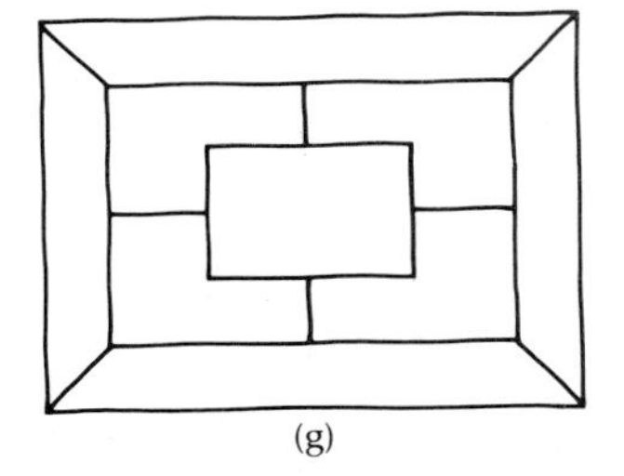
(g)

FIGURE 9.35

EXERCISE SET 9.4

Trace and color the section of a map of the United States shown in Figure 9.35 so that neighboring states do not have the same color. How many colors are required?

PEDAGOGY FOR TEACHERS

ACTIVITIES

1. Engage students at a given level in an experiment with the Moebius strip.
   a. Did the students get involved and interested in the experiment?
   b. Did the students make any conjectures? That is, did they say, "I wonder if this might be true?"
   c. Did the students ask any questions about properties of the strip other than questions on mechanical procedure?
   d. Did the students have a tendency to want to carry the experiment further, or were they satisfied with the initial experience?
   e. How would you change the experiment to make it more challenging or interesting?

2. At what age can students understand topological equivalence? Investigate this question by devising a mini-lesson that would develop this concept. Include some exercises that you could present to students to see if they have understood the idea. You may want to present this mini-lesson to individual students at different grade levels.

3. Consult puzzle books such as *Creative Puzzles of the World* (van Delft and Botermans, 1978) and make at least three topologically oriented puzzles you feel would be appropriate for for students at designated grade levels. Have children try the puzzles and report on the results.

4. Create a sequence of concept cards that use examples and nonexamples to develop the idea of topological equivalence. Use the cards with students and evaluate their effectiveness. Revise the cards for classroom use.

5. Study Chapter 16 of *How Children Learn Mathematics* (Copeland, 1979). On the basis of this study, create some activities that you could use with preschool and primary school children to check their developmental stage with regard to topological notions. Try these activities with children and summarize your findings.

6. Investigate the topological type games on pp. 146–163 of the book *Mathematical Puzzlements* (Kohl, 1987), and select a game for children at a selected grade level to try. Observe them playing the game and write a one-page analysis of the experience. Focus on the children's attitudes, reasoning abilities, and successes and difficulties with the game.

COOPERATIVE LEARNING GROUP DISCUSSION

Discuss the following questions in your group. Summarize your conclusions for the class.

1. Is topology an important topic for children? Defend your answer. If your answer is yes, give some instructional objectives that would be met by studying topological ideas.

2. What topological ideas would be appropriate at what grade levels, if any? If appropriate, give some examples and some ideas about how you might introduce them in your class.

CHAPTER REFERENCES

Bergamini, David, and Editors of Life Magazine. 1963. *Mathematics*. New York: Life Science Library, Time Inc.

Bolt, Brian. 1989. *Mathematical Funfare*. Cambridge, MA: Cambridge University Press.

Copeland, Richard W. 1979. *How Children Learn Mathematics, 3rd Ed.* New York: Macmillan Publishing Co.

Gardner, Martin. 1969. *The Unexpected Hanging*. New York: Simon and Schuster.

Johnson, D.A., and W.H. Glenn. 1960. *Topology, The Rubber Sheet Geometry*. St. Louis: Webster Publishing Co.

Kohl, Herbert. 1987. *Mathematical Puzzlements—Play and Invention with Mathematics*. New York: Schocken Books.

Koziakin, Vladimir. 1972. *Mazes 2*. New York: Grosset and Dunlap Publishers.

van Delft, Pieter, and Botermans, Jack. 1978. *Creative Puzzles of the World*. New York: Harry N. Abrams, Inc., Publishers.

SUGGESTED READINGS

Consult the following sources listed in the Bibliography for related readings: Hirsch, 1988; Richardson, 1987; Shyers, 1987; and Steen, 1990.

CHAPTER TEN Number Patterns in Geometry

INTRODUCTION

The important concept of "pattern" is so broad that any attempt to define it will likely restrict its meaning. And yet, nearly everyone has an intuitive understanding of the concept. In this chapter we think of pattern in the sense of being a recurring configuration or relationship. In a report to the nation on the future of mathematics education (National Research Council, 1989), it was stated that ". . . Mathematics is a science of pattern and order." A later publication (Mathematical Sciences Education Board/National Research Council, 1990) asserts that ". . . A person engaged in mathematics gathers, discovers, creates, or expresses facts and ideas about patterns." In order to restrict this rich subject to a "bite-sized portion," the discussion in this chapter is limited primarily to certain patterns of points and lines and the number patterns that emerge from these geometric settings.

PATTERNS OF POINTS

Thinking of point in the physical sense, we observe that there are both natural and human-made settings in which patterns of points occur. For example, natural settings can be seen in X-ray diffraction patterns (Fig. 10.1), in a magnification of a crystal of virus (Fig. 10.2), or in a photograph of portions of the solar system (Fig. 10.3). The computer printout of prime numbers (Fig. 10.4) represents human-made patterns of points.

Some of these patterns may be useful to the scientist for laboratory analysis, while others are simply aesthetically pleasing. Regardless of our purpose, if we are sensitive to patterns in the world around us, even the physical notion of point can take on new meaning. Investigation 10.1 will provide an opportunity for a closer look at some simple patterns of points.

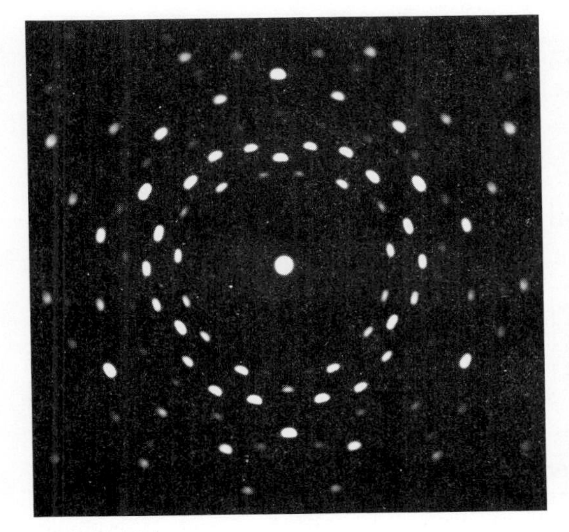

FIGURE 10.1

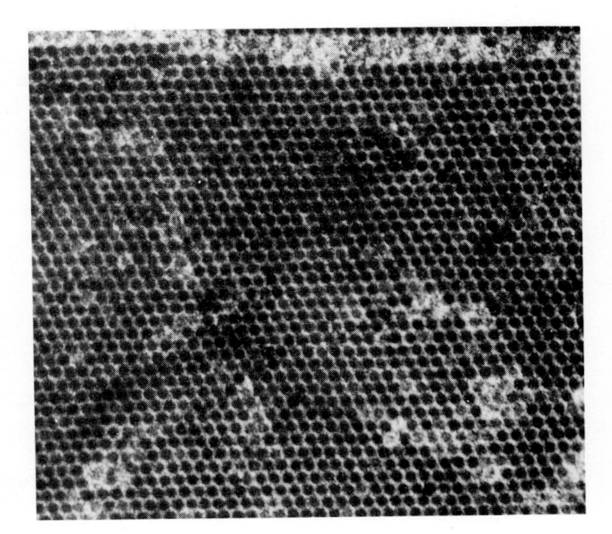

FIGURE 10.2

FIGURE 10.3

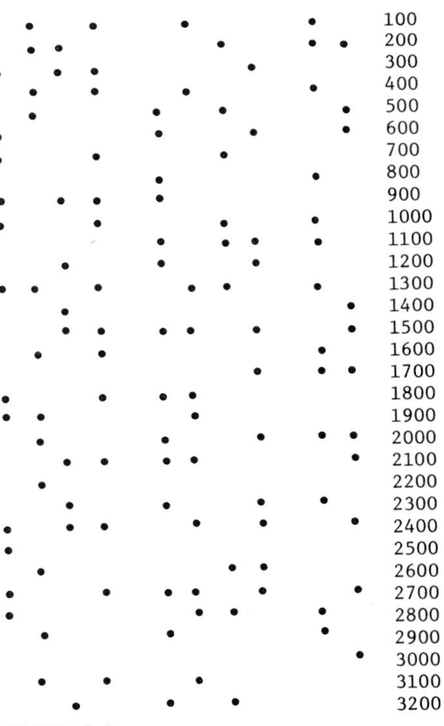

FIGURE 10.4

INVESTIGATION 10.1

A geoboard is a board with a square array of nails (points) on it. A few different-sized geoboards are shown in Figure (a).

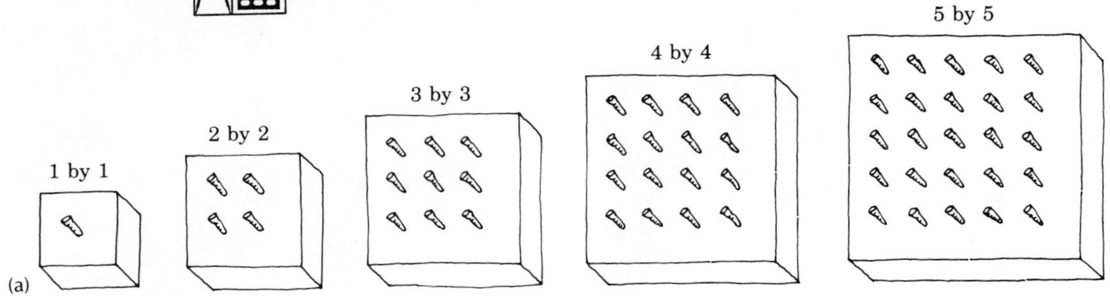

A. If you place rubber bands on these geoboards in special ways and count the nails inside them, you produce a sequence of numbers Figure (b). Write the first 8 numbers in the sequence produced in each situation presented in Figures (c) and (d).

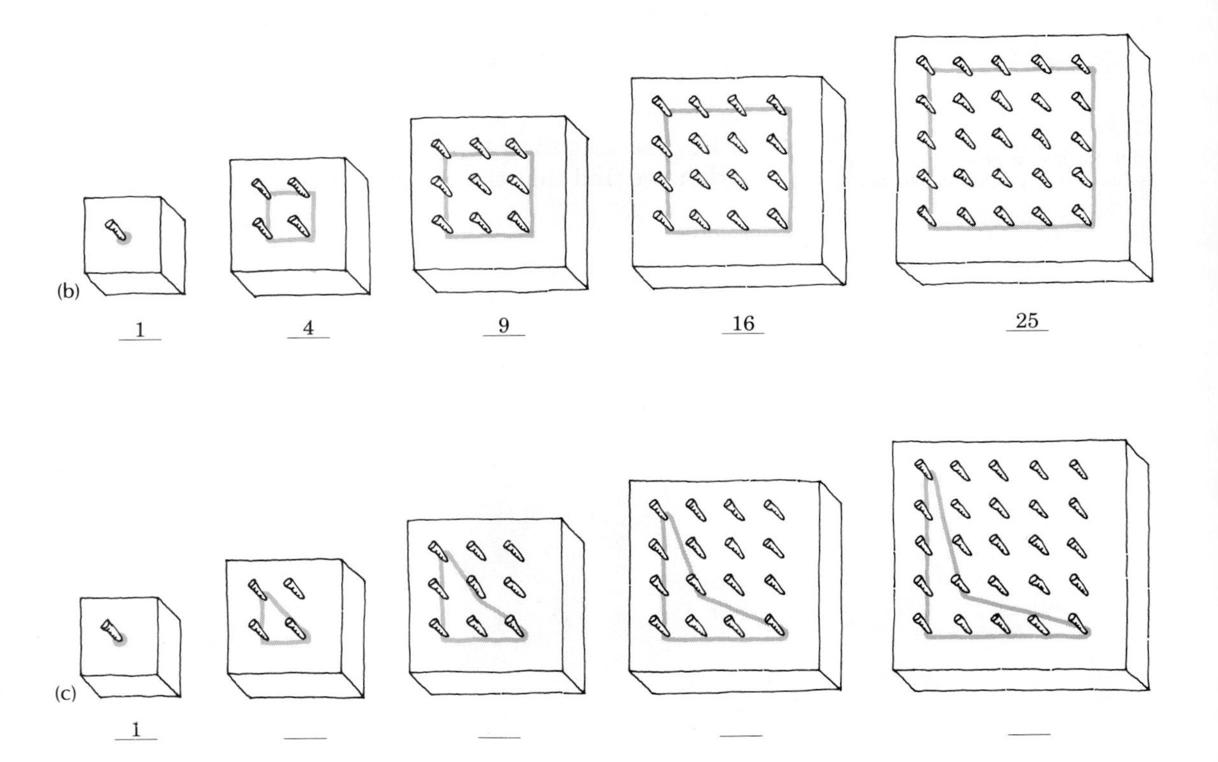

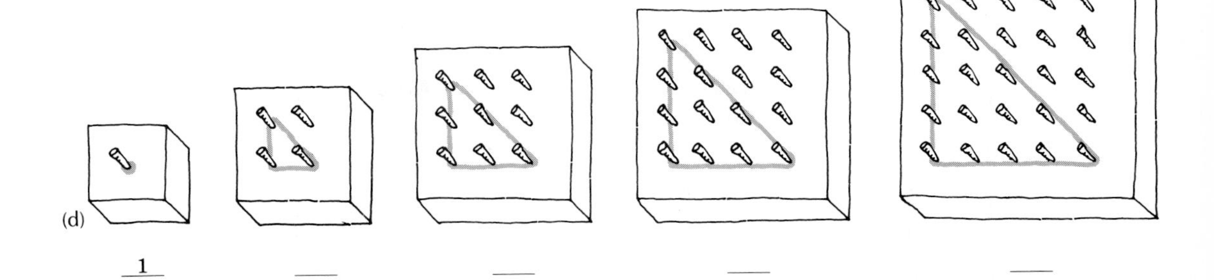

B. For each situation, can you figure out how to find the number of nails inside the rubber band on a 100 × 100 geoboard? On an n by n geoboard? (Write a formula.)

C. Now, make some more geoboard sequences of your own and answer the same questions about them.

TEACHING NOTE

If the teacher sees mathematical patterns as a source of beauty and interest, then the student may also come to recognize this beauty and look for it in a new situation. The geoboard is a valuable device for creating situations where children can discover patterns. The teacher might be interested in the books on geoboards available from Creative Publications (Goodnow, 1991, and Hoogeboom, 1988).

Investigation 10.1 of simple point patterns on geoboards demonstrates that number generalizations can be illustrated and clarified by viewing geometric patterns. For example, the pattern produced by enclosing all nails on each geoboard suggests the sequence of numbers called **square numbers**. Table 10.1 describes the situation. By observing the geometric situation, we find that the generalization "the *nth* square number is n^2" takes on a fuller meaning.

TABLE 10.1

Geometric Situation			Numerical Situation	
Size of Geoboard	Total Number of Nails Enclosed		n	nth Square Number
1 by 1	•	1	1	1
2 by 2	•• ••	4	2	4
3 by 3	••• ••• •••	9	3	9
⋮		⋮	⋮	⋮
n by n		?	n	?

TABLE 10.2

Geometric Situation			Numerical Situation	
Size of Geoboard	Total Number of Nails Enclosed		n	nth Triangular Number
1 by 1	•	1	1	1
2 by 2		3	2	3
3 by 3		6	3	6
4 by 4		10	4	10
.		.	.	.
.		.	.	.
.		.	.	.
n by n		?	n	?

Similarly, the pattern produced by enclosing the triangular array of nails on the geoboard suggests the sequence of numbers called **triangular numbers**. Table 10.2 describes the situation. (Note that the black dots indicate the nails that were enclosed on the geoboard. The gray dots have been included to display rectangular arrays of dots that are helpful in forming the generalization. Observe that each triangular array contains half the number of dots in the rectangular array.)

By observing the arrays of dots we see that the number of dots enclosed on a geoboard with two nails on a side is $(2 \times 3)/2$ (i.e., width of array $\times$ length of array $\div$ 2), the number on a geoboard with three nails on a side is $(3 \times 4)/2$, and the number on a geoboard with four nails on a side is $(4 \times 5)/2$. This geometric discovery suggests the numerical generalization that "the nth triangular number is $[n(n + 1)]/2$."

In the preceding situations, we associated numbers with a geometric situation. This practice is common throughout the study of geometry and results in generalizations involving both numbers and geometric content. Generalizations are valuable in all areas of mathematics because they allow us to economize our effort. Rather than dealing with each of several specific cases as if they were different, we form a generalization that gives us a way of dealing with any of the several cases in the same way.

TEACHING NOTE

Searching for patterns might well be the most important experience the student will have in mathematics. These experiences develop habits and skills of pattern seeking that can be extremely valuable as the student encounters new ideas. The publication on *Geometry for Grades K–6* (NCTM, 1987) suggests several geometric situations rich in patterns.

TABLE 10.3

Dot Pattern	Numerical Relationship
•	$1 = \mathbf{1}^2$
(2 × 2 array)	2 addends: $1 + 3 = \mathbf{2}^2$
(3 × 3 array)	3 addends: $1 + 3 + 5 = \mathbf{3}^2$
(4 × 4 array)	4 addends: $1 + 3 + 5 + 7 = \mathbf{4}^2$
⋮	⋮

TEACHING NOTE

Patterns can often be described using mathematical functions. A function machine has an input, operates according to a rule, and produces an output. A record of the machine's operation can be kept in a table where the input and output are listed as ordered pairs. A function machine such as the one shown can be used to help students discover relationships utilizing geometric figures as well as numbers.

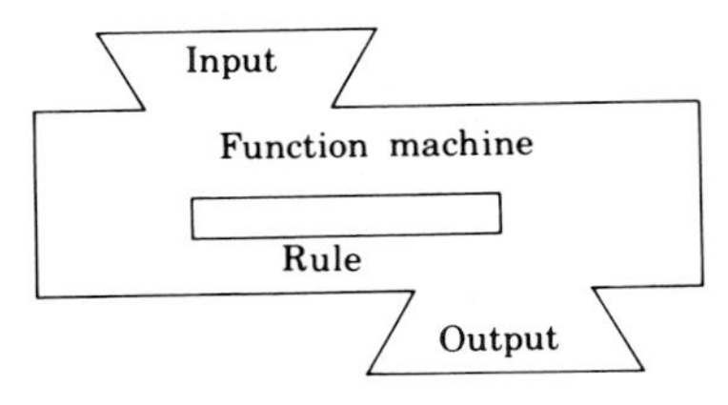

It is important, then, to gain experience in recognizing patterns and finding generalizations, so we continue exploring these patterns. For example, consider the display in Table 10.3, which involves another way of looking at square arrays of dots.

By studying the dot pattern, the generalization "the sum of the first n odd numbers is n^2" becomes clear. In this case, an interesting way of viewing a square array of dots suggests a valuable shortcut for finding the sum of consecutive odd integers, starting with 1. Another similar example involves triangular arrays of dots. See Table 10.4. The reader should study the pattern in Table 10.4 and complete the following generalization: "The sum of the first n whole numbers is ____."

Whether in the context of simple patterns of dots or of complicated patterns of points, lines, and other geometric figures, the search for generalizations in geometry continues. Whenever these geometric generalizations can clarify or suggest numerical generalizations, so much the better. Whenever a numerical generalization can shed light on our understanding of geometry, we experience the value of the close relationship between number and space.

TABLE 10.4

Dot Pattern	Numerical Relationship
•	$\mathbf{1} = \frac{\mathbf{1} \times 2}{2} = 1$
• • •	$1 + \mathbf{2} = \frac{\mathbf{2} \times 3}{2} = 3$
• • • • • •	$1 + 2 + \mathbf{3} = \frac{\mathbf{3} \times 4}{2} = 6$
• • • • • • • • • •	$1 + 2 + 3 + \mathbf{4} = \frac{\mathbf{4} \times 5}{2} = 10$
⋮	⋮

EXERCISE SET 10.1

1. a. What is the 50th square number? The 100th square number?
   b. What is the 50th triangular number? The 100th triangular number?
2. a. What is the sum of the first 50 odd numbers? The first 100 odd numbers?
   b. What is the sum of the first 50 whole numbers? The first 100 whole numbers?
3. A rectangular number is a number that, like 6, can be illustrated with a rectangular array of dots. The number 5 is *not* a rectangular number since neither a single row of dots nor the array with a single extra dot is considered a rectangular array, as shown in Figure 10.5.
   a. Write the numbers up to 35 that are rectangular numbers.
   b. Use the idea of "square" numbers and "rectangular" numbers to write a definition of prime numbers.

APPLICATION

Monument Problem: A sculpture is building a monument using cubical blocks of stone. He has decided to build a pedestal for a statue that is a pyramid with a block on top, 4 in the next layer, and so on. He has 210 blocks and needs to know how many blocks to place on the bottom layer so as to use as many of his stone blocks as possible. What would you tell the sculpture? Is there more than one possible solution?

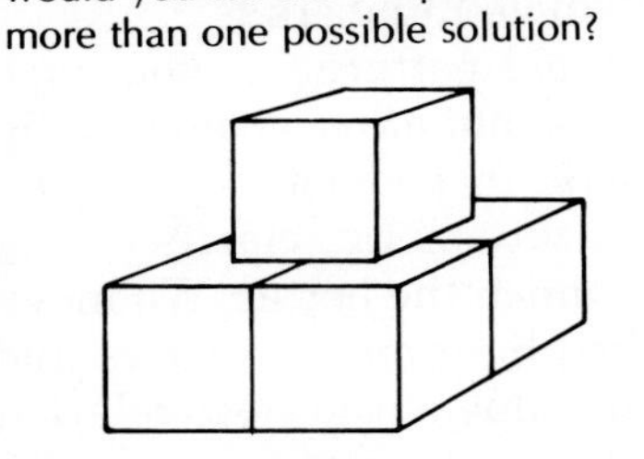

5 6 5

FIGURE 10.5

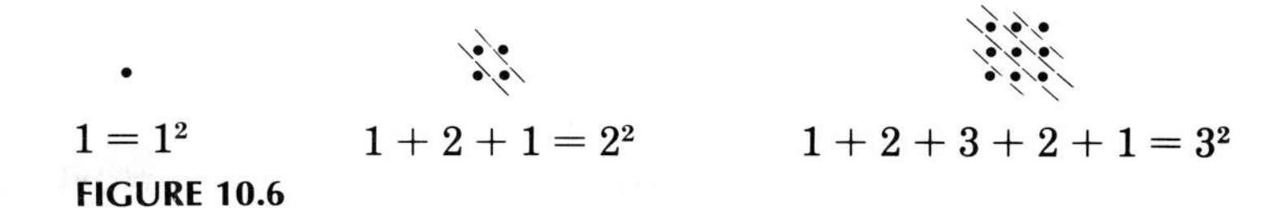

$1 = 1^2$ $\quad$ $1 + 2 + 1 = 2^2$ $\quad$ $1 + 2 + 3 + 2 + 1 = 3^2$

FIGURE 10.6

$1 + 3 = 2^2$ $\quad$ $3 + 6 = 3^2$ $\quad$ $6 + 10 = 4^2$

FIGURE 10.7

4. Refer to Figure 10.6. If you start with the square arrays of dots and draw lines like this, it seems natural to write the equations shown.
   a. Write the next three equations in the sequence.
   b. State a generalization about this relationship.
5. Refer to Figure 10.7. If you start with square arrays of dots and draw lines like this, a pattern emerges.
   a. Write the next three equations.
   b. State a generalization about this relationship. [*Hint:* Note that 1, 3, 6, 10, etc., are triangular numbers.]
6. Suppose you enclose nails on geoboards like those in Figure 10.8 and count all nails enclosed on each board.
   a. Write the first 8 terms in the sequence of numbers determined in this situation.
   b. How many nails are enclosed on a 100×100 geoboard? An $n \times n$ geoboard?

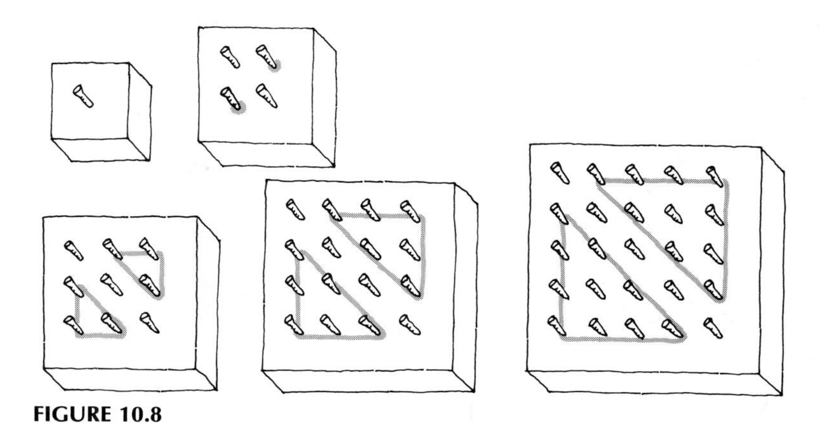

FIGURE 10.8

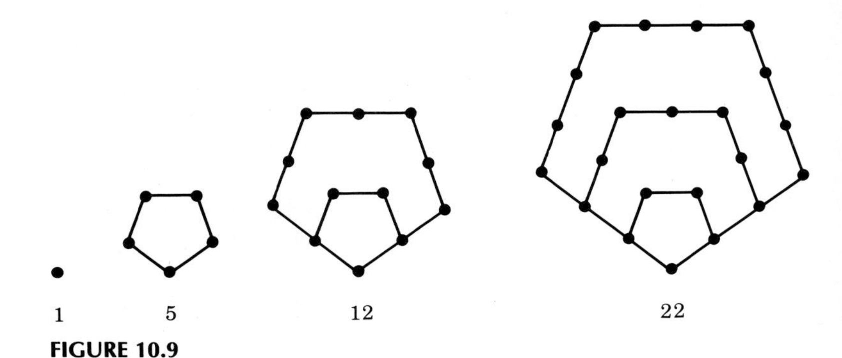

FIGURE 10.9

7. Pentagonal numbers can be described in a manner similar to that used to describe triangular and square numbers. Figure 10.9 shows the first 4 pentagonal numbers. What are the next 3 pentagonal numbers? Show the dot patterns for these numbers.
8. Use the idea suggested in Exercise 7 to find the first 7 hexagonal numbers and show their dot patterns.

*9. Express each pentagonal number and hexagonal number as a sum in such a way that a "pattern of sums" emerges similar to those we observed for triangular and square numbers.

*10. Study the following explanation: Algebraic methods can be used to find the formula for a relationship shown by a table. Such a method is illustrated using Table 10.5 for triangular numbers.

TABLE 10.5

n	nth Triangular Number $T(n)$		
0	0		
		1	
1	1		1
		2	
2	3		1
		3	
3	6		1
		4	
4	10		1
		5	
5	15		1
		6	
6	21		

Column of second differences $(2-1, 3-2, 4-3, 5-4, 6-5$, etc.)

Column of first differences $(1-0, 3-1, 6-3, 10-6, 15-10, 21-15$, etc.)

Since the second difference is constant (always equal to the same number), it can be shown that the equation of $T(n)$ is a quadratic equation of the form

$$T(n) = an^2 + bn + c$$

We proceed to solve simultaneous equations to find values for a, b, and c. Note that

$$T(0) = a \cdot 0^2 + b \cdot 0 + c = 0$$

$$T(1) = a \cdot 1^2 + b \cdot 1 + c = 1$$

$$T(2) = a \cdot 2^2 + b \cdot 2 + c = 3$$

Simplifying, we have the equations

$$c = 0 \quad a + b = 1 \quad 4a + 2b = 3$$

Solving these simultaneously, we have

$$a = 1/2 \quad b = 1/2 \quad c = 0$$

Hence,

$$T(n) = 1/2n^2 + 1/2n + 0$$

or

$$T(n) = \frac{n^2 + n}{2}$$

Use the method explained here and Table 10.6 of pentagonal numbers to find the equation that can be used to find the nth pentagonal number, $P(n)$.

TABLE 10.6

n	$P(n)$
0	0
1	1
2	5
3	12
4	22
5	35
6	51

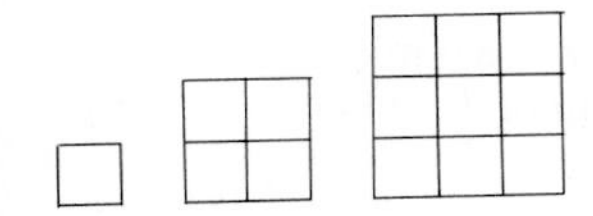

FIGURE 10.10

TABLE 10.7

Number of Rows	1	2	3	4	5	6	7	8	...
Number of Small Squares	1	4	9						
Total Number of Squares	1	5	14						

TABLE 10.8

n	$P(n)$		
0	0		
		1	
1	1		3
		4	
2	5		3
		7	
3	12		3
		10	
4	22		3
		13	
5	35		3
		16	
6	51		

11. Complete Table 10.7 and determine the pattern (Fig. 10.10).
12. If n represents the number of dots in a row and $P(n)$ represents the number of different pairs of dots, construct a table showing the relationship between n and $P(n)$.

*13. The numbers inside the loop in Table 10.8 of pentagonal numbers can be used in a shortcut method for finding the equation that can be used to find the nth pentagonal number. Use the equation

$$P(n) = an^2 + bn + c$$

where: a = (constant second difference) ÷ 2 = 3/2

b = (1st first difference) − a = 1 − 3/2 = −1/2

$c = P(0) = 0$

Hence, the equation is

$$P(n) = \frac{3}{2}n^2 + \frac{-1}{2}n + 0$$

The tables in Table 10.9 express numerical relationships that might describe some geometric situation. Use the shortcut explained here to write an equation for each that can be used to find the $f(n)$ entry in the table for any value of n.

TABLE 10.9

(a) n	$f(n)$	(b) n	$f(n)$	(c) n	$f(n)$
0	3	0	2	0	1
1	4	1	5	1	5
2	9	2	9	2	11
3	18	3	14	3	19
4	31	4	20	4	29
5	48	5	27	5	41
6	69	6	35	6	55

TABLE 10.10

Triangular	1	3	6	16	15	21	28	36	45	55	66
Square	1	4	9	16	25	36	49	64	81	100	121
Pentagonal	1	5	12								
Hexagonal	1										

*14. a. Figure out a way to use the rows of triangular and square numbers in Table 10.10 to find the pentagonal numbers. How did you do it?

b. Use the rows of square and pentagonal numbers in Table 10.10 to complete the row of hexagonal numbers.

15. Complete Table 10.11 below by observing number and formula patterns.

16. **Critical Thinking**. You have studied the concepts of triangular numbers, square numbers, and pentagonal numbers based on polygons in the plane. Generalize from these concepts and describe what you would call *cube numbers*. Explain why your description is appropriate.

TABLE 10.11

Polygonal Numbers	First	Second	Third	Fourth	Fifth	···	*n*th
Triangular	1	3	6	10	15	···	$\frac{n(1n - {}^{-}1)}{2}$
Square	1	4	9	16	25	···	$\frac{n(2n - 0)}{2}$
Pentagonal	1	5	12	22	35	···	$\frac{n(3n - 1)}{2}$
Hexagonal	1	6	15	28	45	···	$\frac{n(4n - 2)}{2}$
Heptagonal	1	7	18				
Octagonal	1	8	21				
Enneagonal	1	9	24				
Decagonal	1	10	27				

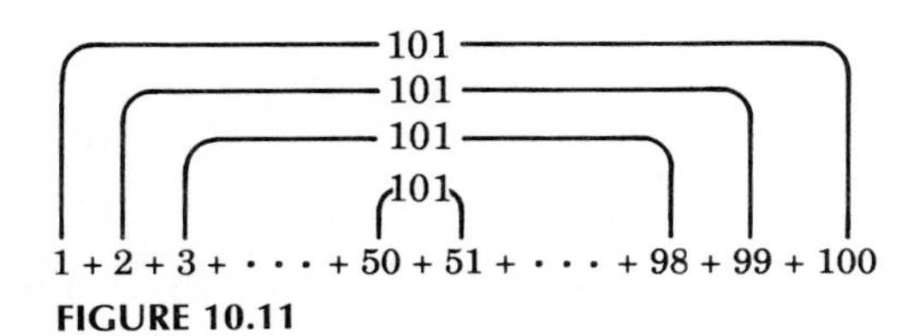

FIGURE 10.11

Problem Solving: Developing Skills and Strategies

An important part of problem solving is **looking back**, analyzing your solution, and checking to see if you can generalize the answer. A famous mathematician, Carl Gauss, is said to have used the diagram in Figure 10.11 as a child to quickly find the sum of the first 100 natural numbers without adding.

1. What is the sum? How do you think he did it?
2. What if there were n natural numbers rather than 100. Can you give a general formula for finding the sum?

PATTERNS OF LINES

Patterns of lines often appear in interesting and exciting ways in everyday situations. Photographs from the world of art and architecture provide illustrations of interesting line patterns.

Many of the familiar mathematical curves and surfaces can be demonstrated with patterns (formally called *envelopes*) of lines. We see examples of the parabola (Fig. 10.12), cardioid (Fig. 10.13), and hyperbolic paraboloid (Fig. 10.14) generated in this way. Patterns of lines of this type are often used by

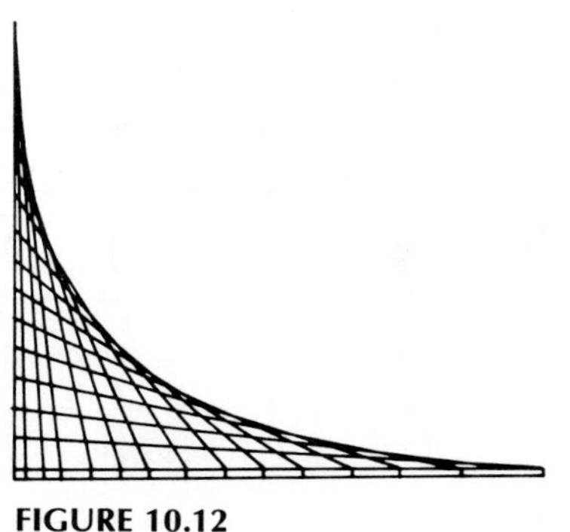

FIGURE 10.12

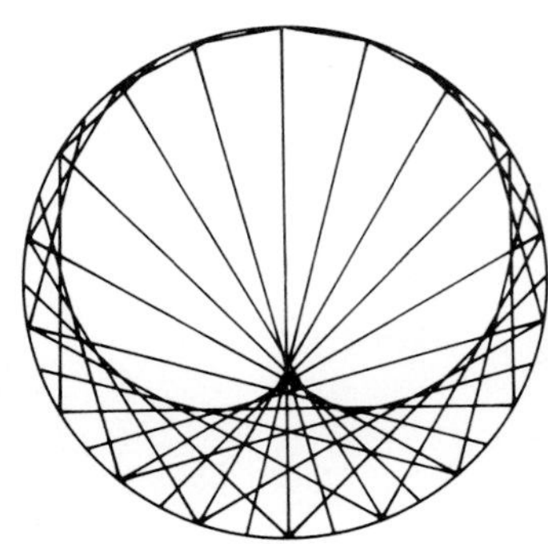

FIGURE 10.13

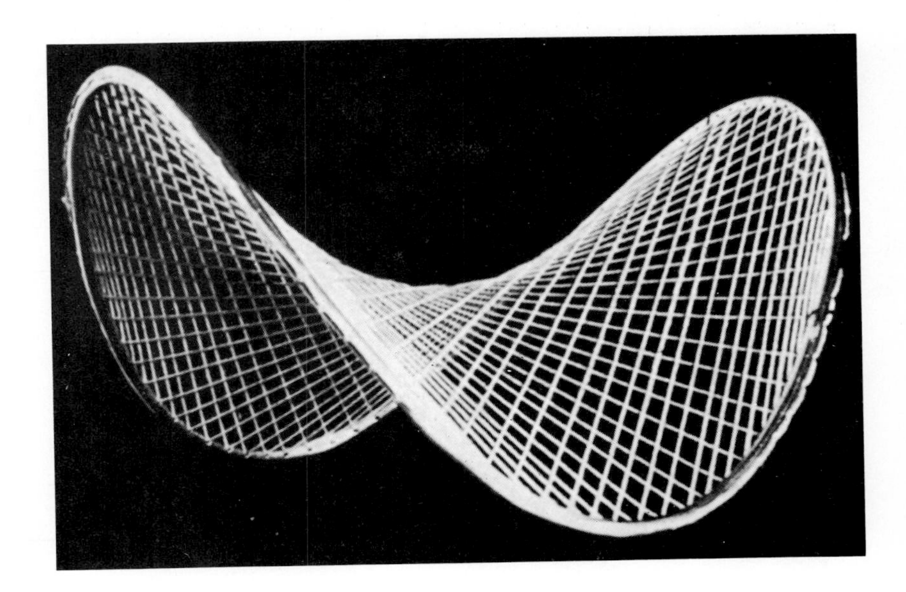

FIGURE 10.14

students involved in constructing curve stitching designs (Figs. 10.15 and 10.16). Investigation 10.2 provides an opportunity to explore line patterns and search for generalizations.

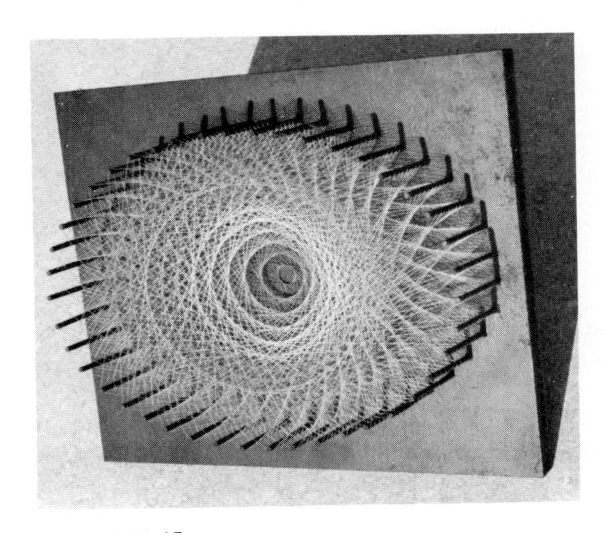

FIGURE 10.15

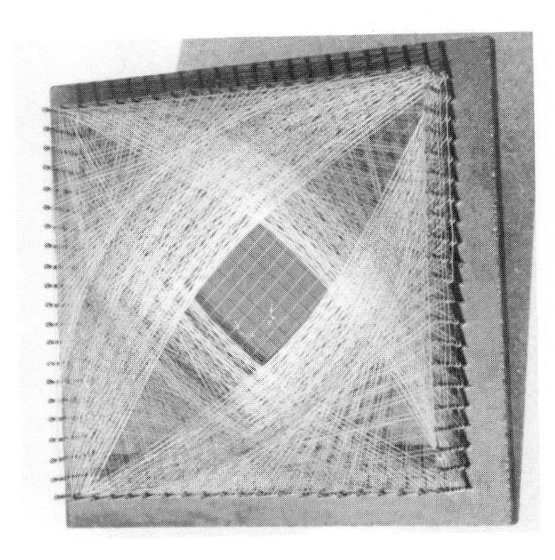

FIGURE 10.16

INVESTIGATION 10.2

Draw pictures to help you complete 5 rows of each table and write an equation expressing the relationship involved.

A.

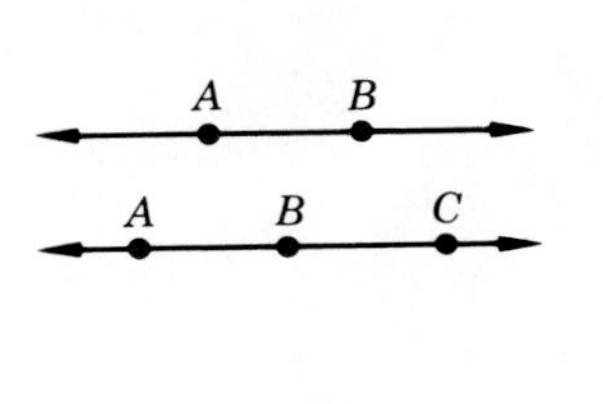

Number of Points Marked on a Line	Number of Different Segments That Can Be Named
→ 2	1 ($\overline{AB}$)
→ 3	3 ($\overline{AB}$, $\overline{BC}$, $\overline{AC}$)
4	
5	
6	
.	
.	
.	
n	?

B.

Number of Points (no 3 on the same line)	Number of Different Lines Determined by These Points
2	1
3	3
4	
5	
6	
.	
.	
.	
n	?

C.

Number of Lines	Maximum Number of Intersections of These Lines
2	1
3	3
4	
5	
6	
.	
.	
.	
n	?

Investigation 10.2 illustrates that a single numerical relationship can describe several seemingly unrelated geometric situations. Also, once a relationship such as that for finding the nth triangular number, $T_n = [n(n + 1)]/2$, is known, it can be useful in finding other relationships. For example, consider the following situation and Table 10.12 of number pairs that describe the relationship.

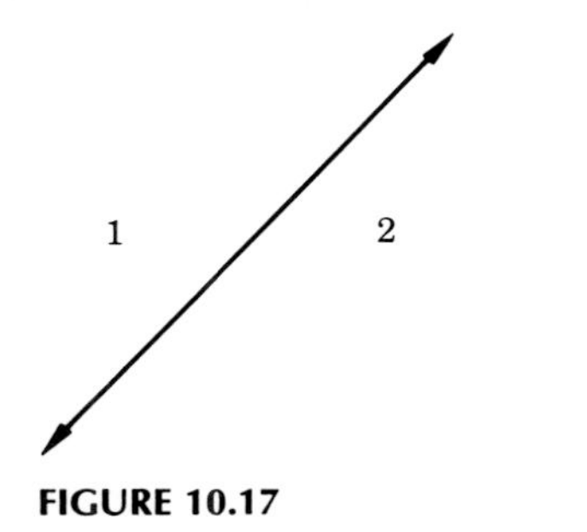

FIGURE 10.17

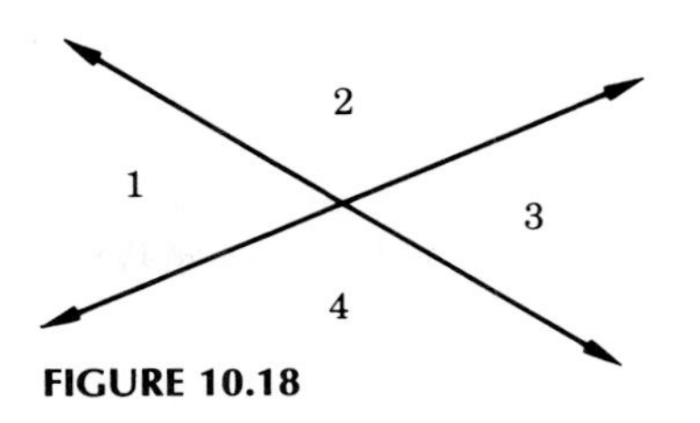

FIGURE 10.18

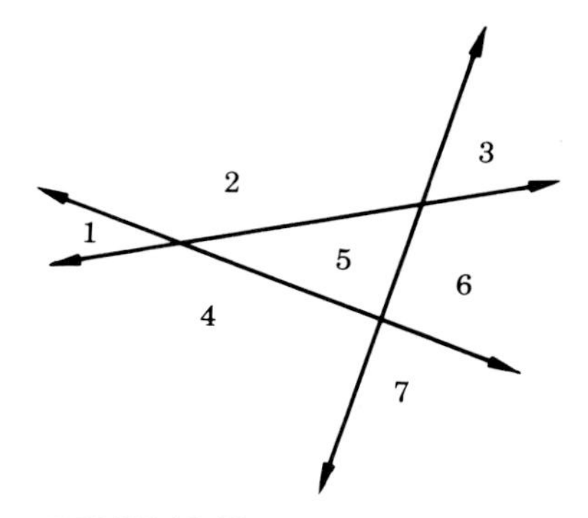

FIGURE 10.19

TABLE 10.12

Number of Lines (n)	Maximum Number of Regions (R_n)
1	2
2	4
3	7
4	11
5	16
.	
.	
.	
n	?

1. One line divides the plane into two regions (Fig. 10.17).
2. Two lines divide the plane into four regions (Fig. 10.18).
3. Three lines divide the plane into a maximum of seven regions (Fig. 10.19).
4. Four lines divide the plane into a maximum of 11 regions and 5 lines divide the plane into a maximum of 16 regions. (The reader should make drawings to show that this is true).

When the table for this relationship is compared with the table for the nth triangular number, we have the results shown in Table 10.13. Note that the maximum number of regions, R_n, is one more than the triangular number for a given value of n.

TABLE 10.13

n	T_n		n	R_n
1	1	⟶	1	⟶ 2
2	3	⟶	2	⟶ 4
3	6	⟶	3	⟶ 7
4	10	⟶	4	⟶ 11
5	15	⟶	5	⟶ 16
.			.	
.			.	
.			.	
n	$\frac{n(n+1)}{2}$	⟶	n	⟶ ?

EXPLORING WITH THE COMPUTER 10.1

Use LOGO software to complete the following.

1. Here is a procedure for a "pattern of lines" design. Predict what the design will look like. Then run the procedure.

```
TO LINE DESIGN
REPEAT 36 [REPEAT 4 [FD 50 RT 90]] RT 10
```

2. How can you change the procedure to create a line design by rotating a triangle instead of a square?
3. How are your line designs in Exercise 2 like the ones in Exercise 1? How are they different?

Thus, the formula for the maximum number of regions is found as follows:

$$R_n = \frac{n(n+1)}{2} + 1 = \frac{n(n+1)}{2} + \frac{2}{2}$$

$$= \frac{n(n+1)+2}{2} = \frac{n^2+n+2}{2}$$

The following exercises present interesting line patterns.

EXERCISE SET 10.2

1. If you draw closed figures with 3, 4, and 5 segments as sides, you can draw 0, 1, and 2 diagonals from a selected vertex (Fig. 10.20).
   a. Complete 3 more rows of Table 10.14. What patterns do you see?
   b. Write an equation to show this relationship.

TABLE 10.14

Number of Sides	Number of Diagonals from One Vertex
3	0
4	1
5	2
6	
7	
8	
.	
.	
.	
n	?

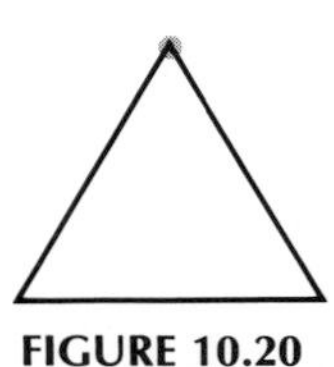

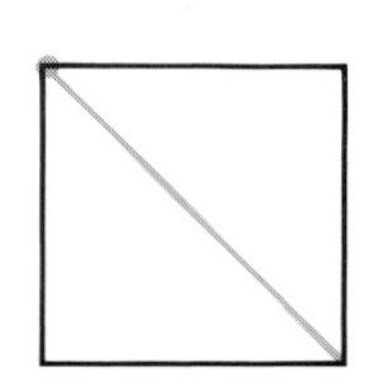

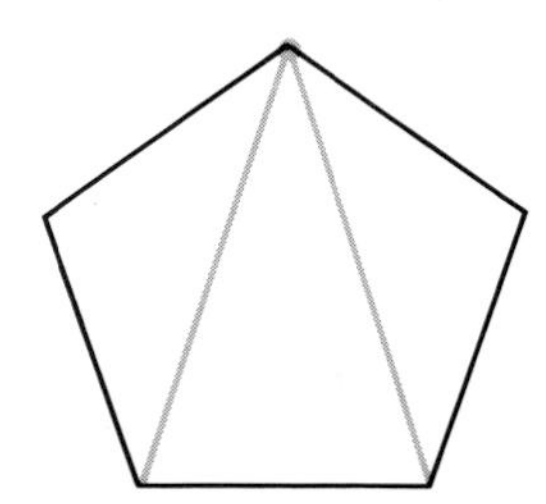

FIGURE 10.20

FIGURE 10.21

TABLE 10.15

Number of Sides	Number of Diagonals
3	0
4	2
5	
6	
7	
8	
.	
.	
.	
n	?

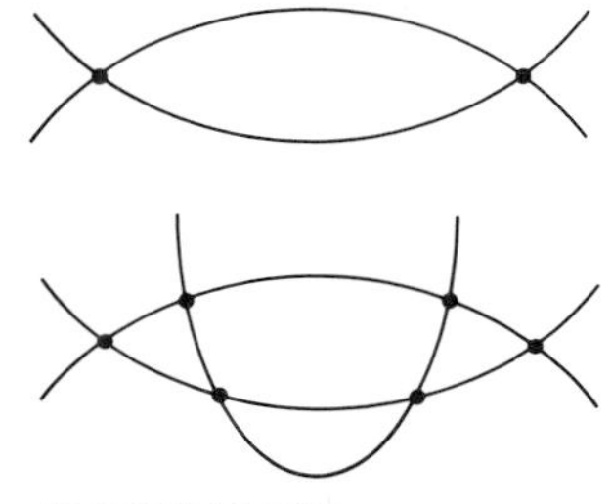

FIGURE 10.22

2. A triangle has no diagonals. A square has a total of 2 diagonals (Fig. 10.21).
   a. How many diagonals does a 5-sided figure have?
   b. Complete 4 more rows of Table 10.15. What patterns can you find in this relationship?
3. If curves lines are drawn and it is agreed that *every line shall cut every other line twice*, we get the pictures in Figure 10.22 and Table 10.16. Continue the drawings to find 3 more entries in the table. What patterns can you find in this relationship?
4. A region of the plane that is bounded on all sides by a line is called a *bounded region*. When 3 lines intersect, one bounded region is formed (Fig. 10.23). When 4 lines intersect, 3 bounded regions are formed (Fig. 10.24). (Note that a region like ABC is not counted since it overlaps with regions 1 and 2.)
   a. Complete the next two rows of Table 10.17.
   b. Write an equation, if possible, to describe this relationship.

TABLE 10.16

Number of Lines	Number of Crossings
2	2
3	6

TABLE 10.17

Number of Lines	Maximum Number of Bounded Regions
3	1
4	3
5	
6	
.	
.	
.	
n	?

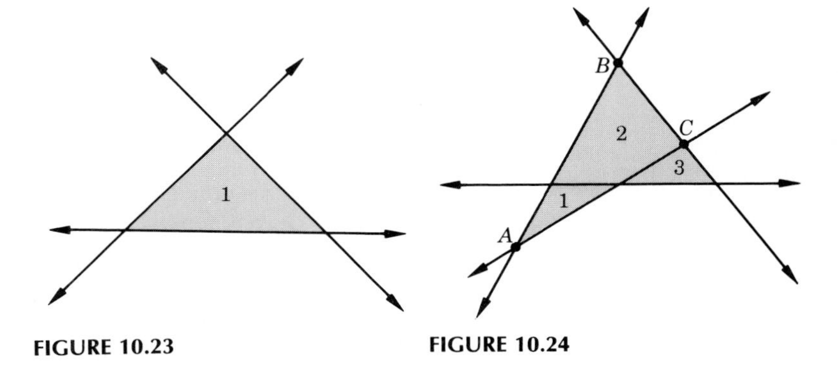

FIGURE 10.23 FIGURE 10.24

TABLE 10.18

Number of Lines	Maximum Number of Triangular Regions
3	1
4	2
5	5
6	7
7	11
8	15
9	21
.	.
.	.
.	.
n	?

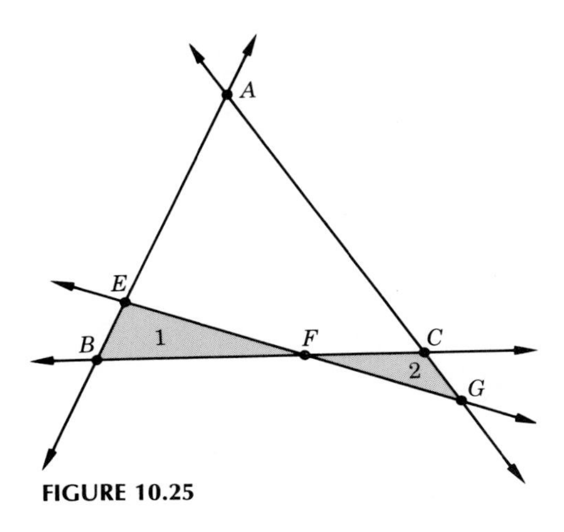

FIGURE 10.25

*5. Instead of counting the total number of regions as in Exercise 4, count the maximum number of *triangular* regions (Fig. 10.25). Four lines determine 2 triangular regions ($\triangle BEF$ and $\triangle CFG$). We do not count $\triangle ABC$ nor $\triangle AEG$ since they overlap with $\triangle BEF$ and $\triangle CFG$. Can you draw diagrams for each number of lines to show the number of regions indicated in Table 10.18?

*6. Make 5 rows of a table showing the maximum number of quadrilateral regions formed by intersecting lines in the plane. (See Exercise 5.)

7. **Critical Thinking**. We have seen that geometrical patterns often suggest numerical patterns. Similarly, number patterns often suggest geometrical patterns. When these points are joined by yarn, this process is known as curve stitching. Suppose 36 points evenly spaced around a circle (and numbered 1 through 36) are joined by the pattern 1 to 2, 2 to 4, 3 to 6, . . . , and x to $2x$ remembering that in this context 37 is equivalent to 1, 38 to 2, and so on (Fig. 10.26). This pattern of lines suggests a curve known as an *epicycloid*. Make a second and third drawing and explore what curve results if

   a. x is joined to $3x$ b. x is joined to $4x$

Problem Solving: Developing Skills and Strategies

Solving some problems requires **special insight**. The usual strategies just don't seem to work for these problems, and you often need to take a new point of view to find a solution. Try the following problems:

1. A waitress had always used 4 straight cuts to cut a cylinder of cheese into 8 identical pieces. One day, she suddenly realized that she could do it with only 3 straight cuts! How did she do it? When

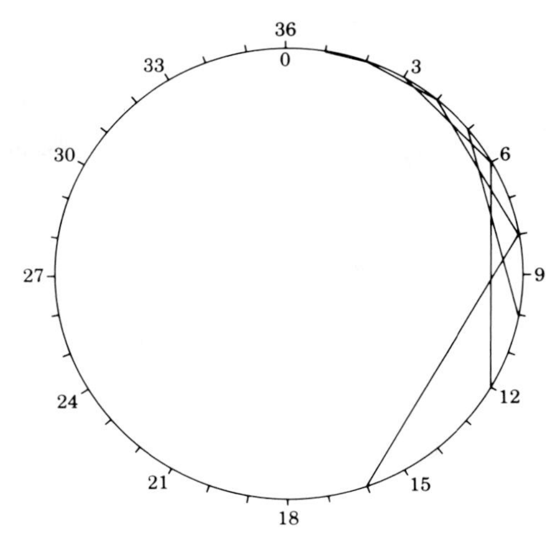

FIGURE 10.26

looking for a solution to the problem, you might first think about cutting the cheese with 4 cuts, as in Figure 10.27a. Then you might think about ways to make 3 straight cuts across the top, as in Figure 10.27b. After several unsuccessful tries, you might try to think about the problem in a different way. Perhaps, suddenly, you will decide to try horizontal cuts as well as vertical cuts and find the solution.

2. Copy the 9 dots in Figure 10.28. Draw 4 connected straight lines through all 9 dots without lifting your pencil or retracing any part of a line.
3. Place the 12 toothpicks of Figure 10.29 together to form 6 squares. Make a drawing of your answer.

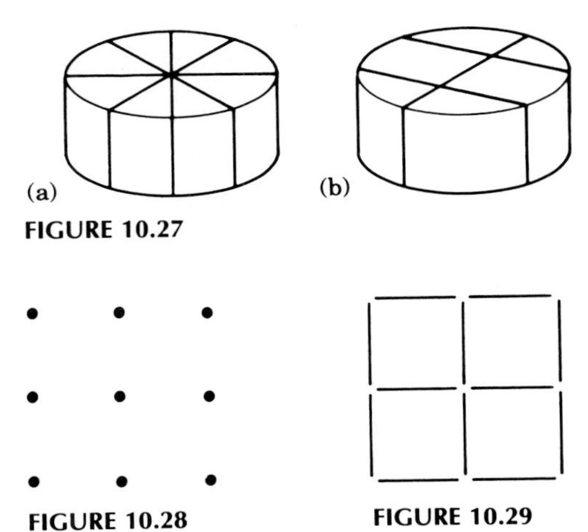

FIGURE 10.27

FIGURE 10.28

FIGURE 10.29

INVESTIGATION 10.3

A segment connecting 2 points on a circle is called a **chord** of the circle, as shown in Figure (a). Two points on a circle determine a chord that divides the interior of the circle into 2 regions. Beginning with 3 points, 3 chords and 4 regions are determined, as shown in Figure (b).

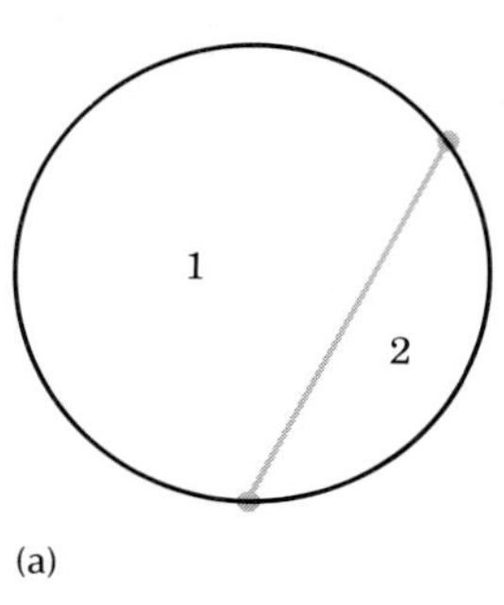

(a)

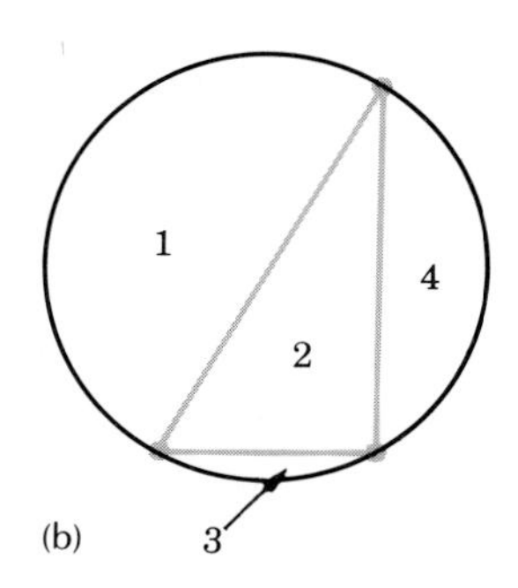

(b)

Number of Points	Number of Regions
2	2
3	4
4	
5	

A. Continue this procedure by marking the specified number of points (unevenly spaced) on the circle, constructing all chords determined by the points, and complete the table.
B. Guess the number of regions generated when beginning with 6, 7, and 8 points.
C. Check the guesses made in part B by going through the construction in each case. For a related Critical Thinking activity, see Activity 12, Appendix D.

PATTERN AND PROOF

The process of inferring how a pattern continues from the consideration of a few examples is called **inductive reasoning**. Investigation 10.3 provides an interesting setting for considering this process.

The lesson we learn from Investigation 10.3 is that we must proceed with a degree of caution when seeking patterns by inductive processes. What we may think at first glance is an obvious pattern may turn out to be a quite subtle one, as illustrated in the investigation. This pattern is referred to as the pattern of the "Lost Region." It illustrates that *we cannot be certain that our inductive conclusion about a particular pattern is correct unless it has been proven*. The process of proof is called a **deductive** process, or **deductive reasoning**. We must recognize the importance of the deductive process, based on accepted assumptions and utilizing a system of logic, as an ultimate means of verifying our conjectured generalizations. While this process has not been emphasized in this book, the reader is referred to Eves (1972) as an excellent resource for further exploration of geometry as a deductive mathematical system.

Even though the potential for incorrect inferences exists, it is desirable that we explore and make inductive conjectures about possible patterns and relationships. All mathematicians engage in this type of inductive process, for it is an integral part of the development of an intuition about mathematical ideas and a method for discovering possible generalizations to be proven. We must "live dangerously," willing to risk making mistakes as we investigate new mathematical situations.

EXERCISE SET 10.3

1. Draw chords of a circle, crossing whenever possible. Then count the number of regions into which these chords divide the circle as shown in Figure 10.30.

   a. Study Table 10.19, *guess* the pattern, and complete 3 more rows.
   b. Draw circles to check your guess. Were you correct?

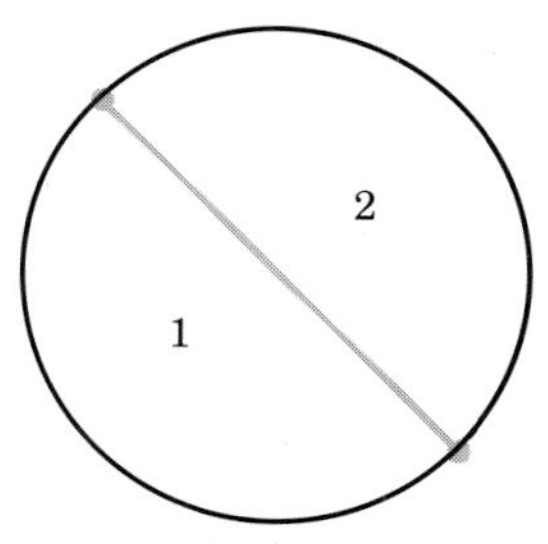

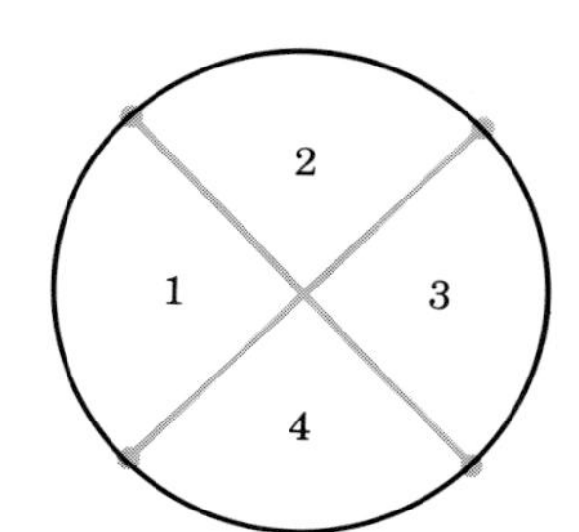

FIGURE 10.30

TABLE 10.19

Number of Segments	Number of Regions
0	1
1	2
2	4
.	.
.	.
.	.

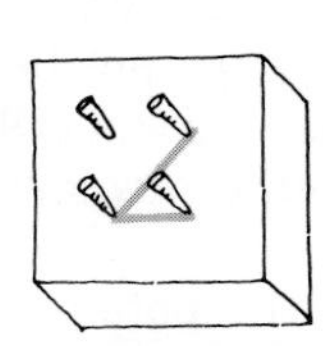
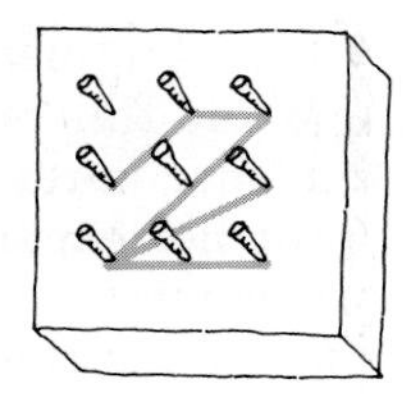

FIGURE 10.31

2. If we use rubber bands to represent segments, a geoboard with 2 nails per side has segments of 2 different lengths, and one with 3 nails per side has segments of 5 different lengths (Fig. 10.31).
   a. Use a geoboard or dot paper and complete the next 2 rows of Table 10.20, studying the pattern developed.
   b. *Guess* the numbers in the next row of Table 10.20.
   c. Check your guess by drawing the different-length segments on 6 × 6 dot paper. (Be careful not to repeat same-length segments in different positions. For example, recall that the hypotenuse of a triangle with legs 3 and 4 units long is 5 units long.) Was your guess correct?
3. **Critical Thinking**. Explore the pattern of differences (see Exercises 10 and 13 in Exercise Set 10.1) for the number pattern discovered in Investigation 10.3. Use the pattern of differences to conjecture about the number of regions generated by 9 and 10 points. Do you think it is possible to find a formula to show this relationship? Explain.

TABLE 10.20

Number of Nails on Each Side	Number of Different-Length Segments
2	2
3	5
4	
5	
6	

PASCAL'S TRIANGLE AND RELATED PATTERNS

A **polygonal path** is the union of a set of line segments, $\overline{AB}$, $\overline{BC}$, $\overline{CD}$, $\overline{DE}$, etc., with common endpoints (Fig. 10.32). Investigation 10.4 involves a search for the number of different possible polygonal paths from one point to another on the geoboard. For this investigation we shall agree to consider only those paths that always move either "up" or "to the right." In Figure 10.33, neither of the paths from A to B would be counted.

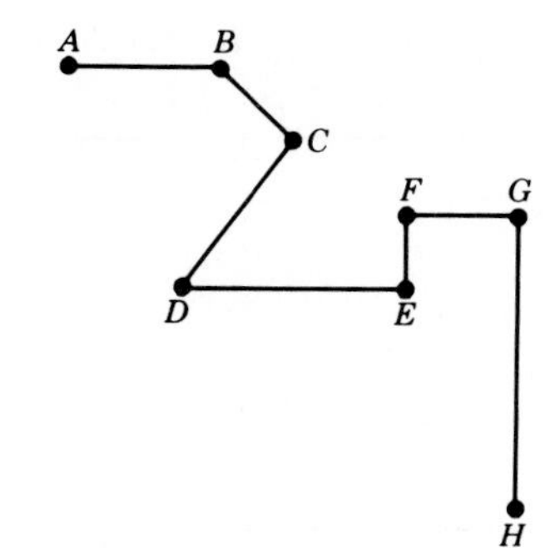

FIGURE 10.32

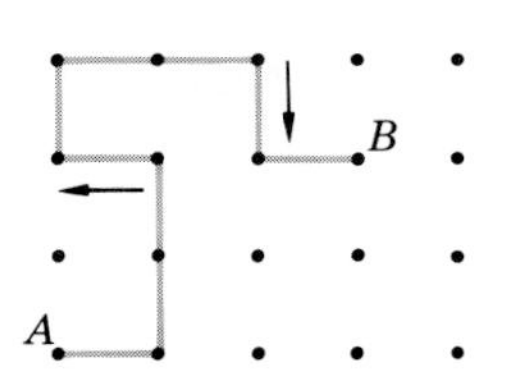

This path goes "to the left" and "down"

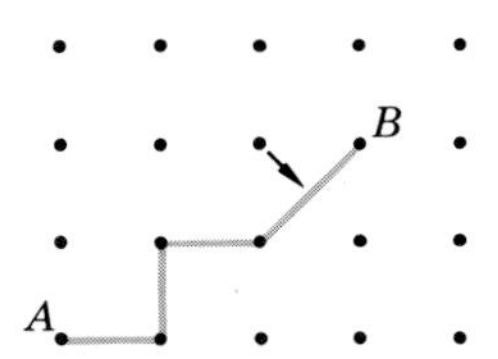

This path goes along a diagonal

FIGURE 10.33

INVESTIGATION 10.4

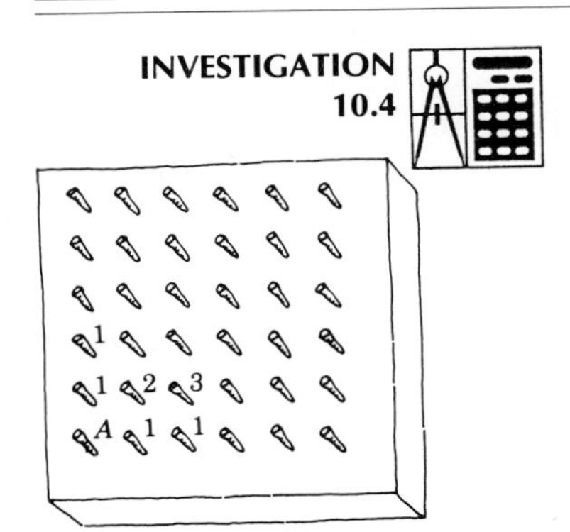

For each nail on the geoboard shown, count the number of paths from A to that nail. Record the number above the nail as is illustrated. If you can discover some useful number patterns as you begin to complete the array, your work will become much easier. Be careful!

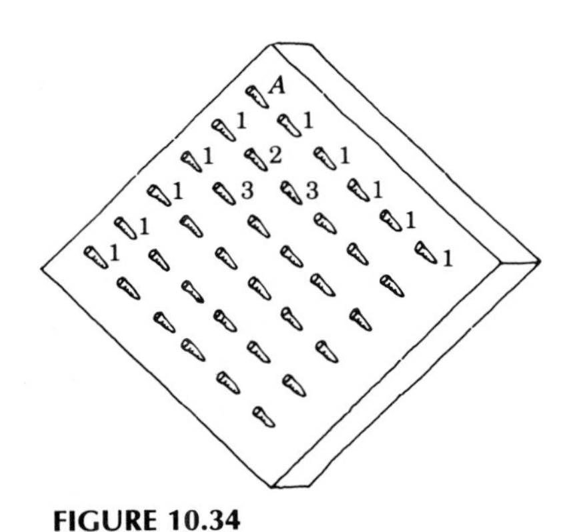

FIGURE 10.34

If the geoboard in Investigation 10.4 is turned as in Figure 10.34, the resulting completed array of numbers appears as a part of a special triangular array of numbers called **Pascal's Triangle**. This array was named after the French philosopher mathematician Blaise Pascal (1623–1662), who discovered many interesting patterns in the array. The following exercises provide an opportunity for the reader to search for patterns in Pascal's Triangle.

TEACHING NOTE

Older students will enjoy a follow up of the idea of Pascal's Triangle and the related Fibonacci numbers. These ideas are beautifully described in Chapters 4, 10, 11, and 12 of *The Divine Proportion* (Huntley, 1970).

EXERCISE SET 10.4

1. One arrangement for Pascal's Triangle is

$$\begin{array}{ccccccccc} &&&&1&&&& \\ &&&1&&1&&& \\ &&1&&2&&1&& \\ &1&&3&&3&&1& \\ 1&&4&&6&&4&&1 \\ &&&&\cdot&&&& \\ &&&&\cdot&&&& \\ &&&&\cdot&&&& \end{array}$$

   Observe the relationship between each entry and the two entries directly above it. Use this relationship and complete an additional 5 rows of the array.

2. What interesting number sequences can you find in the diagonals of Pascal's Triangle?

TABLE 10.21

Row Number	Sum of the Numbers in This Row
1	1
2	2
3	
4	
5	
6	
.	
.	
.	
n	

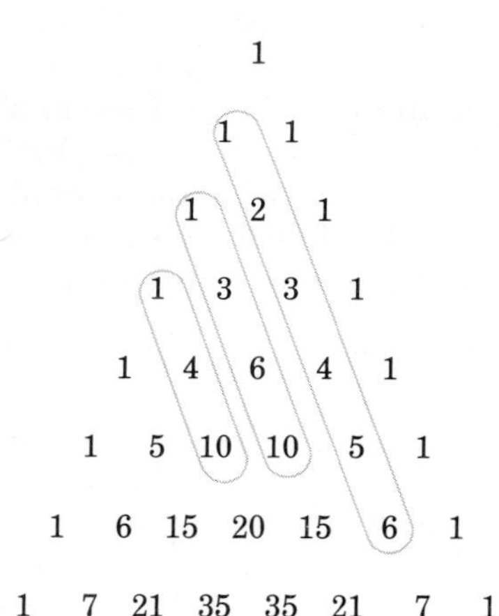

FIGURE 10.35

TABLE 10.22

n	Sum of the First n Fibonacci Numbers
1	1
2	2
3	4
4	7
5	
6	
7	

3. An interesting pattern emerges when the numbers in each row of Pascal's Triangle are added.
   a. Complete 4 more rows of Table 10.21.
   b. Without adding the numbers, can you determine the sum of the numbers in the 10th row?
   c. Write a formula for determining the sum of the numbers in the nth row of Pascal's Triangle.
4. It is also easy to find the sum of numbers along a diagonal of Pascal's Triangle.
   a. Find the sum of the numbers in each gray loop of Figure 10.35.
   b. Do you find these sums in the triangle?
   c. Form a generalization.
5. From the array in Exercise 1, observe that the 1st, 2nd, and 4th rows of Pascal's Triangle contain only odd numbers.
   a. What is the next row that contains only odd numbers?
   b. Can you find a pattern for determining which rows contain only odd numbers?
6. Draw 12 rows of Pascal's Triangle for each part of this exercise.
   a. Circle all even numbers in an array. What patterns do you see?
   b. Circle all multiples of 5 in an array. Do you see any patterns?
   c. Circle all multiples of 3. What patterns emerge?
7. The sums of the numbers in the first 7 special diagonals of Pascal's Triangle are given in Figure 10.36.
   a. Find a pattern in this sequence of numbers and guess the next 3 numbers in the sequence. [*Hint:* Consider the two numbers preceding any number in the sequence.]
   b. Draw more rows of Pascal's Triangle and check your guesses in part a.
8. The sequence of numbers you discovered in Exercise 7 is called the **Fibonacci Sequence**, named after the great medieval mathematician Leonardo of Pisa (called Fibonacci).
   a. Write the first 15 Fibonacci numbers.
   b. Study these equations for finding the sums of Fibonacci numbers:

$$1 = 2 - 1 = 1$$
$$1 + 1 = 3 - 1 = 2$$
$$1 + 1 + 2 = 5 - 1 = 4$$
$$1 + 1 + 2 + 3 = 8 - 1 = 7$$
$$1 + 1 + 2 + 3 + 5 = 13 - 1 = 12$$

   c. Complete the next 3 lines of Table 10.22.

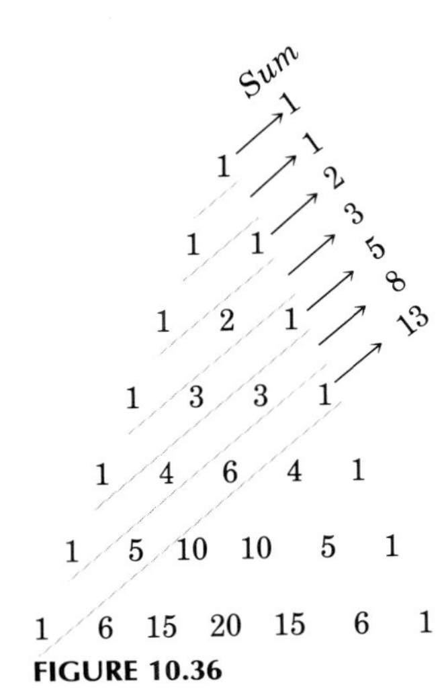

FIGURE 10.36

9. As the second number in the row, the number 5 is a divisor of each other number in its row (except the number 1).
   a. What other second numbers in rows have this property?
   b. Describe a way to determine, simply from observing the second number, whether or not the number divides each other number in the row (except the number 1) evenly.

*10. Refer to Figure 10.37. Given banks of light switches with 2, 3, 4, 5, and 6 switches each, an interesting question is, How many different ways are 2 switches on? 3 switches on? Record in the appropriate entry in Table 10.23 the number of different ways of having the required number of "on" switches.

*11. Suppose you have a set of 5 elements: a, b, c, d, and e.
   a. How many subsets containing 0 elements does this set have?
   b. How many different subsets containing 1 element does the set have?
   c. How many different 2-element subsets does the set have? 3-element subsets? 4-element subsets? 5-element subsets?
   d. Do the numbers you found in parts a, b, and c have any relationship to Pascal's Triangle?
   e. Form a generalization about situations like these.

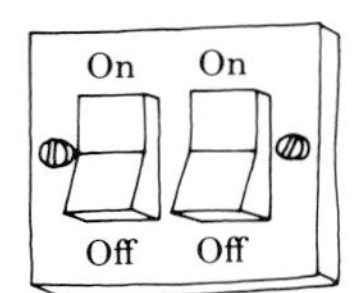

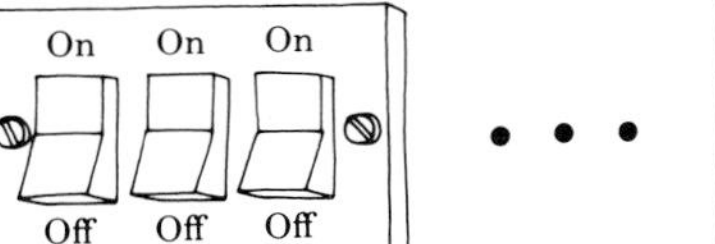

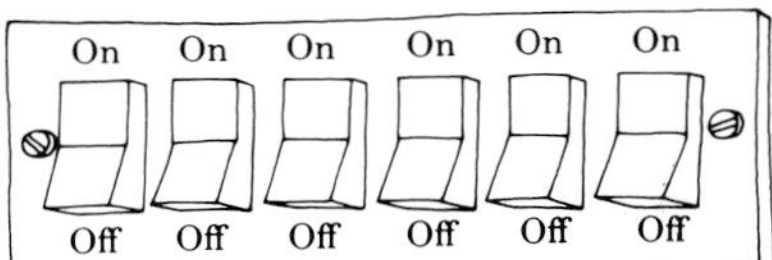

FIGURE 10.37

TABLE 10.23

Number of Switches In Bank	Number of Ways per Number of Switches On				
	1	2	3	4	5
1	1	0	0	0	0
2	2	1			
3	3				
4					
5					
6					

PEDAGOGY FOR TEACHERS

ACTIVITIES

1. Create an investigation card that is appropriate for a level of your choice and that helps a group of students discuss and discover some pattern related to the triangular and square numbers.

2. Read a book such as *Curve Stitching* (Millington, 1989) and (a) write directions for the process of curve stitching, (b) make some special patterns that could be used by students to make interesting curve stitching designs, and (c) present techniques you use to motivate a group of students to become involved in a curve stitching project.

3. A notebook or card file of ideas for teaching mathematics is a valuable aid to teachers at every level. Not only should teachers continue to collect such ideas, but persons preparing to be teachers should start this collection as soon as possible. To this end, consult the geoboard books available from Creative Publications (Goodnow, 1991, and Hoogeboom, 1988) and select at least 3 patterns that students at a given level can discover using a rectangular or circular geoboard.

COOPERATIVE LEARNING GROUP DISCUSSION

Study the recommendations on patterns from the NCTM Standards (NCTM, 1989) on pp. 60–62 and pp. 98–101.

1. Discuss why patterns are important, and brainstorm some useful patterns in geometry.
2. Work together to develop an activity, in the spirit of the Standards, that would help a small group of students explore and discover a pattern in geometry.

CHAPTER REFERENCES

Eves, Howard. 1972. *A Survey of Geometry, Rev. Ed.* Boston: Allyn and Bacon.

Goodnow, Judy. 1991. *Junior High Cooperative Problem Solving with Geoboards*. Sunnyvale, CA: Creative Publications.

Hill, Jane, Ed. 1987. *Geometry for Grades K–6—Readings from the Arithmetic Teacher*. Reston, VA.: National Council of Teachers of Mathematics.

Hoogeboom, Shirley. 1988. *Moving on With Geoboards*. Sunnyvale, CA: Creative Publications.

Huntley, H.E. 1970. *The Divine Proportion*. New York: Dover Publications.

Millington, Jon. 1989. *Curve Stitching*. Norfolk, England: Tarquin Publishing. (Available from Dale Seymour Publications.)

NCTM. 1989. *Curriculum and Evaluation Standards for School Mathematics*. Reston, VA: National Council of Teachers of Mathematics.

Pohl, Victoria. 1986. *How to Enrich Geometry Using String Design*. Reston, VA: National Council of Teachers of Mathematics.

SUGGESTED READINGS

Consult the following sources listed in the Bibliography for related readings: Bruni and Silverman, 1987; Chinn, 1988; Garland, 1987; Miller, 1990; Norman, 1991; and Steen, 1990.

APPENDIX A Using LOGO in Geometry

INTRODUCTION

This appendix presents an introduction to LOGO. LOGO is a computer language. Just as we communicate with other people in English, French, or Spanish, we can communicate with the computer in LOGO. Using LOGO, we can work with pictures, designs, arithmetic, words, and lists. It is a procedural language that is very powerful, but even the beginner can easily work with LOGO successfully. The roots of LOGO are found in artificial intelligence and LISP. The ideas of Piaget can also be found in the language. This makes LOGO an ideal language with which to explore geometry.

From the beginning, the user is in control of the computer. Because it is a friendly language, the user can explore ideas without worrying whether the result is right or wrong. You may not get what you expect, but the result will still be interesting. The philosophy of LOGO is that if it didn't work the first time, fix! This gives the user an ideal environment in which to work. As you work, just think of the screen as a blank slate on which you are going to draw. Remember, you can always erase your work and start again.

LOAD LOGO

LOGO is not a language that the computer knows when you turn on the machine. It must be loaded into the memory of the computer from the Apple LOGO Language disk before you start working.

Use the following steps to load LOGO on machines such as Apple IIe or IIc:

1. Be sure that the computer is turned off.
2. If you are using an Apple IIe, be sure that the caps lock is down or the computer will not understand your commands.
3. Open the door on disk drive number one and insert the Apple LOGO Language disk.
4. Close the door of the disk drive.
5. Turn on the computer and the monitor.

6. Do not touch the disk drive while the red light is on.
7. After the disk drive whirs for a bit and the opening screen appears, hit RETURN.
8. When the red light on the disk drive goes out, open the drive door and carefully remove the Apple LOGO disk. Put the disk in its protective cover and store it in its proper place. You have now completed what is called "booting the disk."

If you are using software such as Logowriter on the Apple Macintosh, simply insert the disk and double click the Logowriter icon.

GETTING STARTED

The screen should now show what is called the Nodraw Picture of the Welcome to LOGO screen with a cursor and prompt. You are in the *text mode*.

This is where the instructions you type appear. To see what happens, let's start by working with the PRINT statement. This command tells the computer to write something on the screen. The question mark on the screen is called the *prompt*. It tells you that the computer is waiting for you to give it a command. The blinking square is called the *cursor*. It tells you where you are on the screen. It moves as you type. You are now in what is called *immediate mode*. After you type a command, simply hit the RETURN key, and the computer will immediately process your instructions.

Now you are ready to try the first LOGO investigation. As LOGO Investigation 1 shows, you can print words like "Hello," lists like [Hi there], and numbers like 3, and perform computations like [3 + 4]. Each of these PRINT statements can be manipulated in several ways, as you will see in later sections.

DRAW MODE

The *draw mode* allows you to draw pictures and designs. To enter this mode, type SHOWTURTLE. This can be abbreviated as ST. The command requires no input. Now your screen should have a splitscreen, with four text lines at the bottom. The little triangular object in the middle of the screen is called the *turtle*. The turtle always has two attributes—position and heading. The *position* is the place on the screen occupied by the turtle. The *heading* is the direction the turtle is pointing. The middle

LOGO INVESTIGATION 1

The following statements show some of the things that the computer will print. Try each one to see what happens. *One very important thing to remember about* LOGO *is that a space is required between a command and any input. A space is also required between commands.*

A. Type each of the commands as they are written.

`PRINT"HELLO` (This is called a word in LOGO.)

`PRINT HI`

`PRINT"HI THERE`

`PRINT 3 + 4`

`PRINT[3 + 4]` Be sure to use the square brackets. (Brackets are made using Shift M and N on the Apple II+.)

`PRINT 3`

`PRINT 3`

`PRINT [HI THERE]` (This is called a list in LOGO.)

B. Now try a few PRINT statements of your own. Be sure to keep track of what happens.

of the screen is the turtle's home position. The original heading here is zero.

There are four primitive statements that change the turtle's position and heading. FORWARD (FD) and BACK (BK) change the turtle's position, while RIGHT (RT) and LEFT (LT) change the turtle's heading. Each of these requires an input value. Remember to put a space between the command and the input! Figure A.1 shows some examples of the effects of these commands on the turtle.

The command CLEARSCREEN (CS) sends the turtle home, sets the heading back to zero, and clears the graphics screen. CS can be used whenever you need a clean slate on which to draw.

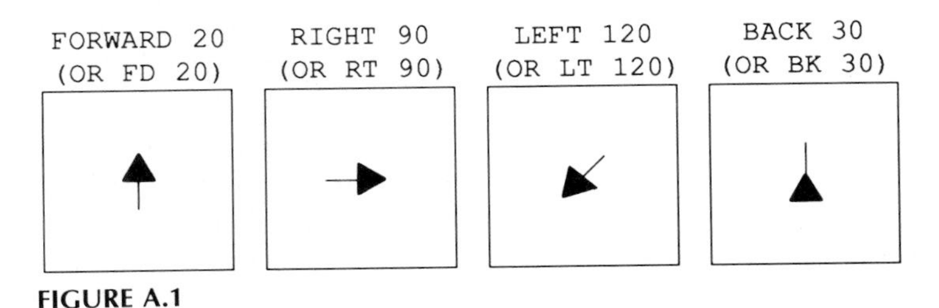

FIGURE A.1

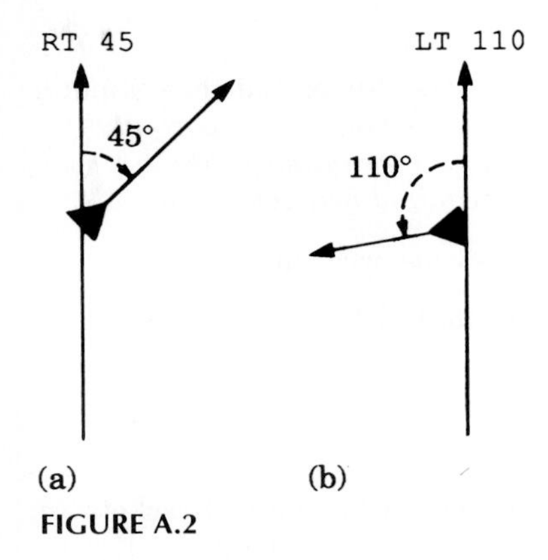

(a) (b)

FIGURE A.2

The RIGHT AND LEFT commands tell the computer to turn the turtle the entered number of degrees. A RIGHT turn (with a positive input) moves the turtle in a clockwise direction the entered number of degrees away from the direction the turtle was heading (Fig. A.2a). The LEFT turn moves the turtle in a counterclockwise direction the entered number of degrees away from the direction the turtle was heading (Fig. A.2b). These commands change only the heading of the turtle—not the position.

The FORWARD and BACK commands change the position of the turtle, but not its heading. Since the turtle draws a line as though a pencil were tied to its tail as it changes position, we need to erase the screen intermittently. As you type, you may make some errors. When you hit RETURN, you will receive an error statement. No harm is done. Just continue when you see the prompt. You can correct errors before you hit RETURN. The ← and → keys move the cursor along the line. When the cursor is over the error, hold down the CONTROL key and hit D (CTRL-D). This will delete the character under the cursor. Control-K will delete the character under the cursor *and* everything to the right of the cursor. To insert a character, just use the arrow keys to move the cursor to the letter following the insertion point. Now hit the key you want, and it will be inserted in the space to the left of the cursor. (This comes in handy when you forget a space between a command and its input!) Try making and correcting some errors. Then try inserting a character somewhere in the line.

LOGO Investigation 2 will help you explore the effects of some of these commands. For example, the turtle can disappear and then reappear on the screen at another point. This is the *wrap mode*. The turtle wraps around the back of the screen and reappears on the other side. If we want to keep the turtle on the screen, the command FENCE will work. If a value is too big to keep the turtle on the screen, an error statement will appear if you are in the *fence mode*. The command WRAP takes the fence down.

You can also type more than one command on a line if you place a space between them. Nothing will be done until you press the RETURN key. The number of characters that you can type before hitting RETURN requires more than one line. If you keep typing characters, LOGO will automatically wrap the information onto the next line. An exclamation point (!) will be placed at the end of the line so you know that the computer put the information on that line and that you did not press RETURN.

LOGO INVESTIGATION 2

A. Move the turtle on the screen to find the dimensions of the screen in turtle steps. Use small inputs at first.
B. What happens when you type each of the following? Clear the screen between the statements.

```
BK -30
RT -45
FD 25.7
BK 28/3
LT 300-120
RT 30 FD 500
LT 120 BK 350
```

Type these on one line before you hit RETURN.

C. Experiment with the commands on your own by creating a design on the screen.

So far, every time the turtle changed position, it drew a line from the original position to the new one. At times you would like to move the turtle without leaving a trail. This can be done by using the command PENUP (PU). This is like picking up your pencil so you can move it to another spot to start writing. If you use the command PENDOWN (PD), the turtle will start leaving a line again. You can also make the turtle disappear by typing HIDETURTLE (HT). The turtle will still move around and draw lines, you just can't see it. ST makes the turtle reappear.

POLYGONS

To draw a polygon, a combination of position and heading changes must be made. If you want to draw an equilateral triangle, you must break it down into the turtle's movements.

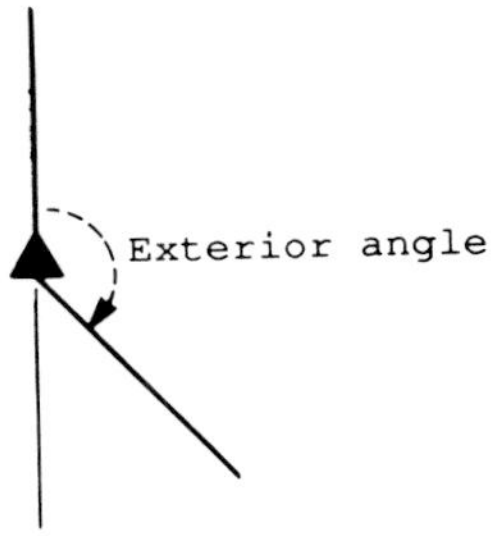

1. First, you need the turtle to draw a side: FD 50.
2. Now the turtle must turn before the next side is drawn. You need to find the measures of the angles to determine the turn. The turn is the measure of the *exterior angle* of the polygon (Fig. A.3). Note the turtle's position and the necessary turn: RT 120.
3. Side two would be drawn next: FD 50.
4. Now, complete the second turn: RT 120.

Exterior angle

FIGURE A.3

LOGO INVESTIGATION 3

A. Modify the combined, final set of commands given in this section so the equilateral triangles shown here are drawn with the turtle starting in the home position each time.
B. Write instructions to draw a square with sides of length 80.
C. Write instructions to draw a regular hexagon with sides of length 20.

5. The third side is now drawn: FD 50.
6. Finally, the last turn is completed: RT 120.

Note that the triangle is drawn without this last turn, but it is always a good idea to return the heading to zero in case you plan to do more drawing.

You will notice that there were three turns of 120° each: $3 \times 120 = 360°$. In order to have a closed figure, the turtle must turn 360° or a multiple of 360°. This is called the *total turtle trip*. As you construct designs, *remember the turns must add to 360°, or a multiple of it, and the distance moved forward must be the same in each direction in order for the design to be closed.* The combined, final set of commands for drawing the equilateral triangle are FD 50 RT 120 FD 50 RT 120 FD 50 RT 120.

In LOGO Investigation 3 you can try out and extend these ideas.

REPEAT STATEMENTS

By now you have noticed that some of the instructions are the same. Blocks of the same instructions that are used in succession can be put into one statement called REPEAT. A REPEAT statement would look like this for our equilateral triangle:

```
REPEAT 3[FD 50 RT 120]
```

This tells the turtle to do the instructions in the square brackets (be sure to use the square bracket [ ] and not the set brackets { }) the number of times specified. Therefore, the turtle will do FD 50 RT 120 three times before going to the next instruction.

Try This. Try this statement to see what happens:

```
REPEAT 50[FD 20 RT 30]
```

(You can stop the turtle at any time with CONTROL-G.)

The REPEAT statement is very useful when instructions are

used more than once. In any design, start your instructions and then look to see if some of it repeats. The instructions in the brackets can be two steps, like the steps for the equilateral triangle, another REPEAT statement, or a series of statements.

To draw a rectangle, you know that all the turns will be 90° and that opposite sides will be equal. Therefore, a REPEAT statement that will draw a rectangle is REPEAT 2[FD 20 RT 90 FD 40 RT 90].

Try This. Try this statement and record what happens:

```
REPEAT 5[REPEAT 4[FD 40 RT 90] RT 72]
```

As you have seen, this is a statement that "spins squares."

THE CONTROL KEY

So far you have seen two screens in LOGO, the textscreen and the splitscreen. At times you want to use the whole graphics screen without the four text lines. The command FULL-SCREEN will hide the text lines. The computer puts your instructions on the textscreen, even though you can't see them. To get back, just type SPLITSCREEN. If you type TEXT-SCREEN when you are in the draw mode, you get the full text-screen. Each of these can be used in a program, as well as in the immediate mode. Each of these screens also has a CON-TROL key:

```
FULLSCREEN CTRL-L
SPLITSCREEN CTRL-S
TEXTSCREEN CTRL-T
```

Try These.

1. Draw a rectangle that is 150 × 100. Try CTRL-L, CTRL-S, and CTRL-T.
2. Draw a design that will fill the screen. Use the commands FULLSCREEN, TEXTSCREEN, and SPLITSCREEN.

CIRCLES

You can use the REPEAT statement for making the polygons you made earlier, and you can also use it to draw circles. In LOGO, a circle is approximated by a polygon with many sides. The more sides, the more circular your figure will look.

LOGO INVESTIGATION 4

A. Try the following.

```
REPEAT 20[FD 2 RT 18]
REPEAT 360[FD 1 RT 1]
REPEAT 75[FD 2 RT 5]
REPEAT 180[FD 1 LT 2]
```

B. As you can see, different-sized circles are made. Now try some combinations of your own. Can you make a small circle? Can you make a large circle?

The key to drawing a circle is the fact that the number of times the instructions are repeated times the angle turned must be 360° so that the turtle makes a complete trip. The size of the angle and the length moved will determine the size of the circle. LOGO Investigation 4 will help you improve your skill in drawing circles.

EXERCISE 1

Use REPEAT statements to produce each of the designs shown in Figure A.4.

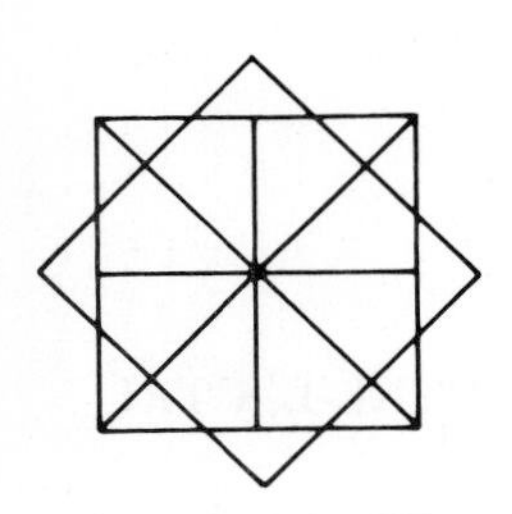

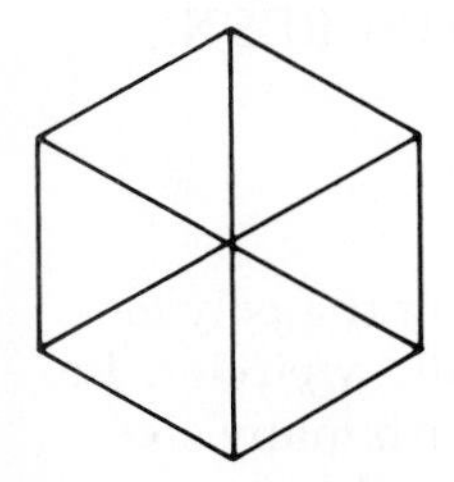

FIGURE A.4

COLOR

If you are not using a color monitor, you may skip this section.

LOGO also allows you to change the background color and the color the turtle uses to draw. When you start, the turtle draws white lines on a black screen. To change these, you use the commands SETPC and SETBG. Each of these commands requires an integer input from zero to five. The following colors are available:

0	Black
1	White
2	Green
3	Violet
4	Orange
5	Blue

The instructions SETPC 5 SETBG 1 would give blue lines on

a white screen. The computer uses the last color values specified until you change the values. Even on black and white, green, or amber screens, you can tell the colors are different even though you cannot see the real colors.

Try These.

1. Draw a green tree on a white background.
2. Draw an orange triangle on a violet background.
3. Draw a black circle on a white background.
4. Create any blue design on a white background.

When you draw a line you do not want, Apple LOGO allows you to erase it without clearing the entire screen. This can be done by retracing the line after typing PENERASE (PE). PU or PD will lift the eraser. PENREVERSE (PX) causes the colors of the points the turtle crosses to reverse—the change depends on the background and pen colors that you are using.

Now you have the basics to make all kinds of designs and geometric figures. The exercises that follow will allow you to explore.

EXERCISE 2

Clear the screen between problems.

1. Draw a regular pentagon and then erase it using PE.
2. Draw a rectangle and then move the turtle around the screen so it intersects the rectangle, its interior, and its exterior using PX.
3. Move the turtle from home to the upper left corner without leaving a trail.
4. Draw a square with sides 50 and another with sides 20.
5. Draw a rectangle that is 40 by 70 and then draw a square with sides 40 on top of it (Fig. A.5).
6. Move the turtle to the left side of the screen leaving no trail. Now, have the turtle draw a dashed line to the right side.
7. Write instructions to draw figures like those in Figure A.6.

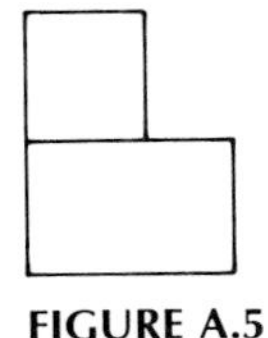

FIGURE A.5

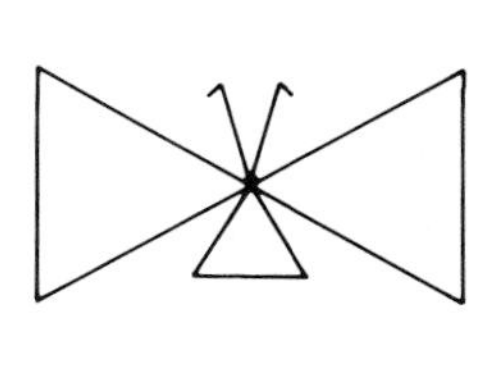

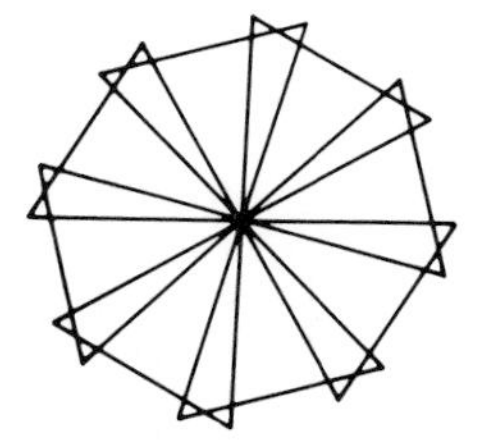

FIGURE A.6

8. Predict the outcome of each of the following. Use the computer to check your answers.

```
REPEAT 5[FD 60 LT 72]
REPEAT 4[FD 30 REPEAT 4[FD 10 RT 90] BK 30 RT 90]
REPEAT 10[FD 5 RT 36]
REPEAT 8[RT 45 REPEAT 4[FD 20 RT 90] FD 40 RT 90]
REPEAT 10[FD 10 PU FD 20 PD]
```

9. Create the following letters: C, I, G, Z, H, N, E, and T.
10. Draw your first name or initials on the screen.
11. Draw 2 circles that are tangent at the turtle's home position.
12. Write a statement using a REPEAT that will draw a nonagon. Use the nonagon in another REPEAT statement that will spin the nonagon and draws it 9 times evenly spaced around the circle.

PROCEDURES

Until now you have been working in the immediate mode. As soon as you type your instructions and press RETURN, the computer performs your commands. This gives you immediate feedback, but you must keep retyping many of the same commands.

Programs in LOGO are called *procedures*. The command TO allows you to define a procedure. This is done by giving a command like TO SQUARE. This tells the computer that what follows is to be saved in memory and is to have the name SQUARE. You can use anything you like for a name as long as it is one word that contains no primitive—like BACK that means something special in LOGO—and is not all numbers. It is usually best to use names that identify the procedure, but these names should not be too long. Here is an example of a procedure:

```
TO SQUARE
REPEAT 4[FD 30 RT 90]
END
```

This procedure defines a square. The END statement tells the computer that the procedure is complete, and it takes you out of the definition mode. Now the computer remembers how TO SQUARE, and it will draw a square with side 30 whenever you type the name SQUARE. The name SQUARE now works like the primitive terms FORWARD and BACK, and the procedure will be performed whenever you type the name.

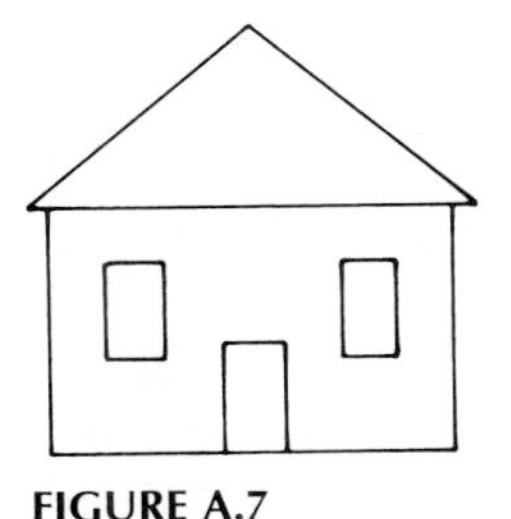

FIGURE A.7

Superprocedures

Procedures make a complicated problem easier to handle since you can break that problem into parts and then write independent procedures for each part. After each piece has been written and tested, you can write one superprocedure that uses all the parts to complete the problem.

For example, you can make a picture of the house in Figure A.7 by writing procedures for BASE, DOOR, ROOF, and WINDOWS. The base of the house is just a rectangle:

```
TO BASE
REPEAT 2[FD 40 RT 90 FD 50 RT 90]
END
```

The roof is an equilateral triangle, but you need to position it correctly:

```
TO ROOF
RT 30
REPEAT 3[FD 50 RT 120]
END
```

The door is also a rectangle:

```
TO DOOR
REPEAT 2[FD 20 RT 90 FD 10 RT 90]
END
```

Finally, you can use two squares for the windows, but you must position them:

```
TO WINDOWS
REPEAT 4[FD 10 RT 90]
PU RT 90 FD 30 LT 90 PD
REPEAT 4[FD 10 RT 90]
END
```

Now the superprocedure TO BUILD can be produced by putting these pieces together. The procedures BASE, ROOF, DOOR, and WINDOW are used, and each is put in the correct place:

```
TO BUILD
BASE FD 40
ROOF LT 30 BK 40 RT 90 FD 20 LT 90
DOOR PU FD 25 LT 90 FD 15 RT 90 PD
WINDOWS
HT
END
```

When you type BUILD, the house will be constructed—first

comes the base, then the roof, followed by the door, and finally, the windows. The parts should be as simple as possible, and the superprocedure should put the parts together. Remember as you write the parts to keep track of the turtle's heading so you can get the parts to fit!

If you make a mistake in a procedure, you can correct it in the edit mode. EDIT (ED) takes procedure names with one set of quotes or a list of procedures in brackets as input and displays the procedures in the edit mode. For example, EDIT"WINDOWS would show the commands making up the procedure WINDOWS in the edit mode. EDIT [DOOR ROOF] would show both procedures.

If no name is given, the last procedure in the edit mode will be displayed, if possible. Corrections and changes can be made to a procedure in the edit mode. The arrow keys CTRL-D and CTRL-K work in this mode as well as the immediate mode. To move from line to line in the edit mode, use CTRL-N for the *next* line and CTRL-P for the *previous* line. The key press CTRL-O will *open* a line wherever the cursor is positioned. You can exit the edit mode with either CTRL-C or CTRL-G. CTRL-C will process the edited procedures, while CTRL-G aborts the action and leaves the procedures in their original form.

Besides making changes in the edit mode, you can also enter procedures. More than one procedure may be typed in the edit mode as long as they are separated by END statements. All the procedures will be defined when you hit CTRL-C.

When you define procedures, they are stored in memory, often called the workspace. To see the contents of this workspace, just type POTS. This means Print Out Titles. To see the listing of a procedure, type PO"NAME or PO[NAME1 NAME2]. The printout command shows a listing. To see the listings of the entire contents of the workspace, type POALL. All the listings will scroll up the screen. The command CTRL-W will stop the scrolling. Using it again will advance the display, while any other key will return the display to its normal form.

As you work with your procedures, you will find that you have improved some and, therefore, would like to take some of the procedures out of memory. The ERASE(ER) command does this. If you have the procedure SQUARE in memory, you can erase it by typing ER"SQUARE. Now when you type SQUARE, you will receive an error statement since SQUARE is no longer in memory.

The command ERALL clears the workspace. Be careful

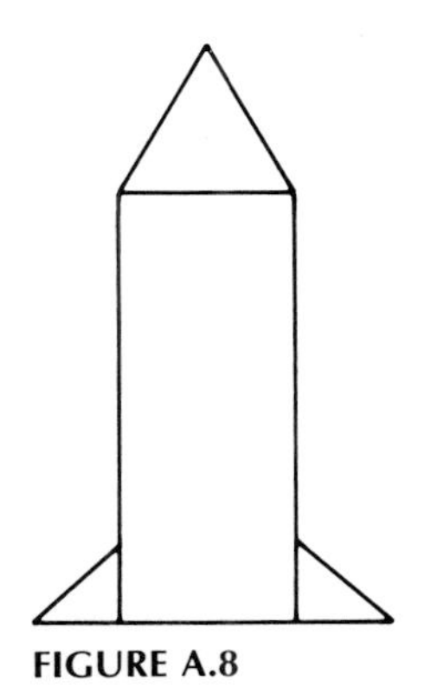

FIGURE A.8

with ERALL, as all your work will disappear with this command. When you have finished a session and have saved your work, you can use it to clear memory for the next user.

EXERCISE 3

1. Write procedures to create the shape shown in Figure A.8.
2. a. Use your set of tangrams to create a design.
   b. Write a procedure to create each of the tangram pieces.
   c. Write a superprocedure to reproduce your design.
3. Write a procedure to draw a rectangle. Practice your editing by doing the following:
   a. Change the width to your value plus 10.
   b. Change the length to your value minus 20.
   c. Open a line and add a REPEAT statement to draw a square on top of the rectangle.

VARIABLES

All the procedures you have created so far draw unique figures. The lengths and turns are fixed unless you edit them. The addition of variables will allow you to change values without editing.

A procedure to draw a fixed rectangle is

```
TO RECT
REPEAT 2[FD 30 RT 90 FD 50 RT 90]
END
```

Try This. Predict what the following commands will draw and check your answer.

```
TO RECT:L:W
REPEAT 2[FD:L RT 90 FD:W RT 90]
END
```

The two values that can vary here are the length and the width. The angle measure must be fixed in order to have a rectangle. The inclusion of variables lets you change the size and appearance. The colon (:) tells the computer that what follows is a variable. Putting them in the title tells the computer that the values will be entered when you call the procedure. The call RECT 30 50 would draw your original rectangle. If you fail to enter values, an error message appears, and the procedure is not processed. When you have more than one variable, the values are assigned to the variables in the order in which they are given. LOGO Investigation 5 provides practice in working with variables in a program.

LOGO INVESTIGATION 5

A. Try the following values in the RECT program. Note what happens with the different inputs. Clear the screen between rectangles.

```
RECT 50 30
RECT 30 50
RECT -30 50
RECT 30(-50)
RECT -30(-50)
```

(For RECT 30(-50): You need the parentheses here or the computer will subtract.)

B. Write and try a TRI procedure for drawing a triangle that uses variables.

EXERCISE 4

1. Write a procedure using a REPEAT statement that will draw an asterisk. Include a variable so you can make them different sizes.
2. Write a superprocedure that uses one variable to pass the value to asterisk that will place an asterisk in each of the 4 corners of the screen.
3. Write a procedure that will draw polygons of various sizes. Use 2 variables—one for the type of polygon and one for the length of a side.
4. Use the RECT procedure that was written with variables to create a superprocedure that will draw a skyline.
5. Use the RECT procedure to spin 6 rectangles. Draw 3 "spinners" of different sizes.
6. Write a procedure that will draw circles, with the input value being the number of units moved forward.
7. Using the procedure that draws the circles, write a superprocedure that draws 3 circles that are tangent at the turtle's home position—they are to be internally tangent.

LOGO FILES

As you work with the procedures, you will want to save them so they can be used again. The procedures in the workspace can be saved on an initialized disk using the command SAVE"NAME. Everything in the workspace at that time will be saved under that name.

To read the file from the disk into the memory of the computer, type LOAD"NAME. The file with that name will then be moved from the disk into the workspace, provided there is room. Reading a file from the disk does not erase the procedures that are already in the workspace. All procedures that were in the file will still be on the disk after they are loaded.

To erase a file from a disk, simply instruct the computer to ERASEFILE"NAME. This will remove the file from the disk so those procedures will no longer be on permanent storage. To see the contents of the disk, ask for CATALOG.

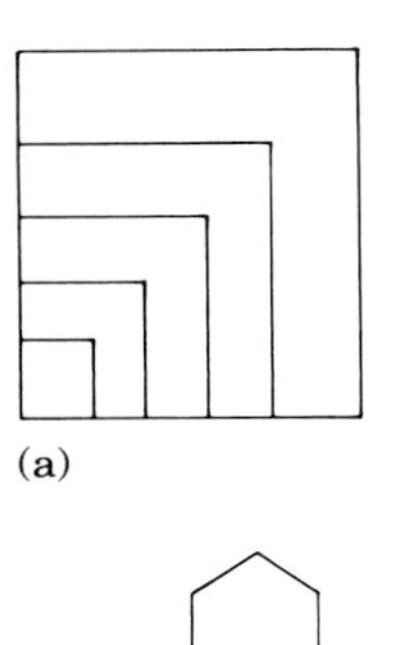

(a)

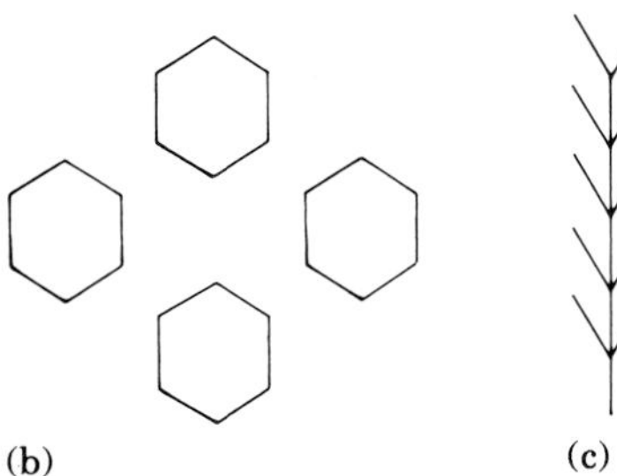

(b)

(c)

FIGURE A.9

EXERCISE 5

1. Write 2 procedures with variables called SQ and TRI. Write another procedure that will put them together to form a house.
2. Write a procedure that uses the house and draws a street with houses of different sizes.
3. Write a procedure that draws a stick person using different polygons and lines.
4. Write a procedure that will draw the numeral 8.
5. Write a procedure that draws a maze. Have the turtle go back and try to travel through it.
6. Write a procedure that will draw coordinate axes and mark off 5 equally spaced points on each axis.
7. Write a procedure that will draw a flag. Write another procedure that draws 8 spinning flags.
8. Write a procedure that draws a snowflake. Remember that a snowflake is symmetric and forms a regular hexagon.
9. Write procedures to draw the figures shown in Figure A.9.

This appendix has provided just an introduction to LOGO. As you work with it, you may find that you would like to explore it in more detail. Check the Bibliography for books about LOGO.

APPENDIX B Constructions

CONSTRUCTION ACTIVITIES

In order to discover relationships in geometry, it is often helpful to draw accurate pictures. The Greek philosopher Plato used only the *compass* and *unmarked straight edge* as the tools for making these pictures (Fig. B.1). To Plato, a *ruler-compass construction* was performed when "an accurate picture of a geometric idea was produced using only the unmarked ruler and compass according to agreements about the use of these tools." This restriction is historically interesting, and some very important ideas of mathematics have been discovered as people have attempted to make certain constructions using only these tools.

For the purposes of investigating relationships in geometry, however, we will not restrict ourselves to the ruler and compass, but will also often use a protractor and other special

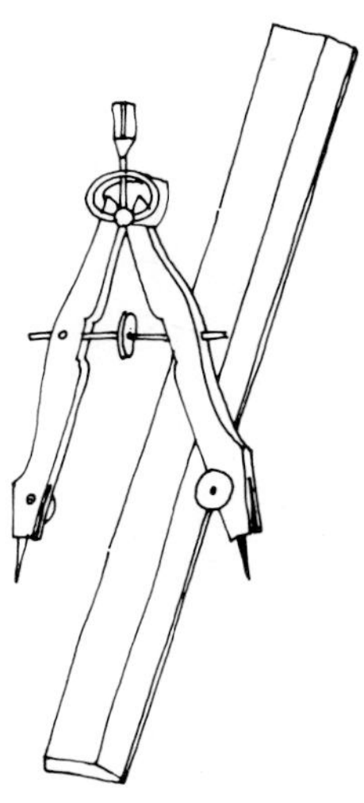

FIGURE B.1

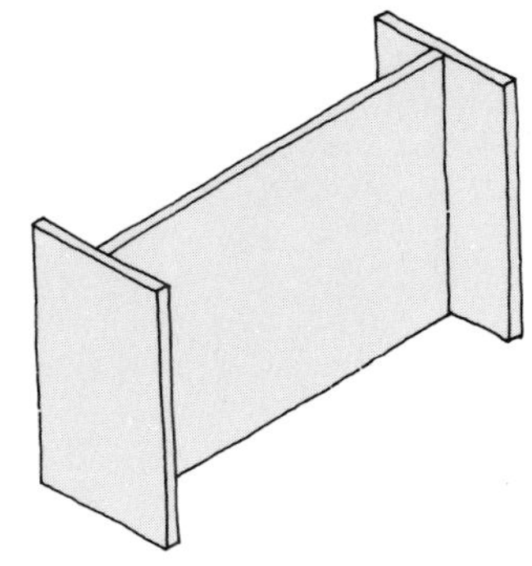

FIGURE B.2

tools. One such tool is called a MIRA and is widely available from producers of commercial aids. The MIRA is made from red plexiglass that gives it the property of a mirror; at the same time, it is transparent (Fig. B.2). The following activities will help you become familiar with the MIRA.

Activity 1

The top of the MIRA is different from the bottom of the MIRA since the bottom has an angled drawing edge. The angled side should always be toward you.

A. Place the MIRA in the correct position and draw a segment along the drawing edge (Fig. B.3). Label it PQ.
B. Then, draw a dot in front of the MIRA about 4 cm from the segment. Label it A.
C. Look through the MIRA. Do you see a reflection of the dot you just drew?
D. Reach around behind the MIRA and mark the image of A on your paper. Label it A'. Draw AA'. Label the point X where AA' intersects PQ.
E. Remove the MIRA. Is AA' perpendicular to the original segment? How do the lengths AX and $A'X$ compare?
F. Would the properties in (E) be true for any point like A that you might draw? Try other points.

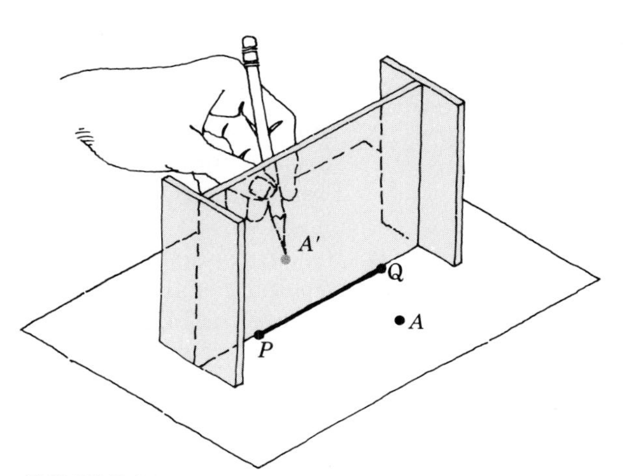

FIGURE B.3

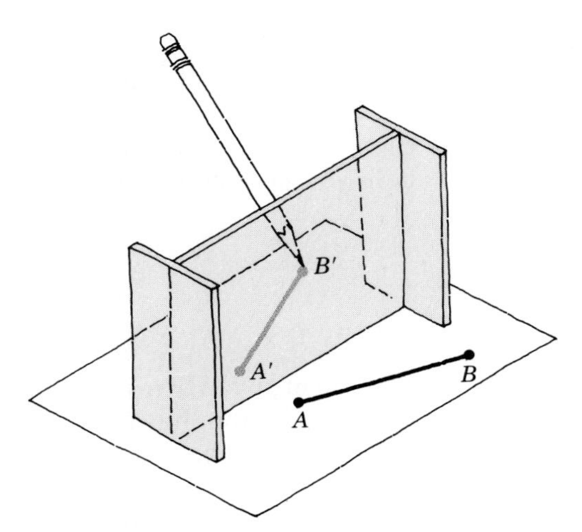

FIGURE B.4

Activity 2

A. Draw a segment AB on the front side of the MIRA (Fig. B.4).
B. Reach around behind the MIRA and draw over the visual image of AB. Label it $A'B'$.
C. Measure the lengths of AB and $A'B'$ to the nearest millimeter. How do the lengths compare?
D. Would your answer in (C) be true for any segment and its image? Try some other segments to be sure.

Activity 3

A. Draw a segment along the drawing edge of the MIRA (be sure the angled edge is toward you). Label it AB (Fig. B.5). From point B, draw a ray BC so that an angle is formed as shown.
B. Reach around behind the MIRA and mark the visual image of point C. Label it C'.
C. Remove the MIRA and draw ray BC'.
D. Use the protractor to measure angles $\angle CBA$ and $\angle C'BA$. How do they compare? Try this with other angles.
E. Draw another angle that does not have one side on the drawing edge of the MIRA. Draw its image. Does the angle and its image have the same measure?

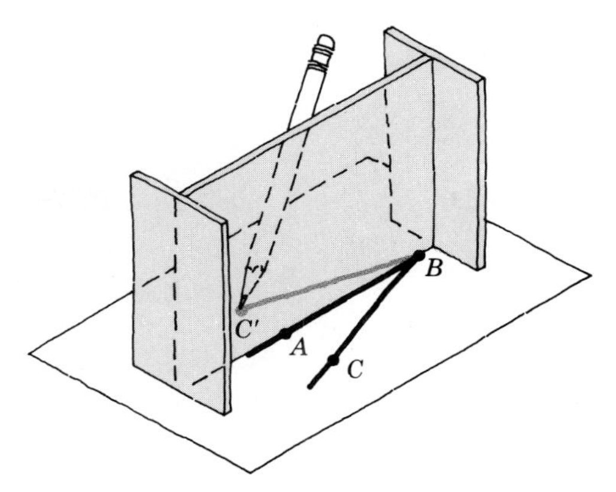

FIGURE B.5

Activity 4

A. Print your name at the bottom of a sheet of paper in front of the MIRA (Fig. B.6). Reach around behind the MIRA and trace the visual image of your name.
B. Cut off your original name so that the visual image remains.
C. Can you place the MIRA in relation to the traced image so that your name reappears?
D. Use this idea to write a "mirror message." Give it to someone to decipher.

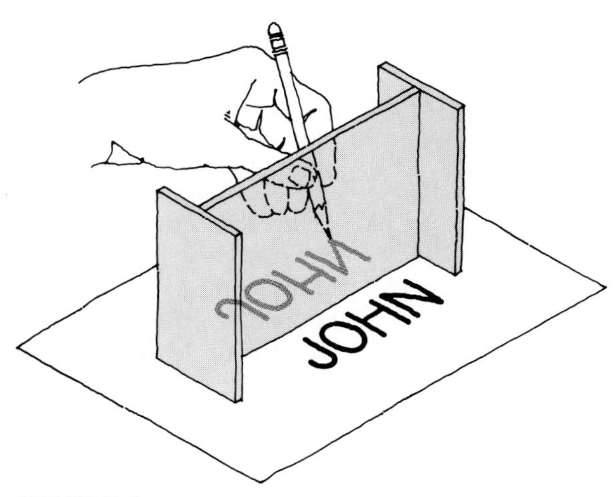

FIGURE B.6

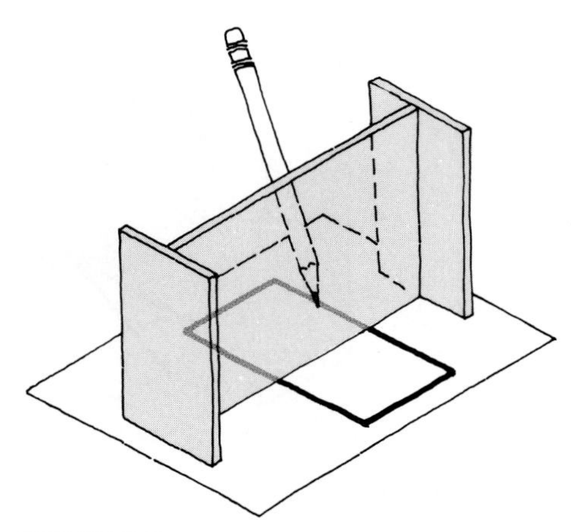

FIGURE B.7

Activity 5

A. Draw three sides of a square in front of the MIRA so that the drawing edge of the MIRA serves as the fourth side. Reach around behind the MIRA and trace the visual image of the square (Fig. B.7).
B. Remove the MIRA. What type of polygon do you see?
C. Can you draw figures in front of the MIRA so that the figure you drew, together with its traced image, forms the following?

1. A square
2. A heart
3. An isosceles triangle
4. An equilateral triangle
5. A trapezoid
6. A regular pentagon
7. A regular hexagon
8. A rhombus
9. An interesting figure

Activity 6

A. Trace the figure in Figure B.8 in the center of a sheet of paper.
B. Move your MIRA away from the figure (Fig. B.9). What happens to the image?
C. Move your MIRA toward the figure. How does the image move?
D. Place the MIRA near the figure and rotate the MIRA clockwise (counterclockwise). How does the image move?
E. Can you place the MIRA so that a familiar figure is formed by the figure and its image?

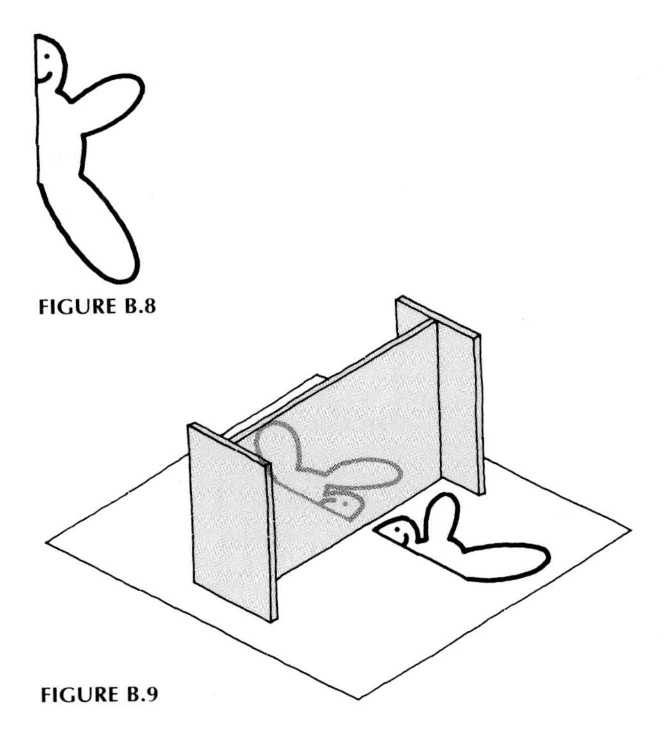

FIGURE B.8

FIGURE B.9

CONSTRUCTIONS

The following pages describe how to use the *ruler*, the *compass*, and a *MIRA* to perform the following basic constructions:

1. Copy a circle.
2. Copy a line segment.
3. Copy an angle.
4. Copy a triangle.
5. Construct the perpendicular bisector of a line segment.
6. Bisect an angle.
7. Given a line l and a point P on l, construct a line through P and perpendicular to l.
8. Given a line l and a point P not on l, construct a line through P and perpendicular to l.
9. Given a line l and a point P not on l, construct a line through P and parallel to l.
10. Construct an equilateral triangle, given the length of one side.

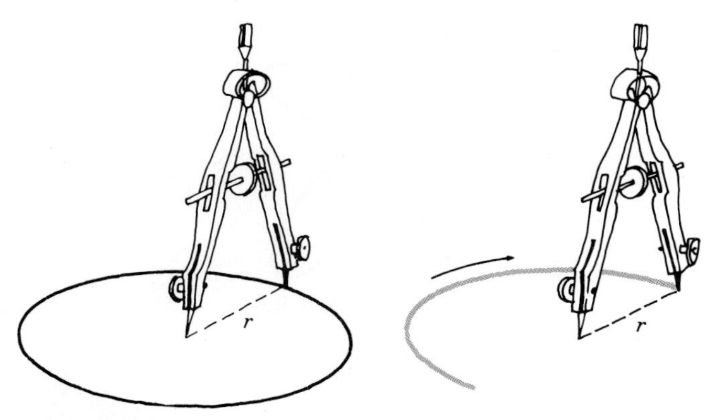

FIGURE B.10

Construction 1

Copy a circle.

A. With a compass (Fig. B.10):

1. Measure the radius of the given circle with the compass.
2. Use this radius to draw a copy of the given circle.

B. With a MIRA* (Fig. B.11):

1. Trace the visual image of the circle.

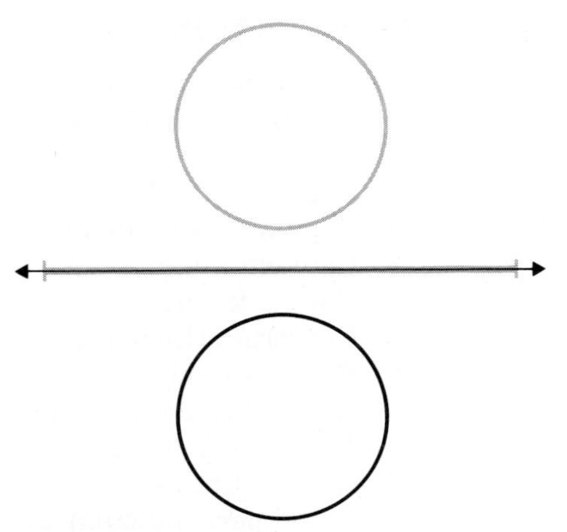

FIGURE B.11

*Note that a MIRA is not very useful for drawing circles. An image of a given circle can be traced freehand if a perfect circle is not required.

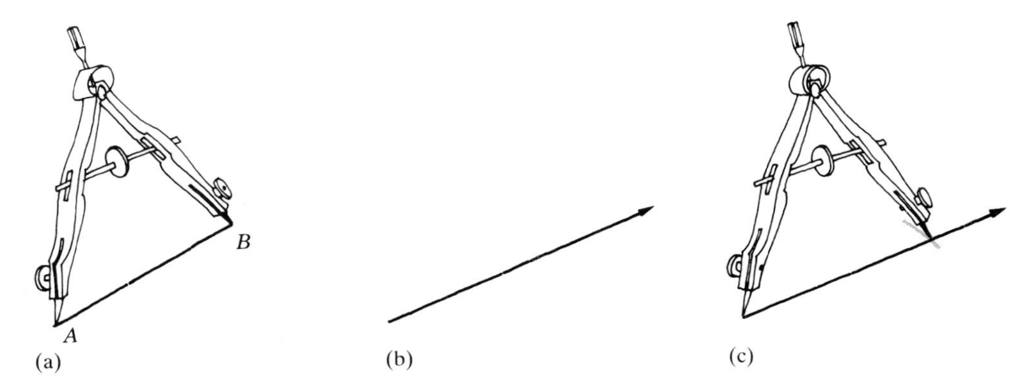

FIGURE B.12

Construction 2

Copy a line segment.

A. With a compass and a ruler (Fig. B.12):

1. Open your compass the length of the given segment (Fig. B.12a).
2. Draw a ray longer than the given segment (Fig. B.12b).
3. Use the same compass to mark a copy of the segment on the ray (Fig. B.12c).

B. With a MIRA* (Fig. B.13):

1. Using the visual image of the segment AB, mark points A' and B'.
2. Use the drawing edge of the MIRA to draw the segment $A'B'$.

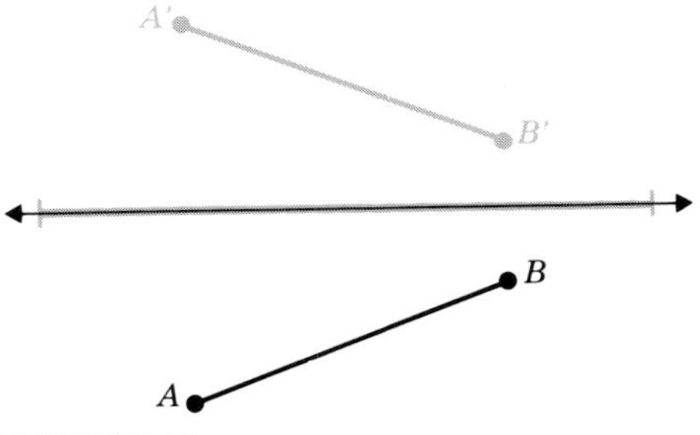

FIGURE B.13

*The possible positions for the copy of a segment, an angle, and a triangle are restricted when using the MIRA. A figure and its copy must be the mirror image of each other unless multiple reflections are used.

Construction 3

Copy an angle.

A. With ruler and compass (Fig. B.14):

1. Draw an arc intersecting both rays of the given angle (Fig. B.14a).
2. Draw a ray to serve as one side of the copy (Fig. B.14b).
3. With the same compass opening as in (1), draw an arc crossing the ray (Fig. B.14c).
4. Open the compass to measure the opening of the given angle (Fig. B.14d).

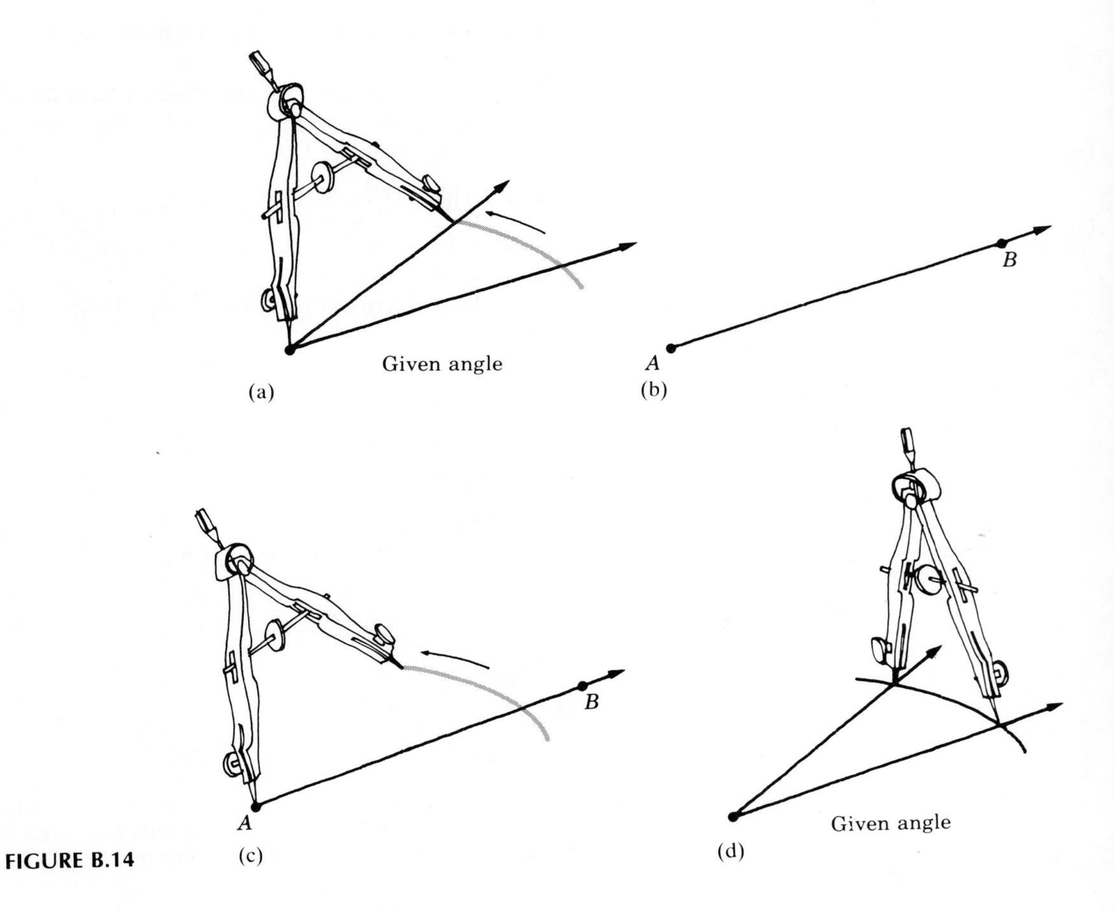

FIGURE B.14

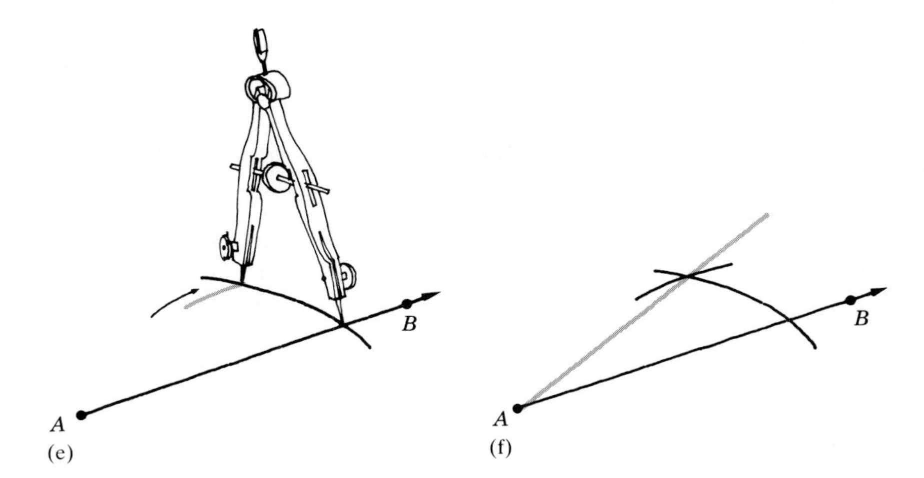

FIGURE B.14 *(continued)*

5. Use the same opening as in step 4 and draw an arc (Fig. B.14e).
6. Draw the second side to complete the copy of the given angle (Fig. B.14f).

B. With a MIRA (Fig. B.15):

1. Using the visual image of $\angle ABC$, mark points A', B', and C'.
2. Use the drawing edge of the MIRA to complete $\angle A'B'C'$.

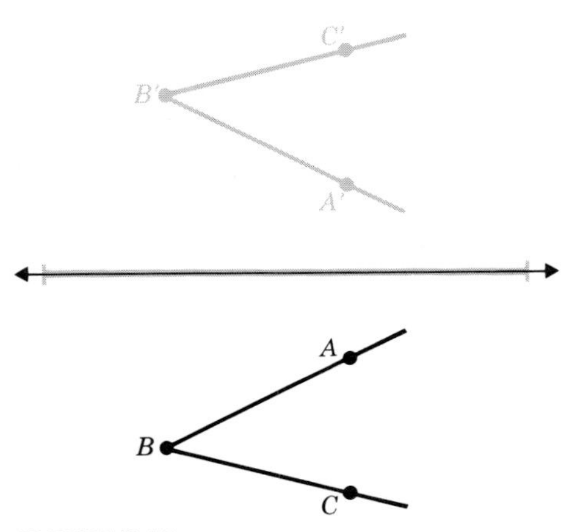

FIGURE B.15

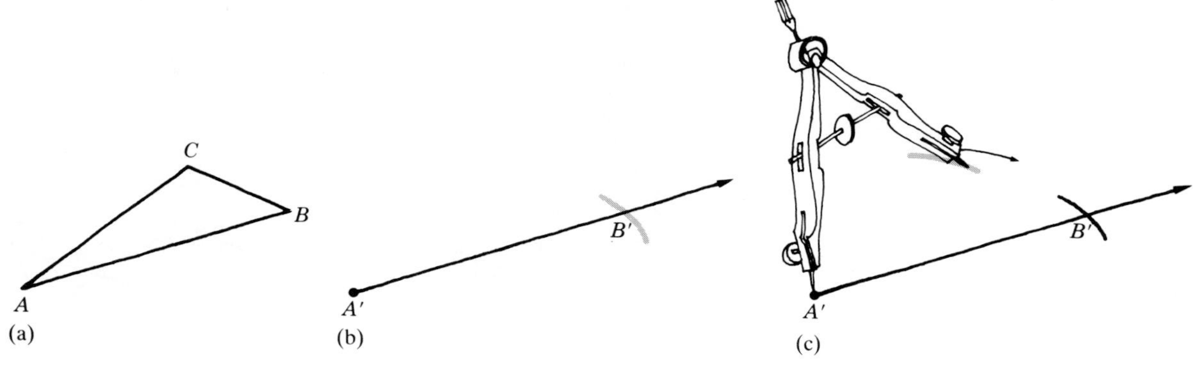

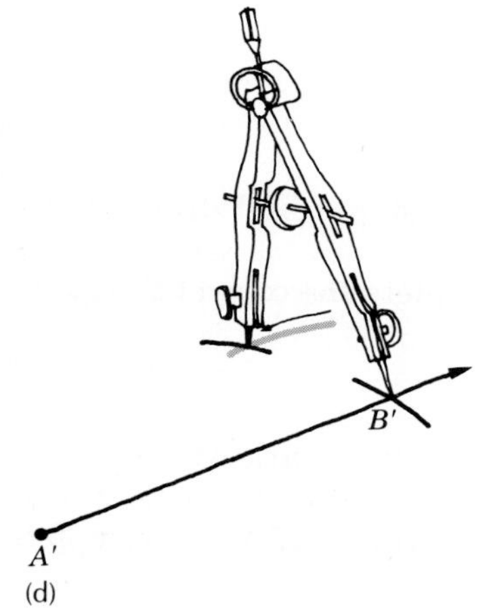

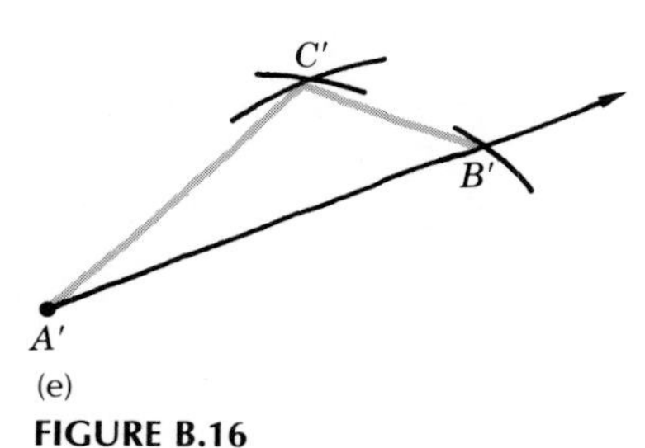

FIGURE B.16

Construction 4

Copy a triangle.

A. Using a ruler and compass (Fig. B.16):

1. Given a triangle ABC (Fig. B.16a), draw a ray and copy segment AB of the triangle (Fig. B.16b).
2. Draw an arc with center A and compass opening AC (Fig. B.16c).
3. Draw an arc with center B and compass opening BC (Fig. B.16d).
4. Use the ruler to complete the drawing of the triangle (Fig. B.16e).

B. Using a MIRA (Fig. B.17):

1. Using the visual image of $\triangle ABC$, mark points A', B', and C'.
2. Use the drawing edge of the MIRA to complete the triangle $A'B'C'$. Note that a copy of $A'B'C'$ will be in the same relative position as $\triangle ABC$.

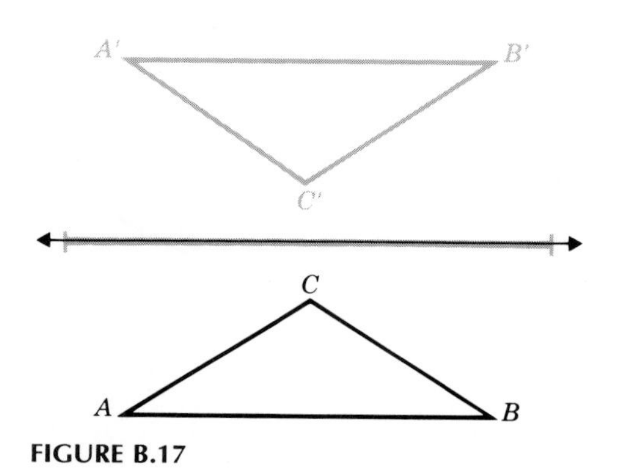

FIGURE B.17

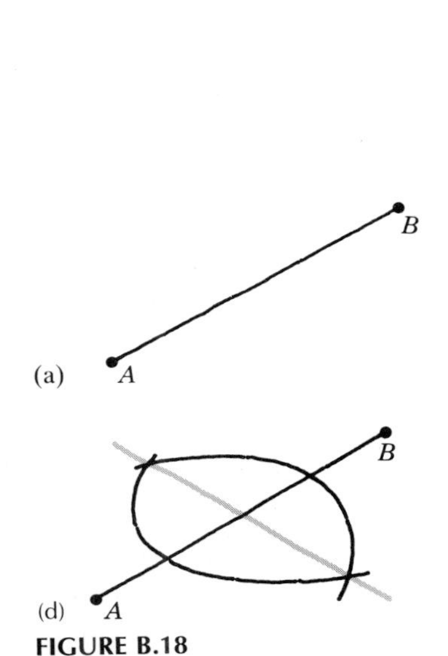

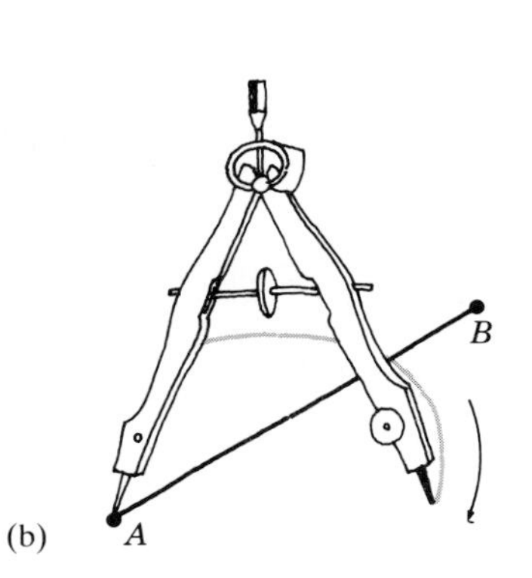

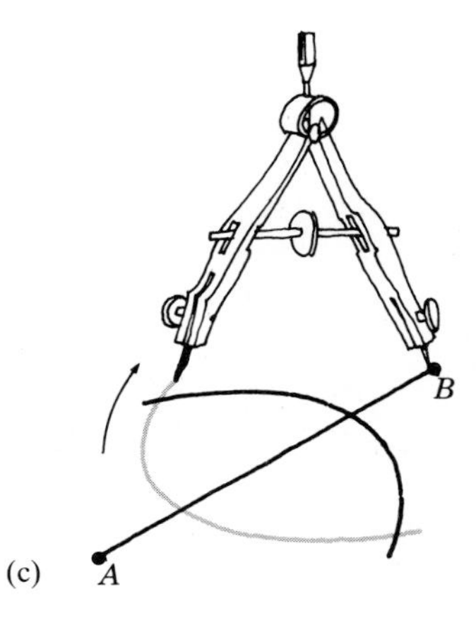

FIGURE B.18

Construction 5

Construct the perpendicular bisector of a line segment.

A. Using a ruler and compass (Fig. B.18):

1. Given line segment AB (Fig. B.18a) and with A as center and a compass opening greater than half of AB, draw a semicircular arc (Fig. B.18b).
2. With B as center and the same opening as in step 1, draw a semicircular arc that intersects the first arc (Fig. B.18c).
3. Connect the two points of intersection to complete the construction of the perpendicular bisector of AB (Fig. B.18d).

B. Using a MIRA (Fig. B.19):

1. Place the MIRA on segment AB in such a way that the visual image of point A is on point B.
2. Draw along the drawing edge of the MIRA to produce the perpendicular bisector.

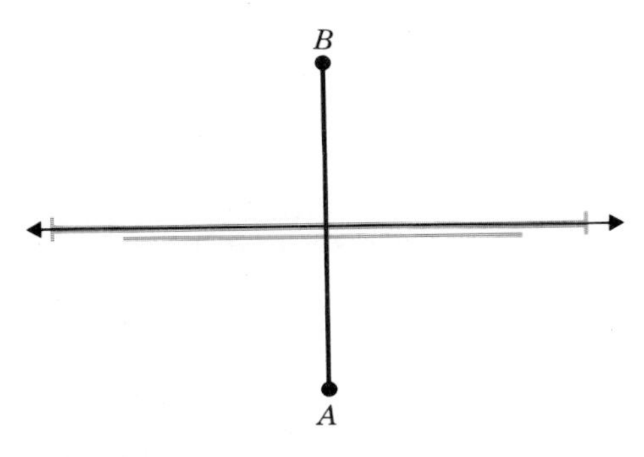

FIGURE B.19

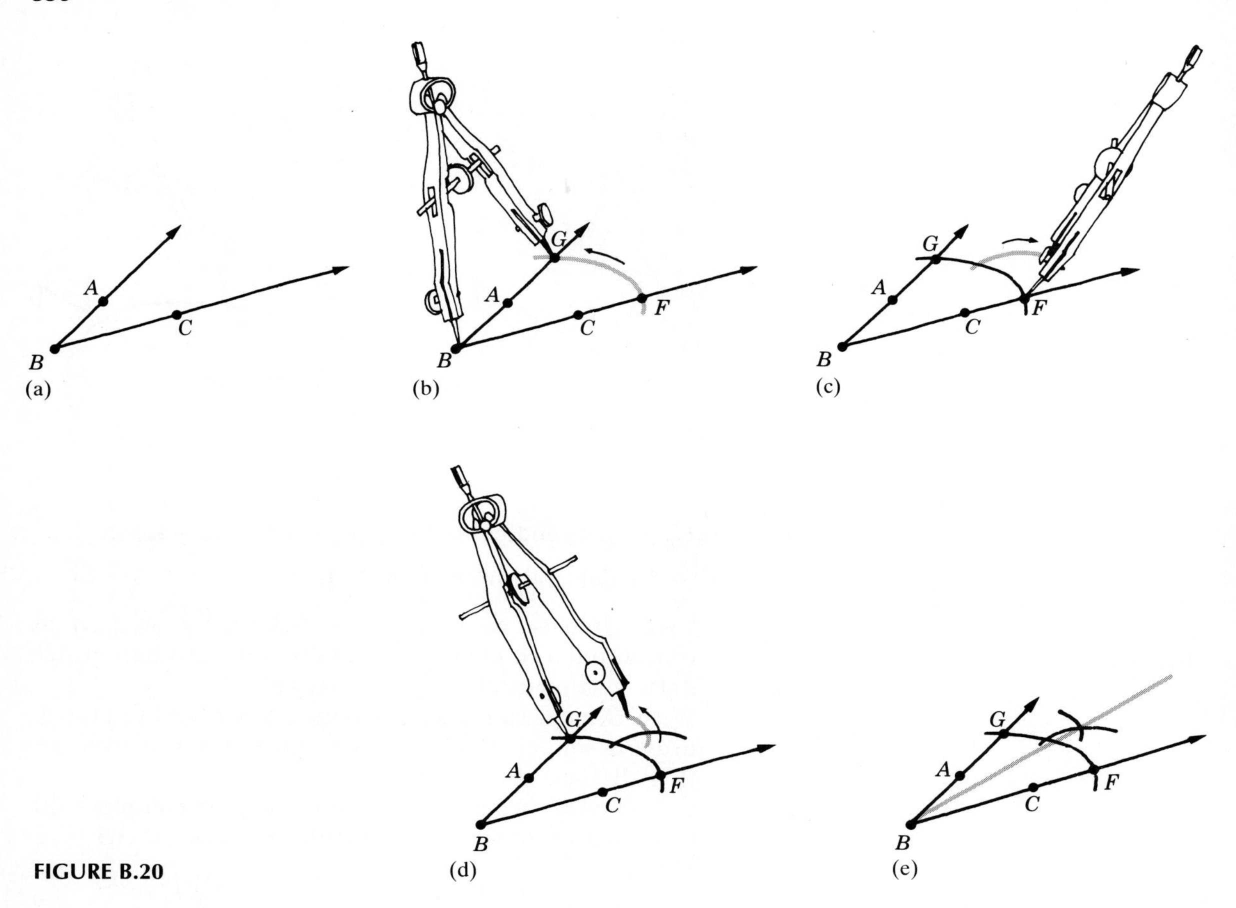

FIGURE B.20

Construction 6

Bisect an angle.

A. Using a ruler and compass (Fig. B.20):

1. Given angle *ABC* (Fig. B.20a) and with *B* as center, draw an arc that intersects both sides of the angle, at *F* and *G* (Fig. B.20b).
2. With *F* as center, draw an arc in the interior of the angle (Fig. B.20c).
3. With *G* as center, and the same opening as in step 2, draw an arc that crosses the first arc (Fig. B.20d).
4. Connect *B* and the point of arc intersection to produce the bisector of the angle (Fig. B.20e).

B. Using a MIRA (Fig. B.21):

1. Place the MIRA with point *B* on the drawing edge in

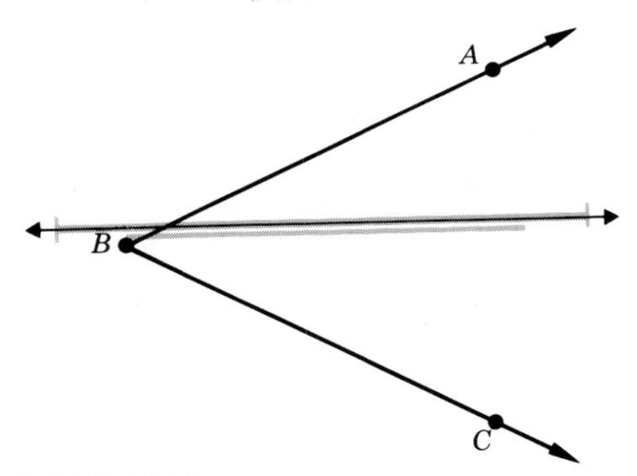

FIGURE B.21

such a way that the visual image of ray BC is on ray BA.

2. Draw along the drawing edge of the MIRA to produce the angle bisector.

Construction 7

Construct a perpendicular to a line through a given point on the line.

A. Using a ruler and compass (Fig. B.22):

1. Given a line l and a point P on l (Fig. B.22a), draw arcs on each side of P (Fig. B.22b).
2. Draw crossing arcs above line l (Fig. B.22c).
3. Draw the perpendicular bisector of line l through P (Fig. B.22d).

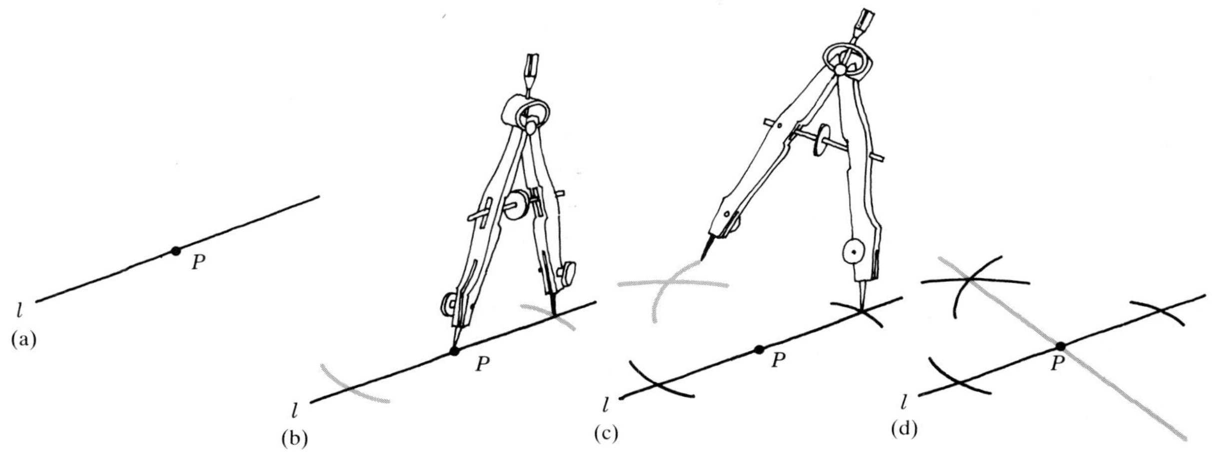

FIGURE B.22

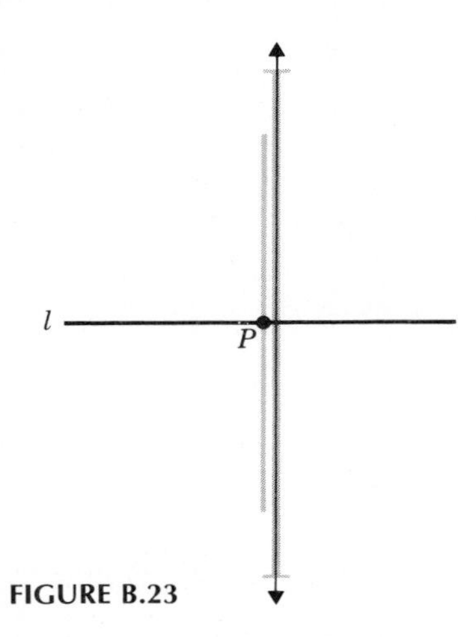

FIGURE B.23

B. With a MIRA (Fig. B.23):

1. Place the MIRA with the drawing edge next to P in such a way that the visual image of half of the line is exactly on the other half of the line.
2. Draw along the drawing edge to produce the perpendicular to line l through point P.

Construction 8

Construct a perpendicular to a line through a given point not on the line.

A. Using a ruler and compass (Fig. B.24):

1. Given a line l and a point P not on l (Fig. B.24a), draw two arcs cutting line l (Fig. B.24b).
2. Draw two crossing arcs below line l (Fig. B.24c).
3. Draw the perpendicular bisector of line l through P (Fig. B.24d).

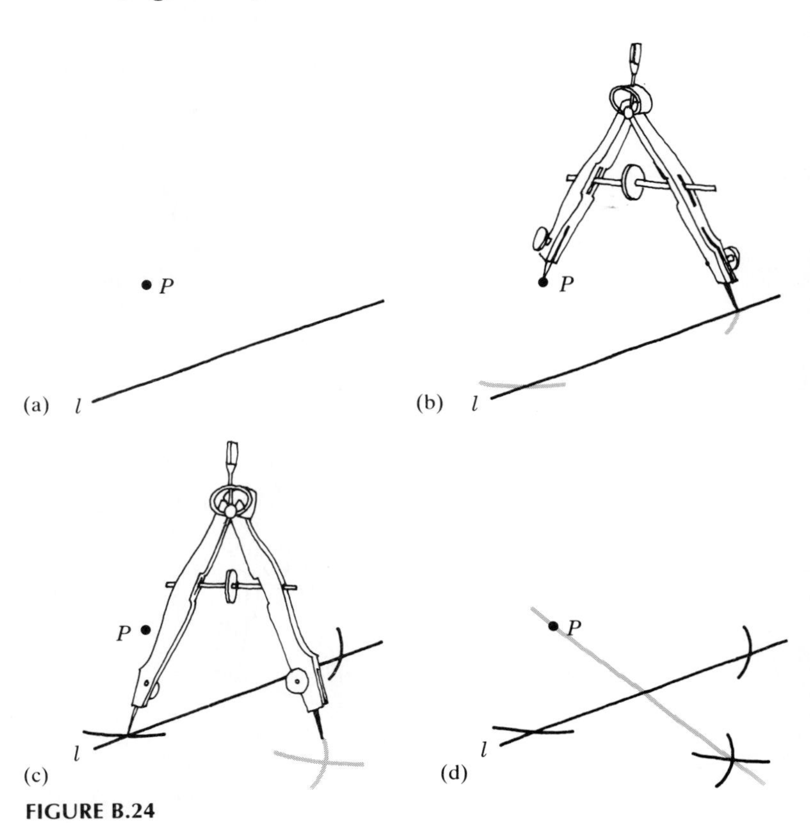

FIGURE B.24

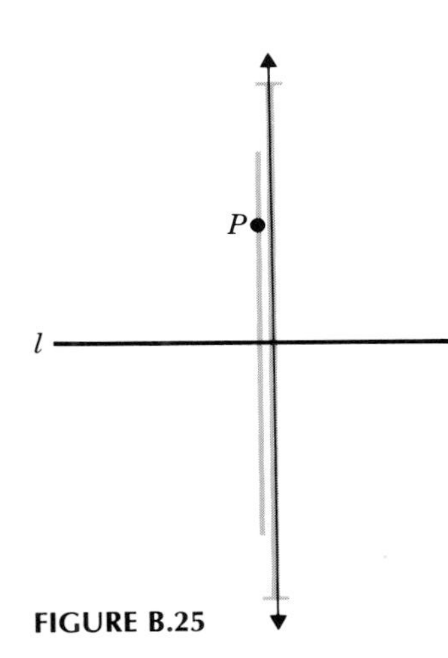

FIGURE B.25

B. With a MIRA (Fig. B.25):

1. Place the MIRA with the drawing edge next to P in such a way that the visual image of the half of line l in front of the MIRA is exactly on the half of line l that is behind the MIRA.
2. Draw along the drawing edge to produce the perpendicular to line l through point P.

Construction 9

Construct a parallel to a line through a point not on the line.

A. Using a ruler and compass (Fig. B.26):

1. Given a line l and a point P not on l (Fig. B.26a), with P as center, draw an arc that crosses line l at point A (Fig. B.26b).
2. With A as center and the same compass opening, draw an arc that crosses line l at point B (Fig. B.26c).
3. With B as center and the same compass opening, draw an arc above line l (Fig. B.26d).

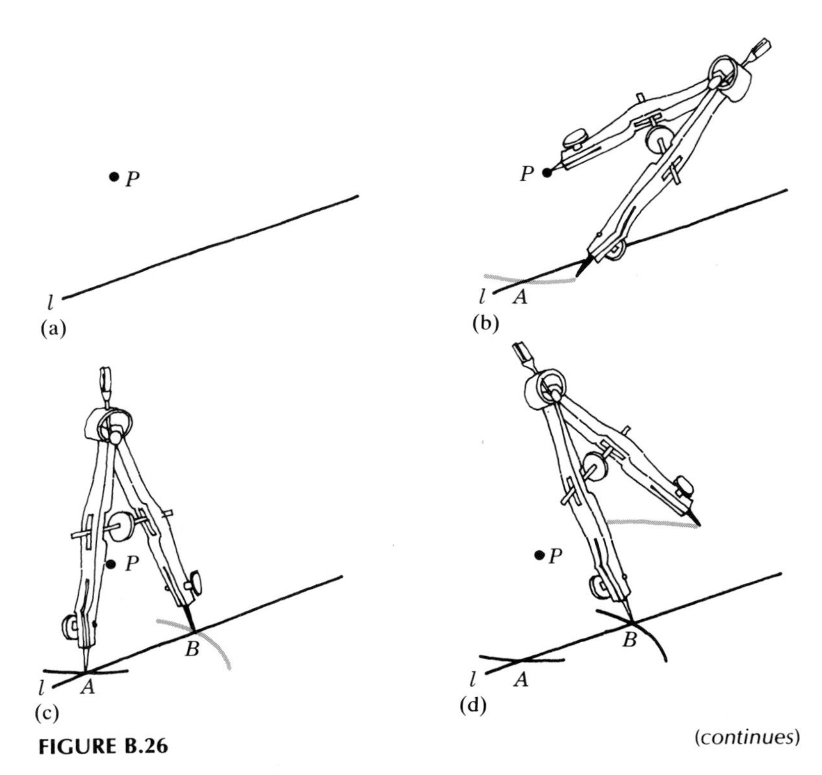

FIGURE B.26

(continues)

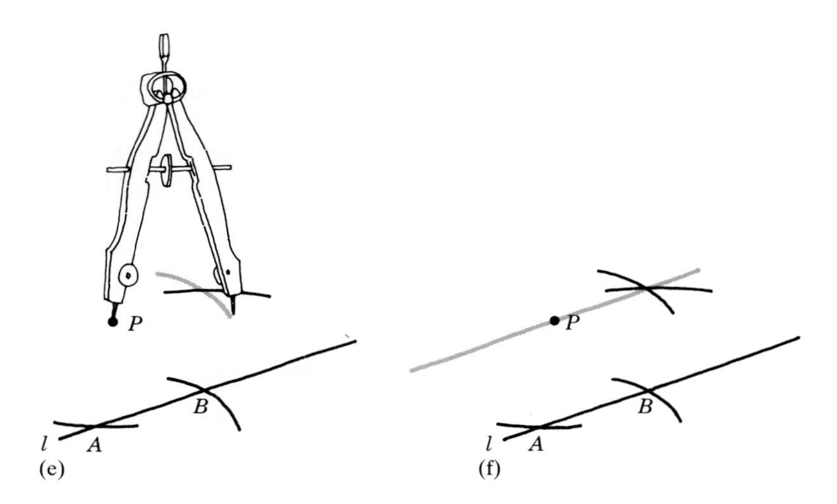

FIGURE B.26 *(continued)*

4. With P as center and the same compass opening, draw an arc crossing the arc you drew in (4) (Fig. B.26e).
5. Draw the line through P and the intersection of the arcs. This line is parallel to line l (Fig. B.26f).

B. With a MIRA (Fig. B.27):

1. Place one end of the MIRA across line l in such a way that the visual image of l falls exactly on l.
2. Then, mark the image P' of point P. When PP' is drawn, it will be parallel to line l.

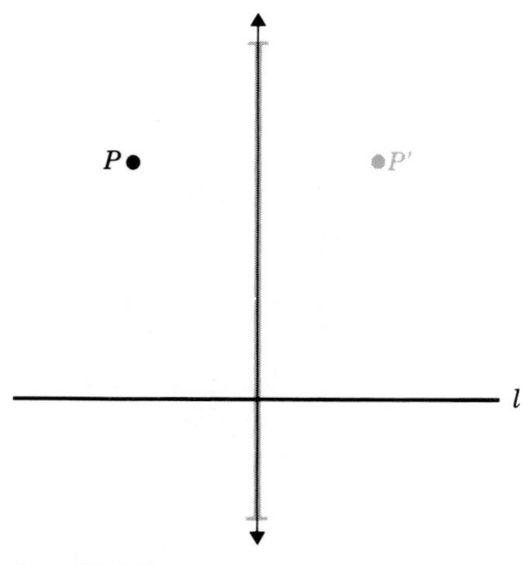

FIGURE B.27

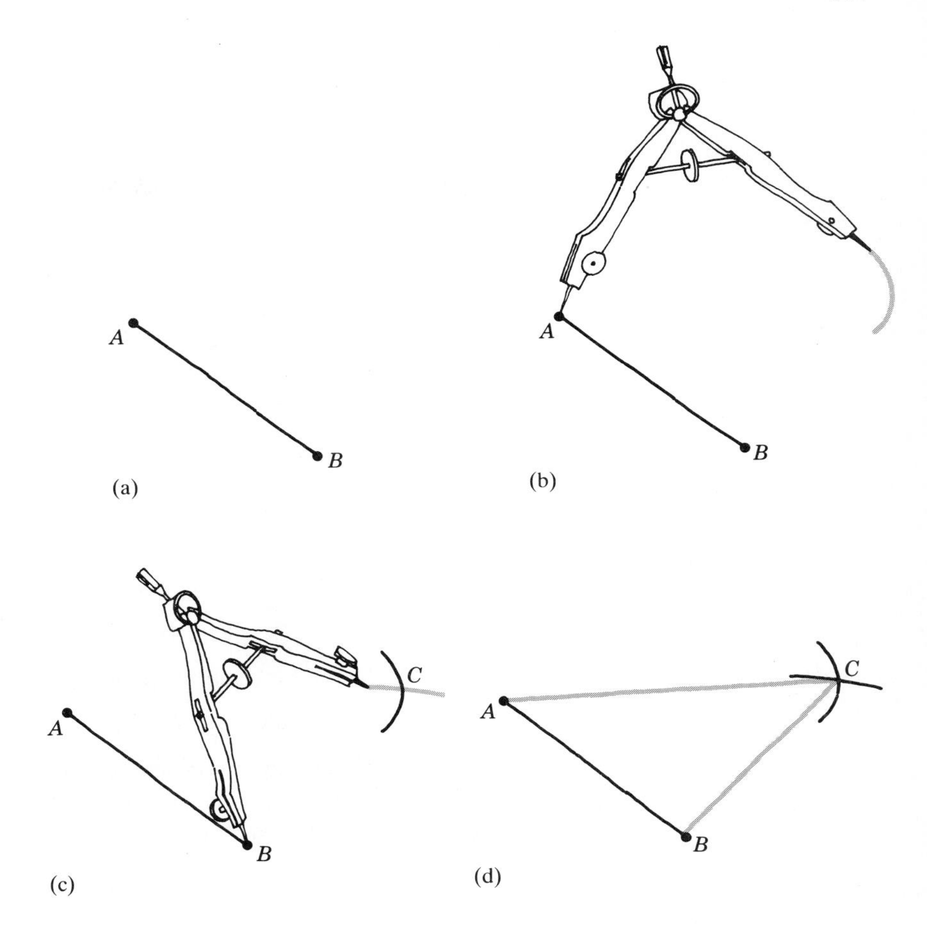

FIGURE B.28

Construction 10

Construct an equilateral triangle, given the length of one side.

A. Using a ruler and compass (Fig. B.28):

1. Given side AB of an equilateral triangle (Fig. B.28a), with A as center and compass opening AB, draw an arc centered above segment AB (Fig. B.28b).
2. With B as center and compass opening BA, draw an arc that crosses the arc you drew in (2) at point C (Fig. B.28c).
3. Draw sides AC and BC to complete the construction of the equilateral triangle (Fig. B.28d).

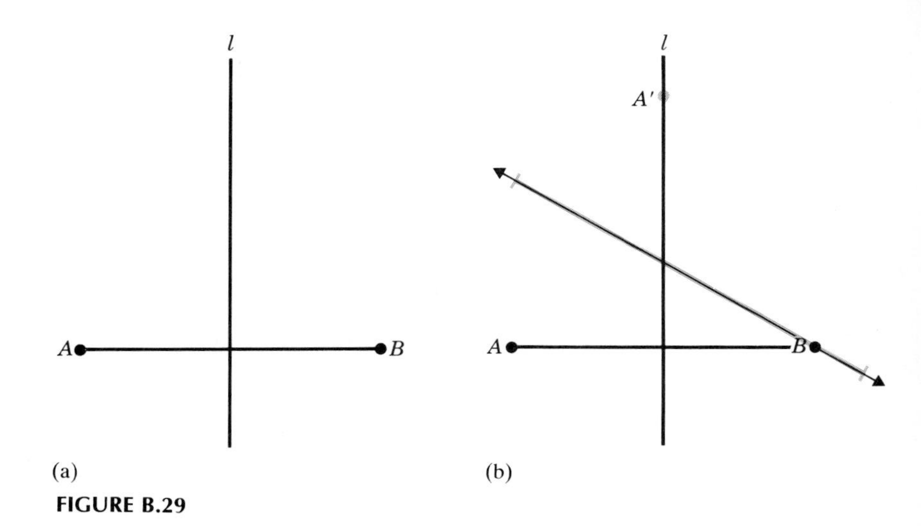

FIGURE B.29

B. With a MIRA (Fig. B.29):

1. Construct the perpendicular bisector l of side AB (Fig. B.29a).
2. Place the drawing edge of the MIRA next to B in such a way that the visual image of A is on l. Mark this image A' and connect AA' and BA' to complete the equilateral triangle (Fig. B.29b).

Other more detailed constructions can be performed using combinations of the 10 basic constructions just described. Here are a few that you might like to try.

1. Given any three points, construct a circle passing through them. *Hint*: The perpendicular bisector of a chord of a circle goes through the center of the circle. Use Construction 5 twice to find the center of the circle.
2. Construct a tangent to a circle at a given point on the circle. *Hint*: A tangent to a circle is perpendicular to the radius of the circle at the point of tangency. Draw a radius, extend it, and use Construction 7.
3. Construct a rhombus, given the length of each of the diagonals of the rhombus. *Hint*: The diagonals of a rhombus are perpendicular. Constructions 2 and 5 should help.

SUGGESTED READINGS

Consult the following sources listed in the Bibliography for activities and ideas about MIRA constructions: Gillespie, 1973, and *MIRA Math* for the Elementary School, 1973.

APPENDIX C Metric Units

SUPPLEMENTARY ACTIVITIES: LENGTH

Activity 1: Body Representation of Basic Units

Materials. Meter stick, centimeter ruler

1. Record your estimates of whether the following body measurements are more or less than the basic metric units given:
   a. Width of fingernail of little finger—1 cm
   b. Length of fingernail on pointer finger (from base to white part)—1 cm
   c. Width of palm plus width of thumb—1 dm
   d. Base of hand to base of ring finger—1 dm
   e. Distance from left side of head to fingertip of outstretched right hand—1 m
   f. Distance from waist to the floor—1 m
2. Make and record the actual measurements; then record the difference between your estimate and the actual estimate.
3. Find at least one other body measurement that is close to each of these basic units.

Activity 2: Body Lengths

Materials. Metric measuring tape, meter stick

1. Record your estimates of the following body lengths to the nearest unit indicated:
   a. Width of thumb—in mm
   b. Length of middle finger—in cm
   c. Length from wrist to fingertip—in cm
   d. Length of span (maximum length from tip of thumb to tip of little finger)—in cm
   e. Length of cubit (elbow to fingertip)—in dm
   f. Height—in m*
   g. Armspan (fingertip to fingertip, arms outstretched)—in m*

*Use a decimal. Note that 1.73 m is 1 m and 73 cm.

2. Measure and record the actual length; then record the difference between your estimate and the actual measurement.
3. Calculate the *percent of error* of your estimate. Percent of error = difference ÷ actual length.

Activity 3: Size Measurements

Materials. Metric tape or string, metric ruler

1. Record your estimates of the following sizes to the nearest centimeter:
   a. Ring size
   b. Wristwatch size
   c. Hat size
   d. Collar size
   e. Belt size
   f. Shoe size
2. Measure and record the actual size; then record the difference between your estimate and the actual measurement.

Activity 4: Estimating Lengths of Objects

Materials. Metric ruler, meter stick

1. Record your estimates of the length of the following objects to the nearest unit indicated:
   a. Piece of chalk (width)—in mm
   b. Width of penny—in mm
   c. Piece of chalk (length)—in cm
   d. Length of this book—in cm
   e. Width of door—in dm
   f. Width of desk—in dm
   g. Length of chalkboard—in m
   h. Height of door—in m
2. Measure and record the actual length, using decimals if appropriate; then record the difference between your estimate and the actual measurement.
3. Practice further estimation of length by choosing certain numbers of centimeters, decimeters, and meters and cutting strings that you estimate to have the chosen length.

Activity 5: Metric Estimation Golf

Materials. Meter stick or ruler

1. Play this game with a partner: Choose 9 objects in the room and make a "scorecard" to record your estimates of the objects' lengths, as shown here. Then measure the objects and find your "score" (the difference between the estimate and the measure) for the "nine holes."

	Object				
	1	2	. . .	9	Total
Player A Estimate					
Actual					
Score					
Player B Estimate					
Actual					
Score					

2. Record the estimates and actual measurements for each player and each object.
3. For each player, find the "score" (the difference between the estimate and the measure) for the nine "holes." Lowest score wins.

Activity 6: Length-of-Pace Estimates

Materials. Meter trundle wheel or metric tape measure

1. Estimate the number of paces it will take you to step off 10 m. Check your estimate by measuring. Pace off the measured 10 m 5 times. Find the average number of paces required and divide by 10 to find the approximate length of your normal pace. Now for each of the following items, record your visual estimate and your calculated pace length measurement.
   a. Length of room
   b. Width of room
   c. Length of longest hallway in building
   d. Outside length of building
   e. Outside width of building

2. Make and record actual measurements; then calculate your percent of pacing error. Percent of pacing error = (actual measure − pace estimate) ÷ actual measure.
3. Estimate some other distances by pacing. Can you reduce your percent of pacing error?

Activity 7: How Much Time for a Kilometer Walk?

Materials. Stopwatch (or regular watch with second hand)

1. A kilometer is 1000 m. Measure off a 25 m distance. (Use a shorter distance if this is not convenient.) Using your normal pace, find how long it takes for you to walk this distance.
2. Record your time for each of four trials and find your average time to walk the distance.
3. Estimate how long it would take you to walk 1 km.

Activity 8: Estimating Longer Distances

Materials. Globe, piece of string, metric ruler

1. Using a globe and a piece of string, choose a scale so that 1 cm on the string is so many kilometers on the globe. (Note that it is approximately 40,000 km around the earth at the equator.) Use your string to estimate the shortest distance between the following cities:
   a. New York to London
   b. Montreal to Paris
   c. San Francisco to Honolulu
   d. Sydney to Rome
   e. Manila to Lisbon
   f. Chicago to Hong Kong
   g. Cairo to Tokyo
   h. Moscow to Washington, D.C.
2. Record your string estimates and the difference between your estimates and the actual distances.*
3. Try this activity using a map of your state or country.

*a. 5532 km; b. 5475 km; c. 3829 km; d. 16,218 km; e. 12,074 km; f. 12,469 km; g. 9496 km; h. 7773 km.

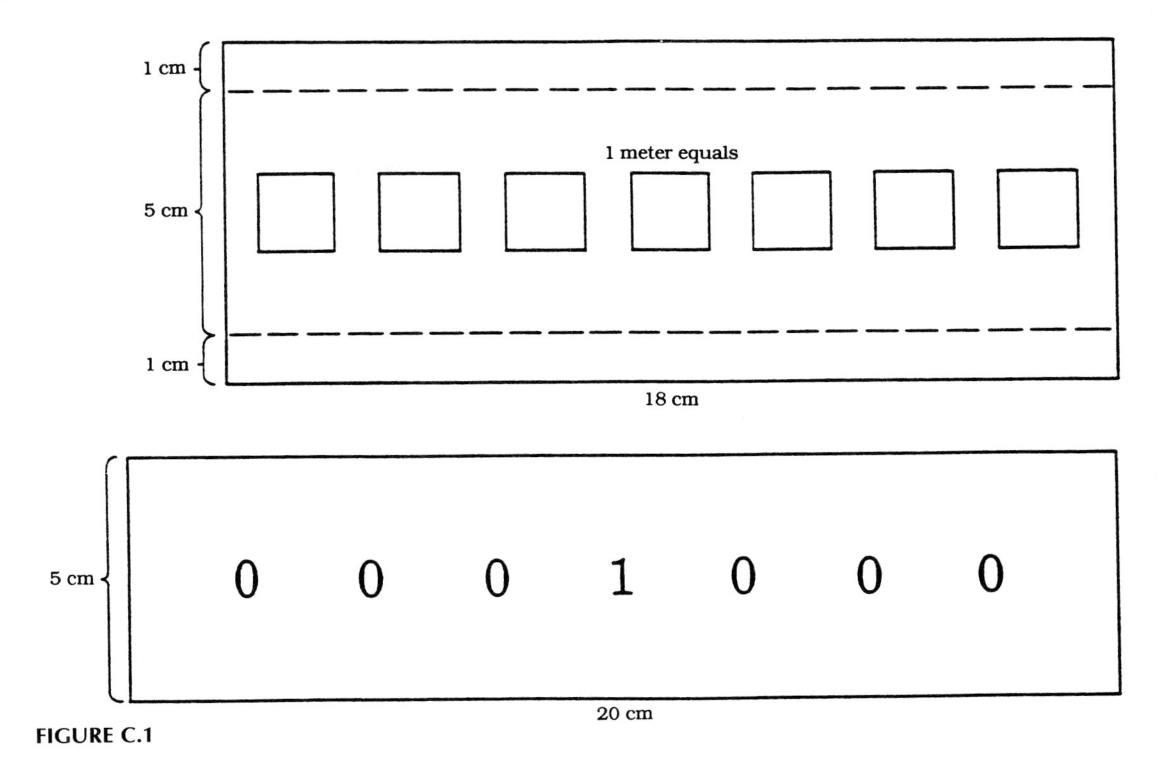

FIGURE C.1

Activity 9: Making a Unit Relationship "Slide Rule"

Materials. Scissors, ruler, paper

Make a model of the pieces in Figure C.1 using the dimensions given. Construct a "slide rule" to show the relationship between the metric units of length. The completed rule looks like Figure C.2. Can you make such a "slide rule" to show the relationships between metric units of area? Of volume?

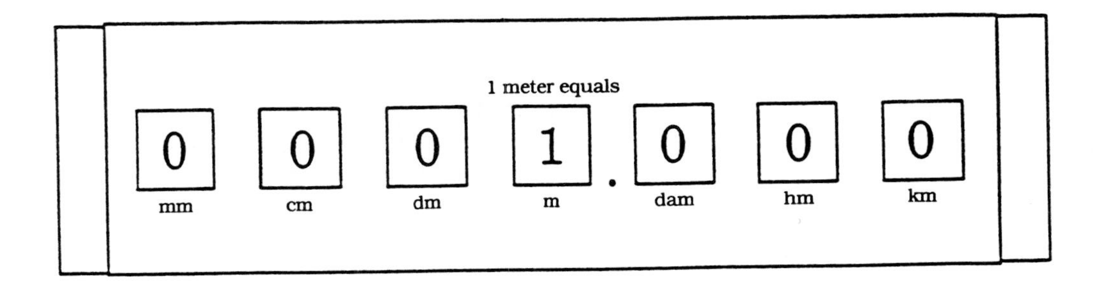

FIGURE C.2

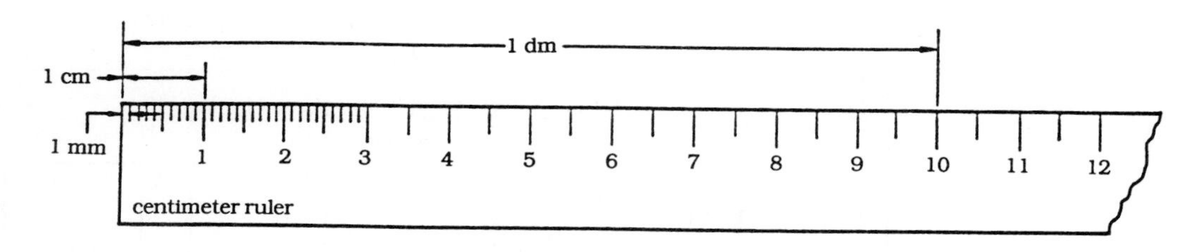

FIGURE C.3

SUPPLEMENTARY EXERCISES: LENGTH

Learning to Think Metric with Metric Units of Length.

The ruler in Figure C.3 shows the millimeter and centimeter units.

Complete the following:

1. 1 meter is ____ decimeters.
2. 1 meter is ____ centimeters.
3. 1 meter is ____ millimeters.
4. Give examples of things 1 meter long.
5. 1 centimeter is ____ meter.
6. 1 centimeter is ____ decimeter.
7. 1 centimeter is ____ millimeters.
8. Give 2 examples of things that measure 1 centimeter.
9. 1 millimeter is ____ meter.
10. 1 millimeter is ____ decimeter.
11. 1 millimeter is ____ centimeter.
12. Give 2 additional examples of things that measure 1 millimeter.
13. 1 kilometer is ____ meters.
14. 1 kilometer is ____ dekameters.
15. 1 kilometer is ____ hectometers.
16. Give examples of 2 distances familiar to you of 1 kilometer.

Complete each statement with the appropriate unit selected from among kilometer (km), meter (m), centimeter (cm), and millimeter (mm):

17. The length of a toothbrush is 15 ____.
18. The length of a picnic table is 2 ____.
19. The piece of wire had a diameter of 2 ____.
20. The distance between Philadelphia and New York City is 150 ____.
21. The diameter of a bicycle wheel is 66 ____.
22. The piece of cardboard is 3 ____ thick.
23. The length of a postcard is 14 ____.
24. The thickness of a window pane is 4 ____.

Match each of the following objects with the correct linear measure:

25. ____ the length of a newborn baby
26. ____ the length of a tennis racket
27. ____ the length of a man's size 10 shoe
28. ____ the width of an adult hand
29. ____ the width of a two-car garage
30. ____ the length of a new pencil
31. ____ the length of an average-sized bedroom
32. ____ the length of a new piece of chalk

a. 8 cm
b. 10 cm
c. 20 cm
d. 30 cm
e. 45 cm
f. 68 cm
g. 120 cm
h. 4 m
i. 7 m
j. 15 m

Choose the correct answer:

33. The width of a card table is approximately

    a. 95 cm b. 250 mm c. 2 m d. 30 cm

34. The diameter of a coffee cup is approximately

    a. 2 cm b. 10 mm c. 8 cm d. 25 mm

35. An average-sized bedroom is about

    a. 10×12 m b. 3.5×4 m
    c. 200×300 cm d. 2×3 km

36. The height of an average living room is about

    a. 150 cm b. 55 cm c. 2.5 m d. 5 m

37. A square cake pan is about ____ on a side.

    a. 20 cm b. 110 cm
    c. 25 mm d. 75 mm

38. The metric building brick is ____ long and ____ wide.

    a. 5 mm, 3 mm b. 25 mm, 12 mm
    c. 80 mm, 40 mm d. 215 mm, 100 mm

    (Even though centimeters may seem to be a more natural unit, metric bricks will probably be produced with millimeter specifications.)

39. The lumber used for wall studs (the familiar 2×4) has a cross-sectional length and width of about

    a. 40 mm $\times$ 85 mm b. 15 mm $\times$ 30 mm
    c. 175 mm $\times$ 350 mm d. 250 mm $\times$ 500 mm

40. A man's shirt size measured according to the circumference of the neck is

    a. 55 cm b. 20 cm
    c. 15 cm d. 38 cm

Compare the kilometer with the unit of mile in the following:

41. Discover the pattern and complete the following table:

Miles	Kilometers
10	10 + 5 + 1 = 16
20	20 + 10 + 2 = 32
30	30 + 15 + 3 = 48
40	40 + 20 + 4 = 64
50	
60	
70	
80	
90	
100	100 + 50 + 10 = 160
110	
120	

42. The pattern discovered in Exercise 41 says that n miles is approximately equal to $n + \frac{1}{2} n + \frac{1}{10} n$ kilometers. Use this pattern to complete the following table:

From	To	Miles	Kilometers
Philadelphia	Baltimore	100	
Philadelphia	New York	95	
New York	Boston	220	
Los Angeles	San Francisco	400	
Chicago	St. Louis	290	
Chicago	Peoria	150	
Denver	Kansas City	600	
Detroit	Toledo		96

43. A man walks 1 km in 10 min. He is walking at a rate of
    a. 2 km/hr b. 6 km/hr
    c. 15 km/hr d. 25 km/hr

44. A girl rides a bicycle about 4 km in 10 min. She is riding at a rate of
    a. 5 km/hr b. 24 km/hr
    c. 48 km/hr d. 75 km/hr

45. A man driving on an interstate highway travels about 15 km in 10 min. He is driving at a rate of
    a. 55 km/hr b. 70 km/hr
    c. 90 km/hr d. 100 km/hr

46. A commercial jet airplane can travel 170 km in 10 min. At this rate the plane is traveling at a speed of
    a. 500 km/hr b. 750 km/hr
    c. 1020 km/hr d. 1500 km/hr
47. During an 8-hour automobile trip on an interstate highway, you can comfortably travel a distance of
    a. 350 km b. 600 km
    c. 850 km d. 1200 km

Solve the following problems:

48. a. If wire costs \$0.55 per meter, how much does it cost per decimeter? Per centimeter?
    b. If wire costs \$0.18 per foot, how much does it cost per inch? Per yard?
49. A telephone company is to lay underground cable.
    a. At a cost of \$3.80 per meter, how much does it cost to lay 1 kilometer? 40 kilometers?
    b. At a cost of \$1.25 per foot, how much does it cost to lay 1 mile? 25 miles?
50. a. If ribbon costs \$1.35 per meter, how much does 1 centimeter cost? 30 centimeters?
    b. If ribbon costs \$1.29 per yard, how much does 1 inch cost? 30 inches?
51. A standard metric brick is 215 mm long, 102.5 mm wide, and 65 mm high. If the mortar joints are 10 mm thick, how high is a wall 50 bricks high?
52. The standard $8\frac{1}{2}'' \times 11''$ sheet of paper will be replaced by metric paper 21 cm by 30 cm. (This is a length-to-width ratio of approximately $\sqrt{2}$ to 1, the standard for metric paper.)
    a. If a 500-sheet package is 50 mm thick, how thick is a single sheet of paper?
    b. How many packages can be stacked on a shelf 1 m long, 30 cm wide, and 27 cm high?
    c. How many centimeters longer would the shelf have to be in order that an additional five packages fit on the shelf?

SUPPLEMENTARY ACTIVITIES: AREA

Activity 1: Estimating Area

Materials. Centimeter graph paper, meter stick, acetate or tracing paper

1. Make a grid of square centimeters on acetate or tracing paper. Outline a square decimeter on this grid. Use your grid to help you estimate the area of the following items:
   a. Face of wristwatch—in mm^2
   b. Light-switch plate—in cm^2
   c. Cover of this book—cm^2
   d. Dollar bill—in cm^2
   e. Window pane—in dm^2
   f. Sheet of notebook paper—in dm^2
   g. Desk top—in dm^2
   h. Door—in m^2
   i. Chalkboard—in m^2
   j. Floor of classroom—in m^2
2. Record your estimates and the actual area measurements.
3. Calculate and record the difference between each of your estimates and the actual measurements.

Activity 2: Estimating Larger Areas

Materials. Meter stick, trundle wheel or metric tape measure

1. An *are* is a unit of area 10 m on a side (1 dam^2). Step off a square with area you estimate to be 1 are. Check your estimate by actual measurement. How far off were you?
2. Estimate the area of your classroom in ares. (Use a decimal if necessary.) Check by actual measurement.
3. Record your estimates of area of the following surfaces in ares or hectares (1 hectare = 1 square kilometer):
   a. Basketball court
   b. Tennis court
   c. Volleyball court
   d. Badminton court
   e. Football field
   f. Soccer field (average size)
   g. Hockey field
   h. Baseball diamond (inside the base paths)
4. For each area, find the difference between your estimate and the actual area.*

*a. 3.78; b. 2.53; c. 1.62; d. 1.79; e. 43.2; f. 69.86; g. 15.3; h. 7.29.

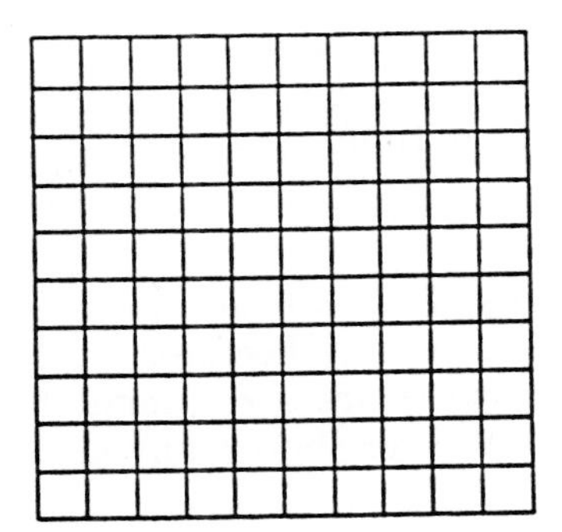

FIGURE C.4

SUPPLEMENTARY EXERCISES: AREA

Learning to Think Metric with Metric Units of Area

Use the square grid pictured in Figure C.4 to complete the following:

1. If each small square on the grid represents a straight pin stuck in a square posterboard, the posterboard is covered with __(a)__ pins and its area is approximately __(b)__ mm^2. The posterboard is 1 cm on each side and has area __(c)__ cm^2. One square centimeter is __(d)__ square millimeters.
2. If each small square represents a thumbtack stuck in a square posterboard, the posterboard is covered with __(a)__ tacks and its area is approximately __(b)__ cm^2. The posterboard is 1 dm on each side and has area __(c)__ dm^2. One square decimeter is __(d)__ square centimeters.
3. If each small square represents a large fruit-juice can sitting on a square table, the table is covered with __(a)__ cans and the area of the table is __(b)__ dm^2. The table is 1 m on a side and has area __(c)__ m^2. One square meter is __(d)__ square decimeters.
4. If the grid represents a restaurant that measures 1 dam by 1 dam and is filled with square tables 1 m × 1 m (i.e., there is no room left for seating people), the room is filled with __(a)__ tables and its area is __(b)__ m^2 or __(c)__ are. The room is 1 dam on each side and has area __(d)__ dam^2. One square decimeter is __(e)__ square meters.
5. If the grid represents a city block measuring 1 hm × 1 hm divided into small garden plots 1 dam × 1 dam, the city block is filled with __(a)__ plots and has area __(b)__ dam^2 or __(c)__ hectare. The block is 1 hm on each side and has area __(d)__ hm^2. One square hectometer is __(e)__ square dekameters. One hectare is __(f)__ ares.
6. If the grid represents a city, the city consists of __(a)__ blocks and its area is __(b)__ hm^2. The city is 1 km on a side and has area __(c)__ km^2. One square kilometer is __(d)__ hectares.

Complete each statement below with the appropriate unit selected from among square millimeter (mm^2), square centimeter (cm^2), square meter (m^2), are (a), or hectare (ha):

7. The picnic table has a surface area of approximately 1.5 ____.
8. The head of a nail might have an area of 30 ____.
9. The area of a living-room floor might be 40 ____.

10. The area of a postcard is approximately 130 ____.
11. The area of a football field is approximately 50 ____.
12. The area of a boating lake might be 1500 ____.

Solve the following problems:

13. A field 1 km on a side has an area of ____ hectares.
14. A field 200 m × 400 m has an area of ____ hectares.
15. The grounds for a large urban airport are a rectangular region 4 km × 10 km. The airport covers how many hectares?
16. How many square tiles, 30 cm on a side, will it take to tile a room 9 m long and 6 m wide?
17. A rectangular region is 750 cm long and 500 cm wide. Find the number of square meters of area of this region in two ways.
    a. First compute the number of square centimeters and then convert to square meters.
    b. First convert the length and width measure to meters and then compute the area in square meters.

    Do you prefer to convert, then compute, or to compute, then convert?
18. A nurse is ordered to give a patient an intravenous solution at a rate of 120 ml per hour. What rate in drops per minute should the nurse establish to maintain a rate of 120 ml per hour? (There are 20 drops per milliliter.)
19. Lumber when cut according to metric specifications will probably be priced by the cubic meter. One standard-size board is 25 mm thick, 100 mm wide, and 3 m long.
    a. How many of these boards are there in a stack of this wood 1 × 1 × 3 m?
    b. If this wood is used as flooring on a sun deck, how many square meters will this 3-cubic-meter stack cover?
20. If a dripping faucet drips 10 drops per minute and there are 20 drops per milliliter, what is the capacity of the water wasted in one day? In one week? In one month?
21. a. *Metric system.* A standard-size drinking straw is 25 cm long and 4 mm in diameter. Find the capacity of the straw. (Recall that the volume of a cylinder is given by the formula volume $= \pi$ ($\frac{1}{2}$ diameter)^2 × height.)
    b. *English system.* A standard-size drinking straw is 10 inches long and $\frac{1}{8}$ inch in diameter. Find the capacity of the straw in fluid ounces.

SUPPLEMENTARY ACTIVITIES: VOLUME AND CAPACITY

Activity 1: Estimating Volume

Materials. Several different-sized boxes, containers, metric ruler, meter stick

1. Use the following categories:
   A. Less than 1 dm^3
   B. More than 1 dm^3 and less than 100 dm^3
   C. More than 100 dm^3 and less than 1 m^3
   D. More than 1 m^3 and less than 100 m^3
   E. More than 100 m^3
2. Match the categories with the following items:
   a. Small match box
   b. Shoe box
   c. File cabinet
   d. Cereal box
   e. Your classroom
   f. Milk carton
   g. Crayon box
   h. Desk drawers
   i. Hallway near your classroom
   j. Toothpaste box
   k. A restroom in your building
3. Record the actual volume for each item by measuring length, width, and height and using the formula $V = l \times w \times h$.
4. Record whether your categorization was correct or incorrect.

Activity 2: Estimating Capacity

Materials. One-liter container marked so milliliters can be measured, containers (or information on capacities of the containers)

1. Use the following categories:
   A. Under 100 ml
   B. Between 100 and 250 ml
   C. Between 250 and 500 ml
   D. Between 500 and 750 ml
   E. Between 750 ml and 1 l
   F. Over 1 l

2. Match the categories with the following items:
   a. Coffee cup
   b. Pop can
   c. 9-cup coffee maker
   d. Saucepan
   e. Small juice can
   f. Teaspoon
   g. Family-sized catsup bottle
   h. Eyedropper
   i. Tea kettle
   j. Tablespoon
   k. Large can of apple juice
3. Record the actual capacities by pouring and measuring.
4. Record whether your categorization was correct.

SUPPLEMENTARY EXERCISES: VOLUME AND CAPACITY

Learning to Think Metric with Metric Units of Volume

Use the cube pictured in Figure C.5 to complete the following:

1. If the box in Figure C.5 is a centimeter on a side, approximately how many granules of salt would fit in it? One cubic centimeter is ____ cubic millimeters.
2. If the box in Figure C.5 is a decimeter on a side, approximately how many sugar cubes could be placed in it? One cubic decimeter is ____ cubic centimeters.
3. If the box in Figure C.5 is 1 meter on a side, approximately how many new softballs would it hold? One cubic meter is ____ cubic decimeters.

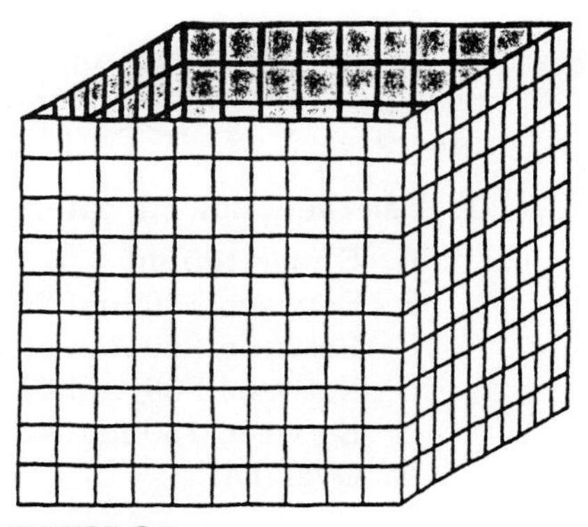

FIGURE C.5

4. If the box in Figure C.5 is 1 meter on a side, approximately how many sugar cubes does it hold? One cubic meter is ____ cubic centimeters.

Learning to Think Metric with Metric Units of Capacity

Complete the following statements:

1. One liter is approximately ____ cups.
2. One cup is approximately ____ ml.
3. One-half cup is approximately ____ ml.
4. One cubic decimeter has a capacity of ____ l.
5. One cubic centimeter has a capacity of ____ ml.
6. One cubic decimeter has a capacity of ____ ml.
7. One cubic meter has a capacity of ____ l.
8. One cubic meter has a capacity of ____ kl.
9. One liter is ____ ml.

Convert from one unit of capacity to another in the following:

	kl		l		ml
10.	0.001 kl	=	____ l	=	1000 ml
11.	____ kl	=	0.1 l	=	____ ml
12.	____ kl	=	1.23 l	=	____ ml
13.	____ kl	=	3.2 l	=	____ ml
14.	1 kl	=	____ l	=	____ ml
15.	____ kl	=	____ l	=	1236 ml
16.	____ kl	=	0.004 l	=	____ ml
17.	476 kl	=	____ l	=	____ ml
18.	____ kl	=	____ l	=	885 ml

Solve the following problems:

19. Find the volume in dm^3 of a box 30 cm long, 20 cm wide, and 10 cm high. Do this problem two ways.
    a. Compute the number of cm^3 and convert to dm^3.
    b. Convert linear units to dm and compute dm^3.

    Do you prefer to convert, then compute, or to compute, then convert?
20. A file cabinet is 40 cm wide, 150 cm high, and 70 cm deep. Find the volume of the cabinet in dm^3. In m^3. Use both of the methods outlined in Exercise 21.
21. a. *Metric system.* A container is 30 cm long, 10 cm wide, and 5 cm high. What is the volume of the container? How many liters of fluid will it hold?
    b. *English system.* A container is 12 inches long, 8 inches wide, and 3 inches high. What is the volume of the

container? How many quarts of fluid does the container hold? (Don't panic if you are not able to do this problem. Most others can't do it either.)

22. A cakepan is 20 cm square and 5 cm deep. How many liters does it hold?
23. A swimming pool 50 m long and 25 m wide is uniformly 2 m deep. How many kiloliters of water does it hold?
24. A 250-ml measuring cup will replace the familiar 2-cup measure used in baking. The ingredients needed to make one batch of Swedish cookies are listed below. Convert this recipe to metric units (rounded to nearest 10 ml):

shortening	¾ cup
sugar	1 cup
eggs	2
nuts, chopped	5 tablespoons
orange rind, grated	1 tablespoon
almond extract	½ teaspoon
flour	2½ cups
baking powder	2 teaspoons

Cream the shortening and sugar. Add eggs, nuts, orange rind and extract. Add flour and baking powder which have been sifted together. Mix. Make into small balls and flatten with a fork. Bake about 12 minutes at 375°F (190°C). Makes about 60 cookies.*

SUPPLEMENTARY ACTIVITIES: MASS

Activity 1: Estimating the Mass of Coins

Materials. Accurate, commercially made balance scale, penny, nickel, dime, quarter, half-dollar, silver dollar (or substitutes)

1. The mass of a dime is 2 grams. Record your estimates of the mass of the following coins:
   a. Penny ____
   b. Nickel ____
   c. Dime ____
   d. Quarter ____
   e. Half-dollar ____
   f. Silver dollar ____

*Copyright 1975. Courtesy of American Metric Journal, Tarzana, CA 91356.

2. Use a scale to check your estimates.
3. For each coin, record the actual mass and the difference between your estimate and the actual measurement.

Activity 2: Guessing the Mass of Persons

Materials. Metric personal scale

1. Conduct a "mass guessing contest" in which each person in your group guesses the mass of 5 volunteer subjects. Check your guesses by finding the mass of each person. Lowest sum of differences wins.
2. Compile a tally sheet recording each participant's estimated mass and the actual mass of each subject. (If a metric scale is not available, use an English scale and multiply the result by 0.45.)
3. Record the differences and percent-of-error values for each.
4. Can you practice estimating the mass of persons and reduce your percent of estimation error?

Activity 3: Estimating the Mass of Objects

Materials. Accurate scale, objects listed in part 2 (or substitutes)

1. Use the following categories:
   A. Under 100 grams
   B. Between 100 and 250 grams
   C. Between 250 and 500 grams
   D. Between 500 and 750 grams
   E. Between 750 grams and 1 kilogram
   F. Over 1 kilogram
2. Match the categories with the following objects:
   a. Chalkboard eraser
   b. This book
   c. Dictionary
   d. Can of pop
   e. Apple
   f. Pencil
   g. Your shoe
3. Check the mass of each object using a scale. Then, record the actual mass and whether your estimate was correct or incorrect.

SUPPLEMENTARY EXERCISES: MASS

Learning to Think Metric with Units of Mass

Complete the following statements:

1. One kilogram is ____ grams.
2. One ton is ____ grams.
3. Name 2 objects whose mass is about 1 g.
4. Name 2 grocery items that are commonly bought in quantities of about ½ kg.
5. Name 2 commodities that are commonly weighed in tons.

Complete each statement below with the appropriate unit selected from among ton (t), kilogram (kg), gram (g), and milligram (mg):

6. A sack of potatoes might have a mass of 5 ____.
7. The net mass of a new tube of lipstick is 5 ____.
8. An aspirin tablet has a mass of 324 ____.
9. A sheet of tablet paper has a mass of 3 ____.
10. An empty Pepsi bottle has a mass of about 440 ____.
11. An adult might have a mass of 75 ____.
12. A canned ham might have a mass of 2 ____.
13. A truckload of coal might have a mass of 2 ____.
14. A new piece of chalk has a mass of 10 ____.
15. A Bic pen has a mass of 7 ____.
16. A new roll of Scotch tape has a mass of 50 ____.
17. A new tablet has a mass of 185 ____.
18. A bucket of water might have a mass of 12 ____.
19. A raisin has a mass of 330 ____.
20. A small car has a mass of about 1 ____.
21. A miniature marshmallow has a mass of about 500 ____.
22. A pair of eyeglasses has a mass of 40 ____.

Choose the correct answer for each of the following:

23. A can of Campbell's vegetable soup has a mass of

    a. 700 mg b. 40 g c. 305 g d. 2 kg

24. A common-sized bag of flour has a mass of

    a. 200 g b. 4.5 kg c. 8 kg d. 20 kg

25. A large coffee can together with its contents has a mass of approximately

    a. 100 g b. 1 kg c. 3 kg d. 5 kg

26. A package of Kool-Aid has a mass of approximately
    a. 7 g b. 70 g c. 700 g d. 7 kg
27. A standard box of Jello has a mass of approximately
    a. 8.5 g b. 85 g c. 850 g d. 8.5 kg
28. A medium-sized tube of toothpaste has a mass of approximately
    a. 14 g b. 140 g c. 800 g d. 2 kg
29. A small can of cinnamon has a mass of about
    a. 40 g b. 300 g c. 800 g d. 1 kg
30. If the kilogram is defined to be the mass of one liter of water at 4°C, the weight of a container of water (or any liquid with the density of water) can be easily determined when the capacity of the container is known. Using this relationship, complete the following.
    a. One milliliter (1 cm^3) of water weighs approximately ____ g.
    b. One liter (1 dm^3) of water weighs approximately ____ kg.
    c. One kiloliter (m^3) of water weighs approximately ____ ton.

SUPPLEMENTARY ACTIVITIES: TEMPERATURE

Activity 1: Estimating Temperature

Materials. Celsius thermometer, container to hold liquid

1. Water freezes at 0°C and boils at 100°C. The average room temperature is about 20°C. Normal body temperature is 37°C. Record your estimates of the following temperatures in °C:
    a. Cold water from a faucet
    b. Hot water from a faucet
    c. Palm temperature (Hold thermometer inside your closed fist for 2 min.)
    d. Light-bulb temperature (Hold the thermometer 2 in from a light bulb for 2 min.)
    e. Today's room temperature
    f. Today's outside temperature
2. Record your estimates and the actual temperatures taken with a celsius thermometer.
3. For each, record the difference between your estimate and the actual temperature.

SUPPLEMENTARY EXERCISES: TEMPERATURE

Learning to Think Metric with Metric Units of Temperature

1. Match the following:

a. 40°C	1. Pleasant summer day
b. 37°C	2. Scorching summer day
c. 35°C	3. Room temperature
d. 30°C	4. Death Valley heat
e. 25°C	5. Body temperature
f. 20°C	6. Hot summer day

Use the thermometer pictured in Figure C.6 to choose the correct answers in the following:

2. The temperature of a sauna is about

   a. 30°C b. 50°C c. 120°C d. 180°C

3. The temperature of a hot cup of tea is about

   a. 30°C b. 80°C c. 120°C d. 150°C

4. A glass of iced cola has a temperature of

   a. 0°C b. 10°C c. 32°C d. 50°C

5. A person ill with a high fever might have a body temperature of

   a. 37°C b. 39°C c. 43°C d. 104°C

6. The temperature on a very warm December day in the midwest might be

   a. 0°C b. 10°C c. 20°C d. 25°C

7. The temperature of the hot-water rinse of a dishwasher is about

   a. 40°C b. 75°C c. 100°C d. 120°C

8. Comparing the 100°C difference versus 180°F difference between freezing point and boiling point of water, it follows that each 9°F rise corresponds to a 5°C rise. Complete the table of corresponding °F and °C temperatures, freezing point and above:

°F	°C
32	0
41	5
50	
.	
.	
.	
113	

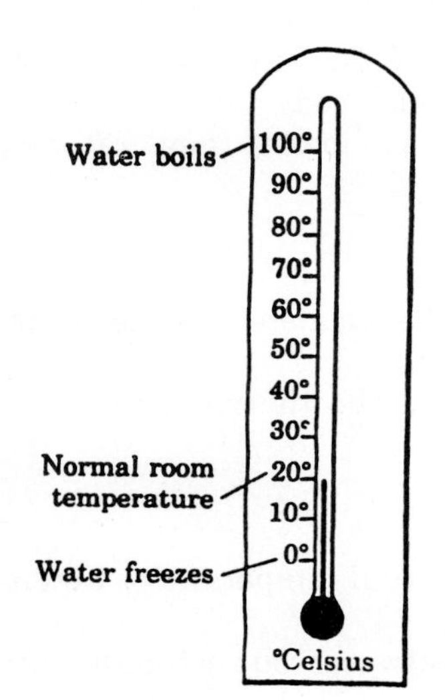

FIGURE C.6

9. Complete the table of corresponding °F and °C temperatures, freezing point and below:

°F	°C
32	0
23	−5
14	
.	
.	
.	
−40	

APPENDIX D Critical Thinking

WHAT IS CRITICAL THINKING?

Critical thinking is an important process that not only helps us make decisions about patterns and relationships in geometry, but also it helps us make decisions in everyday life.

It may be impossible to give a definition of critical thinking that is as clear and precise as the definitions of terms in mathematics. Definitions of such terms as *critical thinking* often give only a general idea about what the term means, but are useful just the same. Here is such a definition: **Critical Thinking** is a process that uses general thinking skills to help one decide what to believe or do in a situation.

Let's make some sense out of the definition. It refers to critical thinking as a *process*. The following gives an idea of what you might need to do to carry out such a process:

1. Understand the situation.
2. Deal with evidence/data/assumptions.
3. Go beyond the data/evidence/assumptions.
4. State and support conclusions/decisions/solutions.
5. Plan/create to act on/apply the conclusion/decision/solution.

The definition above also says that critical thinking uses general thinking skills. Here is a partial list of such skills.

— *Comparing/contrasting:* asking students to look for similarities and differences
— *Classifying/categorizing:* ability to form classifications or categorizations according to chosen attributes
— *Visualizing:* asking students to visualize plane or 3-D figures, drawing, blueprints, etc.
— *Sequencing/ordering:* ability to put items in order in terms of some characteristic; ability to put events or steps in a process in sequence
— *Relating (analogies):* ability to find relationships and form analogies
— *Patterning:* asking students to discover a pattern and extend it

— *Conjecturing/inducing/generalizing/specializing:* ability to look at specific examples with an eye to inducing or discovering something that is true in general; ability to give specific examples for a generalization, including a counterexample that shows that the generalization is false
— *Deducing/concluding:* ability to draw conclusions logically from information given about a situation
— *Validating/proving:* ability to support or validate conclusions by convincing via plausible reasoning with models, inductive reasoning from examples, or proof by logical reasoning
— *Predicting:* ability to make predictions on the basis of given data and relationships
— *Analyzing:* ability to "make sense" out of the parts of a situation as they relate to the whole
— *Evaluating:* ability to evaluate assumptions, evidence, and the reasoning used in a demonstration that something is true

Other important thinking activities include observing, comprehending, identifying, discriminating/distinguishing, perceiving, describing, explaining, justifying, planning, prioritizing, modeling/representing, and creating.

The following list gives an idea of how these general thinking abilities might be involved in the critical thinking process described above.

1. *Understand the situation:* Observe, visualize, comprehend, clarify, and identify to understand meanings of terms, existence of evidence, whether assumptions are stated, and to generally understand the nature of the situation.
2. *Deal with data/evidence/assumptions:* Organize data by classifying/categorizing, ordering, sequencing, patterning, and prioritizing. Evaluate by comparing/contrasting, discriminating (relevant from irrelevant evidence, fact from opinion, reliable sources from unreliable, etc.), identifying (hidden assumptions, probable causes and effects, purposes, inconsistencies, adequacy of data/evidence, questions, etc.), and specializing (looking at specific examples).
3. *Go beyond the data/evidence/assumptions:* Draw conclusions, make decisions, or find solutions by conceptualizing, conjecturing, generalizing, inducing, deducing, solving, inferring, relating, and predicting. Evaluate these processes using ideas of logic and an analysis of emotional/psychological fallacies.

4. *State and support conclusions/decisions/solutions:* Evaluate the conclusions/decisions/solutions and support them by describing, explaining, verifying, convincing, and proving.
5. *Plan/create to act on/apply the conclusion/decision/solution:* Apply the conclusions/decisions/solutions to related situations by planning, modeling/representing, and creating.

The following critical thinking activities focus on the critical thinking process and the skills needed to implement the process.

CRITICAL THINKING ACTIVITY 1

Introducing the Process

Critical thinking has been described as a thinking process used to decide what to believe or do. Consider the following situation.

> Inspector Glueso arrived at the scene of a murder at Motley Motel early in the morning (Fig. D.1). The first thing he found was a broken pocket watch clutched in the victim's hand. Other evidence led Glueso to determine that the murder was committed at the time shown on the watch and that the murderer was one of three people at the motel: Mae East, Mr. Gleem, or Joe Monitor. The hotel operator verified that Mae East, an actress, had been talking on the telephone from 12:32 A.M. to 12:47 A.M. Joe Monitor, the motel manager, had been seen using his computer from about 12:20 A.M. until he shut it off sometime after 12:30 A.M. The computer log showed that it was shut off at 12:34 A.M. Mr. Gleem, the night custodian, had punched out at 12:31 A.M. and had left promptly. The

FIGURE D.1

only fact that Glueso could remember about the watch was that the hour and minute hands were in a straight line when he picked it up. Whom should Glueso arrest as the prime suspect?

The major points below outline the steps in critical thinking. Answer each question as you work, with Glueso, to solve his problem.

1. *Understand the situation.*
   a. What is the situation about?
   b. What conclusion, decision, or solution is required?
   c. Are there any questions you could ask to clarify the situation?
2. *Deal with the data/evidence/assumptions.*
   a. Make a list of the most important data/evidence. Is more needed?
   b. Which evidence is opinion, not fact? Which is not relevant?
   c. What assumptions are made? Can other assumptions be made?
3. *Go beyond the data/evidence/assumptions.* Read the notes Glueso made in his notebook (Fig. D.2). How could you use this and the evidence to solve the mystery? [*Hint*: Since the hour hand travels through 5-min spaces while the minute hand travels through 60-min spaces, the distance the hour hand had traveled was equal to $1/12$ the distance the minute hand had traveled.]
4. *State and support your conclusion.* State your solution to the mystery in writing. Write a paragraph supporting your conclusion.
5. *Apply the conclusion/decision/solution.* The solution of a similar mystery depended upon the fact that a crime occurred the first time after midnight that the hands of a clock were both at the same place. At what time (to the nearest second) did this crime take place?

FIGURE D.2

CRITICAL THINKING ACTIVITY 2

Recognizing and Analyzing Assumptions

A statement of opinion, a discussion, an argument, and a proof are all based on assumptions. Sometimes these assumptions are not stated. You must figure out that assumptions are being made and identify what they are. Consider the following examples.

Example 1. Jeff's teacher said, "Proving theorems in geometry helps you develop more general reasoning abilities." State several assumptions that Jeff's teacher may be making.

Solution:

1. It is important to develop general reasoning abilities.
2. There is a need to justify proving theorems in geometry.
3. The logic used in geometric proof relates to the logic used in everyday reasoning.
4. This statement will motivate students to work harder when proving theorems.

Example 2. Lauren saw a television advertisement showing a marathon runner drinking Sunnyside orange juice. She said, "I'm going to drink that juice before my next aerobics class." State an assumption she may be making.

Solution: Lauren might be assuming that the juice increased the marathon runner's energy level.

Example 3. Ah-hode said, "An altitude from any vertex of a triangle divides it into two triangles, each with an area smaller than the original." State an assumption that Ah-hode is making.

Solution: Ah-hode was making the assumption that the original triangle was an acute triangle.

Applications

1. Accept this story as the depiction of an event that actually happened.

   Little Boy Blue come blow your horn.
   The sheep are in the meadow, the cows in the corn.
   Where is the little boy who looks after the sheep?
   He's under the haystack, fast asleep.

   Which of these assumptions have you accepted as true?

   a. Little Boy Blue's parents are looking for him.
   b. Little Boy Blue looks after the sheep and cows.
   c. The sheep shouldn't be in the meadow and the cows shouldn't be in the corn.

d. Little Boy Blue is covered with hay and can't be seen.
e. If Little Boy Blue would blow his horn, the sheep and cows would do what they are supposed to do.
f. Little Boy Blue is a very young child.

2. Now reread the story. Which of these assumptions are valid, based only on the information in the story?

State an assumption that each speaker is making:

3. "I can't wait until I'm 21 years old!"
4. "I've enjoyed all of this author's other books, so this new one should be good."
5. "Extend the sides of a quadrilateral until they intersect. Call the point of intersection A."
6. "Let line m intersect line n at a point between A and B."
7. "Since $\angle RSV$ is the result of bisecting $\angle RST$, its measure is less than 45."
8. "The product of two positive numbers is always greater than either of the factors."
9. "Since n is a prime number, I know that n is odd."
10. "When evaluated, $-x$ is negative."
11. "I can construct an isosceles triangle with sides congruent to $\overline{AB}$, $\overline{CD}$, and $\overline{EF}$."
12. "I can find the square root of x."
13. Describe a situation that you have recently observed where unstated assumptions were involved.

CRITICAL THINKING ACTIVITY 3

Analyzing and Evaluating Evidence

Sometimes it is possible to jump to a conclusion on the basis of evidence that is unreliable or incomplete. Some evidence is possibly true, other evidence is probably true, and still other evidence is accepted as true. Consider the following examples:

Example 1.

1. Refer to Figure D.3. By observation only, decide if line k, when extended, will meet point A, B, C, or none of these points.
2. Read the following silently and then aloud:

A BIRD IN THE
THE HAND IS WORTH
TWO IN THE BUSH

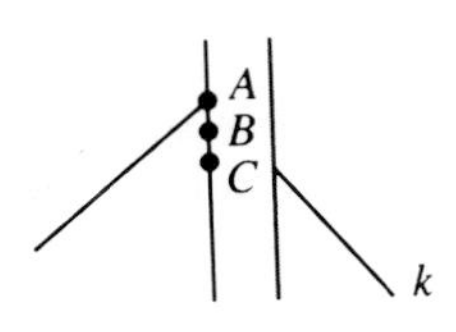

FIGURE D.3

What do you notice? What kind of evidence are you using in these examples?

Solution: In both cases, observational evidence was used. These examples show that we do not always see what we think we see. Often, facts get changed to fit our expectations. These examples show that observational evidence should be treated with caution and classified as possibly or probably true.

Example 2. As Jerry came around a corner, he saw his steady girlfriend, Rhonda, walking into a theater with another boy. Jerry concluded that Rhonda was dating someone else and was very angry. What kind of evidence was Jerry using in his decision?

Solution: Jerry based his conclusion on circumstantial evidence. The truth was that Rhonda was meeting her aunt at the theater and was only walking through the door next to a boy from school. Circumstantial evidence should be considered as possibly true.

Example 3. Mr. Higami was clocked traveling 48 mph in a 35 mph zone by a police officer's radar gun. Two reputable witnesses, who had checked the radar gun before and after Mr. Higami's ticket was issued, testified that the gun was working properly. Another police officer, who was riding in the car with the first officer verified the officer's story. What kind of evidence do you have in this case?

Solution: The evidence here is of the type that is accepted to be true. In a case such as this, many different facts corroborate the fact in contention. Even in such cases where the evidence is considered true beyond any reasonable doubt, there is always a small possibility that it could be false.

Applications

1. After Karen began dating Javier, her grade-point average began to fall. Her counselor checked Karen's study time and found that she spent an average of only 4 hours a week on homework. Karen seemed to be trying to get better grades and did not want to talk about her lower grades. Decide if each of the following statements of evidence is possibly true, probably true, or accepted as true.
   a. Karen was getting better grades before she started dating Javier.

   b. Karen is spending less time on homework since she started dating Javier.
   c. Karen feels guilty or embarrassed about her lower grades.

2. Plan and rehearse a brief incident that a group of 3 or 4 students could act out in class. After a group presents an incident, write a careful "eyewitness report" of exactly what happened. How does your report compare to that of a classmate?
3. Describe a situation in which you drew an incorrect conclusion based on circumstantial evidence.

CRITICAL THINKING ACTIVITY 4

Deciding on the Truth of a Statement

In the process of dealing with data or evidence in critical thinking, you have to decide if a statement is true. Every statement, no matter how simple or complicated, is either true or false. Statements can be represented by letters of the alphabet.

For example, if statement p "You passed geometry" is true, then the statement $\sim p$ (read "not p") "You did not pass geometry" is false, and vice versa. This is summarized in the following truth table.

p	$\sim p$
T	F
F	T

Consider the "p or q" statement below.

p You will pass this geometry course.
q You will be a sophomore again next year.
p or q You will pass this geometry course or you will be a sophomore again next year.

p	q	p or q
T	T	T
T	F	T
F	T	T
F	F	F

The statement is false only when you fail geometry and so you are not a sophomore next year. In the first row p or q is true. You have passed geometry but you could still be a sophomore next year because you failed English!

Check the "p and q" statement below. Do you think the accompanying truth table is correct for the "p and q" statement below?

p	We will review geometry.
q	The geometry test will be easy.
p and q	We will review geometry and the geometry test will be easy.

p	q	p and q
T	T	T
T	F	F
F	T	F
F	F	F

Applications

1. Joe said "It is false that I don't like geometry." Does this mean that Joe doesn't like geometry? Explain. Use symbols to show the statement.
2. Suppose the statement "College students like to go to movies or listen to records" is true. Could a person who likes both movies and records be a college student? Explain.
3. Aaron's doctor said, "You cannot go to work and you must take walks regularly." Aaron thought he must not go to work or he must take walks regularly. Was he correct? Explain.
4. Kwan said, "English majors like movies or records." If this is true and Kwan likes both movies and records, could Kwan be an English major? Explain.
5. Bill's parole officer said, "You can't go out of town and you must be in by 12 o'clock." Bill thought he must not go out of town or he must be in by 12 o'clock. Was he correct? Explain.
6. The salesman said, "You can drive this car home today for nothing down and just $349 a month." Will Corliss be able to drive the car home without having to pay anything on it today?
7. The truth table for "or" statements was defined in a way useful to geometry, but it could be defined in another way. Give a true "or" statement in which both parts cannot be true at the same time.

CRITICAL THINKING ACTIVITY 5

Determining the Truth of a Conditional Statement

To analyze evidence/assumptions and to reason logically when drawing conclusions, it it also necessary to be able to decide when a statement in "if-then" form is true. A statement of the form **If *p*, then *q*** (symbolized as $p \rightarrow q$) is called an **if-then** statement, or **conditional**. The simple statement ***p*** is called the **hypothesis** and the simple statement ***q*** is called the **conclusion**. If an if-then statement, the hypothesis and conclusion are related so that the truth of the hypothesis promises the truth of the conclusion.

Suppose a teacher makes the if-then statement below. For which of the cases would you feel you were treated unfairly and your teacher didn't tell the truth?

If-then statement: If you get A's on all geometry tests, then you will get an A in the course.

Case 1:	You get A's on all tests. (hypothesis true)	You get an A in the course. (conclusion true)
Case 2:	You get A's on all tests. (hypothesis true)	You don't get an A in the course. (conclusion false)
Case 3:	You don't get A's on all tests. (hypothesis false)	You get an A in the course. (conclusion true)
Case 4:	You don't get A's on all tests. (hypothesis false)	You don't get an A in the course. (conclusion false)

The above cases suggest that *an if-then statement is false only when the hypothesis is true and the conclusion is false*. This is summarized in the truth table for an if-then statement.

p	q	If p, Then q
T	T	T
T	F	F
F	T	T
F	F	T

Applications

1. A TV repairman said, "If I don't fix your set, then I won't charge you anything." The repairman didn't charge anything. Assuming that the repairman told the truth, what, if anything, can you conclude?

2. A TV ad said, "If you use Hair Style, then your hair will have a greater shine." Fred didn't use Hair Style, but his hair had a greater shine. Did the ad tell the truth? Explain.
3. Tina's auto mechanic said, "If you don't use a gasoline additive, then you won't get over 30 miles per gallon." Tina used a gasoline additive, but she did not get over 30 miles per gallon. Did Tina's mechanic lie to her? Explain.
4. When asked how she decided if a conditional was true or false, a student said, "I assume that the hypothesis is true and ask myself if the conclusion could possibly be false. If it can, I say that the conditional is false." Does this method work? Explain, using several different examples.

CRITICAL THINKING ACTIVITY 6

Using the Converse, Inverse, and Contrapositive

When drawing conclusions in the process of critical thinking, you will often have to decide upon the truth of statements related to a statement that is known to be true. If we begin with an if-then statement, we can form three types of related statements, called the **converse**, the **inverse**, and the **contrapositive** of the original if-then statement, as shown below.

If-then statement ($p \rightarrow q$): If you live in Chicago, then you live in Illinois.

Converse ($q \rightarrow p$): If you live in Illinois, then you live in Chicago. (Formed by interchanging the hypothesis and the conclusion)

Inverse ($\sim p \rightarrow \sim q$): If you don't live in Chicago, then you don't live in Illinois. (Formed by negating both the hypothesis and the conclusion)

Contrapositive ($\sim q \rightarrow \sim p$): If you don't live in Illinois, then you don't live in Chicago. (Formed by interchanging the hypothesis and the conclusion and negating both)

Assume that Chicago means Chicago, Illinois. Is the if-then statement true? Decide which of the other statements are true.

Truth values for each of the types of statements are given in the truth table below.

p	q	$\sim p$	$\sim q$	Statement $p \rightarrow q$	Converse $q \rightarrow p$	Inverse $\sim p \rightarrow \sim q$	Contrapositive $\sim q \rightarrow \sim p$
T	T	F	F	T	T	T	T
T	F	F	T	F	T	T	F
F	T	T	F	T	F	F	T
F	F	T	T	T	T	T	T

Since the contrapositive is true whenever the statement is true and is false when the statement is false, it is **logically equivalent** to the statement. Note that neither the converse nor the inverse is true in all cases where the statement is true.

Applications

1. Write the converse, inverse, and contrapositive of these statements. Decide which of the statements are true.
   a. If you graduate from college, then you will get a good job.
   b. If a triangle is an equilateral triangle, then it is an isosceles triangle.
2. If you want to be beautiful, then you will buy Smooth-It facial cream. What statement, with negations, has the same meaning as this statement?
3. How would you use the truth table of the preceding discussion to convince someone that the inverse of a statement is logically equivalent to the converse of the statement? Choose a statement and illustrate this idea.
4. Is the converse of the inverse of a statement the same as the contrapositive of the statement? Is the contrapositive of the inverse of a statement the same as the converse? Give a convincing argument to support your conclusions. What other relationships like this can you find?

CRITICAL THINKING ACTIVITY 7

Common Errors in Everyday Reasoning

These situations will help you discover or review some common reasoning errors that involve deciding upon the truth of a statement.

FIGURE D.4

Situation 1. Jessie saw an advertisement on television that stated, "If you want top power and control, then buy a Smasher tennis racket" (Fig. D.4). He assumed that, "If I buy a Smasher, then I'll have top power and control."

1. If the statement in the advertisement is represented by $p \rightarrow q$, how would you represent Jessie's assumption?
2. Do you think Jessie's reasoning was logically correct? Explain why or why not.

Situation 2. Martina's teacher said, "If you don't get an A on this test, then you won't get an A in the course." Martina assumed that, "If I do get an A on this test, then I will get an A in the course."

1. If Martina's teacher's statement is represented by $p \rightarrow q$, how would you represent Martina's assumption?
2. Do you think Martina's reasoning was logically correct? Explain why or why not.

You learned earlier that when the statement is true, the converse and inverse are not necessarily true. In the situation above, Jason was making a very common error in reasoning called *assuming the converse*. Martina was making a common error in reasoning called *assuming the inverse*.

Applications

1. Is there a reasoning error in each situation? If so, explain what error was made.

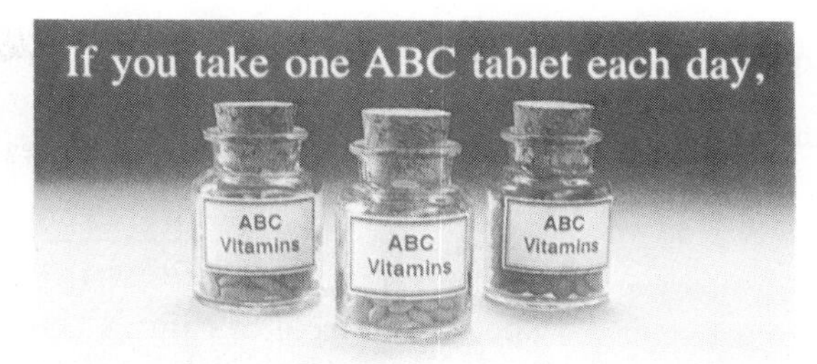

FIGURE D.5

a. Isaac's little brother Ted knew that if a vehicle was a car, then it had 4 wheels. When given a toy wagon with 4 wheels, he concluded that it was a car.
b. Jose's coach said, "If you win the most tryout games, then you will make the tennis team." Jose did not make the team, so he concluded that he did not win the most games.
c. Tanya read that if a dog is show quality, then it has registration papers. She told her friend Marta that a dog that has no registration papers is not show quality.
d. Mr. Wilson read an advertisement that said, "If you take one ABC tablet each day, then you will keep your good health" (Fig. D.5). He began to feel uneasy, thinking, "If I don't take an ABC tablet each day, then I won't keep my good health."

2. Cathy's roommate told her, "If you don't help keep our room clean, then you will have to do our laundry." Cathy kept the room clean and was angry when her roommate asked her to do the laundry. She felt her roommate had broken her promise.
3. Copy a television or magazine advertisement that you think wants the viewer to assume the converse or inverse. Explain your interpretation.

CRITICAL THINKING ACTIVITY 8

Venn Diagrams and Reasoning Patterns

Venn diagrams can be used to test whether an argument is logically correct. An if-then statement can be represented in terms of circles, which help illustrate the validity of arguments based on the conditional. Consider the following examples.

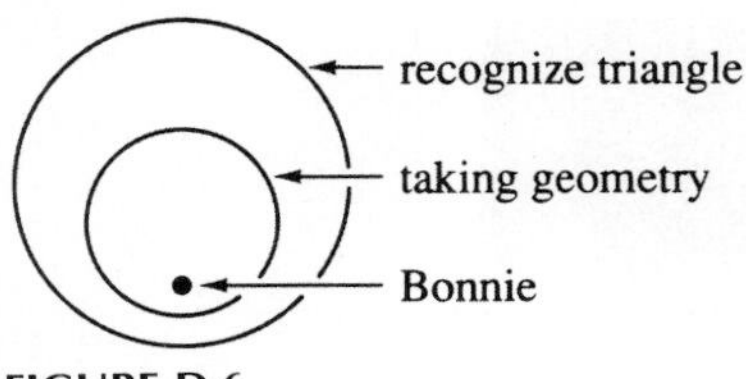

FIGURE D.6

Example 1.

If a student is taking geometry,
then she recognizes triangles. $p \to q$

Bonnie is taking geometry. p (premise)

Bonnie recognizes triangles. q (conclusion)

The outer circle of Figure D.6 defines those students that recognize triangles. The inner circle indicates that all of the students taking geometry recognize triangles. Since Bonnie is in the inner circle, she is automatically among those students who recognize triangles.

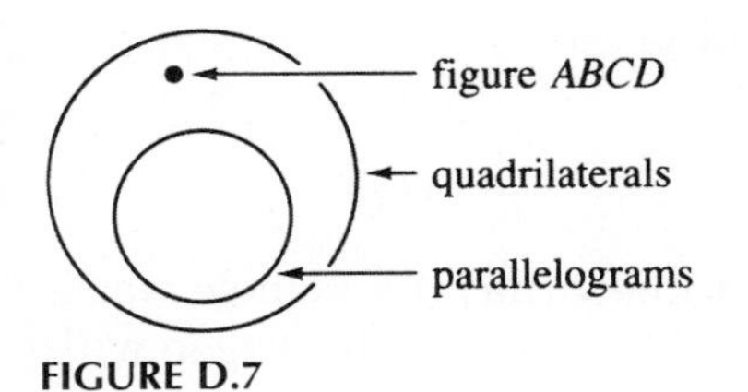

FIGURE D.7

Example 2: Assuming the Converse.

If a figure is a parallelogram,
then it is a quadrilateral. $p \to q$

Figure *ABCD* is a quadrilateral. q

Figure *ABCD* is a parallelogram. p

The Venn diagram (Fig. D.8) shows that a point representing figure *ABCD* can be outside the parallelogram circle, so the reasoning pattern is invalid. Here, the converse of the original conditional is assumed.

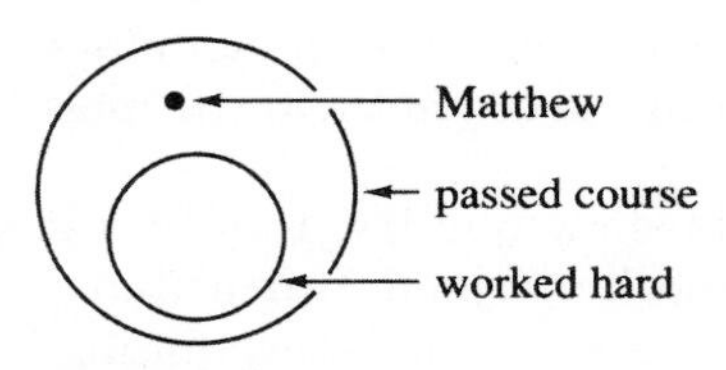

FIGURE D.8

Example 3: Assuming the Inverse.

If a student worked hard,
then he passed the course. $p \to q$

Matthew did not work hard. $\sim p$

Matthew did not pass the course. $\sim q$

The Venn diagram (Fig. D.8) shows that although Matthew did not work hard, the point representing him could still be inside the circle representing those who passed the course. The error in this case is assuming the inverse of the original conditional.

Applications

Draw Venn diagrams to show each reasoning pattern and decide whether it is valid.

1. If you are over 21, then you can vote.

   Marika cannot vote.

   Marika is not over 21.

2. If you do not pay taxes, then you will go to jail.

   Peter paid his taxes.

   Peter went to jail.

3. If you have enough, then you must share.

   Theresa has enough.

   Theresa must share.

4. If you are a citizen, then you want power.

   If you want power, then you must vote.

   If you are a citizen, then you must vote.

5. If you are a doctor, then you have gone to college.

   Pat is not a doctor.

   Pat did not go to college.

6. If x is divisible by 4, then x is even.

   10 is even.

   10 is divisible by 4.

CRITICAL THINKING ACTIVITY 9

Common Errors in Drawing a Conclusion

There are several common reasoning errors that are made every day in the critical thinking process. The ones described below often result in "proofs" that do not support the conclusion. Consider the following examples.

Example 1. Mr. Vasquez said, "All politicians are dishonest." Ms. Tsujimura said, "I agree with you. I've known a lot of politicians in my time, and none of them were honest."

This is an instance of *drawing a conclusion in the absence of a counterexample*. You have seen that a counterexample is a single example that shows a general statement to be false. In Example 1, the reasoning error was deciding a general statement to be true (all politicians are dishonest) just because a counterexample had not been found. This is similar to *drawing a conclusion from selected examples*, illustrated by the freshman who says, "The teachers in this school are strict—look at the ones I have this year."

Example 2. Jackie said, "The neighbors drive a big new car, so they must be rich."

This is an example of *drawing a conclusion that does not follow from the supporting statements.* An expensive-looking car is, of course, not enough evidence to support the conclusion that those who drive it are rich.

Example 3. Rose decided she needed a job. Her friend asked why. Rose said, "I need to work to pay for a car." Her friend asked, "Why do you need a car?" Rose answered, "I need it to drive to my job."

This illustrates *drawing a conclusion by assuming it is true* (or *circular reasoning*). Notice that Rose used the assumption that she needed a job to convince her friend that she needed a job.

Applications

Which reasoning error described in the preceding examples occurs in each situation? Explain.

1. Fred said, "My geometry teacher assigns a lot of homework. She must not like our class."
2. Janelle said to Samantha, "You joined our club because you're strange." Samantha asked, "Why do you think I'm strange?" Janelle responded, "You must be, or you wouldn't have joined our club."
3. Maurice said, "All 4-sided figures are quadrilaterals." Su Ming agreed, "I've never seen a 4-sided figure that wasn't a quadrilateral."
4. Niki said, "I've ordered 2 items from a bike catalog, and they have both been of low quality. That company obviously sells only inferior items."
5. Victoria said, "I'm running for mayor." Marc asked her why. She answered, "I'm quite intelligent." When Marc asked Victoria how she knew she was intelligent, she responded, "I must be, otherwise I wouldn't be running for mayor."
6. Roger drew 3 right triangles, measured their sides, and found that the Pythagorean Theorem held for all of them. He concluded, "The Pythagorean Theorem must be true because I can't draw a triangle for which it doesn't hold."
7. Consider the advertisement in Figure D.9. Which reasoning error occurs in the advertisement?

FIGURE D.9

CRITICAL THINKING ACTIVITY 10

Recognizing Quantifier Errors in Reasoning

Many reasoning errors occur in critical thinking because people do not know how to use the quantifier words *all*, *no*, and *some*. Consider these examples.

Example 1: All. Nino said, "All my friends dislike me." "You're wrong," said Tico, "All your friends like you a lot" (Fig. D.10).

Since *all* means *every* member of a certain set, Tico could have negated Nino's statement by giving a counterexample: "You're wrong—I'm your friend and I like you."

FIGURE D.10

Example 2: No. Jenny said, "No student from the Theatre Club is friendly." "I don't agree," said Sara, "All of the ones I've met are really friendly."

Since *no* means *not one, or none*, Sara found one example to contradict Jenny's statement and could say, "I disagree. I've met Sandy Sloan, and she is very friendly."

Example 3: Some. Tip said, "Some theorems in geometry are easy to prove." "I disagree," said Bill, "Some theorems in geometry are not easy to prove."

Since *some* means *at least one*, Bill would have to say the following to negate Tip's statement: "I disagree. No theorem in geometry is easy to prove."

All, *no*, and *some* statements can be written in one of the following four forms:

I. All A are B.
II. No A is B.
III. Some A are B.
IV. Some A are not B.

Applications

What would you have to do to prove that the statements in 1 through 4 are false?

1. Tom said, "All frogs are green."
2. Katy said, "No compact disc costs less than $10."
3. A teacher said, "Some students in our class are absent today."
4. A geometry student said, "Some triangles have 2 lines of symmetry."

Use statements I–IV of the preceding discussion.

5. Give examples to explain why statements I and II are not negations of each other. Do the same for statements III and IV.
6. Which pairs from statements I, II, III, and IV are negations of each other? Give examples to explain why you think so.
7. What if-then statement means the same as statement I? Statement II?
8. Which statement means the same as "Anytime an A is found to exist, it is guaranteed to be a B"? As "Anytime an A is found to exist, it is guaranteed not to be a B"?

FIGURE D.11

CRITICAL THINKING ACTIVITY 11

Recognizing the Role of Propaganda

Propaganda is any systematic promotion of ideas to further a cause. The term comes from the Latin word meaning "propagate." Sometimes we make incorrect decisions because we react to propaganda that someone uses to persuade us to a point of view. Consider the following examples.

Example 1. Jumping on the Bandwagon. Ben saw the advertisement of Figure D.11 in a magazine.
He felt that if Olympic athletes liked Panthers the best, they must be good running shoes. Ben bought two pairs of Panthers and later decided that he did not really like them.

Analysis: This form of propaganda tries to convince individuals that they will take on characteristics of others if they imitate their behavior.

Example 2. Appealing to Authority or Testimony. A television ad shows Bert Medico, an actor who plays a doctor on a soap opera, reading a report issued by the National Medical Doctors Association (NMDA). It describes the medical value of the pain reliever NoAche. A viewer likes Bert Medico, respects the NMDA, and goes to the store to buy NoAche. Later, she finds it upsets her stomach.

Analysis: In this situation, the fact that Bert plays a doctor on a soap opera carries over, as the viewer evaluates what Bert says during the ad. So it appears that the NMDA report on NoAche is accepted by a well-known doctor.

Example 3. Avoiding the Question. A congressman is asked if he is in favor of a plan to build a controversial highway. He responds, "I'm sure the people of this county will use wise judgment as they decide how to vote on this plan. I have always had great confidence in the decision-making process at our state and local levels."

Analysis: Listeners would think the congressman favors their view, as everyone likes to feel he or she uses good judgment in making decisions. The congressman did not actually state his position on the planned highway.

Example 4. Transferring a Feeling. A television ad for a sportscar shows a beautiful horse racing across a lush pasture as it shows the automobile. Marla watches the ad and tells a friend, "That car is a beauty—it can really fly."

Analysis: Marla equated the speed of the horse with the speed of the car in the advertisement. That is, she transferred her feelings about the horse to the car.

Applications

1. Give another example of each of the above forms of propaganda.

   Identify the type of propaganda involved in each situation:

2. Singh chose to buy a baseball because it had Babe Ruth's name on it.
3. Kara wanted to join a sorority because, "All my friends are joining a sorority."
4. When a store clerk was asked if an item had a guarantee, she said, "I have never had anyone bring one of those back because it did not work properly."
5. Think of another form of propaganda and give an example of its use.

CRITICAL THINKING ACTIVITY 12

Inductive Reasoning

Critical thinking is based on the consideration of evidence, patterns, and relationships. **Inductive reasoning**, which plays an important role, involves forming a conclusion by working from a sequence of specific examples.

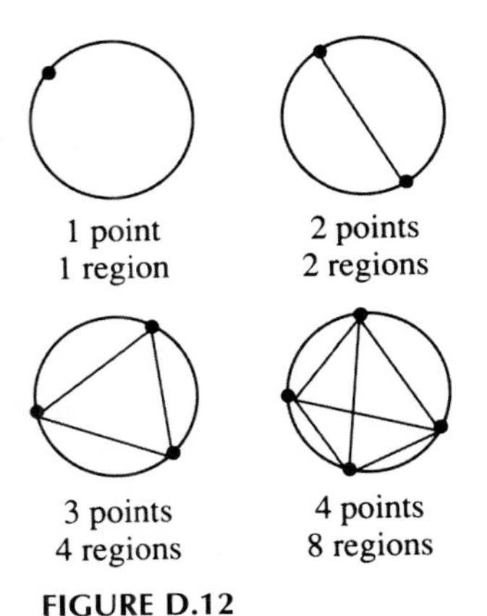

FIGURE D.12

In Figure D.12, consider the diagrams showing circles with interior regions formed by segments connecting points on the circles. What is the maximal number of regions formed in the interior of the circles by a given number of points? From the pattern of regions (1, 2, 4, and 8), you might guess that 5 points will result in 10 regions. You might form the conjecture that n lines cut the interior of a circle into $2n - 1$ regions. An extension of the drawings to the case with 5 points shows that the conjecture appears to hold. With 5 points, 16 regions are formed. However the extension to the case with 6 points shows that only 31 regions are formed. This case gives a counterexample that shows that the pattern formed by inductive reasoning was not valid for all cases.

Consider the pattern in the following sums. What conjecture is appropriate?

$$1 = 1$$
$$1 + 3 = 4$$
$$1 + 3 + 5 = 9$$
$$1 + 3 + 5 + 7 = 16$$
$$1 + 3 + 5 + 7 + 9 = 25$$

The sum for each series of odd numbers appears to be a square number. Thus, one might conjecture: The sum of the first n odd numbers is n^2 or

$$\underbrace{1 + 3 + 5 + 7 + \ldots + (2n - 1)}_{n \text{ numbers}} = n^2$$

This conjecture is valid and can be proved using a method known as mathematical induction. In this case, it led to a valid result. Inductive reasoning is a powerful form of critical reasoning, but the validity of the conjectures must be tested. It is not enough to form a conjecture on the basis of a few cases. If conjectures are to give us a useful generalization, they must be true for all cases.

Applications

Make a 10×10 grid of numbers from 1 to 100 to form conjectures and test if they are valid. If a conjecture is not valid, give a counterexample.

1. Select any 3 × 3 array of numbers from the grid. What can you say about the sum of the numbers on the diagonals?
2. Select any 3 × 3 array of numbers from the grid. What can you say about the relationship of the number in the center to the sum of the numbers on the diagonal?
3. Select any 3 × 3 array of numbers from the grid. What can you say about the sums of numbers in the first row and numbers in the last row?
4. Select any 3 × 3 array of numbers from the grid. What can you say about the sums of numbers from opposite corners?
5. Select any 3 × 3 array of numbers from the grid. What can you say about the sum of the middle column and the sum of the middle row?

CRITICAL THINKING ACTIVITY 13

Making and Supporting Decisions

In previous critical-thinking activities, the focus has been on individual topics, such as the following:

Understanding the situation
Dealing with assumptions
Checking logic

Understanding different patterns of reasoning

Now, consider all of these skills relative to the following situation: A shipment of 8 identical boxes, each assumed to contain the same top-secret equipment, was delivered to Sonia, the manager of the secret equipment warehouse (Fig. D.13).

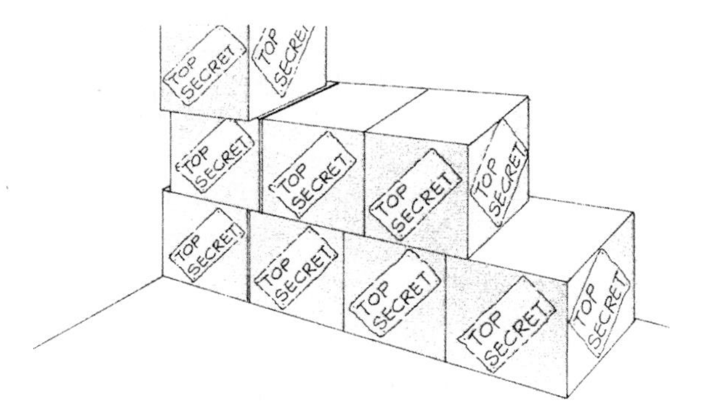

FIGURE D.13

They were marked, "TOP SECRET Do not open under any circumstances." As she was loading them to be taken to their destination, Sonia received a message that one box was missing a very small, but vital, part. Since she could not open the boxes, how could Sonia best determine which box had the missing part?

Answer the questions in each step below to use the process of critical reasoning.

1. Understand the situation.
   a. What is the situation about?
   b. What conclusion, decision, or solution is required?
   c. Are there any questions you could ask to clarify the problem?
2. Deal with the data/evidence/assumptions.
   a. Make a list of the most important evidence and data. Is other data needed?
   b. Which evidence is opinion, not fact? Which is not relevant?
   c. What assumptions are made? What other assumptions could be made?
3. Go beyond the data/evidence/assumptions.
   Sonia decided to use the precision balances at a nearby aerospace laboratory (Fig. D.14). The scale operator allowed 2 free weighings and then charged $50 per weighing thereafter. Furthermore, Sonia did not have any funds at her disposal.
4. State and support your conclusion/decision/solution.
   Write a paragraph detailing your solution to Sonia's problem. Describe one process she could have used to find the desired box without spending any money.

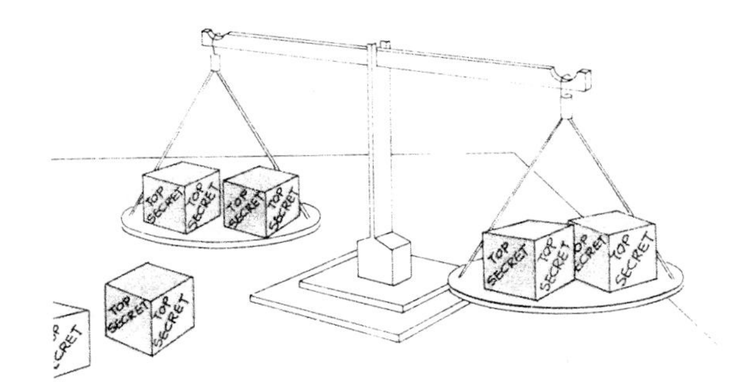

FIGURE D.14

5. Apply the conclusion/decision/solution.
   Two weeks after the above situation occurred, Sonia received a shipment of 12 boxes containing top-secret equipment. Again, one box was missing a very small part. Sonia needed to use the precision balances to determine which box had the missing part. What is the fewest number of weighings necessary in this case?

CRITICAL THINKING ACTIVITY 14

Making and Supporting Decisions

Apply your critical-thinking abilities to solving the following situation: It takes 9 days to cross the Dryasdust Desert on foot. A messenger must deliver a secret written message to the other side of the desert, where no food and water is available, and then return. One person can carry enough food and water to last for 12 days. Food and water may be buried in plastic sacks in the desert and collected for use on the way back. There are 2 messengers available to carry out the task. How quickly can the 2 messengers get the message delivered without either messenger suffering from lack of food and water?

1. Understand the situation.
   a. What is the situation about?
   b. What conclusion, decision, or solution is required?
   c. Are there any questions that would clarify the situation?
2. Deal with the data/evidence/assumptions.
   a. Make a list of the most important evidence/data. Are other data needed?
   b. What evidence is opinion, not fact? Which data are not relevant?
   c. What assumptions are made? What other assumptions could be made?
3. Go beyond the data/evidence/assumptions.
   a. Try the situation with deserts requiring fewer days to cross.
   b. Develop a method of recording the positions of the messengers and the amount of food available.
   c. Is there any sense to having one of the other messengers return to base and then head back out to meet the other messenger returning?

4. State and support your conclusion/decision/solution.
   Write a paragraph, with an accompanying diagram, describing your solution to this situation.
5. Apply the conclusion/decision/solution.
   What is the widest desert that can be crossed by the 2 messengers if they are not allowed to bury food for the return trip?

CRITICAL THINKING ACTIVITY 15

Making and Supporting Decisions

Many critical-thinking problems require the use of visual reasoning. Use these skills to develop a solution for the following situation: In a certain two-person game, the players take turns placing a penny on a circular table top with a green tablecloth (Fig. D.15). The pennies may touch, but are not allowed to rest upon one another. Furthermore, they must be completely on the table—not hanging out over the edge. Once a penny is placed on the table, it cannot be moved. The last player to place a penny on the table is the winner. If each player makes an optimal move on his or her turn, will the player who went first or second be the winner?

Answer the questions in each step below to use the process of critical thinking.

1. Understand the situation.
   a. What is the situation about?
   b. What conclusion, decision, or solution is required?
   c. Are there any questions you could ask to clarify the problem?

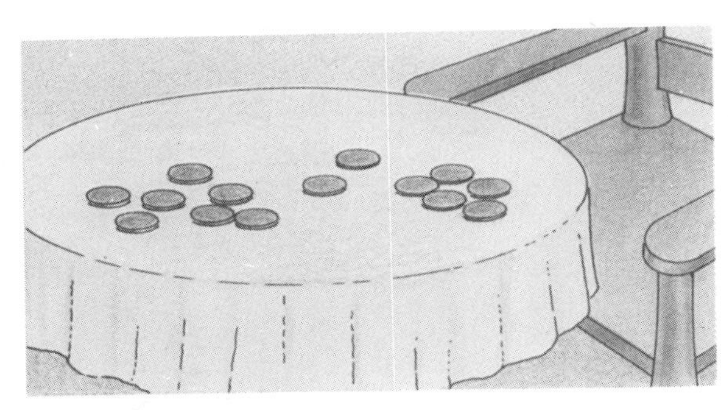

FIGURE D.15

2. Deal with the data/evidence/assumptions.
   a. Make a list of the most important rules and data. Is other information needed?
   b. Which data given in the situation is not relevant?
   c. What assumptions are made? What other assumptions could be made?
3. Go beyond the data/evidence/assumptions.
   a. Try this problem with very small tables.
   b. Consider a table whose diameter is the width of a certain number of pennies.
   c. Does it make a difference if the diameter is a certain number of pennies long?
4. State and support your conclusion/decision/solution.
   Write a paragraph, with an accompanying diagram, describing your solution to this situation.
5. Apply the conclusion/decision/solution.
   A similar situation involved a square table (Fig. D.16). What modifications, if any, would need to be made in the solution strategy to answer the same question about the game played on this table?

Applications

1. Use the above process for the following situation: Lucille found 4 sections of a bracelet. Each piece consisted of 3 links of gold. She wanted to have the 4 pieces joined to form one loop. A jeweler told her it would cost $1 for each link that was opened and welded shut again. When the jeweler wrote a bill for $4, Lucille said she thought the loop could be made with fewer than 4 links being cut. Do you agree?

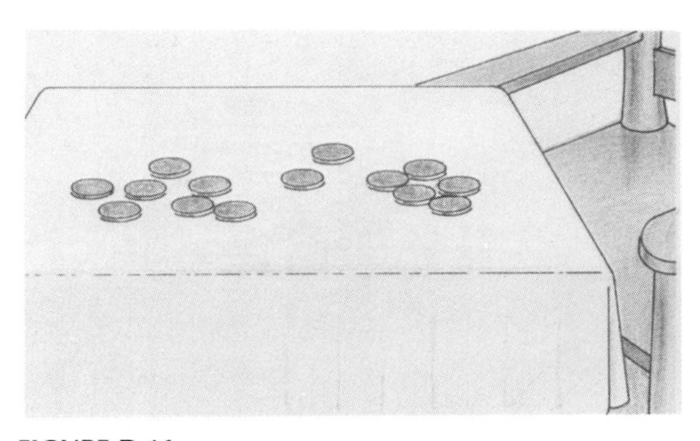

FIGURE D.16

APPENDIX E Computer Explorations in Geometry

Computer geometry exploration software opens exciting possibilities for discovering geometric relationships. The software allows you to draw any type of polygon, mark points, draw segments, and automatically gives the measures of segment length, angles, and area of regions. With these and other capabilities, you can quickly produce and examine a variety of examples. Using inductive reasoning from the examples, you can form and test all kinds of generalizations.

The following list includes data on some currently available geometry exploration software packages for your convenience.

Euclid's Toolbox. Available from Addison-Wesley Publishing Co., Order Department, Reading, MA 01867 (1-800-447-2226).

Rene's Place. Available from Addison-Wesley Publishing Co., Order Department, Reading, MA 01867 (1-800-447-2226).

The Geometric Supposer: Triangles, Quadrilaterals, Circles. Available from Sunburst Communications, 101 Castleton Street, Pleasantville, NY 10570-3498 (800-628-8897 or 914-747-3310).

Geometric Connectors: Transformations, Coordinates. Available from Sunburst Communications, 101 Castleton Street, Pleasantville, NY 10570-3498 (800-628-8897 or 914-747-3310).

GeoDraw. IBM Geometry Series, IBM Corporation (1-800-426-2468).

Geometry Sketch Pad. Available from Key Curriculum Press, 2512 Martin Luther King Jr. Way, Berkley, CA 94704 (415-548-2304).

The 28 explorations in this appendix provide an opportunity for you to explore and discover a number of important relationships in geometry. As recommended in the NCTM Standards, the importance of inductive reasoning as well as the need to establish the truth of theorems using deductive reasoning will become clear.

Computer Exploration 1

Angle Measures of a Triangle

Objective

Determine the sum of the measures of the angles of a triangle.

Investigation

Construct a triangle. Measure its interior angles and record the data. Find the sum of the angle measures. Repeat for other types of triangles.

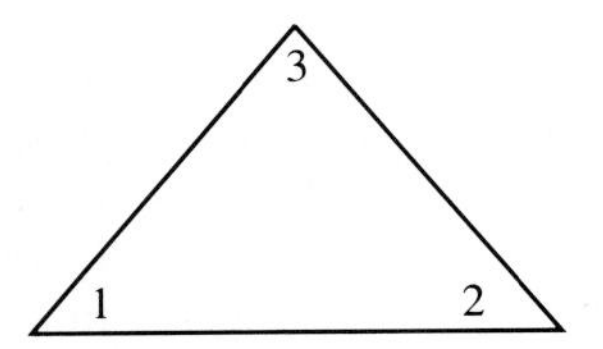

	$m\angle 1$	$m\angle 2$	$m\angle 3$	Sum
Acute				
Acute				
Obtuse				
Obtuse				
Right				
Right				
Isosceles				
Isosceles				
Equilateral				
Equilateral				

Conjectures

1. Make a conjecture about the sum of the measures of the angles of a triangle.
2. Make a conjecture about the angles of an equilateral triangle.
3. Make a conjecture about the acute angles of a right triangle.

Computer Exploration 2

Exterior Angles of a Triangle

Objectives

Determine the sum of the measures of the exterior angles of a triangle.
Determine the relationship between an exterior angle and its two remote interior angles.

Investigation

Construct a triangle and extend each side as shown in the figure. Measure the exterior angles and the interior angles. Record the data. Repeat for the different triangles shown in the charts.

1 2 3

	$m\angle 1$	$m\angle 2$	$m\angle 3$	Sum
Equilateral				
Isosceles				
Right				
Obtuse				
Acute				

	$m\angle 1$	Remote $\angle$s		$m\angle 2$	Remote $\angle$s		$m\angle 3$	Remote $\angle$s	
Equilateral									
Isosceles									
Right									
Obtuse									
Acute									

Conjectures

1. Using your data from the first chart, make a conjecture about the sum of the measures of the exterior angles of a triangle.
2. Using your data from the second chart, make a conjecture about the measure of an exterior angle of a triangle and the measures of its two remote interior angles.

Computer Exploration 3

Isosceles Triangles

Objective

Determine the type of triangle formed by the base of an isosceles triangle and the altitudes drawn from the base angles to the congruent sides.

Investigation

Construct an isosceles triangle. Draw the altitudes from the base angles. Label the intersection of the altitudes as shown. Measure segments AF and BF. Record the data. Repeat for the different triangles shown in the chart.

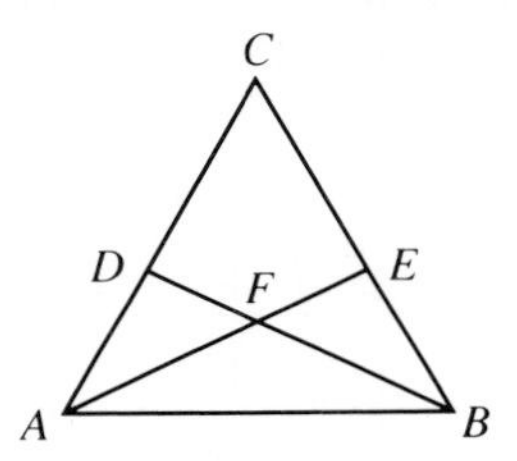

	AF	BF
Isosceles (three acute angles)		
Isosceles (one obtuse angle)		
Equilateral		
Another Isosceles		
Non-isosceles		

Conjecture

1. Make a conjecture about the type of triangle that is formed by the base of an isosceles triangle and the altitudes drawn from the base angles to the congruent sides.

Computer Exploration 4

Diagonals of a Parallelogram

Objective

Investigate the properties of the diagonals of a parallelogram.

Investigation

Draw a parallelogram and its two diagonals. Label the point of intersection of the diagonals. Measure the lengths of the four segments formed by the diagonals.

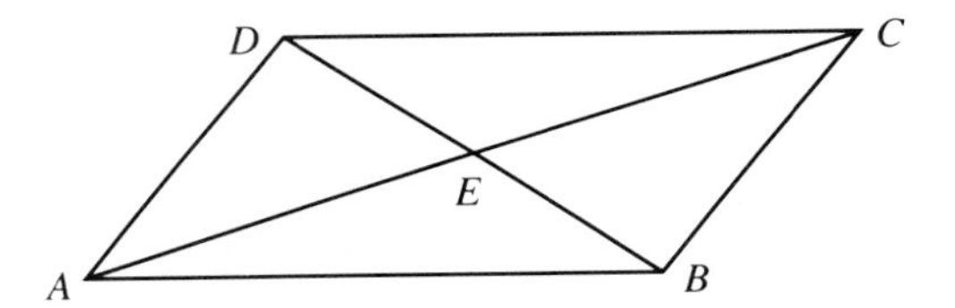

	AE	CE	BE	DE
Parallelogram 1				
Parallelogram 2				
Parallelogram 3				
Parallelogram 4				

Conjecture

1. Make a conjecture about the diagonals of a parallelogram.

Computer Exploration 5

Diagonals of a Quadrilateral

Objective

Determine the type of quadrilateral formed by diagonals that bisect each other.

Investigation

Construct a quadrilateral by drawing diagonal segments that bisect each other. Measure the angles and segments listed in the chart. Record the data. Construct other quadrilaterals as needed to make your conjectures.

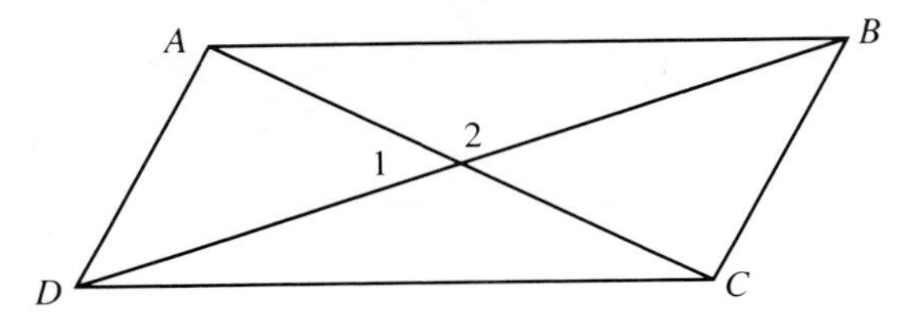

	$m\angle 1$	$m\angle 2$	AC	BD	Type of Quadrilateral
Quadrilateral 1					
Quadrilateral 2					
Quadrilateral 3					
Quadrilateral 4					

Conjectures

1. Make a conjecture about the type of quadrilateral that is formed by diagonals that bisect each other.
2. Make a conjecture about how the angle of the intersection of the diagonals affects the type of quadrilateral that is formed.
3. Make a conjecture about how the lengths of the diagonals affect the type of quadrilateral that is formed.

Computer Exploration 6

Diagonals of a Rhombus

Objective

Investigate the properties of the diagonals of a rhombus.

Investigation

Draw a rhombus and its two diagonals. Measure the angles formed by the diagonals. Measure the angles formed by the diagonals and the sides of the rhombus.

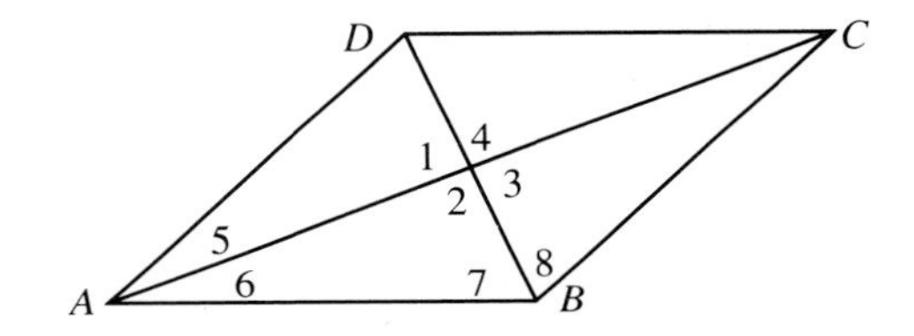

	$m\angle 1$	$m\angle 2$	$m\angle 3$	$m\angle 4$	$m\angle 5$	$m\angle 6$	$m\angle 7$	$m\angle 8$
Rhombus 1								
Rhombus 2								
Rhombus 3								
Rhombus 4								

Conjectures

1. Make a conjecture about the angles formed by the diagonals of a rhombus.
2. Make a conjecture about the pair of angles formed by a diagonal and the consecutive sides of a rhombus.

Computer Exploration 7

Median of a Trapezoid

Objective

Investigate relationships between the median and the two bases of a trapezoid.

Investigation

Draw a trapezoid and its median. Measure the length of the bases and the median. Measure the angles indicated.

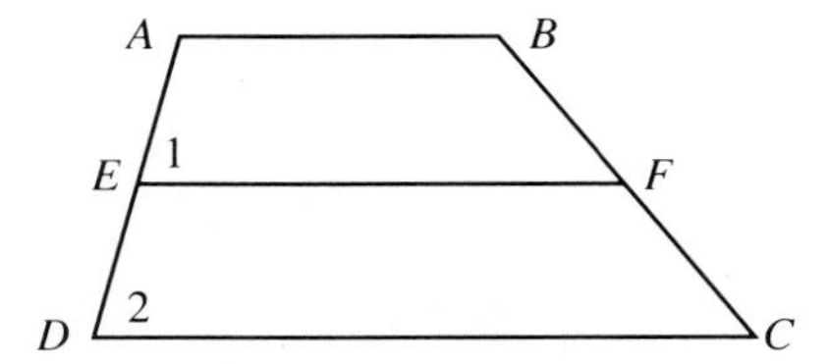

	AB	CD	EF	$m\angle 1$	$m\angle 2$
Trapezoid 1					
Trapezoid 2					
Trapezoid 3					
Trapezoid 4					

Conjectures

1. Make a conjecture about the lengths of a trapezoid's median and its two bases.
2. Make a conjecture about the position of a trapezoid's median in relation to its bases.

Computer Exploration 8

Midpoints of a Quadrilateral

Objective

Determine the type of quadrilateral formed by joining the midpoints of consecutive sides of a quadrilateral.

Investigation

Draw a quadrilateral. Label the midpoints of each side of the quadrilateral. Draw the segments joining the midpoints of consecutive sides of the quadrilateral. Measure the lengths of the sides and the angles of the newly formed quadrilateral.

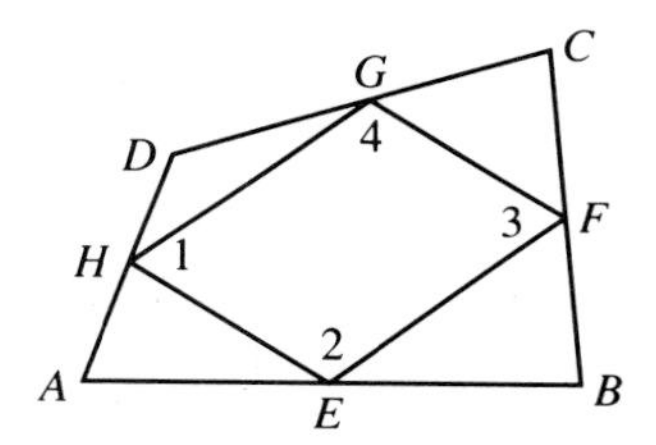

	EF	FG	GH	HE	$m\angle 1$	$m\angle 2$	$m\angle 3$	$m\angle 4$
Quadrilateral 1								
Quadrilateral 2								
Quadrilateral 3								
Quadrilateral 4								

Conjecture

1. Make a conjecture about the type of quadrilateral that is formed by joining the midpoints of consecutive sides of a quadrilateral.

Computer Exploration 9

Midpoints of an Isosceles Trapezoid

Objective

Determine the type of quadrilateral formed by joining the midpoints of consecutive sides of an isosceles trapezoid.

Investigation

Draw an isosceles trapezoid and label the midpoints of each side. Draw the segments joining the midpoints of consecutive sides of the isosceles trapezoid. Measure the lengths of the sides and the angles of the newly formed quadrilateral.

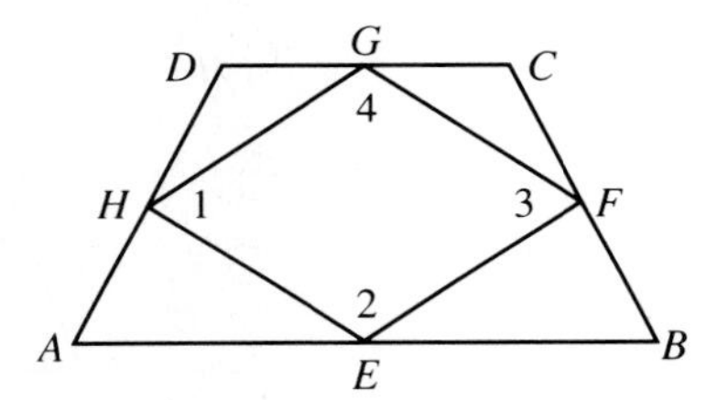

	EF	FG	GH	HE	$m\angle 1$	$m\angle 2$	$m\angle 3$	$m\angle 4$
Isosceles Trapezoid 1								
Isosceles Trapezoid 2								
Isosceles Trapezoid 3								
Isosceles Trapezoid 4								

Conjecture

1. Make a conjecture about the type of quadrilateral formed by joining the midpoints of consecutive sides of an isosceles trapezoid.

Computer Exploration 10

Triangle Midsegments

Objective

Determine special relationships between a midsegment of a triangle and the side that does not contain the endpoints of the midsegment.

Investigation

Construct a triangle and a midsegment as shown. Measure the midsegment and the side of the triangle that does not contain the midsegment's endpoints. Measure angles 1 and 2. Record the data. Repeat for the different triangles shown in the chart.

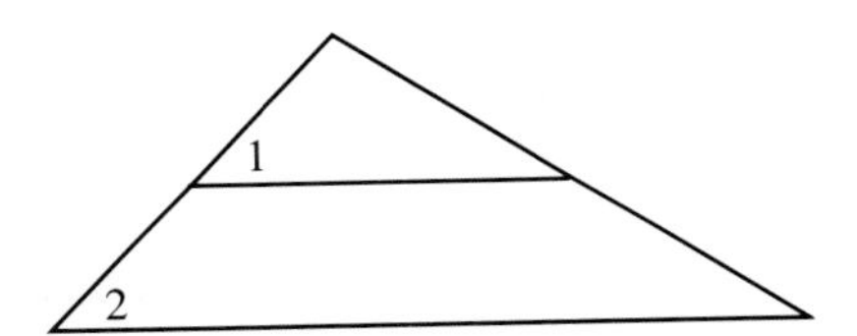

	Length of Midsegment	Length of Side	$m\angle 1$	$m\angle 2$
Equilateral				
Isosceles				
Right				
Obtuse				
Acute				

Conjectures

1. Make a conjecture about the length of the midsegment and the length of the side of a triangle not containing the endpoints of the midsegment.
2. Will the line containing the midsegment intersect the side of the triangle that does not contain the midsegment's endpoints? How do you know?

Computer Exploration 11

Triangle Side Lengths and Angle Measures

Objective

Determine the relationship between the lengths of the sides of a triangle and the measures of the angles of the triangle.

Investigation

Draw a triangle having the side lengths given in the chart for Triangle 1. Measure the angles of the triangle and the lengths of the sides of the triangle. Record your data. Repeat for the other triangles in the chart.

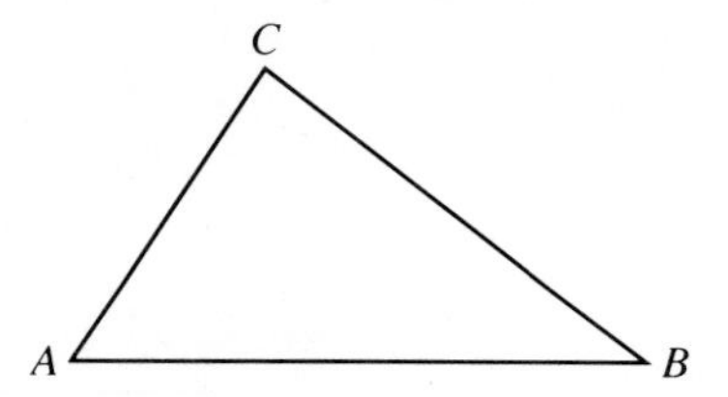

	AB	AC	BC	$m\angle C$	$m\angle B$	$m\angle A$
Triangle 1	10	12	14			
Triangle 2	21	17	25			
Triangle 3	17	20	14			
Triangle 4	18	20	19			
Triangle 5	25	15	20			
Triangle 6	20	18	16			

Conjecture

1. For each triangle, compare the measures of each side and the angle opposite that side. Make a conjecture about the relationship between the measures of the angles of a triangle and the lengths of the sides of the triangle.

Computer Exploration 12

Lines Parallel to a Side of a Triangle

Objective

Determine ratios of the lengths of segments produced by a line that is parallel to one side of a triangle and intersects the triangle's other two sides.

Investigation

Construct a triangle and label a point on a side. Draw a line through the point that is parallel to one of the other sides and intersects the third side. Measure the segments formed. Record the data. Find the ratios shown in the charts. Repeat for different types of triangles.

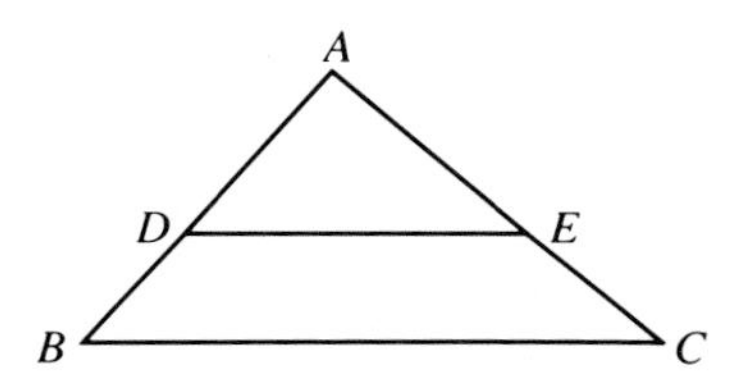

	AD	*DB*	*AE*	*EC*	*DE*	*AB*	*AC*	*BC*
Acute								
Obtuse								
Isosceles								
Equilateral								

	AD:*DB*	*AE*:*EC*	*AD*:*AB*	*AE*:*AC*	*DE*:*BC*
Acute					
Obtuse					
Isosceles					
Equilateral					

Conjectures

1. Make a conjecture about the ratios of the lengths of the segments formed by a line that is parallel to a side of a triangle and that intersects the other two sides.
2. Make a conjecture about the relationships between the ratio of the lengths of the two parallel line segments in your investigation and the ratios of the lengths of the other segments in the triangle
3. Make a conjecture about the relationship between a triangle and the triangle formed by a line that is parallel to one of its sides and that intersects the other two sides.

Computer Exploration 13

Angle Bisector Segments

Objective

Determine relationships between the segments formed by an angle bisector of a triangle and the other two sides of the triangle.

Investigation

Construct a triangle. Construct an angle bisector that intersects the opposite side. Measure segments and record your data. Compare segment ratios as shown in the chart. Repeat for the different types of triangles shown in the chart.

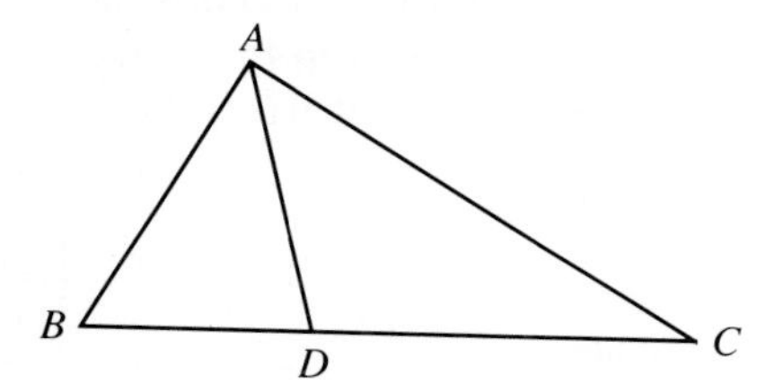

	AB	AC	BD	DC	$AB:AC$	$BD:DC$
Acute						
Obtuse						
Isosceles						
Equilateral						
Right						

Conjectures

1. Make a conjecture about the segments formed by an angle bisector of a triangle and the other two sides of the triangle.
2. Does any class of triangles have special segment ratios? If so, what data supports your conclusion?

Computer Exploration 14

Angle Bisectors of Similar Triangles

Objective

For similar triangles, determine relationships between the lengths of corresponding angle bisectors and the lengths of corresponding sides.

Investigation

Construct a triangle. Construct an angle bisector that intersects the opposite side. Measure segments and record your data. Then draw a triangle similar to the first by giving it a different size (scale), and construct a corresponding angle bisector. Find the segment ratios indicated in the second chart. Repeat for other pairs of triangles.

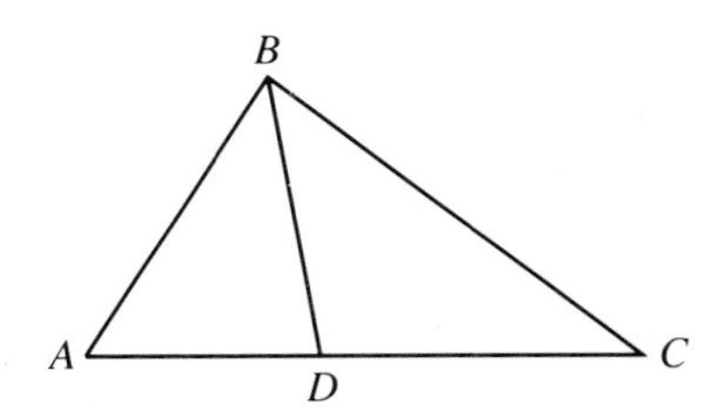

	First Triangle				Scaled Triangle			
	AB	BC	AC	BD	$A'B'$	$B'C'$	$A'C'$	$B'D'$
Acute								
Acute								
Obtuse								
Obtuse								

	$AB:A'B'$	$BC:B'C'$	$AC:A'C'$	$BD:B'D'$
Acute				
Acute				
Obtuse				
Obtuse				

Conjecture

1. Make a conjecture about the relationship between corresponding sides and angle bisectors of similar triangles.

Computer Exploration 15

Altitude to Hypotenuse

Objective

Determine relationships among a right triangle and the two triangles that are formed by the altitude to the hypotenuse.

Investigation

Construct a right triangle. Construct the altitude to the hypotenuse. Measure the angles indicated in the three charts. Record your data in the rows for Right Triangle 1. Repeat for other right triangles.

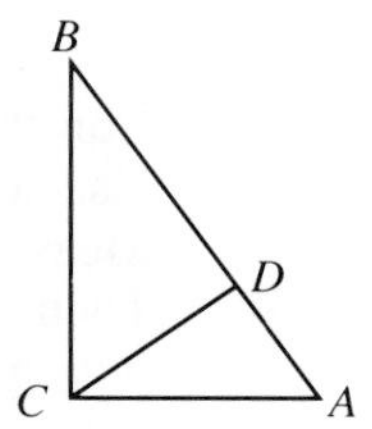

Triangle *ABC*	$m\angle BCA$	$m\angle CAB$	$m\angle ABC$
Right Triangle 1			
Right Triangle 2			
Right Triangle 3			

Triangle *BCD*	$m\angle CDB$	$m\angle BCD$	$m\angle DBC$
Right Triangle 1			
Right Triangle 2			
Right Triangle 3			

Triangle *CAD*	$m\angle ADC$	$m\angle CAD$	$m\angle DCA$
Right Triangle 1			
Right Triangle 2			
Right Triangle 3			

Conjecture

1. Make a conjecture about the relationships among a right triangle and the two triangles that are formed by the altitude to the hypotenuse.

Computer Exploration 16

Length of Altitude to Hypotenuse

Objective

Determine the relationship between the length of the altitude to the hypotenuse of a right triangle and the length of the two segments on the hypotenuse.

Investigation

Construct a right triangle. Construct the altitude to the hypotenuse. Measure the segments. Record your data. Repeat for other right triangles.

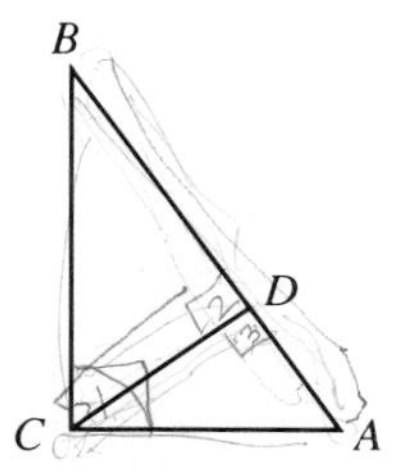

	BD	DA	CD	$BD \cdot DA$
Right Triangle 1				
Right Triangle 2				
Right Triangle 3				
Right Triangle 4				

Conjecture

1. Make a conjecture about the relationship between the length of the altitude to the hypotenuse of a right triangle and the length of the two segments on the hypotenuse.

Computer Exploration 17

Hypotenuse, Legs, and Altitude

Objective

Determine the relationship among the legs, altitude, and hypotenuse of a right triangle.

Investigation

Construct a right triangle. Construct the altitude to the hypotenuse. Measure segments and record your data. Repeat for other right triangles.

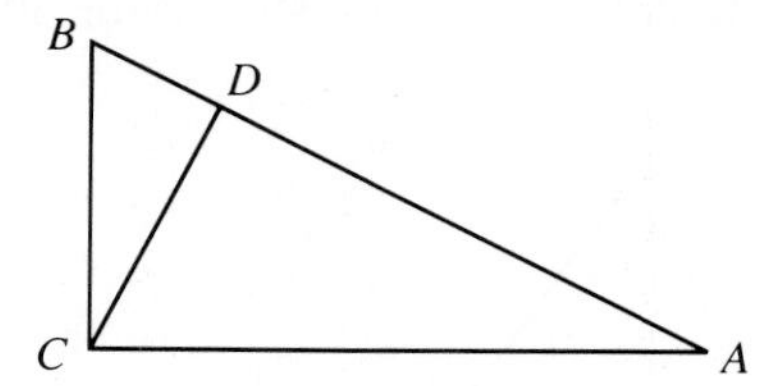

	AC	BC	AB	CD	$AB \cdot CD$
Right Triangle 1					
Right Triangle 2					
Right Triangle 3					
Right Triangle 4					

Conjecture

1. Make a conjecture about the relationship between a right triangle's hypotenuse and altitude, and its two legs.

Computer Exploration 18

Hypotenuse, Hypotenuse Segments, and Legs

Objective

Given an altitude to the hypotenuse of a right triangle, determine the relationship among the hypotenuse, a leg, and a segment on the hypotenuse.

Investigation

Construct a right triangle. Construct the altitude to the hypotenuse. Measure the segments. Record your data. Find the products listed in the chart. Repeat for other right triangles.

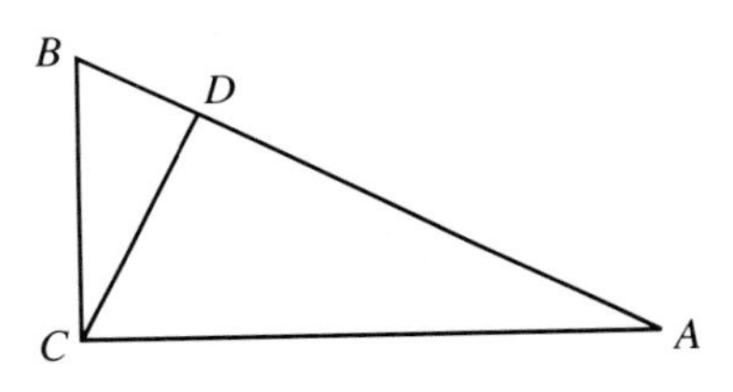

	AC	BC	AB	BD	AD	$BD \cdot AB$	$AD \cdot AB$
Right Triangle 1							
Right Triangle 2							
Right Triangle 3							
Right Triangle 4							

Conjecture

1. Make a conjecture about the relationship among the hypotenuse of a right triangle, the two segments formed by the altitude to the hypotenuse, and the legs.

Computer Exploration 19

Triangles and Side Lengths

Objectives

Find special relationships of side lengths in triangles, and find the type of triangle determined by certain side-length relationships.

Investigation

Construct triangles with sides of different lengths. Label the longest side AB. Measure the lengths of the sides. Record your data. Make the calculations required to complete the chart.

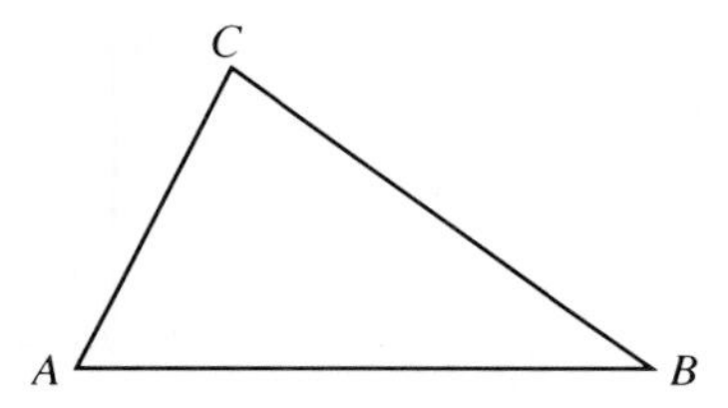

Type of Triangle	AC	BC	AB	AC^2	BC^2	AB^2	$AC^2 + BC^2$

Conjectures

1. For what type of triangle does AC^2 plus BC^2 equal AB^2?
2. For what type of triangle is $AC^2 + BC^2$ less than AB^2?
3. For what type of triangle is $AC^2 + BC^2$ greater than AB^2?

Computer Exploration 20

45°–45°–90° Triangles

Objective

Determine the relationship between the length of the hypotenuse of a 45°– 45°–90° triangle and the length of a leg.

Investigation

Construct a 45°– 45°–90° triangle. Measure the lengths of the sides. Record your data. Calculate the ratio shown in the chart. Repeat for other 45°– 45°–90° triangles.

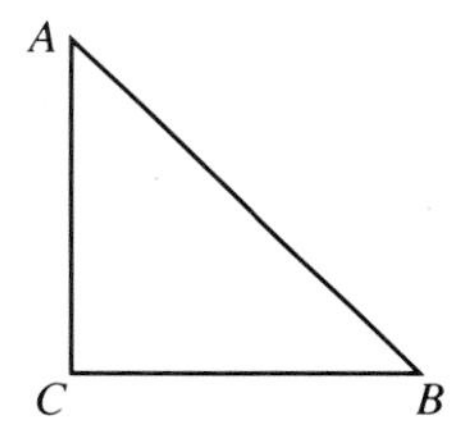

	AC	*CB*	*AB*	*AB*:*CB*
45°– 45°–90° Triangle 1				
45°– 45°–90° Triangle 2				
45°– 45°–90° Triangle 3				
45°– 45°–90° Triangle 4				

Conjecture

1. Make a conjecture about the relationship between the length of the hypotenuse of a 45°– 45°–90° triangle and the length of a leg. Express your conjecture in terms of an exact real number.

Computer Exploration 21

30°–60°–90° Triangles

Objective

Given a 30°–60°–90° triangle, determine the relationship between the lengths of the hypotenuse and the shorter leg, and the relationship between the lengths of the longer leg and the shorter leg.

Investigation

Construct a 30°–60°–90° triangle. Measure the lengths of the sides. Record your data. Calculate the ratios shown in the chart. Repeat for other 30°–60°–90° triangles.

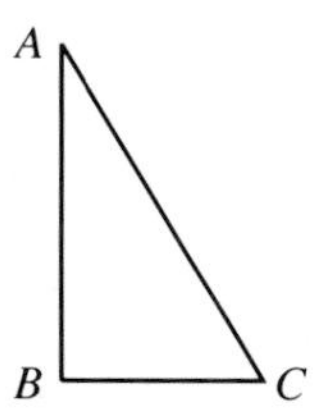

	AC	*AB*	*BC*	*AC*:*BC*	*AB*:*BC*
30°–60°–90° Triangle 1					
30°–60°–90° Triangle 2					
30°–60°–90° Triangle 3					
30°–60°–90° Triangle 4					

Conjectures

1. Make a conjecture about the relationship between the length of the hypotenuse of a 30°–60°–90° triangle and the length of the shorter leg.
2. Make a conjecture about the relationship between the lengths of the legs of a 30°–60°–90° triangle. Express your conjecture in terms of an exact real number.

Computer Exploration 22

The Tangent Ratio

Objective

Investigate the ratio of the lengths of the two legs of similar right triangles.

Investigation

Draw a right triangle. Measure the lengths of the two legs and calculate the ratio of their lengths. Draw a triangle similar to the original one by giving it a different size (scale). Record the data. Repeat for two more similar triangles.

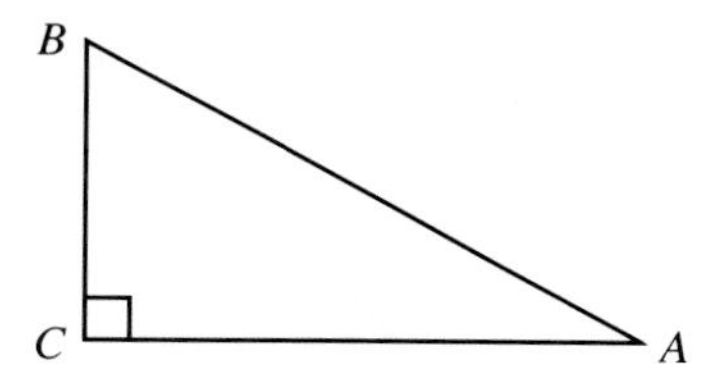

	BC	*AC*	*BC*:*AC*
Right Triangle			
1st Similar Triangle			
2nd Similar Triangle			
3rd Similar Triangle			

Conjecture

1. Make a conjecture about the ratio of the lengths of the legs of similar right triangles.

Computer Exploration 23

Perpendicular Bisectors of the Sides of a Triangle, I

Objective

Investigate the properties of the perpendicular bisectors of the sides of a triangle.

Investigation

Draw a triangle and the perpendicular bisector of each side. Be sure each bisector intersects one of the other sides of the triangle. Do this construction for acute, right, and obtuse triangles.

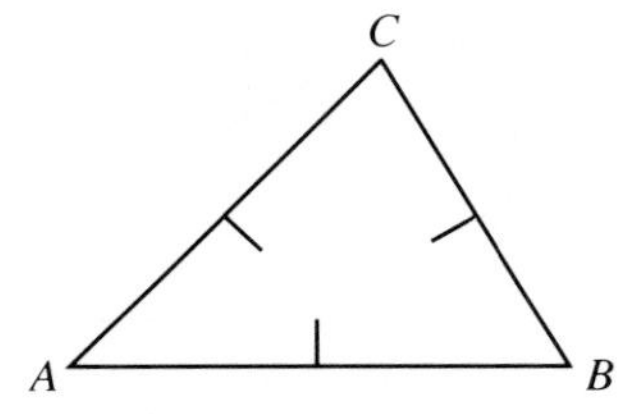

Conjectures

1. Make a conjecture about the three perpendicular bisectors of the sides of a triangle.
2. Make specific conjectures about the perpendicular bisectors of acute, right, and obtuse triangles.

Computer Exploration 24

Perpendicular Bisectors of the Sides of a Triangle, II

Objective

Investigate the properties of the perpendicular bisectors of the sides of a triangle.

Investigation

Draw a triangle and the perpendicular bisector of each side of the triangle. Label the intersection of the perpendicular bisectors. Draw a line segment from the intersection of the perpendicular bisectors to the vertices. Measure each segment. Repeat the construction for the other triangles in the chart.

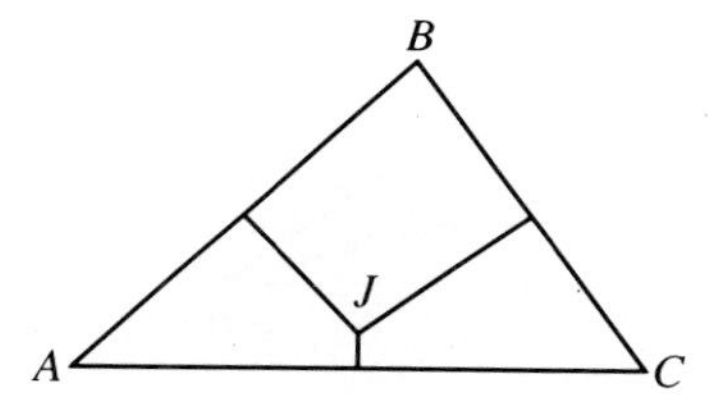

	AJ	BJ	CJ
Acute Triangle			
Obtuse Triangle			
Right Triangle			

Conjecture

1. Make a conjecture about the distance from each vertex of a triangle to the point of intersection of the perpendicular bisectors.

Computer Exploration 25

Angle Bisectors of a Triangle

Objective

Investigate the properties of the bisectors of the angles of a triangle.

Investigation

Draw a triangle. Draw the bisector of each of its angles. Be sure each bisector intersects a side of the triangle. Measure the perpendicular distance from the point of intersection of the angle bisectors to each of the sides of the triangle. Repeat for other triangles.

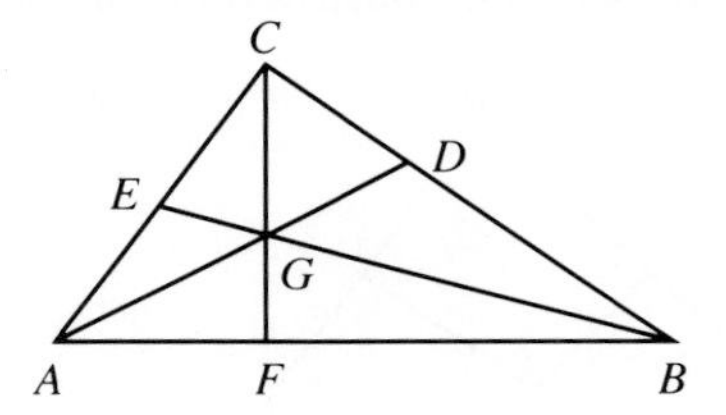

	Distance from G to:		
	AB	BC	CA
Triangle 1			
Triangle 2			
Triangle 3			
Triangle 4			

Conjecture

1. The angle bisectors of a triangle intersect at a point. Make a conjecture about the distance from the point of intersection of the angle bisectors to each of the sides.

Computer Exploration 26

Medians of a Triangle

Objective

Investigate the properties of the medians of a triangle.

Investigation

Draw a triangle and its three medians. Label the point of intersection of the medians. Measure the length of the segments formed by the medians.

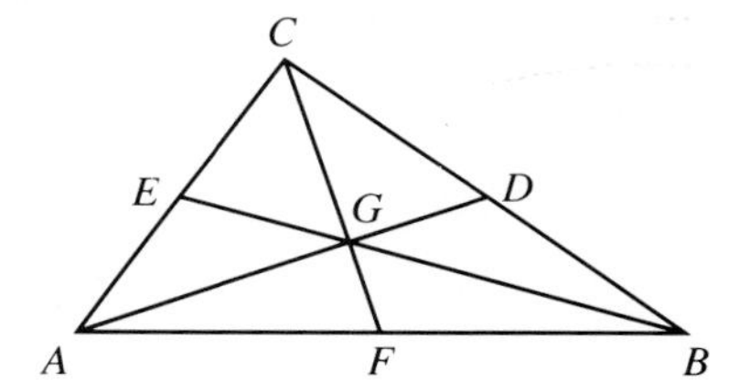

	AG	GD	BG	GE	CG	GF
Triangle 1						
Triangle 2						
Triangle 3						
Triangle 4						

Conjecture

1. The three medians of a triangle intersect at a point. This point divides each median into two segments. Make a conjecture about the relationship between the two segments of each median.

Computer Exploration 27

Similar Triangles and Perimeter

Objective

Determine the relationship between the perimeter and the corresponding sides of similar triangles.

Investigation

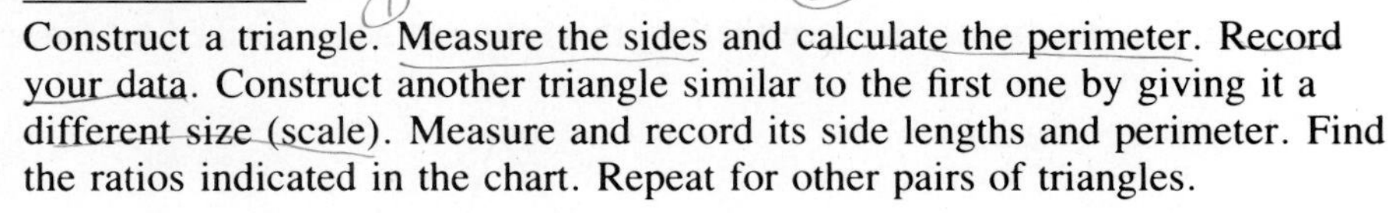

Construct a triangle. Measure the sides and calculate the perimeter. Record your data. Construct another triangle similar to the first one by giving it a different size (scale). Measure and record its side lengths and perimeter. Find the ratios indicated in the chart. Repeat for other pairs of triangles.

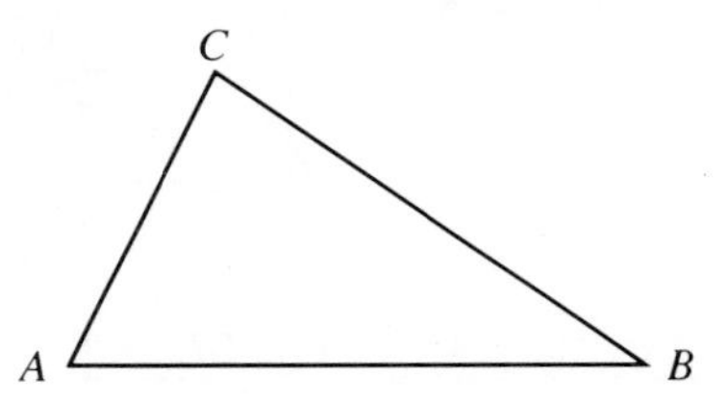

	First Triangle				Scaled Triangle			
	AB	BC	AC	Perimeter	$A'B'$	$B'C'$	$A'C'$	Perimeter
Acute								
Obtuse								
Isosceles								
Right								

	$AB{:}A'B'$	$BC{:}B'C'$	$AC{:}A'C'$	Perimeter of $\triangle ABC$: Perimeter of $\triangle A'B'C'$
Acute				
Obtuse				
Isosceles				
Right				

Conjecture

1. Make a conjecture about the ratio of two similar triangles' perimeters and the ratio of their corresponding side lengths.

Computer Exploration 28

Similar Triangles and Area

Objective

Determine the relationship between the area and the corresponding sides of similar triangles.

Investigation

Construct a triangle. Measure the sides and find the area. Record your data. Construct another triangle similar to the first one by giving it a different size (scale). Measure and record the side lengths and area. Find the ratios indicated in the chart. Repeat for other pairs of triangles.

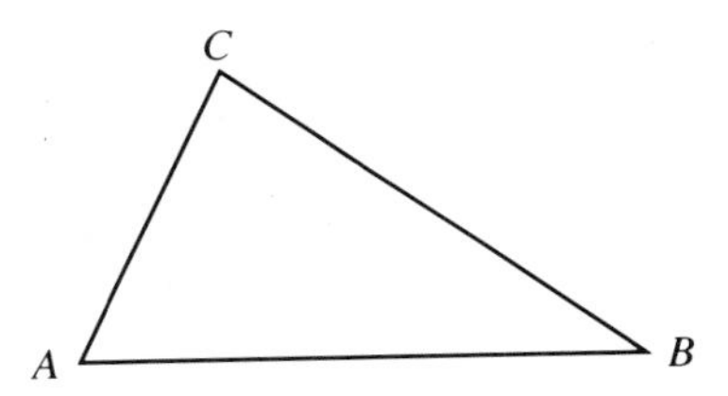

	First Triangle				Scaled Triangle			
	AB	BC	AC	Area	$A'B'$	$B'C'$	$A'C'$	Area
Acute								
Obtuse								
Isosceles								
Right								

	$AB{:}A'B'$	$BC{:}B'C'$	$AC{:}A'C'$	Area of $\triangle ABC$: Area of $\triangle A'B'C'$
Acute				
Obtuse				
Isosceles				
Right				

Conjecture

1. Make a conjecture about the ratio of two similar triangles' areas and the ratio of their corresponding side lengths.

APPENDIX F Glossary, Postulates, and Theorems

GLOSSARY

acute angle An angle with measure less than 90°.

acute triangle A triangle with three acute angles.

alternate exterior angles Two exterior angles with different vertices on opposite sides of a transversal.

alternate interior angles Two interior angles with different vertices on opposite sides of a transversal.

altitude of a triangle A segment from a vertex to a point on the opposite side (perhaps extended) that is perpendicular to that opposite side.

angle The union of two noncollinear rays that have the same endpoint.

axiom A statement that is accepted as true without proof.

bisector of a segment Any point, segment, ray, line, or plane that contains the midpoint of the segment.

bisector of an angle The bisector of $\angle ABC$ is a ray BD in the interior of $\angle ABC$ such that $\angle ABD \cong \angle DBC$.

centroid of a triangle The point of intersection of the medians.

circle The set of all points in a plane that are a fixed distance from a given point in the plane.

circumcenter of a triangle The point that is equidistant from the vertices of a triangle.

circumference of a circle The distance around the circle, represented by the number approached by the perimeters of the inscribed regular polygons as the number of sides of the regular polygons increases.

circumscribed circle A circle that contains the three vertices of a triangle. The center of the circle is the point of intersection of the perpendicular bisectors of the sides of the triangle.

collinear points Points that lie on the same line.

complementary angles Two angles whose measures have a sum of 90°.

concurrent lines Three or more coplanar lines that have a point in common.

cone A solid figure with a vertex and a circular base.

congruent angles Angles that have the same measure.

congruent segments Segments that have the same length.

congruent triangles Two triangles are congruent if there is a correspondence between the vertices such that each pair of corresponding sides and angles are congruent.

conjecture A generalization that is hypothesized to be true.

convex polygon A polygon is convex if all of the diagonals of the polygon are in the interior of the polygon.

coplanar points Points that all lie in one plane.

corresponding angles Two angles on the same side of a transversal. One of the angles is an exterior angle; one is an interior angle.

counterexample A single example that shows a generalization to be false.

cross section of a solid A region common to the solid and a plane that intersects the solid.

cube A regular polyhedron with six square faces.

cylinder A solid figure having congruent circular bases in a pair of parallel planes.

deductive reasoning Starting with a hypothesis and using logic and definitions, postulates, or previously proven theorems to justify a series of statements or steps that lead to the desired conclusion.

degree measure The real number between 0 and 180 that is assigned to an angle.

diameter of a circle A chord that contains the center of a circle.

distance from a point to a line The length of the segment drawn from the point perpendicular to the line.

equiangular triangle A triangle with three congruent angles.

equilateral triangle A triangle with all sides congruent to one another.

Euler line of a triangle The line containing the circumcenter, centroid, and orthocenter of a triangle.

Euler's Formula for a polyhedra The formula $V + F = E + 2$, where V is the number of vertices, F the number of faces, and E the number of edges.

exterior angle of a triangle An angle that forms a linear pair with one of the angles of the triangle.

generalization A statement thought to be true, arrived at through inductive reasoning.

inductive reasoning Observing that an event gives the same result several times in succession, then concluding that the event will always have the same outcome.

intersecting lines Two lines with a point in common.

isosceles triangle A triangle with at least two sides congruent to one another.

kite A quadrilateral with two distinct pairs of congruent adjacent sides.

line of symmetry A line in which a figure coincides with its reflection image.

line reflection A transformation that maps a figure into its reflection image about a line.

linear pair of angles A pair of angles with a common side such that the union of the other two sides is a line.

LOGO A language for a computer that is convenient for exploring ideas of geometry.

median of a triangle A segment joining a vertex to the midpoint of the opposite side.

midpoint of a segment The midpoint of segment AB is a point C between A and B such that $\overline{AC} \cong \overline{CB}$.

MIRA A transparent mirror-like device that can be used to perform geometric constructions and explore reflections.

nonconvex quadrilateral A quadrilateral for which there exists a segment with endpoints on the quadrilateral, but with other points of the segment outside the quadrilateral.

obtuse angle An angle with measure greater than 90°.

obtuse triangle A triangle with an obtuse angle.

orthocenter of a triangle The point of intersection of the lines containing the altitudes.

parallel lines Lines in the same plane that do not intersect.

parallelogram A quadrilateral with both pairs of opposite sides parallel.

Pascal's triangle A triangular array of numbers in which

1's are written along the sides of an imagined isosceles triangle. Each other element filling up the triangle is the sum of the two elements directly above it to its left and its right.

perimeter of a polygon The sum of the lengths of the sides of the polygon.

perpendicular bisector of a segment A line that is perpendicular to the segment and which contains its midpoint.

perpendicular lines Two lines that intersect to form congruent right angles.

polygon A closed plane figure formed by the union of segments meeting only at endpoints such that (1) at most two segments meet at the point, and (2) each segment meets exactly two other segments.

polyhedron A closed solid figure formed by a finite number of polygonal regions called faces. Each edge of a region is the edge of exactly one other region. If two regions intersect, then they intersect in an edge or vertex.

postulates A basic generalization accepted without proof.

prism A polyhedron such that (1) there is a pair of congruent faces that lie in parallel planes, and (2) all other faces are parallelograms.

quadrilateral The union of four segments determined by four points, no three of which are collinear. The segments intersect only at their endpoints.

radius of a circle A segment whose endpoints are the center and point on a circle.

ray A ray, $\overrightarrow{AB}$ is a subset of a line. It contains a given point A and all points on the same side of A as B.

rectangle A parallelogram with four right angles.

reflectional symmetry A figure $\mathcal{F}$ has reflectional symmetry if there is a line l such that the reflection image over l of each point P of $\mathcal{F}$ is also a point of $\mathcal{F}$. The line l is called the line of symmetry.

regular polygon A polygon with all sides congruent to each other and all angles congruent to each other.

regular tessellation A tessellation made of regular, congruent, convex polygons such that each vertex figure is a regular polygon.

remote interior angles Two angles of a triangle with respect to an exterior angle that are not adjacent to the exterior angle.

rhombus A parallelogram with four congruent sides.

right triangle A triangle with a right angle.

rotation A transformation involving a turn about a point O through an angle $x°$ that associates each point P of the plane with an image point P'. The measure of angle POP' is $x°$ and $OP = OP'$.

scalene triangle A triangle with no congruent sides.

segment A segment $\overline{AB}$ is the set of points A and B and all the points between A and B.

semi-regular tessellation A tessellation made of regular polygons of two or more types so that the arrangement of polygons at each vertex is the same.

skew lines Two nonintersecting lines that do not lie in the same plane.

square A rectangle with four congruent sides.

star polygon A polygon formed by joining every dth point of a set of n equally spaced points on a circle ($d < n$).

supplementary angles Two angles whose measures have a sum of 180°.

tangram puzzle A famous puzzle consisting of 7 pieces formed by dissecting a square in a special way.

tessellation, or tiling A complete covering of the plane with polygons with no holes and no overlapping.

theorem A generalization that can be proved to be true using definitions, postulates, and the logic of deductive reasoning.

topological equivalence Two surfaces or curves are topologically equivalent if one can be transformed into the other by distorting, stretching, shrinking or bending.

translation Given an arrow AA', the translation image of a point P for the arrow AA' is the point P', where $AA' = PP'$ and arrows AA' and PP' have the same direction.

transversal A line that intersects two coplanar lines in two different points.

trapezoid A quadrilateral with exactly one pair of parallel sides.

triangle The union of three segments determined by three noncollinear points.

undefined term A basic term at the simplest level that is not defined using other terms.

vertex of a polygon The endpoint of a side of a polygon.

vertex of an angle The endpoint of the two noncollinear rays forming the angle.

vertical angles Two angles that are formed by two intersecting lines, but that are not a linear pair of angles.

volume A measure of the amount of space occupied by a solid. Each solid is assigned a unique positive number called its volume.

POSTULATES*

Postulate 1: *The Distance Postulate.* To every pair of different points there corresponds a unique positive number.

Postulate 2: *The Ruler Postulate.* The points of a line can be placed in correspondence with the real numbers in such a way that

1. To every point of the line there corresponds exactly one real number.
2. To every real number there corresponds exactly one point of the line.
3. The distance between any two points is the absolute value of the difference of the corresponding numbers.

Postulate 3: *The Ruler Placement Postulate.* Given two points P and Q of a line, the coordinate system can be chosen in such a way that the coordinate of P is zero and the coordinate of Q is positive.

Postulate 4: *The Line Postulate.* For every two points there is exactly one line that contains both points.

Postulate 5:

1. Every plane contains at least three noncollinear points.
2. Space contains at least four noncoplanar points.

Postulate 6: If two points of a line lie in a plane, then the line lies in the same plane.

Postulate 7: *The Plane Postulate.* Any three points lie in at least one plane, and any three noncollinear points lie in exactly one plane.

Postulate 8: If two different planes intersect, then their intersection is a line.

*The postulates and theorems from plane geometry are provided for review and reference. They are adaptations of those listed in Moise and Downs (1971).

Postulate 9: *The Plane Separation Postulate.* Given: a line and a plane containing it. The points of the plane that do not lie on the line form two sets such that

1. Each of the sets is convex.
2. If P is in one of the sets and Q is in the other, then the segment $\overline{PQ}$ intersects the plane.

Postulate 10: *The Space Separation Postulate.* The points of space that do not lie in a given plane form two sets such that

1. Each of the sets is convex.
2. If P is in one of the sets and Q is in the other, then the segment $\overline{PQ}$ intersects the plane.

Postulate 11: *The Angle Measurement Postulate.* To every angle $\angle BAC$ there corresponds a real number between 0 and 180.

Postulate 12: *The Angle Construction Postulate.* Let $\overrightarrow{AB}$ be a ray on the edge of the half-plane H. For every number r between 0 and 180, there is exactly one ray $\overrightarrow{AP}$, with P in H, such that $m\angle PAB = r$

Postulate 13: *The Angle Addition Postulate.* If D is in the interior of $\angle BAC$, then $m\angle BAC = m\angle BAD + m\angle DAC$.

Postulate 14: The Supplement Postulate. If two angles form a linear pair, then they are supplementary.

Postulate 15: *The SAS Postulate.* Every SAS correspondence is a congruence.

Postulate 16: *The ASA Postulate.* Every ASA correspondence is a congruence.

Postulate 17: *The SSS Postulate.* Every SSS correspondence is a congruence.

Postulate 18: *The Parallel Postulate.* Through a given external point, there is only one parallel to a given line.

Postulate 19: *The Area Postulate.* To every polygonal region, there corresponds a unique positive real number.

Postulate 20: *The Congruence Postulate.* If two triangles are congruent, then the triangular regions determined by them have the same area.

Postulate 21: *The Area Addition Postulate.* Suppose that the region R is the union of two regions R_1 and R_2. Suppose that R_1 and R_2 intersect in at most a finite number of segments and points. Then, $aR = aR_1 + aR_2$.

Postulate 22: *The Unit Postulate.* The area of a square region is the square of the length of its edge.

Postulate 23: *The Unit Postulate.* The volume of a rectangular parallelepiped is the product of the altitude and the area of the base.

Postulate 24: *Cavalieri's Principle.* Given: two solids and a plane. Suppose that every plane parallel to the given plane, intersecting one of the two solids, also intersects the other and gives cross sections with the same area. Then, the two solids have the same volume.

THEOREMS

Theorem 1: Let $\overrightarrow{AB}$ be a ray, and let x be a positive number. Then, there is exactly one point P of AB such that $AP = x$.

Theorem 2: Every segment has exactly one midpoint.

Theorem 3: If two different lines intersect, their intersection contains only one point.

Theorem 4: If a line intersects a plane not containing it, then the intersection contains only one point.

Theorem 5: Given a line and a point not on the line, there is exactly one plane containing both.

Theorem 6: Given two intersecting lines, there is exactly one plane containing both.

Theorem 7: If two angles are complementary, then both are acute.

Theorem 8: Every angle is congruent to itself.

Theorem 9: Any two right angles are congruent.

Theorem 10: If two angles are both congruent and supplementary, then each is a right angle.

Theorem 11: Supplements of congruent angles are congruent.

Theorem 12: Complements of congruent angles are congruent.

Theorem 13: Vertical angles are congruent.

Theorem 14: If two intersecting lines form one right angle, then they form four right angles.

Theorem 15: Every segment is congruent to itself.

Theorem 16: Every angle has exactly one bisector.

Theorem 17: If two sides of a triangle are congruent, then the angles opposite these sides are congruent.

Theorem 18: If two angles of a triangle are congruent, then the sides opposite them are congruent.

Theorem 19: In a given plane, through a given point of a given line, there is one and only one line perpendicular to the given line.

Theorem 20: The perpendicular bisector of a segment, in a plane, is the set of all points of the plane that are equidistant from the endpoints of the segment.

Theorem 21: Through a given external point, there is at least one line perpendicular to a given line.

Theorem 22: Through a given external point, there is at most one line perpendicular to a given line.

Theorem 23: If M is between A and C on a line L, then M and A are on the same side of any other line that contains C.

Theorem 24: If M is between B and C, and A is any point not on BC, then M is in the interior of $\angle BAC$.

Theorem 25: An exterior angle of a triangle is greater than each of its remote interior angles.

Theorem 26: Every SAA correspondence is a congruence.

Theorem 27: Given: a correspondence between two right triangles. If the hypotenuse and one leg of one of the triangles are congruent to the corresponding parts of the second triangle, then the correspondence is a congruence.

Theorem 28: If two sides of a triangle are not congruent, then the angles opposite them are not congruent, and the larger angle is opposite the longer side.

Theorem 29: If two angles of a triangle are not congruent, then the sides opposite them are not congruent, and the longer side is opposite the larger angle.

Theorem 30: The shortest segment joining a point to a line is the perpendicular segment.

Theorem 31: The sum of the length of any two kinds of a triangle is greater than the length of the third side.

Theorem 32: If two sides of one triangle are congruent, respectively, to two sides of a second triangle, and the included angle of the first triangle is larger than the included angle of the second, then the third side of the first triangle is longer than the third side of the second.

Theorem 33: If two sides of one triangle are congruent, respectively, to two sides of a second triangle, and the third side of the first triangle is longer than the third side of the second, then the included angle of the first triangle is larger than the included angle of the second.

Theorem 34: If B and C are equidistant from P and Q, then every point between B and C is equidistant from P and Q.

Theorem 35: If a line is perpendicular to each of two intersecting lines at their point of intersection, then it is perpendicular to the plane that contains them.

Theorem 36: Through a given point of a given line, there passes a plane perpendicular to the given line.

Theorem 37: If a line and a plane are perpendicular, then the plane contains every line perpendicular to the given line at its point of intersection with the given plane.

Theorem 38: Through a given point of a given line, there is only one plane perpendicular to the line.

Theorem 39: The perpendicular bisecting plane of a segment is the set of all points equidistant from the endpoints of the segment.

Theorem 40: Two lines perpendicular to the same plane are coplanar.

Theorem 41: Through a given point, there passes one and only one plane perpendicular to a given line.

Theorem 42: Through a given point, there passes one and only one line perpendicular to a given plane.

Theorem 43: The shortest segment to a plane from an external point is the perpendicular segment.

Theorem 44: Two parallel lines lie in exactly one plane.

Theorem 45: Two lines in a plane are parallel if they are both perpendicular to the same line.

Theorem 46: Let L be a line and let P be a point not on L. Then, there is at least one line through P, parallel to L.

Theorem 47: If two lines are cut by a transversal, and one pair of alternate interior angles are congruent, then the other pair of alternate interior angles are also congruent.

Theorem 48: Given: two lines cut by a transversal. If a pair of alternate interior angles are congruent, then the lines are parallel.

Theorem 49: Given: two lines cut by a transversal. If a pair of corresponding angles are congruent, then a pair of alternate interior angles are congruent.

Theorem 50: Given: two lines cut by a transversal. If a pair of corresponding angles are congruent, then the lines are parallel.

Theorem 51: If two parallel lines are cut by a transversal, then alternate interior angles are congruent.

Theorem 52: If two parallel lines are cut by a transversal, each pair of corresponding angles are congruent.

Theorem 53: If two parallel lines are cut by a transversal, the interior angles on the same side of the transversal are supplementary.

Theorem 54: In a plane, if two lines are each parallel to a third line, then they are parallel to each other.

Theorem 55: In a plane, if a line is perpendicular to one of two parallel lines, it is perpendicular to the other.

Theorem 56: For every triangle, the sum of the measures of the angles is 180.

Theorem 57: Each diagonal separates a parallelogram into two congruent triangles.

Theorem 58: In a parallelogram, any two opposite sides are congruent.

Theorem 59: In a parallelogram, any two opposite angles are congruent.

Theorem 60: In a parallelogram, any two consecutive angles are supplementary.

Theorem 61: The diagonals of a parallelogram bisect each other.

Theorem 62: Given: a quadrilateral in which both pairs of opposite sides are congruent. Then, the quadrilateral is a parallelogram.

Theorem 63: If two sides of a quadrilateral are parallel and congruent, then the quadrilateral is a parallelogram.

Theorem 64: If the diagonals of a quadrilateral bisect each other, then the quadrilateral is a parallelogram.

Theorem 65: The segment between the midpoints of two sides of a triangle is parallel to the third side and half as long.

Theorem 66: If a parallelogram has one right angle, then it has four right angles, and the parallelogram is a rectangle.

Theorem 67: In a rhombus, the diagonals are perpendicular to one another.

Theorem 68: If the diagonals of a quadrilateral bisect each other and are perpendicular, then the quadrilateral is a rhombus.

Theorem 69: The median to the hypotenuse of a right triangle is half as long as the hypotenuse.

Theorem 70: If an acute angle of a right triangle has measure 30, then the opposite side is half as long as the hypotenuse.

Theorem 71: If one leg of a right triangle is half as long as the hypotenuse, then the opposite angle has measure 30.

Theorem 72: If three parallel lines intercept congruent segments on one transversal T, then they intercept congruent segments on every transversal T' that is parallel to T.

Theorem 73: If three parallel lines intercept congruent segments on one transversal, then they intercept congruent segments on any other transversal.

Theorem 74: If a plane intersects two parallel planes, then it intersects them in two parallel lines.

Theorem 75: If a line is perpendicular to one of two parallel planes, it is perpendicular to the other.

Theorem 76: Two planes perpendicular to the same line are parallel.

Theorem 77: Two lines perpendicular to the same plane are parallel.

Theorem 78: Parallel planes are everywhere equidistant.

Theorem 79: All plane angles of the same dihedral angle are congruent.

Theorem 80: If a line is perpendicular to a plane, then every plane containing the line is perpendicular to the given plane.

Theorem 81: If two planes are perpendicular, then any line in one of them, perpendicular to their line of intersection, is perpendicular to the other plane.

Theorem 82: If a line and a plane are not perpendicular, then the projection of the line into the plane is a line.

Theorem 83: The area of a rectangle is the product of its base and its altitude.

Theorem 84: The area of a right triangle is half the product of its legs.

Theorem 85: The area of a triangle is half the product of any base and the corresponding altitude.

Theorem 86: The area of a trapezoid is half the product of its altitude and the sum of its bases.

Theorem 87: The area of a parallelogram is the product of any base and the corresponding altitude.

Theorem 88: If two triangles have the same base b and the same altitude h, then they have the same area.

Theorem 89: If two triangles have the same altitude h, then the ratio of their areas is equal to the ratio of their bases.

Theorem 90: In a right triangle, the square of the hypotenuse is equal to the sum of the squares of the legs.

Theorem 91: If the square of one side of a triangle is equal to the sum of the squares of the other two sides, then the triangle is a right triangle, with its right angle opposite the longest side.

Theorem 92: In an isosceles right triangle, the hypotenuse is $\sqrt{2}$ times as long as each of the legs.

Theorem 93: If the base of an isosceles triangle is $\sqrt{2}$ times as long as each of the two congruent sides, then the angle opposite the base is a right angle.

Theorem 94: In a 30-60-90 triangle, the longer leg is $\sqrt{3}/2$ times as long as the hypotenuse.

Theorem 95: If a line parallel to one side of a triangle intersects the other two sides in distinct points, then it cuts off segments that are proportional to these sides.

Theorem 96: If a line intersects two sides of a triangle and cuts off segments proportional to these two sides, then it is parallel to the third side.

Theorem 97: Given: a correspondence between two triangles. If corresponding angles are congruent, then the correspondence is a similarity.

Theorem 98: If $\Delta ABC \sim \Delta DEF$, and $\Delta DEF \cong \Delta GHI$, then $\Delta ABC \sim \Delta GHI$.

Theorem 99: Given: a correspondence between two triangles. If two pairs of corresponding sides are proportional, and the included angles are congruent, then the correspondence is a similarity.

Theorem 100: Given: a correspondence between two triangles. If corresponding sides are proportional, then the correspondence is a similarity.

Theorem 101: In any right triangle, the altitude to the hypotenuse separates the triangle into two triangles that are similar to each other and to the original triangle.

Theorem 102: Given a right triangle and the altitude to the hypotenuse:

1. The altitude is the geometric mean of the segments into which it separates the hypotenuse.
2. Each leg is the geometric mean of the hypotenuse and the segment of the hypotenuse adjacent to the leg.

Theorem 103: If two triangles are similar, then the ratio of their areas is the square of the ratio of any two corresponding sides.

Theorem 104: Two nonvertical lines are parallel if and only if they have the same slope.

Theorem 105: Two nonvertical lines are perpendicular if and only if their slopes are negative reciprocals of each other.

Theorem 106: The distance between the points (x_1, y_1) and (x_2, y_2) is $\sqrt{(x_2 - x_1)^2 + (y_2 - y_1)^2}$.

Theorem 107: Given: $P_1 = (x_1, x_2)$ and $P_2 = (y_1, y_2)$. The midpoint of P_1P_2 is the point

$$P = \left(\frac{x_1 + x_2}{2}, \frac{y_1 + y_2}{2}\right).$$

Theorem 108: If P is between P_1 and P_2, and $\frac{P_1P}{PP_2} = r$, then

$$P = \left(\frac{x_1 + rx_2}{1 + r}, \frac{y_1 + ry_2}{1 + r}\right)$$

Theorem 109: Let L be a line with slope m, passing through the point (x_1, y_1). Then every point (x, y) of L satisfies the equation $y - y_1 = m(x - x_1)$.

Theorem 110: The graph of equation $y - y_1 = m(x - x_1)$ is the line that passes through the point (x_1, y_1) and has slope m.

Theorem 111: The graph of the equation $y = mx + b$ is the line that passes through the point $(0, b)$ and has slope m.

Theorem 112: The intersection of a sphere with a plane through its center is a circle with the same center and the same radius.

Theorem 113: A line perpendicular to a radius at its outer end is tangent to the circle.

Theorem 114: Every tangent to a circle is perpendicular to the radius drawn to the point of contact.

Theorem 115: The perpendicular from the center of a circle to a chord bisects the chord.

Theorem 116: The segment from the center of a circle to the midpoint of a chord is perpendicular to the chord.

Theorem 117: In the plane of a circle, the perpendicular bisector of a chord passes through the center.

Theorem 118: In the same circle or in congruent circles, chords equidistant from the center are congruent.

Theorem 119: In the same circle or in congruent circles, any two congruent chords are equidistant from the center.

Theorem 120: If a line intersects the interior of a circle, then it intersects the circle in two and only two points.

Theorem 121: A plane perpendicular to a radius at its outer end is tangent to the sphere.

Theorem 122: Every tangent plane to a sphere is perpendicular to the radius drawn to the point of contact.

Theorem 123: If a plane intersects the interior of a sphere, then the intersection of the plane and the sphere is a circle. The center of this circle is the foot of the perpendicular from the center of the sphere to the plane.

Theorem 124: The perpendicular from the center of a sphere to a chord bisects the chord.

Theorem 125: The segment from the center of a sphere to the midpoint of a chord is perpendicular to the chord.

Theorem 126: If B is a point of $\widehat{AC}$, then $m\widehat{ABC} = m\widehat{AB} + m\widehat{BC}$.

Theorem 127: The measure of an inscribed angle is half the measure of its intercepted arc.

Theorem 128: In the same circle or in congruent circles, if two chords are congruent, then so are the corresponding minor arcs.

Theorem 129: In the same circle or in congruent circles, if two arcs are congruent, then so are the corresponding chords.

Theorem 130: Given: an angle with its vertex on a circle, formed by a secant ray and a tangent ray. The measure of the angle is half the measure of the intercepted arc.

Theorem 131: The two tangent segments to a circle from a point of the exterior are congruent and determine congruent angles with the segment from the exterior point to the center.

Theorem 132: Given: a circle C, and a point Q of its exterior. Let L_1 be a secant line through Q, intersecting C in points R and S; and let L_2 be another secant line through Q, intersecting C in points U and T. Then, $QR \times QS = QU \times QT$.

Theorem 133: Given: a tangent segment $\overline{QT}$ to a circle, and a secant line through Q, intersecting the circle in points R and S. Then, $QR \times QS = QT$.

Theorem 134: Let $\overline{RS}$ and $\overline{TU}$ be chords of the same circle, intersecting at Q. Then, $QR \times QS = QU \times QT$.

Theorem 135: The graph of the equation $(x - a)^2 + (y - b)^2 = r^2$ is the circle with center (a,b) and radius r.

Theorem 136: Every circle is the graph of an equation of the form $x^2 + y^2 + Ax + By + C = 0$.

Theorem 137: The graph of the equation $x^2 + y^2 + Ax + By + C = 0$ is (1) a circle, (2) a point, or (3) the empty set.

Theorem 138: The perpendicular bisectors of the sides of a triangle are concurrent. Their point of concurrency is equidistant from the vertices of the triangle.

Theorem 139: The three altitudes of a triangle are always concurrent.

Theorem 140: The bisector of an angle, minus its endpoint, is the set of all points of the interior of the angle that are equidistant from the sides.

Theorem 141: The angle bisectors of a triangle are concurrent in a point that is equidistant from the three sides.

Theorem 142: The medians of every triangle are concurrent, and their point of concurrency is two-thirds of the way along each median, from the vertex to the opposite side.

Theorem 143: Given: two circles of radius a and b, with c as the distance between their centers. If each of the numbers a, b, and c is less than the sum of the other two, then the circles intersect in two points, on opposite sides of the line through the centers.

Theorem 144: The ratio of the circumference to the diameter is the same for all circles.

Theorem 145: The area of a circle of radius r is πr^2.

Theorem 146: If two arcs have equal radii, then their lengths are proportional to their measures.

Theorem 147: If an arc has measure q and radius r, then its length is

$$L = \frac{q}{180} \cdot \pi r$$

Theorem 148: The area of a sector is half the product of its radius and the length of its arc.

Theorem 149: If a sector has radius r and its arc has measure q, then its area is

$$A = \frac{q}{360} \cdot \pi r^2$$

Theorem 150: All cross sections of a triangular prism are congruent to the base.

Theorem 151: All cross sections of a prism have the same area.

Theorem 152: The lateral faces of a prism are parallelogram regions.

Theorem 153: Every cross section of a triangular pyramid, between the base and the vertex, is a triangular region similar to the base. If h is the altitude, and k is the distance from the vertex to the cross section, then the area of the cross section is equal to k^2/h^2 times the area of the base.

Theorem 154: In any pyramid, the ratio of the area of a cross section to the area of the base is k^2/h^2, where h is the altitude of the pyramid, and k is the distance from the vertex to the plane of the cross section.

Theorem 155: If two pyramids have the same base area and the same altitude, then cross sections equidistant from the vertices have the same area.

Theorem 156: The volume of any prism is the product of the altitude and the area of the base.

Theorem 157: If two pyramids have the same altitude and the same base area, and their bases lie in the same plane, then they have the same volume.

Theorem 158: The volume of a triangular pyramid is one-third the product of its altitude and its base area.

Theorem 159: The volume of a pyramid is one-third the product of its altitude and its base area.

Theorem 160: Every cross section of a circular cylinder is a circular region congruent to the base.

Theorem 161: Every cross section of a circular cylinder has the same area as the base.

Theorem 162: Given: a cone of altitude h and a cross section made by a plane at a distance k from the vertex. The area of the cross section is equal to k^2/h^2 times the area of the base.

Theorem 163: The volume of a circular cylinder is the product of its altitude and the area of its base.

Theorem 164: The volume of a circular cone is one-third the product of its altitude and the area of its base.

Theorem 165: The volume of a sphere of radius r is $4\pi r^2/3$.

Theorem 166: The surface area of a sphere of radius r is $A = 4\pi r^2$.

BIBLIOGRAPHY

Abbott, Edwin A. 1963. *Flatland—A Romance of Many Dimensions*. 5th Rev. Ed. New York: Barnes and Noble Books.

Abbott, Janet S. 1968. *Learn to Fold—Fold to Learn*. Pasadena, CA: Franklin Publications.

Abbott, Janet S. *Mirror Magic*. Pasadena, CA: Franklin Publications.

Abelson, Harold. 1989. *Apple Logo*. Peterborough, NH: BYTE Publications, Inc.

Adler, Irving. 1966. *A New Look at Geometry*. New York: Signet Science Library Books.

Association of Teachers of Mathematics. 1970. *Mathematical Reflections*. New York: Cambridge University Press.

Association of Teachers of Mathematics. 1968. *Notes on Mathematics in Primary Schools*. New York: Cambridge University Press.

Ball, P.G. 1972. "A Different Order of Reptile." *Mathematics Teaching* 60(September):44–45.

Ball, W.W.R., and H.S.M. Coxeter. 1967. *Mathematical Recreations and Essays*. New York: The Macmillan Company.

Bannon, Thomas J. 1991. "Fractals and Transformations." *The Mathematics Teacher* (NCTM), 84(3):178–185.

Baravalle, H.V. "The Geometry of the Pentagon and the Golden Section." *The Mathematics Teacher* (NCTM), 41(1):22–23.

Barr, Stephen. *Experiments in Topology*. 1964. New York: Thomas Y. Crowell.

Bassetti, Fred, Hy Ruchlis, and Daniel Malament. 1968. *Math Projects: Polyhedral Shapes*. Brooklyn, NY: Book-Lab.

Bassetti, Fred, Hy Ruchlis, and Daniel Malament. 1961. *Solid Shapes Lab*. New York: Science Materials Center.

Beamer, James E. 1989. "Using Puzzles to Teach the Pythagorean Theorem." *The Mathematics Teacher* (NCTM), 82(5):336–341.

Beard, Robert S. 1973. *Patterns in Space*. Palo Alto, CA: Creative Publications.

Beck, Anatole, Michael N. Bleicher, and Donald W. Crowe. 1969. *Excursions in Mathematics*. New York: Worth Publishers.

Bergamini, David. 1970. *Mathematics*. Life Science Library. New York: Time-Life Books.

Bezuszka, Stanley, Margaret Kenney, and Linda Silvey. 1977. *Tessellations: The Geometry of Patterns*. Palo Alto: CA: Creative Publications.

Blockwell, William. 1984. *Geometry in Architecture*. New York: John Wiley & Sons.

Boorman, Phil. 1971. "More About Tessellating Hexagons (2)." *Mathematics Teaching* 55(Summer):23.

Bourgoin, J. 1973. *Arabic Geometrical Pattern and Design*. New York: Dover Publications.

Brandes, Louis G. *Geometry Can Be Fun*. 1958. Portland, ME: J. Weston Walch.

Bright, George W. 1989. "Teaching Mathematics with Technology: Logo and Geometry." *The Arithmetic Teacher* (NCTM), 36(5):32–34.

Bruni, James, and Helen Silverman. 1987. "Making Patterns with a Square." In Jane Hill, Ed. *Geometry for Grades K–6—Readings from the Arithmetic Teacher*. Reston, VA: National Council of Teachers of Mathematics.

Brydegaard, M., and J.E. Inskeep, Jr., Eds. 1970. *Readings in Geometry from the Arithmetic Teacher*. Washington, DC: National Council of Teachers of Mathematics.

Budden, F.J. 1962. *The Fascination of Groups*. London: Cambridge University Press.

Burke, Mike, and Ron Genise. 1987. *Logo and Models of Computation*. Menlo Park, CA: Addison-Wesley Publishing Company.

Caldwell, J.H. 1966. *Topics in Recreational Mathematics*. New York: Cambridge University Press.

Cameron, A.J. 1966. *Mathematical Enterprises for Schools*. New York: Pergamon Press.

Carrol, William M. 1988. "Cross Sections of Clay Solids." *The Arithmetic Teacher* (NCTM), 35(7):6–11.

Chazan, Daniel. 1990. "Students' Microcomputer-Aided Exploration in Geometry: Implementing the Standards." *The Mathematics Teacher* (NCTM), 83(8):628–635.

Chinn, Phyllis Zweig. 1988. "Inductive Patterns, Finite Differences, and a Missing Region." *The Mathematics Teacher* (NCTM), 81(6):446–449.

Clauss, Judith Enz. 1991. "Pentagonal Tessellations." *The Arithmetic Teacher* (NCTM), 38(5):52–56.

Clemens, S.R., P.G. O'Daffer, T.J. Corney, and J.A. Dossey. 1990. *Addison-Wesley Geometry*. Menlo Park, CA: Addison-Wesley Publishing Company.

Clemens, Stanley R. 1985. "Applied Measurement—Using Problem Solving." *The Mathematics Teacher* (NCTM), 78(3).

Clemens, Stanley R. 1984. "Systematic Experimentation—A Frequently Used Problem-Solving Strategy." *The Mathematics Teacher* (NCTM), 77(3).

Clemens, Stanley R. 1974. "Tessellations of Pentagons." *Mathematics Teaching* 67(June).

Cohen, Donald. 1967. *Inquiry in Mathematics Via the Geoboard: Teacher Guide*. New York: Walker Educational Book Corporation.

Courant, Richard, and Herbert Robbins. 1941. *What is Mathematics?* New York: Oxford University Press.

Coxeter, H.S.M. 1961. *Introduction to Geometry*. New York: John Wiley & Sons.

Coxeter, H.S.M. 1959. "The Four-Color Map Problem, 1890–1940." *The Mathematics Teacher* (NCTM), 59(4):283–289.

Coxeter, H.S.M. 1948. *Regular Polytopes*. London: Methuen.

Coxeter, H.S.M., and S.L. Greitzer. *Geometry Revisited*. 1967. New York: Random House, The L.W. Singer Co.

Cundy, H.M., and A.P. Rollett. 1961. *Mathematical Models*. London: Oxford University Press.

Del Grande, John. 1972. *Geoboards and Motion Geometry for Elementary Teachers*. Glenview, IL: Scott Foresman.

DeSimone, Daniel V. 1971. *A Metric America: A Decision Whose Time Has Come*. Washington, DC: National Bureau of Standards.

Dunkels, Andrejs. 1990. "Making and Exploring Tangrams." *The Arithmetic Teacher* (NCTM), 37(6):38–42.

Dunn, James A. "More About Tessellating Hexagons (1)." *Mathematics Teaching* 55(Summer):23–24.

Ehrenfeucht, A. 1964. *The Cube Made Interesting*. New York: A Pergamon Press Book, Macmillan.

Elliott, H.A., James R. MacLean, and Janet M. Jordan. 1968. *Geometry in the Classroom—New Concepts and Methods*. Canada: Holt, Rinehart and Winston of Canada, Limited.

Escher, Maurits C. 1960. *The Graphic Work of M.C. Escher*. New York, Hawthorn Books, Inc.

Eves, Howard. 1972. *A Survey of Geometry*. Rev. Ed. Boston: Allyn and Bacon.

Fielker, David S. 1969. *Topics from Mathematics—Cubes*. New York: Cambridge University Press.

Fujii, John N. 1966. *Puzzles and Graphs*. Washington, DC: National Council of Teachers of Mathematics.

Gardner, Martin. 1989. *Penrose Tiles to Trapdoor Ciphers*. New York: W.H. Freeman and Company.

Gardner, Martin. 1978. *Aha! Insight*. New York/San Francisco: Scientific American, Inc./W.H. Freeman and Company.

Gardner, Martin. 1969. *The Unexpected Hanging*. New York: Simon and Schuster.

Gardner, Martin. 1966. *New Mathematical Diversions from Scientific American*. New York: Simon and Schuster.

Gardner, Martin. 1964. *The Ambidextrous Universe*. New York: Basic Books.

Gardner, Martin. 1961. *The Second Scientific American Book of Mathematical Puzzles and Diversions*. New York: Simon and Schuster.

Gardner, Martin. 1959. *The Scientific American Book of Mathematical Puzzles and Diversions*. New York: Simon and Schuster.

Garland, Trudi H. 1987. *Fascinating Fibonaccis*. Palo Alto, CA: Dale Seymour Publications.

Ghyka, Matelo. 1946. *The Geometry of Art and Life*. New York: Sheed and Ward.

Gibbs, Richard A. 1973. "Euler, Pascal, and the Missing Region." *The Mathematics Teacher* (NCTM), 66(1):27–28.

Giganti, Paul, Jr., and Mary Jo Cittadino. 1990. "The Art of Tessellation." *The Arithmetic Teacher* (NCTM), 37(7):6–16.

Gillespie, N.J. 1973. *MIRA Activities for Junior High School Geometry*. Palo Alto, CA: Creative Publications.

Glenn, John. 1968. "The Quest for the Lost Region." *Mathematics Teaching* 43:23–25.

Golomb, Solomon W. 1965. *Polyominoes*. New York: Charles Scribner's Sons.

Goodnow, Judy. 1991. *Junior High Cooperative Problem Solving with Geoboards*. Sunnyvale, CA: Creative Publications.

Grunbaum, Branko, and G.C. Shephard. 1989. *Tilings and Patterns—An Introduction*. New York: W.H. Freeman and Company.

Grunbaum, Branko, and G.C. Shephard. 1987. *Tilings and Patterns*. New York: W.H. Freeman and Company.

Haag, Vincent H., Clarence E. Hardgrove, and Shirley A. Hill. 1970. *Elementary Geometry*. Reading, MA: Addison-Wesley Publishing Company.

Hallerberg, Arthur E. 1973. "The Metric System: Past, Present—Future?" *The Arithmetic Teacher* (NCTM), 20(4):247–255.

Hilbert, D., and S. Cohn-Vossen. 1952. *Geometry and the Imagination*. New York: Chelsea.

Hill, Jane, Ed. *Geometry for Grades K–6—Readings from the Arithmetic Teacher*. Reston, VA: National Council of Teachers of Mathematics.

Hilton, Peter, and Jean Pederson. 1988. *Build Your Own Polyhedra*. Menlo Park, CA: Addison-Wesley Publishing Company.

Hirsch, Christian R. 1988. "Graphs, Games, and Generalizations." *The Mathematics Teacher* (NCTM), 81(9):741–745.

Holden, Alan. 1971. *Shapes, Space, and Symmetry*. New York: Columbia University Press.

Holiday, E. 1970. *Altair Design*. London: Pantheon.

Hoogeboom, Shirley. 1988. *Moving on with Geoboards*. Sunnyvale, CA: Creative Publications.

Huntley, H.E. 1970. *The Divine Proportion*. New York: Dover Publications.

Izard, John. 1990. "Developing Spatial Skills with Three-Dimensional Puzzles," *The Arithmetic Teacher* (NCTM), 37(6):44–47.

Jeger, Max. 1969. *Transformation Geometry*. New York: American Elsevier.

Jencks, Stanley M., and Donald M. Peck. 1987. *Beneath Rules: An Approach to Mathematical Thinking and Problem Solving for Elementary School Teachers*. Menlo Park, CA: Benjamin/Cummings.

Jennings, Donald E. 1970. "An Intuitive Approach to Pierced Polygons." *The Mathematics Teacher* (NCTM), 63(4):311–312.

Johnson, D.A., and W.H. Glenn. 1960. *Topology, The Rubber Sheet Geometry*. St. Louis: Webster Publishing Company.

Johnson, Donovan A. 1957. "Paper Folding for the Mathematics Class." Washington, DC: National Council of Teachers of Mathematics.

Kaiser, Barbara. 1988. "Explorations with Tessellating Polygons." *The Arithmetic Teacher* (NCTM), 36(4):19–24.

Keedy, Mervin L., and Charles W. Nelson. 1965. *Geometry—A Modern Introduction*. Reading, MA: Addison-Wesley Publishing Company.

Kelley, John L., and Donald Richert. 1970. *Elementary Mathematics for Teachers*. San Francisco, CA: Holden Day.

Kingston, Maurice. 1957. "Mosaics by Reflection." *The Mathematics Teacher* (NCTM), 50(4):280–286.

Klarner, David, Ed. 1981. *The Mathematical Gardner*. Boston: Weber & Schmidt.

Kohl, Herbert. 1987. *Mathematical Puzzlements—Play and Invention with Mathematics*. New York: Schocken Books.

Koziakin, Vladimir. 1972. *Mazes 2*. New York: Grosset and Dunlap Publishers.

Lappan, Glenda, William Fitzgerald, Elizabeth Phillips, and Mary Jean Winter. 1986. *Similarity and Equivalent Fractions*. The Middle Grades Mathematics Project Series. Menlo Park, CA: Addison-Wesley Publishing Company.

Laycock, Mary. 1984. *Bucky for Beginners: Synergetic Geometry*. Howard, CA: Activity Resources Company, Inc.

Laycock, Mary. 1970. *Straw Polyhedra*. Palo Alto, CA: Creative Publications.

Lietzmann, W. 1969. *Visual Topology*. London: Chatto and Windus.

Lindgren, Harry. 1972. *Recreational Problems in Geometric Dissections and How to Solve Them*. New York: Dover Publications.

Lindgren, Harry. 1964. *Geometric Dissections*. Princeton, NJ: D. Van Nostrand Company.

Lindquist, Mary M., Ed. 1980. *Selected Issues in Mathematics Education*. Berkeley, CA: McCutchan Publishing Company.

Lindquist, Mary M., and Albert P. Schulte, Eds. 1987. *Learning and Teaching Geometry, K–12*. 1987 Yearbook. Reston, VA: National Council of Teachers of Mathematics.

Loeb, Arthur L. 1976. *Space Structures, Their Harmony and Counterpoint*. Reading, MA: Addison-Wesley Publishing Company.

Loomis, Elisha Scott. 1940. *The Pythagorean Proposition*. Reston, VA: National Council of Teachers of Mathematics. © 1968 NCTM.

MacGillavry, Caroline H. 1965. *Symmetry Aspects of M.C. Escher's Periodic Drawings*. Utrecht: A Oosthoek's Uitgeversmaatschappig NV.

Mann, J.E. 1969. "Discovering Special Polyhedra." *Mathematics Teaching* 49(Winter):48–49.

McWeeny, R. 1963. *Symmetry: An Introduction to Group Theory and Its Applications*. New York: Pergamon Press.

Meyer, Rochelle Wilson. 1971. "Mutession: A New Tiling Relationship Among Planar Polygons." *Mathematics Teaching* 56(Autumn):24–27.

Miller, William A. 1990. "Polygonal Numbers and Recursion." *The Mathematics Teacher* (NCTM), 83(7):555–562.

Millington, Jon. 1989. *Curve Stitching*. Palo Alto, CA: Dale Seymour Publications.

Millman, Richard S., and Ramona R. Speranza. 1991. "The Artist's View of Points and Lines." *The Mathematics Teacher* (NCTM), 84(2):133–138.

Minnesota Mathematics and Science Teaching Project. 1968. *Introducing Symmetry*. Minneapolis: University of Minnesota.

MIRA Math for the Elementary School. 1973. Palo Alto, CA: Creative Publications.

Moise, Edward, and Floyd Downs. 1971. *Geometry*. Reading, MA: Addison-Wesley Publishing Company.

Moise, Edwin E. 1974. *Elementary Geometry from an Advanced Standpoint*. Reading, MA: Addison-Wesley Publishing Company.

Mold, Josephine. 1969a. *Topics from Mathematics—Solid Models*. New York: Cambridge University Press.

Mold, Josephine. 1969b. *Topics from Mathematics—Tessellations*. New York: Cambridge University Press.

Monie, H.M.J. 1971. "Tessellating Hexagons." *Mathematics Teaching* 54(Spring):26–27.

Moore, Charles. 1986. "Pierced Polygons." *The Mathematics Teacher* (NCTM), 61(6):31–35.

Moore, Margaret L. 1984. *Geometry Problems for Logo Discoveries*. Palo Alto, CA: Creative Publications.

Morrow, Lorna J. "Geometry Through the Standards." *The Arithmetic Teacher* (NCTM), 38(8):21–25.

Myer, Walter. 1972. "Garbage Collecting, Sunday Strolls, and Soldering Problems." *Mathematics Teaching* 65(April):307–309.

NCTM. 1989. *Curriculum and Evaluation Standards for School Mathematics*. Reston, VA: National Council of Teachers of Mathematics.

NCTM. 1987a. *Geometry in Our World*. (Includes slides.) Reston, VA: National Council of Teachers of Mathematics.

NCTM. 1987b. *Learning and Teaching Geometry, K–12*. 1987 Yearbook. Washington, DC: National Council of Teachers of Mathematics.

NCTM. 1976. *Measurement in School Mathematics*. Reston, VA: National Council of Teachers of Mathematics.

NCTM. 1973. *Geometry in the Mathematics Curriculum*. Thirty-Sixth Yearbook. Reston, VA: National Council of Teachers of Mathematics.

NCTM. 1948. *Twentieth Yearbook on National Council of Teachers of Mathematics—The Metric System of Weights and Measures*. New York: Columbia University, Bureau of Publications.

NCTM. 1945. *Multisensory Aids in the Teaching of Mathematics*. Eighteenth Yearbook. Washington, DC: National Council of Teachers of Mathematics.

Neufeld, K. Allen. 1989. "Body Measurement." *The Arithmetic Teacher* (NCTM), 36(9):12–15.

Newman, James R. 1956. *World of Mathematics*. Vols. 1, 2, 3, 4. New York: Simon and Schuster.

Norman, Alexander F. 1991. "Figurate Numbers in the Classroom." *The Arithmetic Teacher* (NCTM), 38(7):42–45.

Nuffield Mathematics Project. 1968a. *Environmental Geometry*. New York: John Wiley & Sons.

Nuffield Mathematics Project. 1968b. *Shape and Size*. New York: John Wiley & Sons.

O'Daffer, Phares G. 1990. "Inductive and Deductive Reasoning." *The Mathematics Teacher* (NCTM), 83(5):378–384.

Olsen, Alton T. 1975. *Mathematics Through Paper Folding*. Palo Alto, CA: Dale Seymour Publications.

Ore, Oystein. 1963. *Graphs and Their Uses*. New York: Random House, The L.W. Singer Company.

Pearce, Peter, and Susan Pearce. 1978. *Polyhedra Primer*. New York: Van Nostrand Reinhold Company.

Phillips, J.P. 1965. "The History of the Dodecahedron," *The Mathematics Teacher* (NCTM), 58(3):248–250.

Piaget, J., H. Inhelder, and Alina Szminska. 1964. *The Child's Conception of Geometry*. New York: Harper and Row.

Pohl, Victoria. 1986. *How to Enrich Geometry Using String Designs*. Palo Alto, CA: Dale Seymour Publications.

Posamentier, Alfred. 1983. *Investigations in Geometry*. Reading, MA: Addison-Wesley Publishing Company.

Raab, Joseph A. 1962. "The Golden Rectangle and Fibonacci Sequence, As Related to the Pascal Triangle." *The Mathematics Teacher* (NCTM), 55(11):538–543.

Rademacher, H., and O. Toeplitz. 1957. *The Enjoyment of Mathematics*. Princeton, NJ: Princeton University Press.

Ranucci, Ernest R., and Joseph L. Teeters. 1977. *Creating Escher-Type Drawings*. Palo Alto, CA: Creative Publications.

Ranucci, Ernest R. 1974. "Master of Tessellations: M.C. Escher, 1898–1972." *The Mathematics Teacher* (NCTM), 67(4):229–306.

Ranucci, Ernest R. 1971. "Space Filling in Two Dimensions." *The Mathematics Teacher* (NCTM), 64(11):587–593.

Ranucci, Ernest R. 1968a. "Tiny Treasury of Tessellations." *The Mathematics Teacher* (NCTM), 61(2):114–117.

Ranucci, Ernest. 1968b. "Jungle-Gym Geometry." *The Mathematics Teacher* (NCTM), 61(1):25–28.

Read, Ronald C. 1965. *Tangrams—330 Puzzles*. New York: Dover Publications, 1965.

Reys, Robert E. 1989. "Some Discoveries with Right Rectangular Prisms." *The Mathematics Teacher* (NCTM), 82(2):118–123.

Richardson, Lloyd I. 1987. "The Mobius Strip: An Elementary Exercise Providing Hypothesis Formation and Perceptual Proof." In Jane Hill, Ed. *Geometry for Grades K–6—Readings from the Arithmetic Teacher*. Reston, VA: National Council of Teachers of Mathematics.

Roper, Ann. 1991. *Junior High Cooperative Problem Solving with Pattern Blocks*. Sunnyvale, CA: Creative Publications.

Runion, Garth E. 1990. *The Golden Section and Related Curiosa*. Palo Alto, CA: Dale Seymour Publications.

Runion, Garth E. 1972. *The Golden Section and Related Curiosa*. Glenview, IL: Scott Foresman.

Sanok, Gloria. 1987. "Living in a World of Transformations." In Jane Hill, Ed. *Geometry for Grades K–6—Readings from the Arithmetic Teacher*. Reston, VA: National Council of Teachers of Mathematics.

Schaaf, William L. 1970. *A Bibliography of Recreational Mathematics*. Reston, VA: National Council of Teachers of Mathematics.

School Mathematics Project. 1971. *The School Mathematics Project*. Books A, B, C, D, E, and F. New York: Cambridge University Press.

Senk, Sharon, and Daniel B. Hirschhorn. 1990. "Multiple Approaches to Geometry: Teaching Similarity." *The Mathematics Teacher* (NCTM), 83(4):274–279.

Seymour, Dale, and Jill Britton. 1988. *Introduction to Tessellations*. Palo Alto, CA: Dale Seymour Publications.

Shaw, Jean M., and Mary Jo Puckett Cliatt. 1989. "Developing Measurement Sense," In *New Directions for Elementary School Mathematics*. Reston, VA: National Council of Teachers of Mathematics.

Shengle, Carol Elizabeth. 1972. "A Look at Regular and Semi-Regular Polyhedra." *The Mathematics Teacher* (NCTM), 65(12):713–718.

Shimabukuro, Gini. *Thinking in Logo: A Sourcebook for Teachers of Primary Students*. Menlo Park, CA: Addison-Wesley Publishing Company.

Shroyer, Janet, and William Fitzgerald. 1986. *Mouse and Elephant: Measuring Growth*. The Middle Grades Mathematic Project Series. Menlo Park, CA: Addison-Wesley Publishing Company.

Shubnikov, A.V., and N.V. Belov, et al. 1964. In W. Holser, Ed. *Colored Symmetry*. Oxford: Pergamon Press.

Shyers, Joan H. 1987. "You Can't Get There from Here—An Algorithmic Approach to Eulerian and Hamiltonian Circuits." *The Mathematics Teacher* (NCTM), 80(2):95–98.

Siemens, David F. 1965. "The Math of the Honeycomb." *The Mathematics Teacher* (NCTM), 58(4):334–337.

Smith, Lyle R. 1990. "Areas and Perimeters of Geoboard Polygons." *The Mathematics Teacher* (NCTM), 83(5):392–398.

Spikell, Mark A. 1990. "Equilateral Triangles on an Isometric Grid." *The Mathematics Teacher* (NCTM), 83(9):740–743.

Steen, Lynn Arthur, Ed. 1990. *On the Shoulders of Giants—New Approaches to Numeracy*. Section on Pattern. Washington, DC: National Academy Press.

Steinhaus, H. 1938. *Mathematical Snapshots*. New York: G.E. Stechert.

Stevens, Peter S. 1988. *Handbook of Regular Patterns*. Cambridge, MA: MIT Press.

Stevens, Peter S. 1974. *Patterns in Nature*. Boston, MA: Little, Brown.

Teeters, Joseph L. 1974. "How to Draw Tessellations of the Escher Type." *The Mathematics Teacher* (NCTM), 67(4):307–310.

Teppo, Anne. 1991. "Van Hiele Levels of Geometric Thought Revisited." *The Mathematics Teacher* (NCTM), 84(3):210–221.

Thiessen, Diane, and Margaret Matthias. 1989. "Selected Children's Books for Geometry." *The Arithmetic Teacher* (NCTM), 37(4):47–51.

Thompson, D'Arcy W. 1961. *On Growth and Form*. London: Cambridge University Press.

Thornburg, David. 1986a. *Beyond Turtle Graphics*. Menlo Park, CA: Addison-Wesley Publishing Company.

Thornburg, David. 1986b. *Exploring Logo Without a Computer*. Menlo Park, CA: Addison-Wesley Publishing Company.

Thornburg, David. 1983. *Discovering Apple Logo: An Introduction to the Art and Pattern of Nature*. Reading, MA: Addison-Wesley Publishing Company.

Tietze, Heinrich. 1965. *Famous Problems of Mathematics*. Baltimore, MD: Graylock Press.

Toth, L. Fejes. 1964. *Regular Figures*. Oxford: Pergamon Press.

Trigg, Charles W. 1972. "Collapsible Models of Regular Octahedrons." *The Mathematics Teacher* (NCTM), 65(10):530–533.

Van Hiele, P.M. 1985. *Structure & Insight*. San Diego: Academic Press.

van Delft, Pieter, and Jack Botermans. 1978. *Creative Puzzles of the World*. New York: Harry N. Abrams, Inc., Publishers.

Wahl, John, and Stacey Wahl. 1976. *I Can Count the Petals on a Flower*. Reston, VA: National Council of Teachers of Mathematics.

Walter, Marion. 1985. *The Mirror Puzzle Book*. New York: Parkwest Publications.

Walter, Marion. 1966. "An Example of Informal Geometry: Mirror Cards." *The Arithmetic Teacher* (NCTM), 13(6):448–452.

Wardcop, R.F. 1970. "A Look at Nets of Cubes." *The Arithmetic Teacher* (NCTM), 17(2):127–128.

Watt, Dan, and Molly Watt. *Teaching with Logo*. Menlo Park, CA: Addison-Wesley Publishing Company.

Weibe, Arthur. 1985. *Symmetry with Pattern Blocks*. Fresno, CA: Creative Teaching Associates.

Wells, Peter. 1971. "Symmetries of Solids." *Mathematics Teaching* 55(Summer):48–52.

Wenninger, Magnus. 1970. *Polyhedron Models*. New York: Cambridge University Press.

Wenninger, Magnus J. 1966. *Polyhedron Models for the Classroom*. Washington, DC: National Council of Teachers of Mathematics.

Weyl, Hermann. 1952. *Symmetry*. Princeton, NJ: Princeton University Press.

Williams, Robert. 1979. *The Geometrical Foundation of Natural Structure*. New York: Dover Publications.

Winter, John. 1972. *String Sculpture*. Palo Alto, CA: Creative Publications.

Winter, Mary Jean, Glenda Lappan, Elizabeth Phillips, and William Fitzgerald. 1986. *Spatial Visualization*. The Middle Grades Mathematics Project Series. Menlo Park, CA: Addison-Wesley Publishing Company.

Woods, Jimmy C. 1988. "Let the Computer Draw the Tessellations That You Design." *The Mathematics Teacher* (NCTM), 81(2):138–141.

Yates, R.C. 1949. *Geometric Tools: A Mathematical Sketch and Model Book*. St. Louis, MO: Educational Publishers.

Zaslavsky, Claudia. 1990. "Symmetry in American Folk Art." *The Arithmetic Teacher* (NCTM), 38(1):6–12.

Selected Answers to Exercises

Chapter 1

Exercise Set 1.2

2. All "lines" through point P intersect line ℓ.
3. (a) valid, (b) invalid, (c) invalid, (d) invalid

Exercise Set 1.3

8. It would work, since $5^2 + 12^2 = 13^2$.

Chapter 2

Exercise Set 2.1

6. (a) true, (b) false, (c) true, (d) true
7. The lines ℓ and m are parallel, since they are coplanar and they do not intersect.

Exercise Set 2.2

1. (a) line segment, (b) ray, (c) circle, (d) angle, (e) angle, (f) parallel lines, (g) perpendicular lines
9. (a) never, (b) sometimes, (c) never, (d) never, (e) sometimes, (f) sometimes, (g) never

Exercise Set 2.3

1. (a) four angles of 90° each, (b) infinitely many, (c) no others through point P
7. The lines are parallel.
9. The angles are supplementary.

Exercise Set 2.4

1. (a) $m\angle 1 = 138°$, $\angle 1$ forms a linear pair with the given angle.

   (b) $m\angle 2 = 138°$, $\angle 2$ forms a linear pair with the given angle.

   (c) $m\angle 3 = 42°$, $\angle 3$ and the given angle are vertical angles.
3. (a) $m\angle 1 = 62°$, $\angle 1$ and the given angle are complementary.

   (b) $m\angle 2 = 90°$, $\angle 2$ is made by two perpendicular lines.

   (c) $m\angle 3 = 62°$, $\angle 3$ and $\angle 1$ are vertical angles.

   (d) $m\angle 4 = 28°$, $\angle 4$ and the given angle are vertical angles.
5. Let the other interior angle where the transversal meets line r be $\angle 4$, so $\angle 4$ is supplementary to both $\angle 2$ and $\angle 3$. But $\angle 4$ is a corresponding angle to $\angle 1$, so $\angle 4$ and $\angle 1$ are congruent.

Exercise Set 2.5

1. (a) sometimes, (c) always, (e) always, (g) sometimes, (i) sometimes, (k) sometimes, (m) sometimes
2. (a) always, (c) always, (e) always, (g) sometimes, (i) sometimes, (k) sometimes, (m) sometimes
11. six exterior angles
15. three altitudes

Chapter 3

Exercise Set 3.1

1. (a) any scalene triangle, (b) an isosceles triangle, (c) not possible, (d) an equilateral triangle, (e) not possible, (f) an isosceles triangle, (g) not possible

Exercise Set 3.2

1. triangle, 120°; square, 90°; pentagon, 72°; hexagon, 60°; octagon, 45°
3. six; a triangle, square, hexagon, octagon, dodecagon, and 24-gon
5. (a) heptagon, $51\frac{3}{7}$°; octagon, 45°; nonagon, 40°; decagon, 36°; dodecagon, 30°; 100-gon, 3.6°

   (b) If the central angle is small, the vertex angle is close to 180°.
9. (a) pentagon, 5; hexagon, 6; octagon, 8

   (b) n-gon, n

Exercise Set 3.3

3. $\{^{10}_{3}\}$ is the only 10-sided star polygon.
5. n and d must be relatively prime and $d > 1$.
7. There are four 11-sided star polygons.
9. (a) You can find a trapezoid, rhombus, kite, isosceles triangles, nonconvex quadrilaterals, and other geometric figures.

Exercise Set 3.4

1. 90°
3. It is twice as far from the intersection of medians to the vertex than it is from the intersection to the midpoint of the side.
5. The four angles are right angles.
7. The two ratios are equal.
9. $\overleftrightarrow{AD} \parallel \overleftrightarrow{BC}$ and $\overleftrightarrow{AB} \parallel \overleftrightarrow{CD}$
11. The triangle is equilateral.
13. The segments are perpendicular.
17. (a) The nine points lie on a circle.

    (b) The center of the circle is on the Euler line.

Chapter 4

Exercise Set 4.1

4. Pythagorean theorem
6. (a) When two parallel lines are cut by a transversal, corresponding angles are congruent.

   (b) The line joining the midpoints of two sides of a triangle is parallel to the third side.

   (c) The sum of the angles in a triangle is 180°.

   (d) If corresponding sides of two triangles are parallel, the triangles are similar.
9. (a) The dual of 4^4 is another 4^4.

   (b) The dual of 3^6 is 6^3.

Exercise Set 4.2

1. The others are (4,6,12), (3,3,3,3,6), (3,4,3,3,4), (3,4,6,4), (3,3,3,4,4), (3,12), (3,3,6).
7. The sum of the angles around a (5,6,8) vertex is 363°, not 360°.

Chapter 5

Exercise Set 5.1

1. (a) triangular prism, (b) pentagonal pyramid, (c) rectangular pyramid, (d) rectangular prism
4. $n + 1$ faces, $2n$ edges, $n + 1$ vertices
5. $n + 2$ faces, $3n$ edges, $2n$ vertices
6. In all cases, $F + V = E + 2$.

Exercise Set 5.2

1. The matchsticks are the edges of a tetrahedron.
3. (a) AB and HG; AD and FG; AE and CG; BF and DH; BC and EH; CD and FE

   (b) A and G; B and H; C and E; D and F
5. Cube (4.3); tetrahedron (3.3); octahedron (3.4); dodecahedron (5.3); icosahedron (3.5)
8. (a), (c) and (d) are nets for a cube.
9. (a) and (c) are nets for a tetrahedron.
10. (a) The dual of a cube is an octahedron. The dual of an octahedron is a cube. The dual of a tetrahedron is a tetrahedron. The dual of a dodecahedron is an icosahedron. The dual of an icosahedron is a dodecahedron.

    (b) The dual of an (x, y) is a (y, x).

Exercise Set 5.3

2. (a) six, (b) three
3. (a) four, (b) six, (c) three
4. (a) four, (b) six, (c) five, (d) infinite number, (e) nine, (f) two
7. The answer is 5 if two faces are square; the answer is 3 if no faces are square.

Exercise Set 5.4

1. (a) There is one axis of rotational symmetry.

There are three planes of symmetry.

(b) The answer depends on which two corners are truncated. Consider three cases: (1) opposite corners, (2) adjacent corners, and (3) corners that are opposite vertices of one square face.

4. A truncated octahedron has 24 vertices, 36 edges, and 14 faces. A truncated dodecahedron has 60 vertices, 90 edges, and 32 faces.

Chapter 6

Exercise Set 6.1

5. 40½ inches is more precise.
7. Ascertain the number of units used.
11. (d)
13. (a) true, (b) false, (c) true
15. A tire pump could be used to measure "expandibility" of a balloon. For example, one balloon might hold "ten pumps" of air before bursting while another can hold only "five pumps."

Exercise Set 6.2

3. The prefixes of the metric system are based on powers of ten, like our numeration system. Also, the same prefixes are applied to units measuring different properties: length, capacity, and mass.
5. (a) larger, (b) smaller, (c) smaller

Exercise Set 6.3

1. (a) 15 mm or 1.5 cm, (c) 50 mm or 5 cm, (e) 93 mm or 9.3 cm
9. (a) 1661 km, (c) 2464 km, (e) 1176 km
11. (a) 8.67 m, (c) 632 m, (e) 8.67 hm
13. $c = 6\pi$ cm ≈ 18.85 cm
15. πr

Exercise Set 6.4

1. (a) 2 units, (b) 8 units, (c) 12 units
3. (a) cm^2, (b) m^2, (c) cm^2, (d) km^2
6. (a) 65, (c) 63, (e) 149.5
7. Small triangle area is ½; square, middle triangle, and parallelogram areas are 1; large triangle area is 2.
10. 300π $mm^2 \approx 942.48$ $mm^2 \approx 9.42$ cm^2

Exercise Set 6.5

5. (a) 1,000,000, (c) 1,000,000,000, (e) 1,000,000, (g) 1,000,000
7. (a) 94.7 cl, (c) 0.947 l, (e) 649.6 dal
9. (a) 90, (c) 58⅓, (e) 75π, (g) 75π, (i) 1980
14. The volume of the larger sphere is eight times that of the smaller sphere.

Exercise Set 6.6

5. 1.87 kg is more.
8. (a) 1000 g, (c) 1000 cm^3, (e) 1 kg, (g) 0.001 kg
12. (a) (i) approximately 458 ft^3, (ii) approximately 3426 gallons, (iii) approximately 28,594 pounds

    (b) (i) approximately 29.218 m^3, (ii) approximately 29,218 l, (iii) approximately 29,218 kg
17. One mile per gallon equals approximately .425 km per liter. If you know miles per gallon, about 40% of that number will be km per liter.

Chapter 7

Exercise Set 7.1

8. A' is (5,3), B' is (13,8), C' is (11,0).
13. (a) (2,3), (b) (7,0), (c) (10,7), (d) $(x + 3, y + 2)$

Exercise Set 7.2

1. (a) Point A', (b) Point B', (c) $\overline{A'B'}$
3. (a) A, (c) E, (e) $\overline{CD}$, (g) $\overline{DH}$
5. (a) $(1, -2)$, (b) $(3, -4)$, (c) (2,1)
9. $(4, -2)$

Exercise Set 7.3

1. (a) H, (b) B, (c) $\overline{CB}$, (d) $\triangle HGF$
4. (a) point A' $(-1, 4)$, (c) point E' $(-2, -2)$, (e) $\overline{B'C'}$ from B' $(-3, 5)$ to C' $(-2, 3)$
9. Each letter of TIMOTHY remains the same after a reflection. This is not true of the letters in REBECCA.
11. Have you reflected on this message?

Exercise Set 7.4

1. (a) K, (b) O, (c) D, (d) B

3. (a) D, (b) A, (c) C, (d) O, (e) B, (f) L
5. (a) The rotation is centered at the intersection of the lines, through a clockwise angle of 60°.
7. (b) The single motion is a translation in the direction of $\overrightarrow{XY}$ with a distance of ZXY.

Exercise Set 7.5

3. Flipping the triangle through p, then q, is a translation. The combination of q, then p, is the opposite translation. The two results are the same when $p = q$.
4. If p and q intersect, the final position is a rotation about the point of intersection. One of (p, then q) or (q, then p) is clockwise and the other is counterclockwise. The results are the same when p and q meet at right angles.

Chapter 8

Exercise Set 8.1

5. (a) G, (b) $EFGH$, (c) $ABCD$, (d) L, (e) $IJKL$, (f) $IJKL$

Exercise Set 8.2

1. (a) 2, (c) $2\sqrt{5}$, (e) $3\sqrt{5}$, (g) ½, (i) 3⁄2
3. (a) 8⅓, (b) 18⅓, (c) 5, (d) 13⅓
6. Construct a line p through B' parallel to $\overleftrightarrow{AB}$. A' is the intersection of lines p and l.
9. Each image line is parallel to the original line.

$$\overleftrightarrow{A'B'} \parallel \overleftrightarrow{AB} \text{ and } \overleftrightarrow{D'C'} \parallel \overleftrightarrow{DC}, \text{ so } \overleftrightarrow{A'B'} \parallel \overleftrightarrow{D'C'}.$$

$$\overleftrightarrow{A'D'} \parallel \overleftrightarrow{AD} \text{ and } \overleftrightarrow{B'C'} \parallel \overleftrightarrow{BC}, \text{ so } \overleftrightarrow{A'D'} \parallel \overleftrightarrow{B'C'}.$$

Exercise Set 8.3

3. (c) $A \to X, B \to Y, C \to Z, D \to W$
4. (a) $BC = 15^{11}/_{14}$, $CF = 3^{3}/_{14}$
   (b) $OD = 4$, $CD = 6$
6. 53 feet
9. (a) not necessarily; for example, if $A'B' = 3$ and $A'E' = 3$
   (b) They are similar because sides are also proportional.

Chapter 9

Exercise Set 9.1

2. This array of numbers completes the table:

2	2	4	4
1	1	2	2
2	2	4	4
1	1	2	2

Exercise Set 9.2

1. (a) the third and fourth figures; (b) the second, third, and fourth figures; (c) the third figure; (d) the fourth figure
5. (a) the middle figure, (b) the right figure, (c) the right figure, (d) the right figure

Exercise Set 9.3

1. The network in Figure 9.23 has more than two odd vertices, so it is not traversable.
3. For type 2, add 1, 2, and 2 edges, respectively. For type 1, add 2, 3, and 3 edges, respectively.

Chapter 10

Exercise Set 10.1

1. (a) 2500 and 10,000, (b) 1275 and 5050
3. (a) 4, 6, 8, 10, 12, 14, 15, 16, 18, 20, 21, 22, 24, 26, 27, 28, 30, 32, 33, 34, 35
   (b) A prime number is any number greater than 1 that is neither a square nor a rectangular number.
5. (a) $10 + 15 = 5^2$, $15 + 21 = 6^2$, $21 + 28 = 7^2$
   (b) The sum of two consecutive triangular numbers is a square number.
7. 35, 51, 70

Exercise Set 10.2

1. (b) Number of diagonals = Number of sides − 3
3. The next two pairs are (4, 12) and (5, 20).

Exercise Set 10.3

1. (a) The next 3 pairs are (3, 7), (4, 11), and (5, 16).
2. (a) The next 2 rows of the table are (4, 9) and (5, 14).

(c) There are two different ways to get the length of 5 units on 6 × 6 dot paper, so the expected pattern is not found.

Exercise Set 10.4

1. The next 5 rows are

1 5 10 10 5 1
1 6 15 20 15 6 1
1 7 21 35 35 21 7 1
1 8 28 56 70 56 28 8 1
1 9 36 84 126 126 84 36 9 1

3. (a) The next 4 rows are (3, 4), (4, 8), (5, 16), (6, 32).

   (c) The sum of the numbers in row $n = 2^{n-1}$.

5. (a) 8th

   (b) Any row that is a power of 2 (1, 2, 4, 8, 16, etc.) will have all odd entries.

7. (a) The next 3 sums are 21, 34, and 55.

9. (a) Rows 3 and 7 also have this property.

   (b) If the second number is prime, it will divide every other number in the row (except 1).

Photo Credits

Fig. 1.1 (starfish) Grant Heilman & Associates; Fig. 1.5 (tree on horizon) LeRoy Troyer & Associates; Fig. 1.6 (bees) Grant Heilman & Associates; Fig. 1.7 (Parthenon) Editorial Photocolor Archives; Fig. 1.8 (starfish) Patty Benner; Fig. 1.9 (snail) Grant Heilman & Associates; Fig. 1.12 (electronic pen) Lockheed Missiles & Space Co.; Fig. 1.14 (arrowheads) Peabody Museum, Harvard University; Fig. 1.15 (basket/pots) From Tom Bahti, *Southwestern Indian Tribes*, Las Vegas: KC Publications; Fig. 1.17 (pyramids at Giza) Editorial Photocolor Archives; Fig. 1.25 (apple on a brick wall) Technical Texts; Fig. 1.26 (DaVinci, St. Jerome) Scala Fine Arts Publishers; Fig. 1.27 (Transamerica Bldg.) Transamerica Corporation; Fig. 1.28 (dec. ironwork) Technical Texts; Fig. 2.1a (spiral pattern...) Los Alamos Scientific Laboratory, University of California; Fig. 2.1b (pattern/circle) From Ruchlis & Engelhardt, *The Story of Mathematics*, NY: Harvey House; Fig. 2.1c (Alltair Design) From Ensor Holiday, *Alltair Designs*, © 1970 Alltair, NY: Pantheon Books; Figs. 4.1, 4.2 (Greek & Egyptian patterns) From L. Fejes-Toth, *Regular Features*, NY: Pergamon Press. Fig. 4.3 (Chinese patterns) From L. Fejes-Toth, *Regular Features*, NY: Pergamon Press; Fig. 4.4 (Alahambra drawings) M.C. Escher, Escher Foundation, Haags Gemeente museum, The Hague; Fig. 4.5 (Day & Night) M.C. Escher, Escher Foundation, Haags Gemeente museum, The Hague; Fig. 4.6 (Mitchell Gardens) Daniel Brody/Stock, Boston; Fig. 4.53 (Horses, 1946) M.C. Escher, *Symmetry Aspects*, Escher Foundation, Haags Gemeente museum, The Hague; Fig. 5.1 (flower) Technical Texts; Fig. 5.2 (Saturn) Hale Observatories; Fig. 5.3 (pyramids) Scala, Editorial Photocolor Archives; Fig. 5.4 (bricks) From H. Steinhaus, *Mathematical Snapshots 3rd Edition*, © 1950, 1960, 1968, 1969 Oxford University Press, Inc.; Fig. 5.24 (Radiolarians) From D'Arcy, *On Growth and Form*, NY: Cambridge University Press; Fig. 5.25 left (cubic crystal) Walter Dawn; Fig. 5.25 right (fluorite crystal) Lester Bergman & Associates; Fig. 5.29 (cube/octahedron) From H. Steinhaus, *Mathematical Snapshots 3rd Edition*, © 1950, 1960, 1968, 1969 Oxford University Press, Inc.; Fig. 5.41 (flower) Walter Dawn; Fig. 5.42 (snowflake) Patty Benner; Fig. 5.55 (blocks) From Alan Holden, *Shapes, Space & Symmetry*, 1971, NY: Columbia University Press; Fig. 5.59 (semi-irreg. solids) From Alan Holden, *Shapes, Space &*

Symmetry, 1971, NY: Columbia University Press; Fig. 5.77 (two photos) From H. Steinhaus, *Mathematical Snapshots 3rd Edition*, © 1950, 1960, 1968, 1969 Oxford University Press, Inc.; Fig. 7.24 (cartoon "You're fired") Drawing by Richter; © 1957 The New Yorker Magazine, Inc.; Fig. 7.43 (cartoon, Barber/mirrors) Drawing by Charles Addams, © 1957 The New Yorker Magazine, Inc.; Fig. 7.49 (Magic Mirror) M.C. Escher, Escher Foundation, Haags Gemeente museum, The Hague; Fig. 8.1 (two photos, auto assembly) Chevrolet Motor Division, General Motors Corporation; Fig. 8.2 (two photos, aircraft) Lockheed-California Company; Fig. 8.3 (two aerial views) Itek Corporation; Fig. 10.1 (x-ray diffraction...) Dr. B.E. Warren; Fig. 10.2 (Virus Crystal) The Public Health Research Institute of the City of New York; Fig. 10.3 (Orion) Hale Observatories; Fig. 10.14 (hyperbolic paraboloid) From H. Steinhaus, *Mathematical Snapshots 3rd Edition*, © 1950, 1960, 1968, 1969 Oxford University Press, Inc.; Figs. 10.15/10.16 (string sculptures) From John Winter, *String Sculpture*, Palo Alto, CA: Creative Publications, Inc.

Index

Activities. *See also* Construction activities; Teaching activities
  critical thinking, 364–390
  metric units, 343–363
Alhambra, 86, 87
Angle, 32
  acute, 35
  bisector of, 40
  obtuse, 35
  right, 35
Angles
  complementary, 41
  corresponding, 42
  linear pair of, 41
  relationship between pairs of, 42
  supplementary, 41
  vertical, 41
Appel, K., 277
Applications
  box pattern problem, 62
  box problem, 185
  brass band problem, 170
  dock problem, 220
  duct choice problem, 186
  Earth problem, 177
  factory problem, 50
  flagpole problem, 253
  hungry cow problem, 179
  monument problem, 288
  old wheel problem, 43
  pen problem, 177
  pencil problem, 187
  power plant problem, 51
  salesman problem, 272
  sprinkler problem, 36
  suitcase problem, 19
  tiling problem, 99
  wiring problem, 121
Archimedes, 147
Area, measuring, 172–176

Bhaskara, 13
Bolyai, Johann, 11, 13
Bounded region, 299

Carroll, Lewis, 122
Cayley, Arthur, 277
Chord, 33
Circle, 33
Circumference, 168
Common notions, 10
Computer explorations, 16, 18, 29, 34, 42, 48, 49, 68, 70, 76, 79, 92, 106, 110, 120, 168, 176, 200, 209, 218, 226, 240, 245, 251, 271, 298, 391–419
Congruence
  of angles, 35
  of segments, 34
Congruent, 230
Construction activities, 324–342
  with MIRA, 325–328
  with ruler, compass, and MIRA, 329–342

Contiguous regions, in topology, 276–277
Cooperative learning group discussion, 22, 54, 85, 117, 157, 195, 234, 255, 281, 308
Critical thinking, 10, 21, 31, 38, 44, 46, 52, 64, 74, 78, 84, 101, 112, 122, 132, 140, 161, 164, 171, 180, 187, 194, 203, 212, 222, 229, 233, 254, 259, 272, 293, 300, 304, 364–390
  activities and applications, 366–390
Cross section
  of a cube, 139
  of a solid, 183
Cube, 124
  dual of, 129
Curve stitching designs, 19, 295
Cut point, 265

Decimal system of numeration, 166–167
Deductive reasoning, 302, 373, 377, 379, 381
Degree measure, 34
Deltahedron, 130
Distance, 41
Distortions, 260

Einstein, 11
Endpoints, 32
Equivalence, in topology, 259–264
Escher, Maurits, 86, 88, 114, 215
Euclid, 8, 9, 21
  The Elements of Geometry, 8
  fifth postulate, 11
Euler, Leonard, 80
Euler line, 80
Euler's Formula, 123

Fedorov, 86
Fermat, Pierre de, 71
Fermat prime, 71
Flips, 213–216. *See also* Motions in Geometry
Four-color problem, in topology, 277–279

Garfield, James, 17
Gauss, Carl Friedrich, 11, 71
Geometric figures and their measures, 31–35
Geometric forms in nature, 5
Geometrical patterns, 24
Geometry. *See also* Geometry in three dimensions; Motions in geometry aesthetic and recreational, 2, 12–18
  as an art, 12–14
  basic ideas of, 24–49
  as a formal axiomatic structure, 9–12
  figures and their measures, 31–35
  line and angle relationships, 38–42
  as a mathematical system, 2, 8
  overview, 1–21
  in the physical world, 2–6
  points, lines, and their relationships, 25–29
  for recreation, 16–18
  symbols for basic figures, 32–33
  theorems about intersecting lines, 45
  triangles and quadrilaterals, 47–49
Geometry in three dimensions, 118–154
  regular polyhedra, 122–127
  semiregular polyhedra, 140–147
  symmetry in space, 132–137
  tesselations of space, 148–154
Geomotoon, 37
Golden Ratio, 3, 4, 21, 74
Golden Rectangle, 3
Guthrie, Francis, 277

Haken, W., 277
Hardy, G.H., 12
Hexaminoe, 97

Inductive reasoning, 302, 384
Investigations, 1, 8, 10, 25, 31, 47, 56, 57, 65, 68, 69, 90, 103, 105, 114, 119, 122, 134, 136, 141, 149, 159, 167, 172, 182, 198, 217, 223, 230, 248, 256, 270, 273, 275, 276, 279, 284–285, 296, 302, 305, 311, 313, 314, 316, 322

Jordan Curve Theorem, 272–274

Kaleidoscope, three-mirror, 112–113
Konigsberg bridge question, 267–268

Length measure, 33
Length, measuring, 165–169
Line, 25, 26, 27
  and angle relationships, 38–42
Lines
  concurrent, 28
  intersecting, 28
  parallel, 28
  patterns of, 294–298

Lobachevsky, 11
Logical reasoning, 8, 371, 373–374, 377, 379, 381
LOGO, using, in geometry, 309–323

Magnification, 236–244
  with center O and scale factor $k > 0$, 239
  properties of, 242–244
Mass and weight, measuring, 188–191
Mathematical reasoning, 373, 377, 379, 381, 384
Measurement, 158–191
  area, 172–176
  English system, 164
  length, 165–169
  mass and weight, 188–191
  metric system, 163
  process of, 158–159
  standard units in, 162
  volume and capacity, 180–185

Metric system, 163
Metric units
  activities and exercises, 343–363
MIRA, 39
  construction activities with, 324–342
Moebius band, 256–257
Moors, 86, 88
Motion geometry, 21
Motions in geometry, 196–231
  combining slides, turns, and flips, 222–225
  flips or reflections, 213–216
  motions and congruence, 229–231
  product of two motions, 224–225
  slide or translation, 197–199
  turns or rotations, 204–208

Networks, in topology, 267–269
  traversable, 268–269
Non-Euclidean geometry, 11, 13
n-gon, 59
  regular, 65, 107
Number patterns, 282–305
  Pascal's Triangle, 304–305
  pattern and proof, 302–303
  patterns of lines, 294–298
  patterns of points, 282–287
Parpolygon, 63
Pascal's Triangle, 304–305
Patterns
  of lines, 294–298
  Pascal's Triangle, 304–305
  of points, 282–287
  of polygons, 86–115
  and proof, 302–303
Pentagram, 74
Pentiamond, 100
Pentominoe, 63
Perigal, Henry, 16
Perimeter, 167–168
Perpendicular, 39
Piaget, 26, 309
Pick's Formula, 175
Plane, 26, 27
Planes
  intersecting, 29
  parallel, 29
  skew, 29
Plato, 1, 118
Point, 25
Points
  collinear, 28
  coplanar, 28
  and lines, 25–29
  patterns of, 282–287
Polygon, 33
  angle of, 57
  central angle of, 67
  convex, 57
  diagonal of, 57
  nonconvex, 57
  nonsimple, 57
  pierced, 64
  relationships, 56–80
  simple, 56–57
Polygons
  about, 56–59
  classification of, 58, 66–67
  construction of regular, 68–71
  patterns of , 86–115
  regular, 65–67
  star, 21, 74–77
  symmetry of, 59–62
  tesselations of, 90–106
  theorems about, 78–81
Polyhedra, 120
  regular, 122–127
  semiregular, 140–147

Postulates, 9–10, 425–427 (listed)
Problem-solving strategies
consider a simpler related problem, 101
draw a picture, 31, 53
formulating a problem, 161
guess and revise, 7
look for a pattern, 267
looking back, 294
make an organized list, 65, 254
make a table, 267
special insight, 300
use data from a diagram, 188
use logical reasoning, 222, 242
using space visualization, 132, 140
Ptolemy, 21
Pythagorean Theorem, 13, 17
paper-and-scissors proof of, 16

Quadrilateral, 33, 48, 49

Ray, 32
Reflections, 213–216. *See also* Motions in geometry
Rhombicosidodecahedron, 146
Rhombicuboctahedron, 143, 144
Rhombohedron, 150
Rotations, 204–208. *See also* Motions in geometry
Russell, Bertrand, 13

Schlegel diagram, 129
Segment, 32
bisector of, 40
midpoint of, 40
Similarity, 248–251
Slide, 197–199. *See also* Motions in geometry
Snub cube, 147
Space, 26
Square numbers, 285
Star polygons, 74–77
regular, 75
Symmetry
plane of, 133
of polygons, 59–62
reflectional, 59, 60, 133
rotational, 59, 61, 134–135, 136
in space, 132–137

Tangram puzzle, 16
Teaching activities, 22, 54, 84, 116, 155–157, 194–195, 233–234, 254–255, 280–281, 308
Teaching notes, 1, 2, 3, 8, 11, 14, 17, 21, 26, 31, 38, 47, 58, 59, 60, 67, 69, 78, 90, 95, 102, 104, 113, 114, 119, 120, 122, 124, 126, 133, 140, 162, 165, 167, 169, 173, 180, 182, 184, 189, 196, 197, 204, 213, 222, 236, 239, 240, 242, 256, 259, 260, 263, 269, 274, 277, 285, 286, 287, 305
Tesselations, 21, 86–115
of curved patterns, 114–115
dual, 99
of polygons, 90–97
with regular polygons, 101–106
semiregular, 102, 106–111
of space, 148–154
three-mirror kaleidoscope, 112–113
Thales of Miletus, 8, 13
Theaetetus, 127, 132
Theorems, 427–436 (listed)
about intersecting lines, 45
Topological transformations, 260–263
Topology, 21, 256–279
contiguous regions, 276–277
equivalence, 259–264
four-color problem, 277–279
Jordan Curve Theorem, 272–274
networks, 267–269
Translation, 197–199. *See also* Motions in geometry
Traversable networks, 268–269
Triangle, 32
altitude of, 48
equilateral, 47
isosceles, 47
median of, 48
Triangles and quadrilaterals, 47–49
Triangular numbers, 286
Truncating, 141
Turns, 204–208. *See also* Motions in geometry

Venn diagram, 50
Vertex, 32
Vinci, Leonardo da, 196
Volume and capacity, measuring, 180–185

Weyl, Herman, 132

CURRICULUM STANDARDS FOR GRADES 5–8

STANDARD 1: Mathematics as Problem Solving

In grades 5–8, the mathematics curriculum should include numerous and varied experiences with problem solving as a method of inquiry and application so that students can—

- use problem-solving approaches to investigate and understand mathematical content;
- formulate problems from situations within and outside mathematics;
- develop and apply a variety of strategies to solve problems, with emphasis on multistep and nonroutine problems;
- verify and interpret results with respect to the original problem situation;
- generalize solutions and strategies to new problem situations;
- acquire confidence in using mathematics meaningfully.

STANDARD 2: Mathematics as Communication

In grades 5–8, the study of mathematics should include opportunities to communicate so that students can—

- model situations using oral, written, concrete, pictorial, graphical, and algebraic methods;
- reflect on and clarify their own thinking about mathematical ideas and situations;
- develop common understandings of mathematical ideas, including the role of definitions;
- use the skills of reading, listening, and viewing to interpret and evaluate mathematical ideas;
- discuss mathematical ideas and make conjectures and convincing arguments;
- appreciate the value of mathematical notation and its role in the development of mathematical ideas.

STANDARD 3: Mathematics as Reasoning

In grades 5–8, reasoning shall permeate the mathematics curriculum so that students can—

- recognize and apply deductive and inductive reasoning;
- understand and apply reasoning processes, with special attention to spatial reasoning and reasoning with proportions and graphs;
- make and evaluate mathematical conjectures and arguments;
- validate their own thinking;
- appreciate the pervasive use and power of reasoning as a part of mathematics.

STANDARD 4: Mathematical Connections

In grades 5–8, the mathematics curriculum should include the investigation of mathematical connections so that students can—

- see mathematics as an integrated whole;
- explore problems and describe results using graphical, numerical, physical, algebraic, and verbal mathematical models or representations;
- use a mathematical idea to further their understanding of other mathematical ideas;
- apply mathematical thinking and modeling to solve problems that arise in other disciplines, such as art, music, psychology, science, and business;
- value the role of mathematics in our culture and society.

STANDARD 5: Number and Number Relationships

In grades 5–8, the mathematics curriculum should include the continued development of number and number relationships so that students can—

- understand, represent, and use numbers in a variety of equivalent forms (integer, fraction, decimal, percent, exponential, and scientific notation) in real-world and mathematical problem situations;
- develop number sense for whole numbers, fractions, decimals, integers, and rational numbers;
- understand and apply ratios, proportions, and percents in a wide variety of situations;
- investigate relationships among fractions, decimals, and percents;
- represent numerical relationships in one- and two-dimensional graphs.

STANDARD 6: Number Systems and Number Theory

In grades 5–8, the mathematics curriculum should include the study of number systems and number theory so that students can—

- understand and appreciate the need for numbers beyond the whole numbers;
- develop and use order relations for whole numbers, fractions, decimals, integers, and rational numbers;
- extend their understanding of whole number operations to fractions, decimals, integers, and rational numbers;
- understand how the basic arithmetic operations are related to one another;
- develop and apply number theory concepts (e.g., primes, factors, and multiples) in real-world and mathematical problem situations.